AF788571

Handbook of Poultry Feed from Waste

Handbook of Poultry Feed from Waste

Processing and Use

Second edition

by

Adel R.Y. El Boushy
Prof. Emeritus of Poultry Nutrition

and

Antonius F.B. van der Poel
Senior Lectuer of Feed Science & Technology

Department of Animal Sciences, Group Animal Nutrition,
Wageningen University,
Wageningen, The Netherlands

SPRINGER-SCIENCE+BUSINESS MEDIA, B.V.

Library of Congress Cataloging-in-Publication Data

ISBN 978-90-481-5523-1 ISBN 978-94-017-1750-2 (eBook)
DOI 10.1007/978-94-017-1750-2

Printed on acid-free paper

All Rights Reserved
© 2000 Springer Science+Business Media Dordrecht
Originally published by Kluwer Academic Publishers in 2000
No part of the material protected by this copyright notice may be reproduced or utilized in any form or by any means, electronic or mechanical, including photocopying, recording or by any information storage and retrieval system, without written permission from the copyright owner.

Dedicated to Professors

M.T. RAGAB
To the Memory of the Former Head of Department of Animal Science, Faculty of Agriculture, Cairo University, Cairo, Egypt.

M. VAN ALBADA
To the Memory of the Former Head of Department of Poultry Science, Wageningen Agricultural University, Wageningen, The Netherlands.

C.C. OOSTERLEE
Professor Emeritus of Department of Animal Science and Former Vice-Chancellor of Wageningen Agricultural University, Wageningen, The Netherlands.

... whose teachings will be found reflected in many pages of this book,

and to my wife Thérèse
Adel El Boushy

Contents

Foreword

In developed market economics with intensive animal production systems, such as The Netherlands, many new feedstuffs have been introduced as part of the diets of ruminant and monogastric animals. These new feedstuffs are often by-products of human food processing. It is important that these by-products and also the by-products from wastes are properly evaluated with regard to the possibilities of incorporating them into livestock diets.

Research on the subject of feed from waste, its processing and its use in the nutrition of poultry has increased considerably during the last decade. The Department of Animal Sciences, Animal Nutrition Group of Wageningen University (WU), Wageningen, The Netherlands in close co-operation with the Poultry Feeding and the Processing Industry has been active in this field.

In order to update research and to expedite further work in this field a comprehensive review of the literature on this subject of feed from waste was made. Such a study would not only bring the industry up to date on the subject but could also indicate specific topics which may be of great value for developing market economics. Poultry scientists and technologists suggested that a review would fill a need as a reference and textbook not only for the industry but also for undergraduates and graduates of agricultural colleges and extension services all over the world.

The theories developed and results obtained from the research in our Department are included in the various chapters of this book. A list of relevant references, with emphasis on papers and reviews giving a comprehensive citation of literature, has been provided for students and investigators who may wish to make a detailed study on various aspects of feed from waste as a feedstuff. Particular consideration has been given to the evaluation of research results obtained in the past and to the problems remaining to be solved.

The continuous efforts of Prof. Dr. ir. A.R. El Boushy, Prof. Emeritus in poultry nutrition and Dr. ir. A.F.B. Van der Poel, senior lecturer in animal feed science and technology, who are both members of the Department of Animal Sciences, Animal Nutrition Group, Wageningen University, Wageningen, The Netherlands, in writing this book are greatly appreciated.

On behalf of the Wageningen University (WU), Wageningen, The Netherlands, we are indebted to research workers throughout the world who have conducted the research from which the results and the principles outlined in this book are derived.

Professor Dr. ir. M.W.A. Verstegen
Monogastric Nutrition

Professor Dr. ir. S. Tamminga
Ruminant Nutrition

Preface

Much of the world's human population suffers from malnutrition. In the developing market economics in particular this is a great problem and the gap between the developed and developing regions tends to increase rather than to decrease. Accordingly, much effort is being made to study the possibilities of utilizing agricultural, animal and industrial waste in the nutrition of poultry. As a consequence, this can be a reduction in the use of traditional feed ingredients such as maize, wheat and soybeans that can also be consumed by humans. By using converted biological waste as an animal feed, a new industry and market can be established and pollution can be lowered in developing as well as developed countries. This may be reflected in the national income. Beside this, low-cost poultry meat and eggs will be available and will assist in reducing hunger by lowering the competition between humans and poultry for food. In recent years several books have been published on various aspects of non-traditional feed. However, a comprehensive study in this field of processing and poultry nutrition has not been published previously.

This book has been written to describe the potential of nutrient recovery from wastes such as poultry manure, slaughter waste, sludge recovered from wastewater, tannery waste, municipal waste, fruit and vegetable wastes and their integration in poultry feeding systems.

Numerous processing methods, technologies and systems (dehydrations, chemical and mechanical treatments, biodegradation to improve the nutritive value by using insects and earthworm cultures and other complex recycling systems) are described as potential methods to manufacture feed from waste depending on the origin of the waste material. Results are described and evaluated according to their chemical analysis, nutritive value, reliability, applicability and use as a feedstuff for poultry. Industrial processes involving the regular application of treatments of the original product as well as the conversion of the wastes into protein or carbohydrate sources are described.

Since the sensory quality of feed ingredients has a considerable influence on feed consumption in poultry, a separate chapter deals with the palatability of feed from waste. In addition to the appearance and texture of feed ingredients as components of sensory quality, special attention has been paid to taste. The ability of the birds to taste with a physiological and histological background, has been discussed in detail. The use of high intensity sweeteners to improve palatability of waste is also discussed.

This book has been written to serve as a text-book for undergraduates, a ready reference and text book for graduate students, and a source of information for the research worker, the practical nutritionist and technologist. This book can also be of considerable value for teaching, in extension work and providing advice concerning the technical aspects to the feed industry and related agribusiness.

The selected lists of references at the end of each chapter provide the reader with a maximum entry to the literature covering the extensive research reports on processing and nutrition. Comprehensive compilation of the literature was minimized and the presentation of literature data was based on interpretation or correlation of research findings.

The authors are indebted to research workers throughout the world who have initiated the experimental and practical research from which the principles outlined in this book are derived.

We wish to thank and acknowledge our colleagues of Wageningen University, Wageningen, The Netherlands for their kind help during the writing of this book. We like to express our sincere appreciation to Mrs T.M.S. El Boushy for typing the whole manuscript and to Mrs Y.S. Abeln, member of the Centre for Agricultural Publishing and Documentation, for her continuous supply of literature and documentation. Special thanks to the head of the Sector of Animal Production Mr J. Wien and his staff members, Mr W.J.A. Valen for taking the photographs and Mr K. Boekhorst for the excellent drawings and lay-outs delivered. For the library facilities the efficient support of Mrs A.M. Zijlmans and Mrs L.M.T. Zeeuwen is acknowledged. The computing work of the F.A.O. data was made possible with the continuous support of Ing. P.L. van der Togt.

The authors acknowledge the publication of this book and are convinced that it can serve as a text-book, a research reference and a useful guide to feed from waste, and its processing and use as a feedstuff for all classes of poultry.

A.R. El Boushy
A.F.B. van der Poel

The Authors

Adel R.Y. El Boushy is Prof. Emeritus in Poultry Nutrition at the Wageningen University, Wageningen, The Netherlands. He received his B.Sc. degree from the Faculty of Agriculture, Cairo University, Egypt in 1955. His M.Sc. was from the same institution in 1961 and he obtained his Ph.D. from the Wageningen University, Wageningen, The Netherlands in 1966.

For 8 years, he was employed at the Ministry of Agriculture, Cairo, Egypt, where he conducted research on husbandry and nutritional problems of poultry. He moved to The Netherlands in 1969, where he has continued his teaching and research. He also served as a senior lecturer in poultry nutrition for undergraduate and graduate students. Major areas of investigation have been the use of treated waste as a poultry feedstuff to reduce the competition of food between humankind and poultry. He has authored and co-authored over 150 scientific international articles. He has participated as a chairman on numerous international congresses.

In 1990 Prof. El Boushy received the Egyptian Feeding Industry Research Award from the Ministry of Agriculture for outstanding research in poultry nutrition, particularly feed from waste. He received an honourable Ph.D. from Mexico, University of Guadalajara and was awarded with a guest Professorship from The Netherlands, Egypt, Pakistan and the USA. In The Netherlands he received the highest Royal Award for his scientific distinction, for introducing the new theory in this book for the struggle of hunger campaign in the third world. He received this award from her Majesty the Queen of The Netherlands and he obtained an order of civil reward in the widest sense, as a proof for who gain special merits in sciences (Knight of Order of the Dutch Lion). He is member of the World Poultry Science Association, advisory member of International Agricultural Centre, Ministry of Agriculture, The Hague, The Netherlands and the Dutch Feed Industry and he is a member of Lions International.

Antonius F.B. van der Poel is an associated Professor in animal feed science and technology at the Animal Nutrition Group, Wageningen University, Wageningen, The Netherlands. He received his ir. (B.Sc. & M.Sc.) degree in 1979 from the above mentioned University.

He formerly joined Cyanamid Benelux B.V. at Louvain-La-Neuve, Belgium as a Business Manager for five years. In 1984 he joined the Animal Nutrition Group, Department of Animal Sciences, Wageningen University, where he started teaching and research on feedstuffs' concentrates and animal feed technology. Major areas of investigation are dealing with the reduction of antinutritional factors in feedstuffs. This speciality led to the organization of the third International Workshop on Antinutritional Factors in Legume Seeds held in Wageningen, 1998. This was also the subject of his Ph.D. research.

He obtained his Ph.D. degree from the above mentioned University in 1990. At this moment, he is involved in research in the fields of antinutritional factors, (recycling of feed from waste) and alternatives to conventional feed manufacturing systems. He has authored and co-authored several 50 scientific international articles and booklets on antinutritional factors in legume seeds, on expander processing of animal feeds and on liquid addition of feed additives.

1. THE BENEFIT OF FEED FROM WASTE

Contents

1.1 GENERAL INTRODUCTION

Advances in technology have resulted in an improved standard of living in many parts of the world. Because of differences in rate of population growth, developed countries have benefited most from these developments. Developing countries are also demanding an improved standard of living and, for these countries, the first condition to be improved is the supply of an adequate diet. However, the failure of crops owing to adverse weather conditions and invasions of predators as well as a rapid rate of population growth may cause widespread misery and even death. Unfortunately, this situation represents an unresolved problem in the third world today (Rolfe, 1976). Emphasis, therefore, is put on increasing food supplies in traditional as well as other ways. Traditional ways include the exhaustion of soil and sea to yield more food with the help of high-yielding crop varieties (hybrids) as well as the use of fertilisers and irrigation.

In addition to these traditional ways, alertness to the problem of waste disposal and its utilization has been regularly postulated (Inglett, 1973; Birch *et al.*, 1976; El Boushy, 1986; Boucqué and Fiems, 1988; Boďa, 1990; El Boushy, 1990).

Wastes are generated particularly by the agricultural, industrial and municipal segments of the population, including wholesalers and consumers (Figure 1.1). Nowadays, a confrontation with the challenge of the processing and disposal of these by-products as a result of the modern industrialization is taking place. The utilization of such waste needs urgent investigation because the recycling and reduction of waste can reduce the pollution and ameliorate the present situation by creating new feeds from waste.

The increasing costs and pressures concerned with waste disposal stress the need for a reappraisal of the utilization of waste, either directly (as a diet ingredient, such as citrus pulp or feather meal) or indirectly (upgrading by micro-organisms such as algae, larvae house-fly, earthworm *etc.*), for livestock and poultry feeding. Present-day investigations therefore must include studies on the management of waste, its technology and subsequent feeding value for livestock. The rapid change in modern animal and poultry farming implies scrutiny in the studies on nutritional evaluation with respect to the target animal species and a low-technology approach. Studies on plant and process design of specific waste streams or effluent treatment of wastewater involve nutritional evaluation on digestibility, feeding value, bio-hazards and feasibility for waste management and utilization. Consequently, scientific increase and constructive action in view of these objectives has to be generated from various disciplines of scientific investigation.

Agricultural waste originates from the primary agricultural production (plant products such as straw, culls, leaves, press cakes), from intensive farm production (animal and poultry by-products - as manure) and from livestock processing plants (by-products of slaughtered animals and tannery). All these materials merit consideration to be converted to animal and vegetable feedstuffs from which poultry and animals will produce eggs, milk, meat, *etc.* for the welfare of humankind. The upgraded animal waste can be considered as relevant new feedstuffs to be included in the category of agro-industrial by-products. Alternatives are its use as fuel from manure (methane), as fertilizer, or its use as a substrate for microbial fermentation processes (wastewater treatments of citrus, winery, tomato and potato to generate [activated] sludge).

Especially the use of animal or vegetable and fruit wastes in animal and poultry nutrition represents a valuable means of the indirect production of feed from waste. Despite of the obvious potential, the utilization of feed from waste in diet formulation until now has been negligible owing to constraints imposed by several nutritional and technical considerations. Present investigations, therefore, have to concentrate on the factors limiting the use of these feedstuffs, originating from waste in animal feed formulation. The nutritional and technical aspects are outlined in Table 1.1.

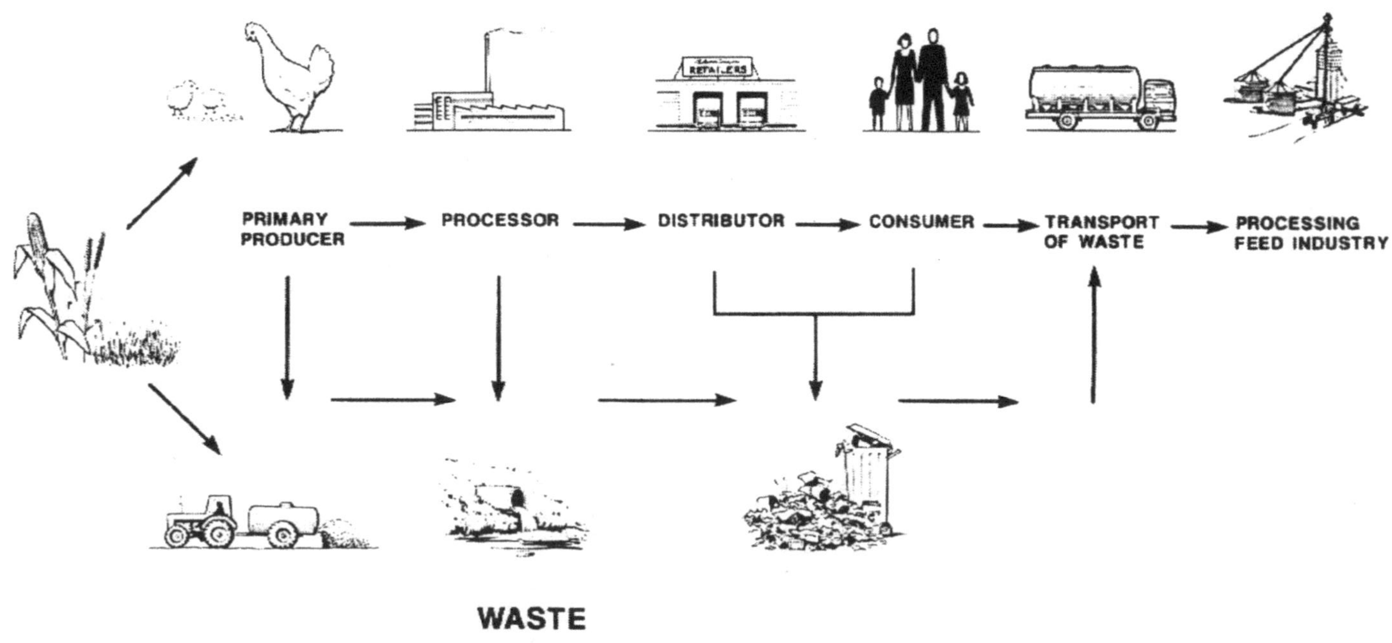

Figure 1.1. Animal, industrial and municipial waste streams and their manufacture into feedstuffs.

Table 1.1 Factors limiting the use of feed from waste in feed formulation. (Modified after Ravindran and Blair, 1991)

Nutritional aspects

- Variability in nutrient level and quality
  (soil, climate [temperature; rain], variety, harvest method, processing)
- Presence of naturally occuring antinutritional and/or toxic factors
  (alkaloïds, non-starch polysaccharides, glycosids, tartarates, heavy metals)
- Presence of pathogenic micro-organisms
  (*Salmonellae*; present if waste is not processed/sterilized properly)
- Need for supplementation
  (minerals, most limiting essential amino acids)

Technical aspects

- Seasonal and unreliable supply (need for storage)
  (winery, apple, dates; duration of transport)
- Bulkiness, wetness and/or powdery texture (need for pelleting)
  (Brewers' spent grains; poultry manure, sludge; potato starch)
- Processing requirements (drying, detoxification)
  (availability of machinery; knowledge of processing; energy source)
- Lack of research and development efforts (feed industries)
  (co-operation developed/developing countries; transfer of knowledge)

Waste disposal routes

Because of their fertilizing properties, animal wastes have been traditionally disposed by spreading them on the land. However, since the advent of chemical fertilizers, there has been a significant decline in the use of organic fertilizers, mainly owing to the costs of transport compared with that of more concentrated fertilizers. However, other modes of conversion of wastes are possible such as the conversion into feed by ensilation, dehydration, chemical treatment or fermentation to yield protein biomass. The estimated recovery from these conversion methods varies widely but the value of animal wastes as a feed ingredient, for example, appears to be far superior to their other uses with operating costs for conversion being low (Müller, 1980).

The means of waste disposal include waste treatment or waste utilization and, in general, conversion for use directly into food, feed or upgrading by micro-organisms are efforts to control the disposal of wastes (Rolfe, 1976). Some wastes such as fish muscle tissue can be readily upgraded for their use directly into food (food or precursor for food products). This application draws attention to public concern in general (food grade quality), to microbiological quality as well as to the acceptance by the consumer.

Other wastes can be converted directly into feed. This means the conversion of animal or vegetable protein and of minerals into livestock feed for ruminants, pigs and poultry who will, in turn, provide protein for humans. This application draws attention to the palatability, digestibility and nutritional value of these products when included in livestock diets. In addition, health hazards have been considered for the reservations regarding the feeding of animal wastes are usually based on the potential risks related to several factors. These factors include the nature of the waste biomass, its high bacteriological activity, the accumulation of antimetabolites, medicinal drugs or feed additives and other antinutritional excretory products derived from the wastes (Müller, 1980). These problems, however, do also relate to conventional feedstuffs. The nutritional as well as the toxicological evaluation of wastes as an animal feed is therefore a most relevant one. Finally, when wastes are unsuitable for either of the above uses, they may be converted into products through their conversion by means of micro-organisms. Both fermentation (yielding yeast or single cell protein) and ensilage can be used.

The quantitative and qualitative potency of raw materials that are suitable for the manufacture as by-products varies widely between economic classes and regions throughout the world. This is a particular reason why a given definition of offal or of by-product is often vague. The terms by-product and offal are basically used to denote the parts which are not included in the primary product (often used for human consumption). Such a primary product is the juice from fruits or the dressed carcass in the case of a slaughtered animal or poultry. It is noted for example that a group of organs such as spleen, brains and lungs are edible parts in certain developing countries whereas they are considered rather inedible in developed ones.

Slaughtery by-products can be divided into primary by-products and secondary ones (Mann, 1967). The primary products may include hides and skins, feathers, bones and also hatchery by-products such as infertile eggs and egg shells; the secondary class of by-products includes a wide range of products manufactured from the primary by-products. These secondary products include blood meal, blood albumen, egg albumen *etc.*

For vegetable products, a sound classification is much more difficult owing to the divergence of the primary objectives of processing as well as the processing procedure used (dry versus wet processing) associated with a specific group of fruits/vegetables such as dates, grapes, tomatoes or potatoes.

In the following paragraphs, some features are discussed that stress the need for the conversion of waste materials into feedstuffs. These features include analytical studies on the production and consumption of populations, of livestock products and of crops in several years and estimates for the year 2050. In addition, benefits derived from the use of feeds from waste, the race between food and population, the increase in the production of waste by-products and the role of the developed countries towards the developing ones are outlined. Finally, the fact that using feed from waste as a feedstuff will reduce the competition for food between humans and poultry is discussed.

1.2 ANALYTICAL STUDIES AND ESTIMATES OF FOOD AND WASTE PRODUCTION

1.2.1 Analytical Studies

Consumption and production of grain, and other commodities and wastes derived from them are influenced by variables that determine the quantities supplied and demanded from year to year in developed and developing countries. This section, therefore, has been set up as a discussion on factors that affect the race between food and population by analysing production and consumption items over time, and by partitioning the world's data over developed and developing countries.

1.2.2 Data Handling

The estimation of data on certain crops and livestock wastes, described in the following chapters, is based on data published by the Food and Agricultural Organisation (FAO 1971-1991). In an example of livestock products such as eggs, the data gathered from the FAO production yearbooks will be expressed as eggs^{FAO}. From these data, other characteristics have been calculated and these data are indicated in this study in italic such as *egg waste*.

All FAO data in a particular item (e.g. population^{FAO}) have been put to a data file, when necessary, per country. From this file, additional characteristics have been calculated to give total values for the developed and developing countries and the world, respectively. The used conversion factors for the calculation of wastes from FAO data on, for example, hide production are described below. All data have been used for the estimates of a certain characteristic in the year 2050 and either the characteristics under investigation are expressed in certain years (1962, 1975, 1988, 2050) or an estimate is given on the basis of a time-series approach.

In a time-series approach, projections are made by using a subset of known data in order to obtain a future estimate. For this objective, a variety of methods is available, and such will give different results based, for example on the accuracy of the projection, on the pattern of data and on the time period covered by the known data (Makridakis, Wheelwright and McGee, 1983).

Population, for example, is of course an important determinant for the consumption figures throughout the world, and projections for population numbers are useful for analytical and planning purposes. In order to make population projections, different assumptions can be made, providing more or less optimistic estimates for the forthcoming years. For example, Epperson (1986) gives a low, middle and high population projection based on different sets of assumptions based on lifetime births per woman, life expectancy and yearly net immigration. According to these projections, U.S. population by the year 2050 could be anywhere between 231 million and 429 million. This represents quite a range of possibilities and it is noted that population estimates have been considered relative to past projections. Bogue (1966) projected the world population increase from 1965 till the year 2000 with a high, medium and low variant (Figure 1.2). Judging and comparing these variants with the population in 1975 (Boda, 1990), the most accurate variant for the year 2000 proved to be the one providing a more linear increase and predicting the lowest population increase. In addition, different computer scenarios have been used and reported to illustrate the wide range of possible paths for the variable "population". Some of these scenarios decline, others characterize a society that has achieved a stable population with a high and sustainable standard of living until the year 2100 (Meadows, Meadows and Randers, 1992).

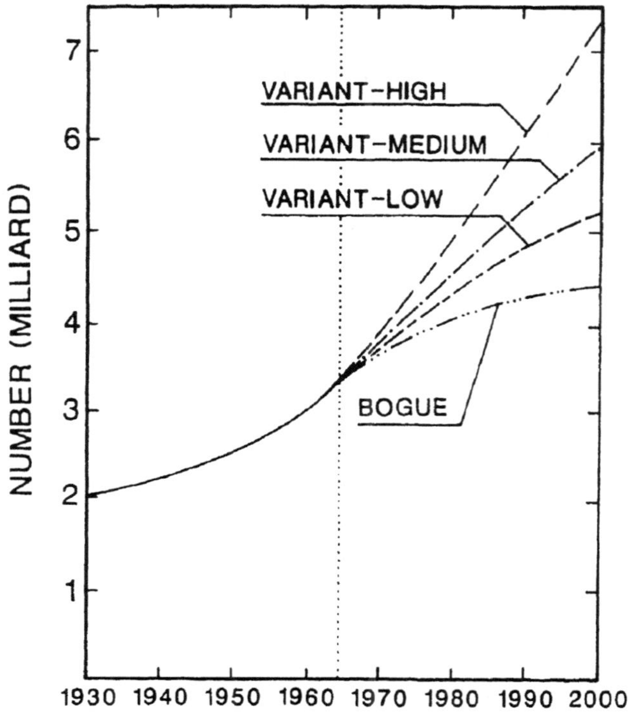

Figure 1.2. Projected estimates of world population. (Reproduced with permission from Bogue, 1966 and Boďa, 1990).

Taking into account the observation of Boďa (1990) as well as the alternative projections for population up to 2100 (Meadows, Meadows and Randers, 1992), the estimates for the several characteristics in the study of this book were based on linear regression, (up to the year 2050). In addition, it is assumed that the variables to be estimated in this study are based on FAO data which are not all specific or well defined enough to obtain a specific forecasting accuracy. For example, it is notable that disparities between the developed and developing countries in the rates of population increase and the growth of food production remain long-term problems as is, for instance, also the change in the make-up of the population characterized by its age distribution (Epperson, 1986). Moreover, the unit of measurements such as the waste percentage of livestock or crop processing, and the partitioning of total produced product into fresh eaten and processed products is not always well defined and estimates are used. Finally, in the long term forecasting when time horizon lengthens to more than fifty years, the trend element will dominate which is the more general objective of this study analysis.

The handling of data, therefore, was as follows:

- Time period: FAO data were used from 1961 till 1990 and prediction towards 2050.
- Variables: *population, protein* and *energy consumption, cereals* (*total cereals, maize and wheat*), *soya, consumption of eggs* and *poultry meat, production of waste products* (*slaughtery waste, manure, tannery waste, vegetables* and *fruit wastes*, respectively). For each item, a partitioning has been made for the total world, the developed countries and the developing ones, respectively.

Regression analyses were performed using the linear regression program (PROC REG) of the Statistical Analysis System (SAS, 1985).

1.2.3 Particular Items Derived from FAO Statistics

(a) Population

The data on population used in this study are generally provided by individual countries and described the time series of the United Nations' *Demographic Yearbook* which are estimating on total population and indicated as "population^{FAO}". For many developing countries, however, further adjustment on available estimates was sometimes needed in order to maintain a reasonable degree of consistency in time period estimates and consistency with data from external sources. The data used in the tables and figures, therefore, are based on a series of estimates for each country covering a fairly long period, particularly for developed countries whose demographic statistics are considered to be reliable.

(b) Food supply

In the FAO data, the total quantity of feedstuffs or foodstuffs produced in a country is added to the total quantity imported and adjusted for any change in stocks, and thus gives the supply available for the reference period. The per caput supply of a food characteristic available for human consumption is obtained by dividing the food supplies by the related data for human consumption. Also, the quantities of food available relate to the quantities of food reaching the consumer but not necessarily to the amounts of food actually consumed. The latter quantities will be lower owing to household losses and quantities fed to animals or thrown away.

(c) Poultry and livestock products

Slaughterings, production of (poultry) meat
The data used from tables concerning slaughterings relate to animals slaughtered within national bounderies, irrespective of their origin. Similarly, data on poultry meat relate to animals slaughtered (both farm and commercial) in the country concerned, regardless of the origin of the animal. Poultry meat includes meat from all domestic birds, wherever possible, based on ready-to-cook weight. Data on poultry meat production reported by national statistical offices are expressed in terms of either liveweight, eviscerated

weight, ready-to-cook weight or dressed weight. The data on total "poultry meat" from the FAO include broiler meat, meat from grand parent stock for meat and eggs, as well as meat from layers that reached one year production or more. Owing to the unspecified nature of the FAO data, the total "poultry meat" production used in the calculation was based on the assumption that all meat was derived from broilers. From the "poultry meat^{FAO}" data, the characteristics *poultry meat consumption*, production of *broiler slaughtery waste* and *production of broiler manure*, were calculated.

Poultry meat consumption was calculated by dividing the data for poultry meat^{FAO} by those of population^{FAO}.

The conversion factor for *broiler slaughtery waste* was estimated to be 53.8% of the broiler meat which is similar to 34.3% of the total live weight (El Boushy, 1990; El Boushy *et al.*, 1990).

For *production of broiler manure*, the calculation was based on the assumption that 1 kg of broiler feed will give a production of 1 kg broiler manure (North, 1978). Broiler meat is estimated to be 63.7% of the total broiler liveweight. Then, a conversion factor of 100/63.7 = 1.57 has been used to estimate the total production of broiler liveweight from the data of meat production from broilers (poultry meat^{FAO}). From these figures one can calculate the total quantity of feed consumed (average feed conversion = 1.8; North, 1978). Accordingly, the total manure produced was estimated. It is noted that it is difficult to estimate the total manure production from broilers without litter because the number of broilers kept in cages and on floor is not classified separately in the FAO data.

Egg production

Some countries have no statistics on egg production, and estimates have been derived from such related data as chicken or total poultry numbers and reported as assumed rates of egg laying. Most of the countries have statistics on egg production report either the total weight of eggs or the number of eggs produced. In addition, it should be noted that this mass of eggs is not divided into hatchery eggs or parent stocks for laying or broiler birds. Data on numbers have been converted into weight using appropriate conversion factors as follows. For the calculation of the characteristics *consumption of eggs*, *layer manure* and *layer slaughtery waste*, the following estimates were used (North, 1978; El Boushy, 1990): 16.4 eggs per kg of eggs; 280 eggs per hen per year (based on hybrid layer production); 34.3% of hen live weight as total waste available for rendering; 2.0 kg body weight on average per hen.

The *consumption of eggs* was estimated as follows:

$$\text{Consumption of eggs} = [\text{eggs produced}^{FAO}\ (\text{kg}) \times 16.4]\ /\ \text{population}^{FAO}$$

For the calculation of the production of manure from laying hens first an estimate of the number of laying hens was made:

$$\text{Number of layers} = [\text{eggs produced}^{FAO}\ (\text{kg}) \times 16.4]\ /\ 280$$

The production of manure was estimated by multiplying the number of layers by 100 (the estimated quantity of manure per hen per day; North, 1978), by 0.5 (assuming that in the world, 50% of the laying hens are kept on batteries), and by 365 (days per year).

For the calculation of the *total slaughtery waste* per layer, the number of layers was multiplied by the average weight of the hen (2.0 kg) and by 34.3% being the estimated slaughtery waste percentage of the total bird live weight for rendering.

Hides and skins

The estimate figures for hide wastes are based on the total production of fresh hides from cattle and buffalo, and skins from sheep and goats (expressed as fresh *total hide production*; FAO, 1971-1991). These FAO data used have been referred to fresh weight of hides and skins. Data for countries reporting production in numbers or expressed in dry, cured, or salted weight have been converted into fresh weight data using appropriate conversion factors. Where no official data are available, estimates based on slaughterings and on other information have been used.

Yields of fresh hide depend on the composition of raw material (animal species), moisture content, fat removal, age of the animals as well as dirt and manure attached to the hide. In general, loss of fleshing during tanning was estimated to be 16% (Ockerman and Hansen, 1988). For the calculation of *hide wastes* from total hide production^{FAO}, therefore, a conversion factor of 0.16 was used.

(d) Crops

Cereals (wheat, maize)

Area and production data of FAO on cereals relate to crops harvested for dry grain only. Cereal crops harvested for hay of harvested green for food, feed or silage or used for grazing are excluded. Wheat and maize were included in the data since these cereals are the most widely used carbohydrate energy sources included in practical poultry diets. For *total cereals* the data include other cereals such as rice, barley, rye, oats, sorghum, millet, coarse and mixed grain and buckwheat. Available data for spelt are included with those for *wheat*, except for Commonwealth of Independent States (FAO data of the formerly USSR).

Pulses (soybean)

The used FAO data are those of the production of the crop harvested for dry beans only as far as this can be ascertained, whether used for feed or food.

Vegetables (tomato, potato)

The FAO data used are those relating to the main vegetable crops (tomato and potato) grown mainly for human consumption. Statistics on vegetables are not available in many countries and the coverage of the reported data differs from country to country. In general, the estimates refer to crops grown in the field and market gardens mainly for sale. It is noted that quite an important part of the estimated total production in certain countries such as France (ca 40%) and Italy (ca 20%) is not included in the FAO data. Tomatoes and potatoes were selected owing to the relevancy of their wastes for use in poultry diets. For tomatoes, data for certain countries, particularly in central and northern Europe, are related to crops grown mainly or totally under glass with, subsequently, large yields per hectare.

Tomato. For the estimate of *tomato waste* the basis was provided by the production of tomatoes^{FAO} data. Tomatoes are consumed either fresh or in a processed form. From the total production of tomatoes, about 50% is processsed in any form (Goose and Binsted, 1973). This processing yields tomato wastes depending on the methods used. Waste yields are estimated to be 3-4.8% (coreing), 4.8% (peeling) while seeds of tomatoes represent about 2% of their fresh weight (Edwards *et al.*, 1952) providing about 9.8-11.6% total waste. Therefore, an average conversion factor of 10.7% has been used to calculate waste from total processed tomatoes.

Potato. According to Talbert (1987), 48% of the total production of potatoes is processed in any form. The waste produced from processed potatoes varies largely and is estimated to range from 2-50% depending on the size of the potato, variety, season, kind of end products and the peeling process. In general, 10% is considered as a reasonable yield for peeling losses using abrasive peeling (Smith and Huxsoll, 1987). To convert total production potatoes^{FAO} into total *potato wastes*, conversion factors of .48 (processed potatoes) and .10 (percentage waste from processing) have been used, respectively.

Grapes and fruits (oranges, dates, wine)
Certain countries do not publish data on total grape production. Estimates for these countries have been based on information available on the production of table grapes, raisins and wine. In most of the major wine-growing countries, FAO wine production has been estimated from the quantity of grapes crushed at harvest time. Consequently, it corresponds to the amount of "grapes for wine" for the same crop year and represents total output at wine presses, irrespective of its final destination. In addition, FAO data for some countries have been based on calculations using tax returns on trade estimates or on the basis of the quantity of grapes crushed for wine when such information was available.

Grape. There are special types of grapes that are suitable for making drinks. These grapes are mostly used for the manufacture of wine. Other types of fresh grapes for direct consumption or preparing of grape juice are not included in the calculation. *Wine waste*, known as "pomace", includes seeds, the skin and the pulp of grapes. It represents about 12.4% of the total wine^{FAO} (Amerine *et al.*, 1980).

Orange. The orange was selected in this study from the citrus family due to its world-wide popularity and its tremendous production. From the oranges, 50% or more is directed towards the fresh market. The orange waste estimate is based on the calculation of the total *orange waste* as peel, pulp, rag and seeds (from 50% of total oranges processed) being 44.6% of the fresh weight equivalent to 11.0% dry feedstuff with 10% moisture (Berry and Veldhuis, 1977; Nagy, Shaw and Veldhuis, 1977).

Date. Dates are pitted to avoid payment of the transportation of the pit and to improve their attractive quality for the consumer. Pits constitute to about 10% of the weight of the whole date (Rygg, 1975). A further estimate is that of the consumption of fresh dates being 50% of the total production (Dowson and Aten, 1962; Rygg, 1975).

The calculated *total waste* is derived as the sum of estimates for *wine*, *tomato*, *potato* and *date wastes*.

(e) Classification of countries by economic classes and regions

The classification of countries has been used as given by the FAO (1990). For the purpose of FAO analytical studies, the world is divided into economic classes, developed and developing as well as regions (Figure 1.3). The geographical coverage of the regions and their countries within the two classes is as follows:

CLASSIFICATION OF COUNTRIES BY ECONOMIC CLASSES AND REGIONS

CLASS I : ALL DEVELOPED COUNTRIES	CLASS II : ALL DEVELOPING COUNTRIES
Region A – North America : Canada, United States.	*Region A* – Africa.
Region B – Europe.	*Region B* – Latin America.
Region C – Oceania : Australia, New Zealand.	*Region C* – Near East : Africa.
Region D – Commonwealth of Independent States. (USSR)	*Region D* – Far East.
Region E – Other developed countries.	*Region E* – Other developing market economies.

Figure 1.3. Classification of countries by economic classes and regions. (Adapted from FAO, 1990).

Class I (All developed countries)

Region A - North America

Canada, United States

Region B - Europe

Albania, Andorra, Austria, Belgium-Luxembourg, Bulgaria, Czechoslovakia, Denmark, Faeroe Islands, Finland, France, Germany (including the former Federal Republic of Germany), Gibraltar, Greece, Holy See, Hungary, Iceland, Ireland, Italy, Liechtenstein, Malta, Monaco, Netherlands, Norway, Poland, Portugal (incl. Azores and Madeira), Romania, San Marino, Spain (incl. Spanish North Africa), Sweden, Switzerland, United Kingdom (incl. Channel Islands and Isle of Man), former Yugoslavia (in classification).

Region C - Oceania

Australia, New Zealand

Region D - Commonwealth of Independent States (Formerly USSR)

Region E - Other developed countries

Israel, Japan (incl. Bonin and Ryukyu Islands), South Africa.

Class II (All developing countries)

Region A - Africa

Algeria, Angola, Benin, Botswana, British Indian Ocean Territory, Burkina Faso, Burundi, Cameroon, Cape Verde, Central African Republic, Chad, Comoros, Zaïre (Congo), Côte d'Ivoire, Djibouti, Equatorial Guinea, Ethiopia, Gabon, Gambia, Ghana, Guinea, Guinea-Bissau, Kenya, Lesotho, Liberia, Madagascar, Malawi, Mali, Mauretania, Mauritius, Morroco, Mozambique, Namibia, Niger, Nigeria, Reunion, Rwanda, Saint Helena, Sao Tome and Principe, Senegal, Seychelles, Sierra Leone, Somalia, Swaziland, Tanzania, Togo, Tunisia, Uganda, Western Sahara, Zaire, Zambia, Zimbabwe.

Region B - Latin America

Anguilla, Antigua and Barbuda, Argentina, Aruba, Bahamas, Barbados, Belize, Bolivia, Brazil, British Virgin Islands, Cayman Islands, Chile, Colombia, Costa Rica, Cuba, Dominica, Dominican Republic, Ecuador (incl. Galapagos Islands), El Salvador, Falkland Islands (Malvinas), French Guyana, Haiti, Honduras, Jamaica, Martinique, Mexico, Montserrat, Netherlands Antilles, Nicaragua, Panama, Paraguay, Peru, Puerto Rico, Saint Kitts and Nevis, Saint Lucia, Saint Vincent and the Grenadines, Suriname, Trinidad and Tobago, Turks and Caicos Islands, Uruguay, US Virgin Islands, Venezuela.

Region C - Near-East

Africa: Egypt, Libya, Sudan.

Asia: Afghanistan, Bahrain, Cyprus, Gaza Strip (Palestine), Islamic Republic of Iran, Iraq, Jordan, Kuwait, Lebanon, Oman, Qatar, Kingdom of Saudi Arabia, Syrian Arab Republic, Turkey, United Arab Emirates, Yemen.

Region D - Far east

Bangladesh, Bhutan, Brunei Darussalam, Cambodia, China, East Timor, Hong Kong, India, Indonesia, Democratic People's Republic of Korea, Republic of Korea, Laos, Macau, Malaysia, Maldives, Mongolia, Myanmar, Nepal, Pakistan, Philippines, Singapore, Sri lanka, Thailand, Vietnam.

Region E - Other developing market economics

America: Bermuda, Greenland, Saint Pierre and Miquelon,

Oceania: American Samoa, Canton and Enderbury Islands, Christmas Islands (Australia), Cocos (Keeling) Islands, Cook Islands, Fiji, French Polynesia, Guam, Johnstone Island, Kiribati, Midway Islands, Nauru, New Caledonia, Niue, Norfolk Island, Pacific Islands (incl. Marshall Islands, Federated States of Micronesia, Northern Mariana Islands and Palau), Papua New Guinea, Pitcairn Islands, Samoa, Solomon Islands, Tokelau, Tonga, Tuvalu, Vanuatu, Wake Islands, Wallis and Futuna Islands.

1.3 BENEFITS DERIVED FROM THE USE OF FEED FROM WASTE

New, local industry will benefit from recycling waste by-products and the benefits from it lead to the following features.

1.3.1 Environmental Sanitation

The sanitary disposal of offal such as the by-products of the slaughterhouse present great difficulties. Not only these offals attract vermin and are causing the danger of spreading disease; it also tends to decompose rapidly forming an ideal substrate for micro-organisms and leads to objectionable odors. Burning or burying of inedible offal or their use as fertilizer leads to a total loss of potential by-products and, unless properly carried out, it may also lead to spread in disease (El Boushy, 1990). From the public health, veterinary and animal husbandry viewpoints, the direct feeding of inedible by-products to livestock or manuring of fields with certain by-products are unsatisfactory (Mann, 1967).

Since agricultural activities cannot be adjusted to the by-product throughput of processing plants, the by-products must be preserved until they are required. A process of concentration, preservation and/or sterilization (rendering) must be applied to products suitable for livestock feeding in view of the prohibition of bacterial multiplication (decreasing the moisture level) and in view of pathogens present.

Further disposal of offal concerns the effluents from slaughter houses and from the processing of potatoes, citrus, grapes and wine as well as from breweries. If these effluents will be purified, it would be possible to reuse wastewater and produce sludge that could be used in the dry form for livestock and poultry feeding purposes. The sludge could also be activated by means of micro-organisms that use the organic material as a substrate to provide a higher nutritional value in comparison with the original sludge. Subsequently, this sludge could be used as a protein supplement for monogastric animals and ruminants.

1.3.2 Livestock Health and Productive Agriculture

Offals can be used either to manufacture fertilizers or livestock feedstuffs depending on the speed with which they are handled and the freshness of the raw materials. Vast areas of less developed territories for example, have soil and pasture which are deficient in calcium and phosphorus. For these areas, treated offals which are very rich in nitrogen, calcium, phosphorus and trace elements may be of great benefit when used as fertilizers. In addition, where these offals are not utilized as animal feed supplements, the wastes of large amounts of offal (citrus, grapes, potatoes, wastewater) rich in nutrients such as protein and minerals, are prodigious (Mann, 1967). The creation of a by-product industry, therefore, is the challenge that can be met by putting the proteins, minerals and vitamins into circulation for the benefit of livestock and consumer rather than exporting them.

1.3.3 Secondary Rural Industries

In this context, effective planning of animal feed sectors, including rural industries for the manufacture of by-products, making use of local feed resources and not relying on imports, would seem desirable. For example, the planning of a poultry feather industry will make it possible to create a market for fillings for upholstery such as pillows and mattresses. Other animal by-products such as extracted fat, dried blood and meat meal, are important basic ingredients for the local soap industry and livestock feed, respectively. Other relevant by-products are produced from the processing of citrus, tomatoes, potatoes, dates, apples, grapes, *etc.* The water purification for citrus, apple and wine processing, and slaughterhouses also produces by-products such as (activated) sludge. Leather and tannery processing also produce their by-products as hide waste or hydrolized hide meal.

1.3.4 Price Structure

The use of by-products and recycling waste will influence the prices of meat and eggs and the price paid in foreign currency (owing to the imports of fish meal or soybean products used as feedstuffs from abroad) to the producer of livestock. Depending on local circumstances, the return derived from the use of by-products (animal, vegetable and fruit wastes) may be used to decrease the prices of meat and eggs to the consumer or to give the livestock producer more gain for his product and by-products.

1.3.5 Creation of New Employment

There is no doubt that the disposal of waste (inedible offal, fruit and vegetables, tannery or municipal refuse) needs little or no manpower in developing market economics. The conversion of offal, however, into valuable by-products creates new employement and skills at the place of production and, in addition, in the secondary industries based on such raw materials. If the scientific knowledge will be transferred from developed to developing countries and the subsidies can be in a form of installation of new factories, this will be reflected in the new employment, on more income and a better life.

1.3.6 Reducing Imports of Feedstuffs (Foreing Currency) by Upgrading Local Waste

If a local industry such as the poultry industry will be able to transfer all the offals by rendering and produce poultry offal meal (blood, feet, heads, intestines and feathers), a product will be created with high nutritive value. Also waste from fruit and vegetables such as citrus pulp, tomato pomace, potato waste, dates and even hide and tannery by-products can be transfered into valuable feedstuffs (El Boushy, 1986). Similar potential feedstuffs are waste products originating from rendering and from wastewater treatments. This application may lower the present imports of feedstuffs, fertilizers, cereals and soya, animal products and total agricultural products that require foreign currency (Table 1.3).

Table 1.2. Values of several relevant imported items in developing market economics (x 10.000 $) per year. (Reproduced from Trade year-books, vols. 26, 30, 35, 39, 43, 50, published by the FAO, 1972-1990 and 1996)

Items/Years	*1971*	*1975*	*1980*	*1985*	*1989*	*1996*
Agricultural Products	1.169.123	3.155.055	6.824.087	6.235.720	8.559.710	16.860.500
Cereals	14.902	29.466	537.663	562.930	570.337	671.352
Soybeans	11.341	27.676	110.018	126.736	180.505	385.562
Poultry meat	5.733	15.127	110.758	82.886	113.663	250.881
Eggs	3.013	3.776	36.228	34.565	31.071	38.751
Feedstuffs	15.427	26.603	103.082	119.436	219.203	-
Lard and fat	4.434	9.004	9.601	11.861	9.374	53.172
Fertilizers crude	27.078	164.093	146.469	99.260	119.116	117.630
Fertilizers manufactured	10.661	42.794	130.789	162.337	257.528	997.698

(a) Municipal refuse

Accumulation of municipal refuse, mainly garbage (known as swill or kitchen refuse) and classified as hotel and restaurants, institutional, military camp and municipal garbage, will cause tremendous pollution, odours and illness owing to spreading of micro-organisms, rats and insects. This is the case in developing countries, where sanitation and hygienic systems are below the standards in comparison with developed countries. In addition, if this kitchen waste was processed and sterilized it will be a very reasonable feedstuff when separately used or when mixed with other wastes such as blood. feather or manure to be used in animal and especially poultry nutrition.

The study of the nutritional composition and the use of kitchen/restaurant waste as a diet ingredient for animals presents a rather recent concept in the trend of using feed from waste in animal nutrition. Under certain conditions, treated waste of this origin appears to be a promising possibility as an animal feed ingredient because of its nutritional potential and availability in large quantities (Yoshida and Hoshii, 1979; Lipstein, 1985). As early as fifty years ago it was already found that, for example, kitchen waste could be used with advantages in poultry diets (Pierson, 1943).

(b) Leather tannery waste

The world production of fresh hides is estimated to be 8.8 million tons per year (FAO, 1991). The hide and tanning industry produces a large amount of waste and polluting materials and has the unevitable reputation of being filthy and evil smelling. This is confined to the present processing methods for the conversion of hides into leather and the use of chromium salts that, unless recovered and recycled, are the polluting chemicals. Dechromed materials derived from modern tannery procedures are potential sources of nutrients and therefore can be used as an animal feedstuff.

(c) Wastewater

Increased regulatory requirements for the reduction of the use of water and the production of wastewater has warranted management attention today. In poultry processing plants as well as in plants producing vegetable waste from potato and citrus, for example, reduction in the polluting load are objectives to establish economic, environmentally friendly and modern treatment procedures.

Since wastewater can be treated by means of flotation, sedimentation and flocculation, the effluent produced will contain proteins and fat that can be recovered and subsequently processed to eliminate undesired ingredients in order to make the waste suitable for livestock feeding.

(d) Fruit, vegetable and brewers' waste

Waste residues from fruit, vegetables and brewers' grains are rarely identified in developing market economics. Although costs are involved for the recovery of a useful feedstuff from this type of processing waste, credit is introduced by the use of the derived feedstuffs in animal diets. The use of feed grade protein or starch from the potato industry is a typical example of recoverable fractions, either as solid or as sludge, that can, after drying and sterilization, be included directly in livestock diets as a sources of protein and carbohydrates.

1.4 INCREASE OF POPULATION AND THE SHORTAGE OF FOOD SUPPLY

1.4.1 Population

At first, the world's population was estimated to grow at about 1% per annum in the periods when the first and second FAO World Food Surveys were published (1946 and 1952). Since the beginning of the last decade, a successful control of epidemics and diseases in large parts of the world has, among other causes, led to a marked lowering of death rates and, consequently, to an accelerated growth in the world's population (FAO, 1963). For example, the annual percentage rates of growth for the world's population in the period 1938-50 and 1950-60 were estimated to be 1.1 and 2.1, respectively (Table 1.3). The rate of growth in the less developed regions in these periods was 1.4 and 2.5% indicating the increasing share of the less developed areas in the world's population up from the sixties. This is also valid for the following years up to the decade of 1980-90.

The trend in the development of the population increase up from 1961 is shown in Figure 1.4. In comparison with the developing countries, the developed countries showed no significant increase in population. In this way, the estimate for the increase in the total world population by 2050 is largely based on the high increase of developing countries. Our expected development for 2050 is based, however, on the linear regression of the period 1961-1990. If no extreme disasters such as earthquakes, epidemic diseases, climatic disorders (e.g. drought, flooding) or war occur, this development seems not a pessimistic one. Developments, however, may change these expectations considerably (Meadows, Meadows and Randers, 1992).

Table 1.3 Population (millions) and population growth (%) in annual percentage in the world and in developed and developing countries. From FAO (1963) and FAO (1971-1997)

Years		*Developed*		*Developing*		*World*	
		Population	*Growth*	*Population*	*Growth*	*Population*	*Growth*
1938	- -	717	-	1.478	-	2.195	-
1950	- -	751	-	1.733	-	2.484	-
-	1938-1950	-	0.4	-	1.4	-	1.1
1960	- -	852	-	2.161	-	3.014	-
-	1950-1960	-	1.3	-	2.5	-	2.1
1970	- -	1.074	-	2.603	-	3.677	-
-	1960-1970	-	2.6	-	2.1	-	2.2
1980	- -	1.169	-	3.268	-	4.437	-
-	1970-1980	-	0.9	-	2.6	-	2.1
1990	- -	1.248	-	4.046	-	5.294	-
-	1980-1990	-	0.7	-	2.4	-	1.9
1997	- -	1.462	-	4.386	-	5.894	-
-	1990-1997	-	1.8	-	3.1	-	5.0

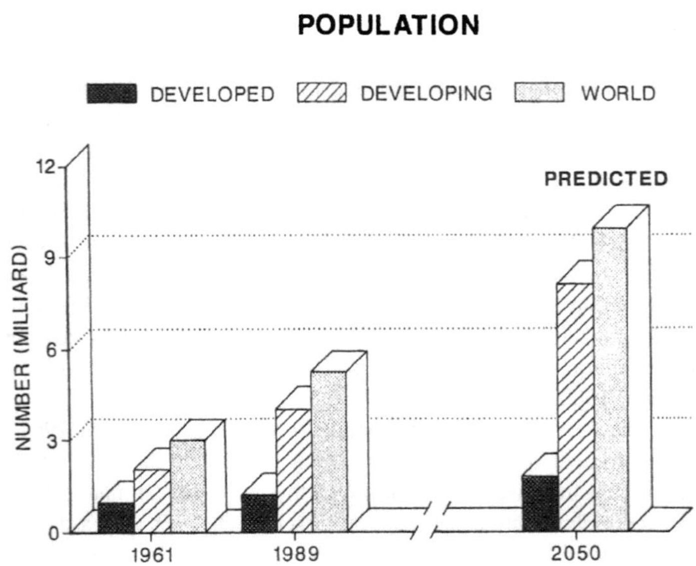

Figure 1.4. The expected development of the increase in population in the world, and in developing and developed countries [(1961, 1989 and estimate for 2050); milliard = 10^9].

1.4.2 Food Supply

For comparison of food supplies it is notable, that disparities in the per caput supplies are best described in terms of the nutritive value of the diet. The quantitative aspects are then measured by the calorie or protein level of the diet; for qualitative aspects, however, no simple measure is available. The FAO data were based on Calories (Cal), therefore we used this unit in this study, although "Joule" (J) is the unit currently used (1 J is 0.239 Cal). Figure 1.5 shows the average levels of calorie and protein supplies and their partitioning for animal and vegetable sources for the periods 1961, 1989 and an estimate for 2050.

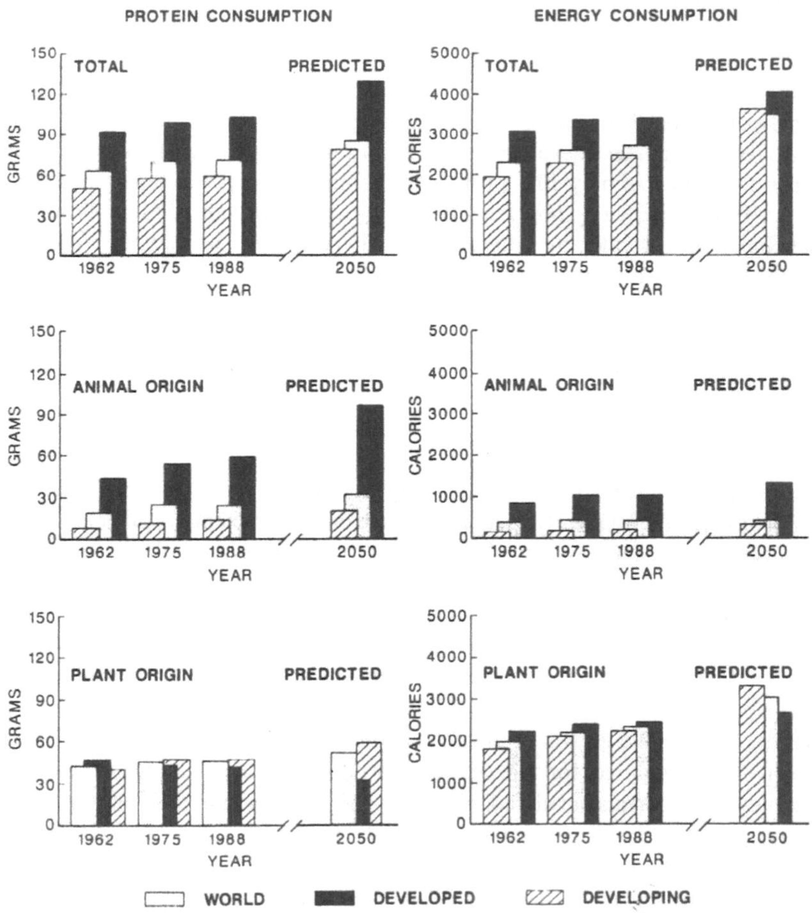

Figure 1.5. Contrasts in the total consumption of protein (g/caput/day) and energy (calories/caput/day) from animal and plant origin in 1962, 1975, 1988 and estimates for 2050.

Calories
For calories, Figure 1.5 shows that the less developed regions consume between 2,000 and 2,500 calories per caput per day in recent times which is only slightly higher compared to the 2,000 calories in the period 1938-1950 (FAO, 1963). The food consumed by the more developed regions provides more than 3,000 calories per caput per day of which about 30% is energy from animal origin. It appears that the current level of calorie supplies of plant origin has reached a stable level in developed regions in contrast with the potential growth in developing regions. Calorie supply of animal origin is still growing for both developed and developing regions. The expectations of high energy consumption will - of course - not continue in an increasing trend since the more modern nutrition is leading to changing eating habits. A lower use of high-energy sources (fats, starch, sugar) is observed with increasing fibre in the diet to avoid obesity, heart and coronary disease problems.

Protein
For protein, a striking contrast can be seen in the consumption of total protein and the supply of protein from animal origin. Although the total protein supplies per caput per day have gone up, less developed regions consume about 10 g of animal protein per caput per day, which is only 15% of the consumption per caput in the more developed regions in 1988. Figure 1.5 shows that the gap in animal protein consumption between the developed and developing countries is still widening. The estimate for the year 2050 shows that, in the developing countries, the amount of plant protein consumed is about twice that of protein from animal origin. The development of cheap sources of energy and of protein of animal origin to replace the traditional feedstuffs is of utmost economic importance.

1.4.3 The Race Between Food and Population

As the selected four feedstuffs under investigation are cereals (maize and wheat, total cereals) and soya, are frequently used in livestock diets. For example, in diets for poultry they are used at inclusion levels up till 60 percent (total cereals) and till 15 percent (soya). If these feedstuffs were eliminated as a diet ingredient, roughly half of the ration could be replaced by carbohydrate wastes such as potato waste, date pits, tomato waste or protein wastes such as by-products from slaughtery houses and products from wastewater treatments. Cereals and soya, then are left for human consumption and the competition between human kind and poultry will be lowered.

Figure 1.6 shows the growth of the population in relation to the production of cereals and soya. It is clear that for the developed countries the situation is rather stable. Accordingly, total cereals are abundant in terms of production in relation to population. In contrast, for the developing countries which show a tremendous increase in their population, there is no compensation by the increase in total cereals or soya, a situation that will be more severe by the year 2050. A similar shortage for cereals and soya then is present in the world's situation, although the differences is somewhat smaller. This example clearly shows the race between food and population, and draws attention to the use of waste materials rather than these major food groups in practical animal diets.

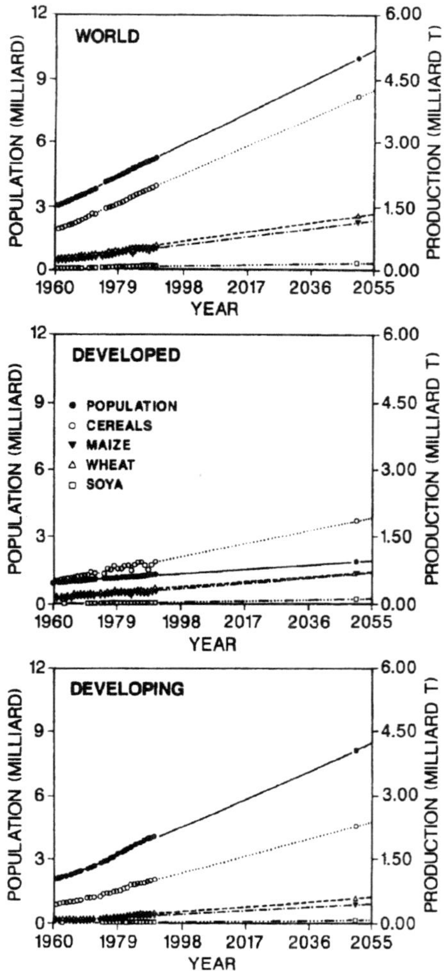

Figure 1.6. World population and production of cereals and soya; milliard = 10^9.

In the calculations, roots, pulses other than soya, certain vegetables and fruits, meat from pork and beef as well as animal products such as milk and fish were not included, because our concern was only to select a special and relevant group of feedstuffs commonly used in poultry diets.

In Figure 1.7 the per caput consumption of poultry products (eggs and poultry meat) is shown for the world and for developed and developing countries.

Egg consumption is low and increasing only slowly in developing countries. In the developed countries, however, it will reach the 450 eggs per caput per year estimated for 2050.

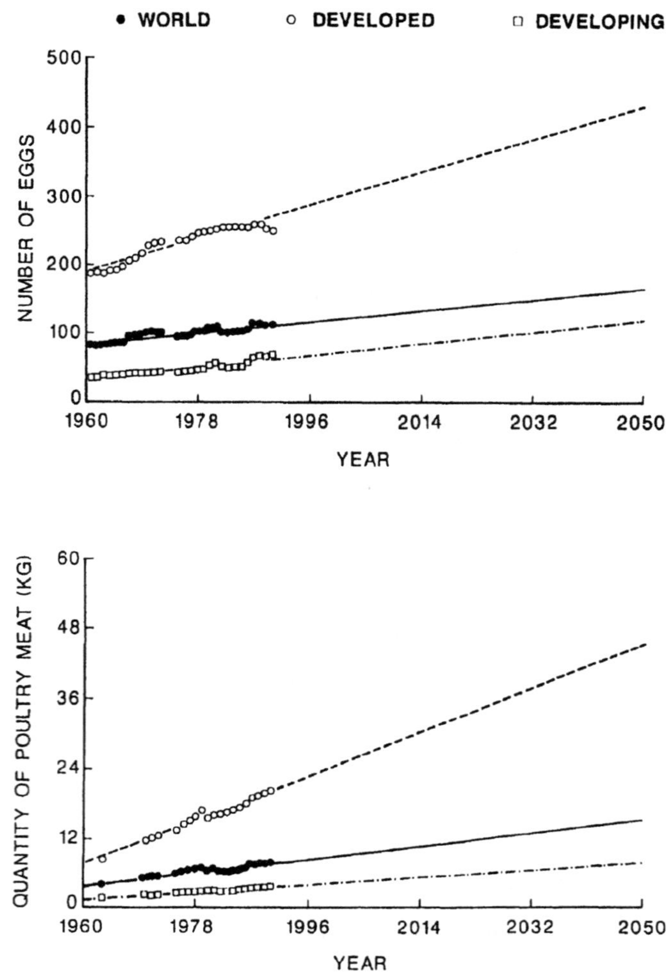

Figure 1.7. Consumption (caput/day) of eggs (human and industrial) and poultry meat (kg) in various years and estimates till 2050.

The figure for consumed eggs includes fresh eggs and industrial ones consumed as salads, cakes, pie *etc.* It is noted that in the developed countries the number of eggs consumed has decreased slightly according to the FAO data in the years after 1985. It is assumed that the problem of cholesterol and the subsequent fear of heart diseases and arterioscleroses may play a role in this consumer behaviour and is responsible for the decreased egg consumption. On the other hand, it is also known that the present egg consumption per caput per year in Israel and Japan is 365 and 355, respectively (FAO, 1991).

For *poultry meat consumption*, a similar trend is noticeable. The developing market economics is not reaching an appreciable level of kilograms of consumed poultry meat per caput per year. On the contrary, in the developed countries the present consumption of poultry meat per caput per year (20.5 kg) is higher than the estimate for its consumption in developing countries in the year 2050 (7.5 kg per caput per year).

The data on egg and poultry meat consumption are in agreement with the animal protein consumption data of Figure 1.5 showing the increase in the consumption of protein in developing countries which is largely based on plant proteins, mainly pulses.

1.4.4 The Increase of Waste

The current trend towards animal wastes recycling is motivated by both economic and environmental considerations. Since feed costs are about 80 percent of the total animal production costs (Boucqué and Fiems, 1988), the substitution of conventional feedstuffs by processed animal or vegetable wastes will lead to a significant reduction in the costs of animal feed and the ultimate derived products. Based on overcoming health hazards, especially with regard to animal wastes (accumulation of macrominerals, trace elements, heavy metals and harmful organisms; Müller, 1980) to ascertain the safe inclusion of wastes in feedstuffs, this substitution will be the key to the economics of the use of wastes in the nutrition of livestock. It is noted, however, that the the problem mentioned above does not relate to wastes alone: conventional feeds may also contain a large number of contaminants and antinutritional factors (Van der Poel, 1990).

The economic potential of the utilization of animal and vegetable wastes as new feed resources is of great importance, and a greater efficiency of this utilization is urged for the future on the basis of the data, illustrated in Figures 1.8 and 1.9. In Figure 1.8, the total estimated production of animal wastes in the world, and the developed and developing countries is shown. It can be seen that the world production of manure from laying hens in 1990 will be doubled by the year 2050 being equally divided over the developed and developing countries. The increase in the production of broiler manure by the year 2050 is even greater with a higher production in the developed countries. Estimates for the increase in slaughter waste from layers and broilers, respectively, are similar to that of the manure of these types of poultry due to the method of calculation: broiler liveweight and the number of laying hens were the basis of the calculation of manure and slaughtery waste, respectively.

The total quantity of hide waste, derived from cattle and buffalo hides and from sheep and goat skins, shows a steady increase in future years to a level of about two million tons by the year 2050. Although the greatest production before 1980 was established in developed countries, this has now turned to the developing market economics which will produce about 60 percent of the total hide wastes in 2050.

Figure 1.9 shows the production of vegetable and fruit wastes (separately and combined) in various years and estimates for 2050. These wastes are nowadays important in both developed and developing countries. Wine is an important product in developed countries while dates are important for developing ones. These countries are fully responsible for the increase in the production of their respective wastes in future years. For orange waste, however, its world production increase can be fully attributed to the increase in the production by developed countries. The production of potato waste in the world is somewhat decreasing combining an increase in production in developing countries and a decrease in the developed ones.

In general, the estimates for the production of total vegetable and fruit wastes reflects the previously discussed individual waste groups and shows a steady increase in the production of waste for the developing countries up to 35 million tonnes in 2050. This production in developed market economics is not increasing and may even slightly decrease as there has been a pronounced decrease of the production of potato as a crop because of the consumers' preferences for other sources of energy such as rice and wheat based pasta. Excluding the potato waste, the total production of vegetable and fruit wastes will also increase in the developed market economics.

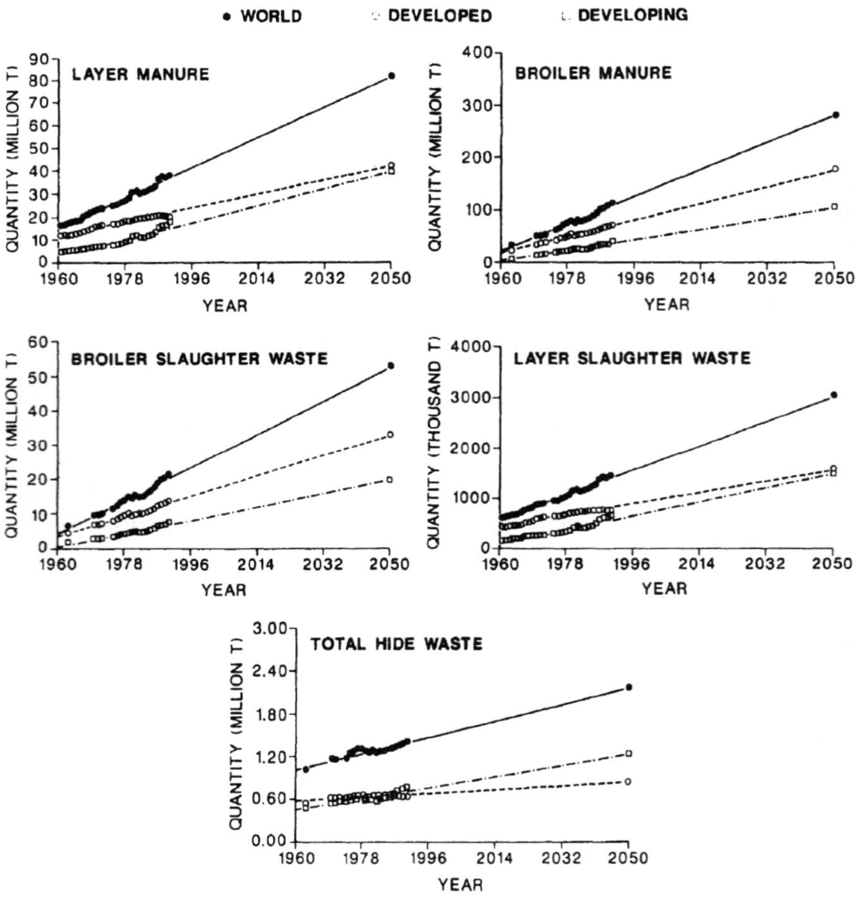

Figure 1.8. Production of poultry manure, poultry slaughter waste and total animal hide waste (cattle, buffalo, sheep and goat) in various years and estimates for 2050.

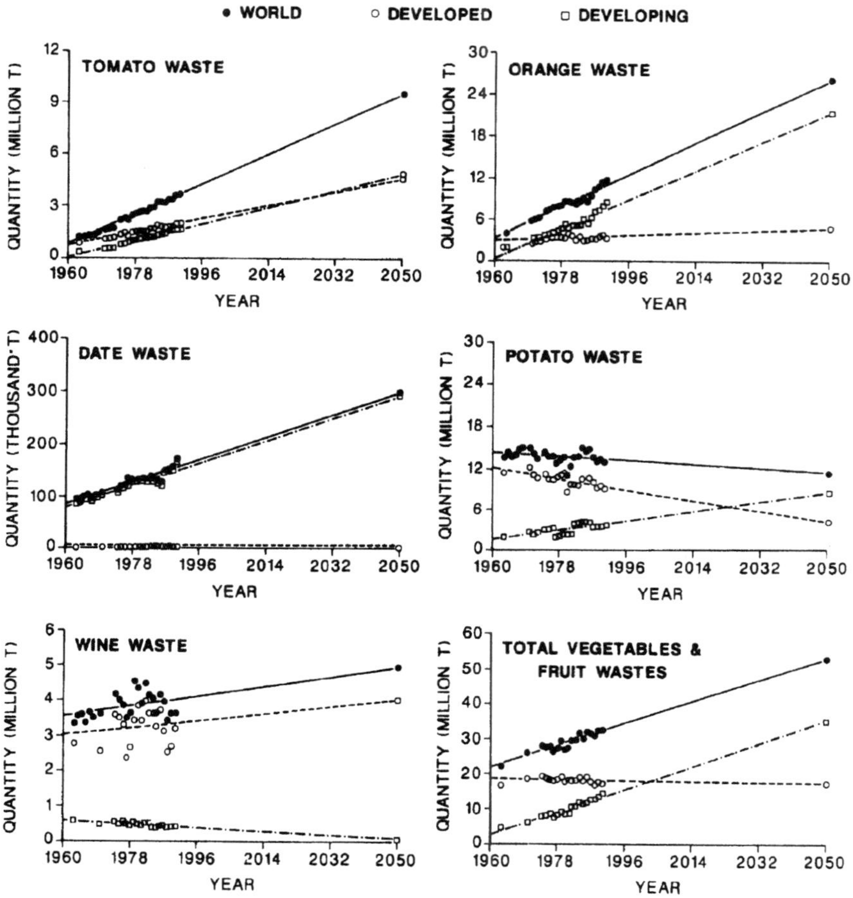

Figure 1.9. Production of vegetable and fruit wastes (separately and combined) in various years and estimates for 2050.

1.5 THE ROLE OF THE DEVELOPED COUNTRIES TOWARDS THE DEVELOPING ONES

Potential food production and the potential use of wastes in animal nutrition may increase if present technical knowledge in the field of animal and poultry feed science and its technology can be spread. This will minimize the factors which are limiting the use of feed from waste at present. This means that, in addition to overcoming, for example, the problem of scarcity of feedstuffs, present knowledge must be given to developing countries concerning:

- Composition of animal and vegetable wastes as potential feedstuffs.
- Timely use of waste materials and practical application.
- Scientific feeding strategies.

This implies, that both nutritional and technical aspects have to be covered, although certain socio-economic aspects also have to be considered (Ravindran and Blair, 1991). According to these authors the latter aspects concern the competition of products as human food, ingredient cost per unit energy and protein, and processing costs as well as fear of loosing market share if a new feedstuff fails to maintain production standards.

In West Africa, Nigeria for example, there is an ample supply of the principal by-product of the brewing industry, brewers' spent grains (BSG). In addition, large quantities of wheat are imported by Nigerian flour mills. Replacement of this relatively expensive imported wheat with spent grains is even desirable for it helps to reduce imports which require foreign exchange; BSG are much cheaper per tonne in comparison with wheat flour and BSG can be used for both poultry feed and bread and cookies. Cookies made with 20% BSG contained 55% more protein, 90% more lysine and 220% greater fibre than control cookies made with wheat flour and were within the organoleptic limitations established for taste and texture (Kissel and Prentice, 1979). This example may illustrate the importance of the use of agro-industrial by-products rather than expensive traditional products.

Transfer of knowledge of agricultural waste processing and waste utilization as an animal feed ingredient is also one of the tasks of the developed countries. It has to include both knowledge of the processing of waste into feedstuffs and of the knowledge of its utilization in diets for poultry and other animals. The extent of the use of waste in poultry diets further depends on the costs of the traditional feedstuffs, the safety for poultry health and the attractiveness of alternative use.

Developed countries may help by gaining and transferring relevant information on items such as:

- Machinery (high pressure cookers for feathers and for leather meal; disc driers for pomace, flash driers for blood), its processing conditions for the treatment of waste or waste mixtures), and the introduction of new methods of processing as well as the use of local, cheap energy sources.
- Methods for the chemical evaluation of antinutritonal factors, micro-organisms, non-starch polysaccharides and palatability.
- Methods for the nutritional evaluation, both in vitro methods and bio-assays. This must also cover information on the possible inclusion level of waste products in diets for different animal categories and the effects on reproductive characteristics and performance.

The use of waste products in animal nutrition may lower animal performance, however, these products may be much cheaper than the traditional (and sometimes imported) feedstuffs currently in use.

For the identification of novel agricultural wastes as an animal feed, for example, an evaluation scheme can be applied covering waste production method and nutritional value (Van der Wal, 1976). Therefore, a detailed description of the chemical composition and of the production process is required and the chemical evaluation must at least include the nutritional value for protein and energy, including the amino acid profile of the waste protein. Moreover, biological testing of the nutritional and toxicological characteristics is necessary, the latter being concerned with the safety for the animals as well as the safety of the animal products (residuals) for human consumption. Digestion of feed ingredients and absorption of nutrients and energy, the requirements for nutrients and the susceptibility to antinutritional factors, pathogenic micro-organisms and toxins are specific for an animal species and may vary with their physiological stage. In addition to the classical laboratory animal tests, the biological evaluation must therefore also include experiments with the main target species during various stages of the life-cycle (Van der Poel *et al.*, 1990). The extent to which such an evaluation scheme has to be applied will then depend on the degree of novelty judged from the data obtained from the different evaluations. As stated previously (Van der Wal, 1976), the criterion for a relevant evaluation is a sound judgement, from the product and process characteristics, of the degree of novelty, and the risks that are deducible from that with regard to toxicity, pathogenic micro-organisms and nutritional value, including digestibility.

1.6 CONCLUSION

In both developed and developing countries, continued growth of the by-product industry is anticipated. In this study, linear regression analysis has been used to estimate developments in the total consumption of animal products and production of waste products. In some cases this may be somewhat optimistic as far as a linear plateau concept (Bogue, 1966 and Boďa, 1990; Figure 1.2) or even alternative, complex scenarios can be expected (Meadows, Meadows and Randers, 1992). The number of slaughter farm animals and, consequently, the quantity of hides produced has been indicated to have reached its maximum in developing countries. However, an increase in the production of waste products can be expected. This means that large quantities of various waste-products, of animal and vegetable origin, will be available. For reason of a better utilization of these by-products and increased consciousness of environmental protection, the use of waste in practical livestock and poultry diets should be increased thereby decreasing the competition for food between humans and poultry.

Since many by-products or wastes have a substantial potential value as a diet ingredient in livestock diets this kind of utilization may be economically worthwhile especially in developing countries where traditional and often expensive feedstuffs can be replaced. At present, the agro-industry in developing countries has still to be developed on the basis of fundamental and applied research in the area of feed science and technology. Fundamental research is especially needed to increase our knowledge concerning the constituents of the various by-products and especially the changes that can take place for these constituents during stages of primary processing, subsequent storage and during secondary processing in the animal feed industry (Mann, 1967; Rolf, 1976; Boucqué and Fiems, 1988; Van der Poel, 1990). Practical research has to be carried out under local circumstances with the help of developed countries providing the information on processing equipment, processing conditions and the chemical/nutritional evaluation of waste products in animal diets.

Owing to modern scientific knowledge and the practical accumulated results from commercial applications in relation to nutrient recovery from animal and vegetable wastes, utilizing these wastes as a feedstuff will result in reduced feed costs and lower prices for animal products. It also contributes to self-sufficiency in nutrients from locally present wastes. Therefore, the system of feed from waste makes possible a vertical, mutually complemented integration of animal production among individual species which can, in turn, solve some problems of local waste disposal and thus some problems of pollution (Müller, 1980). It is noted that feeding certain wastes (such as municipal sewage sludge containing heavy metals) to poultry may be reflected in an increased pollution of the birds. The effluent of wastewater from potato, citrus, or wine processing or slaughter, however, it is not causing this kind of pollution.

Finally, the combined results of the investigations concerning feed from waste will serve as a sound basis for continued technological developments as well as applied research for different products from local wastes. It will help also to ensure that these wastes will be of a real value as an outlet for the main products from which they have been derived.

In the next chapters attention will be paid to the utilization of feed from several types of wastes in the nutrition of poultry, being one of the monogastric species suitable to convert waste nutrients. Firstly, poultry feeds include expensive grains and therefore poultry can be seen as competitors with humans as far as nutrition is concerned. Secondly, in many developing countries (Middle East for example) pork meat is prohibited by religion and poultry husbandry can then be an alternative in the provision of animal protein.

The following chapters describe the potential of nutrient recovery from certain animal and vegetable wastes. These include dried poultry waste, slaughtery products, tannery products and wastewater from slaughter-houses and vegetable and fruit wastes, municipal garbage and wastewater from, among others, the citrus and brewery industries.

The use of these waste products raises the problem of palatability of poultry diets. As poultry have the ability to differentiate between sweet, salt, sour and bitter chemicals, palatability and acceptability of diets based on waste products may be improved by using high-intensity sweeteners in the diets. Some aspects of these feed additives in controlling feed intake will be discussed.

REFERENCES

Amerine, M.A., Berg, H.W., Kunkee, R.E., Ough, C.S., Singleton, V.L. and Webb, A.D. (1980). *The Technology of Wine Making*. AVI Publishing Company, Inc., Westport, Connecticut, p. 663.

Berry, R.E. and Veldhuis, M.K. (1977). Processing of oranges, grapefruit and tangerines. In: *Citrus Science and Technology* (Eds. S. Nagy, P.E. Shaw and M.K. Veldhuis) Vol. 2. AVI Publishing Company, Inc., Westport, Connecticut, pp. 177-252.

Birch, G.G., Parker, K.J. and Worgan, J.T. (1976). *Food from Waste*. Applied Science Publishers Ltd, London, p. 301.

Bod̆a, K. (1990). *Nonconventional Feedstuffs in the Nutrition of Farm Animals*. Developments in Animal and Veterinary Sciences, 23, Elsevier, Amsterdam, p. 258.

Bogue, D.J. (1966). The prospects for population control. In: *New Protein Foods*, 3. Univ. of Chicago Press, pp. 21-43.

Boucqué, Ch. V. and Fiems, L.O. (1988). Vegetable by-products of agro-industrial origin. *Livest. Prod. Sci.*, **19**, 97-135.

Dowson, V.H.W. and Aten, A. (1962). Dates: Handling, Processing and Packing. *FAO Agricultural Department Paper*, no. 72, Rome, Italy, p. 223.

Edwards, P.W., Eskew, R.K., Hoersch, A., Jr., Aceto, N.C. and Redfield, C.S. (1952). Recovery of Tomato Processing Wastes. *Food Technol.*, **6**, 383-6.

El Boushy, A.R. (1986). Local processing industries offer food and beneficial by-products to developing countries. *Feedstuffs*, **58**, (3), 36-7.

El Boushy, A.R. (1990). Using non-traditional feed as a feedstuff will reduce the competition for food between human kind and poultry. *Third International Symposium on Feed Manufacturing and Quality Control*, 7-9 may 1990, Cairo, Egypt, pp. 305-20.

El Boushy, A.R., Van der Poel, A.F.B., Boer, H. and Gerrits, W.J.J. (1990). Effects of processing conditions on quality characteristics of feather meal. *Internal Report 1990/7, Wageningen Agricultural University, Department of Animal Nutrition*, Wageningen, The Netherlands, p. 32.

Epperson, J.E. (1986). Fruit consumption trends and prospects. In: *Commercial Fruit Processing*, 2nd Edition (Eds. J.G. Woodroof and B.S. Luh), AVI Publishing Company, Inc., Westport, Connecticut, pp. 647-72.

FAO (1963). Third World Food Survey. *FFHC Basic Study*, no. 11, Rome, Italy, 102 p.

FAO (1971-1997). *Production yearbook*, vol. **24-44, 51** 1970-1990,1997. *Food and Agricultural Organization of the United Nations*, FAO Statistical Series, Rome, Italy.

FAO (1972, 1976, 1981, 1986, 1990, 1996). *Trade yearbook*, vol. **26, 30, 35, 39, 43, 50.** *Food and Agricultural Organization of the United Nations*, FAO Statistical Series, Rome, Italy.

Goose, P.G. and Binsted, R. (1973). *Tomato Paste and other Tomato Products*. 2nd Edn. Food Trade Press Ltd., p. 266.

Inglett, G.E. (1973). The challenge of waste utilization. In: *Symposium: Processing Agricultural and Municipal Wastes*. AVI Publishing Company, Inc., Westport, Connecticut, pp. 1-5.

Lipstein, B. (1985). The nutritional value of treated kitchen waste in layer diets. *Nutr. Rep. Int.*, **32**, (3), 693-8.

Kissel, L.T. and Prentice, N. (1979). Protein and fiber enrichment of cookie flour with Brewers' spent grain. *Cereal Chem.*, **56**, (4), 261-6.

Makridakis, S., Wheelwright, S.C. and McGee, V. (1983). *Forecasting: Methods and applications*. 2nd edn., John Wiley & Sons, New York, p. 923.

Mann, I., (1967). Processing and utilization of animal by-products. *FAO Agricultural Development Paper* no. **75**, 2nd printing, FAO, Italy, pp. 80-139.

Meadows, D.H., Meadows, D.L. and Randers, J. (1992). *Beyond the limits*. Global collapse or a sustainable future. Earthscan Publications Ltd., London, p. 300.

Müller, Z.O. (1980). Feed from animal wastes: state of knowledge. *FAO Animal Production and Health Paper*, Rome, Italy, p. 195.

Nagy, S., Shaw, P.E. and Veldhuis, M.K. (1977). *Citrus Science and Technology*. Vol. 2. AVI Publishing Company, Inc., Westport, Connecticut, p. 557.

North, M.O. (1978). *Commercial Chicken Production Manual*. 2nd edn. AVI Publishing Company, Inc., Westport, Connecticut, p. 692.

Ockerman, H.W. and Hansen, C.L. (1988). *Animal By-product Processing*. Ellis Horwood Ltd., Chichester, England and VCH Verlagsgesellschaft mbH, Weinheim, Germany, p. 362.

Pierson, K.W. (1943). Preliminary experiments on feeding processed garbage meal to poultry. *Progress*

Notes, Univ. Hawaii, Agric. Exp. Stn., **40**, 1-8.

Ravindran, V. and Blair, R. (1991). Feed resources for poultry production in Asia and the Pacific region. *World's Poultry Sci. J.*, **47**, 213-31.

Rolfe, E.J. (1976). Food from waste in the present world situation. In: *Food from Waste*. (Eds. G.G. Birch, K.J. Parker and J.T. Worgan). Applied Science Publishers Ltd., London, United Kingdom, pp. 1-7.

Rygg, G.L. (1975). Date development, handling, and packing in the United States. Agriculture *Handbook no. 482, Agricultural Research Service*, US Department of Agriculture, Washington DC, pp. 1-56.

SAS, *Statistical Analysis System*, (1985). SAS Institute Inc., Cary NC, p. 956.

Smith, T.J. and Huxsoll, C.C. (1987). Peeling potatoes for processing. In: *Potato Processing*. 4th edn. (Ed. W.F. Talburt, and O. Smith), AVI, Van Nostrand Reinhold Company, New York, pp. 333-69.

Talburt, W.F. (1987). History of potato processing. In: *Potato Processing*. 4th edn. (Ed. W.F. Talburt and O. Smith), AVI, Van Nostrand Reinhold Company, New York, pp. 1-9.

Van der Poel, A.F.B. (1990). Legume seeds: effects of processing on antinutritional factors (ANF) and nutritional value for non-ruminant feeding. In: *Advances in Feed Technology*, Verlag Moritz Schäfer, Detmold, **4**, pp. 22-34.

Van der Poel, A.F.B., Mollee, P.W., Huisman, J. and Liener, I.E. (1990). Variations among species of animals in response to the feeding of heat processed beans (*Phaseolus vulgaris* L.). *Livest. Prod. Sci.*, **25**, 121-150.

Van der Wal, P. (1976). Nutritional and toxicological evaluation of novel feed. In: *Food from Waste* (Ed. G.G. Birch, K.J. Parker and J.T. Worgan), Applied Science Publishers Ltd., London, United Kingdom, pp. 256-62.

Yoshida, M. and Hoshii, H. (1979). Nutritive value of garbage of supermarket for poultry feed. *Japan Poultry Sci.*, **16**, (6), 350-5.

2. DRIED POULTRY WASTE

Contents

2.1 INTRODUCTION

The increased size of poultry units causes a tremendous accumulation of large amounts of manure. For instance, a flock of 100.000 layers, kept in cages will produce more than 12 t daily or 4380 t a year.

Because disposal of poultry manure as a fertilizer is not very promising, increased interest has developed in alternative methods of waste disposal.

Alternative uses of poultry wastes are fuel production, fertilizers, fuel briquets and feedstuff ingredients. Recycling poultry waste after proper sterilization and processing has been advocated for two reasons: its useful nutrients can be used in case of feedstuff shortage and to reduce pollution.

Dry poultry waste varies widely in composition. The nutritive value depends on the type of ration, age and type of birds producing the droppings, extent of feed spoilage, quantity of feathers present, age of droppings before drying, drying temperature and duration.

Although the low available energy content of poultry manure is the main limitation for its use in poultry diets, non-protein nitrogen (NPN) is also a problem.

More than half of the crude protein (N x 6.25) in dry poultry waste is in a form of NPN such as uric acid, ammonia, urea, creatine and creatinine. Such compounds cannot be fully utilized by poultry and are therefore not a very useful type of protein for monogastrics. Uric acid may even be mildly toxic. A possible solution would be the biodegradation of cage layer manure by some natural living organisms, for instance the breakdown of hen excreta by means of house-fly larvae or earthworms. Fermentation or upgrading by aerobic digestion, or oxidation ditch, or through algae culture are also possibilities that lead to a final product rich in true protein, low in NPN and suitable as an animal protein feedstuff.

2.2 DRIED POULTRY WASTE

2.2.1 Nutritive Value of Dried Poultry Waste

After processing the water content of fresh cage layer excreta, has to be reduced from 75-10% and pathogenic organisms have to be killed. The product is an entirely new feed ingredient and is called dry poultry waste (DPW). It may also be classified as dehydrated poultry waste, dry poultry (excreta, faeces, manure, droppings, by-product) recycled nutrients and poultry anaphage (Zindel, 1974).

The nutritive value of DPW varies with: its age before drying; moisture content when fresh; method of storage; type; age; physiological status of layers; composition of ration fed; feed spoilage; environmental temperature; drying temperature; and speed of drying.

The Association of American Feed Control Officials, Inc. defined DPW as: "Dried Poultry Waste DPW is a product composed of freshly collected faeces from commercial laying or broiler flocks not receiving medicaments. It shall be thermally dehydrated to a moisture content of not more than 15%. It shall not contain any substances at harmful levels. It shall be free of extraneous materials such as wire, glass, nails, *etc.* The product shall be labelled to show the minimum percent protein, fat and maximum percent fibre. It may be used as an ingredient in sheep, lamb, beef, dairy cattle, broiler, and layer-chick feeds. Broiler and laying rations shall be limited to 20% and 15% DPW, respectively" (Couch, 1974). Table 2.1 shows the composition of DPW determined by several workers. Their analysis show moderate protein content (N x 6.25) ranging between 24 and 31 percent. The total protein is composed of true protein which accounts for more than one third and the rest is in a form of NPN, mainly uric acid and some ammonia, urea, creatine and creatinine. However, the high contents of ash, Ca and P are creating a problem when rations are formulated (Biely *et al.*, 1972; Price, 1972; Blair, 1973; Fadika, Walford and Flegal, 1973; Shannon, Blair and Lee, 1973; Zindel, 1974 and El Boushy and Vink, 1977).

Young and Nesheim (1972) listed some relevant nutrients in DPW as: Calcium, phosphorus, trace minerals, B-Complex vitamins, metabolizable energy (ME) and some amino acids. The ME was reported variously as 1.8 MJ/kg (Young and Nesheim, 1972), 8.1 MJ/kg (Biely *et al.*, 1972) and 2.3 MJ/kg (El Boushy and Vink, 1977).

El Boushy and Roodbeen (1984) showed that the essential amino acids content is limited (Table 2.2), the average available amino acids being 57%, which is low in comparison with feather meal (68%) and meat meal (83%). Total and available amino acids are, therefore, not giving a high economic value to the DPW.

Table 2.1. Chemical analysis, energy and amino acid composition of DWP on air dry basis in percentages. Fr om [1] Flegal and Zindel (1970a); [2] Biely *et al.* (1972); [3] Blair and Knight (1973) and [4] El Boushy and Vink (1977)

	(1)	*(2)*	*(3)*	*(4)*
Moisture	7.36	9.40	11.40	4.50
Crude Protein	24.21	31.08	28.70	24.28
True Protein	10.84	23.18	10.50	14.73
NPN (Nx6.25)	13.37	7.90	18.20	9.55
Ether extract	2.13	1.62	1.76	4.07
Crude fibre	13.72	10.70	13.84	10.11
Ash	26.90	23.76	26.50	35.79
Calcium	7.78	8.27	7.80	10.61
Phosphorus	2.56	2.00	2.45	2.71
Potassium	1.91	-	-	2.37
ME MJ/kg	-	8.09	2.76	2.34
Amino acids				
Lysine	0.49	0.48	0.39	0.56
Histidine	0.20	0.21	0.23	0.19
Arginine	0.47	0.45	0.38	0.53
Aspartic acid	1.06	1.10	0.71	1.22
Threonine	0.50	0.44	0.35	0.60
Serine	0.52	0.47	0.38	0.72
Glutamic acid	1.54	1.36	1.12	1.69
Glycine	0.82	1.61	1.33	0.93
Alanine	1.06	-	0.61	1.07
Valine	0.62	0.78	0.46	0.83
Methionine	0.09	0.20	0.12	0.29
Isoleucine	0.50	0.42	0.36	0.66
Leucine	0.80	0.69	0.55	0.94
Tyrosine	0.26	0.31	0.27	0.40
Phenylalanine	0.45	0.40	0.35	0.53
Cystine	1.09	-	0.15	0.21
Tryptophan	0.53	-	-	-

NPN = Non-Protein Nitrogen

Table 2.2. Amino acids (AA), expressed as total (TAA), available content (AAA), percentage of total (%) and their limitation in relation to the National Research Council (1977) requirements for dried poultry waste (DPW), feather meal (FM), poultry by-products (PBM), meat meal (MM) and soybean meal (SBM). From El Boushy and Roodbeen (1984)

	DPW			FM			PBM			MM			SBM			
AA	TAA[f]	AAA	%[g]	TAA	AAA	%	TAA	AAA	%	TAA	AAA	%	TAA	AAA	%	F-value
Aspartic acid	0.89	0.59	66^{b}	5.02	2.81	56^{a}	5.46	3.66	67^{b}	5.77	4.62	80^{c}	5.15	4.84	94^{d}	32.5^{**}
Threonine	0.38	0.20^{h}	52^{a}	3.36	2.08	62^{b}	3.22	2.45	76^{c}	2.52	2.07	82^{c}	1.70	1.53	90^{d}	38.0^{**}
Serine	0.37	0.22	59^{a}	6.73	4.31	64^{a}	6.09	4.93	81^{b}	2.83	2.29	81^{b}	2.04	1.86	91^{c}	16.6^{**}
Glutamic acid	1.10	0.65	59^{a}	7.96	4.94	62^{a}	8.00	6.16	77^{b}	7.93	6.66	84^{b}	8.05	7.65	95^{c}	31.0^{**}
Proline	0.59	0.37	62^{a}	9.39	6.67	71^{b}	6.13	4.72	77^{c}	4.45	3.83	86^{d}	2.27	2.11	93^{e}	38.0^{**}
Glycine	2.91	-	-	4.47	-	-	6.59	-	-	6.55	-	-	1.90	-	-	-
Alanine	0.65	0.40	61^{a}	4.85	3.78	78^{b}	4.35	3.39	78^{b}	4.74	4.03	$85^{b,c}$	2.42	2.25	93^{e}	14.7^{**}
Cystine	0.35	0.19	55^{a}	4.26	2.77	65^{a}	2.43	1.15	62^{a}	0.79	0.63	80^{b}	0.88	0.80	91^{b}	15.5^{**}
Valine	0.48	0.27^{h}	57^{a}	6.41	4.81	75^{b}	4.81	3.70	77^{b}	3.53	2.93	83^{b}	2.55	2.42	95^{c}	24.7^{**}
Methionine	0.16	0.10^{h}	61^{a}	0.79	0.51^{h}	65^{a}	1.14	0.88	77^{b}	1.08	0.93	$86^{b,c}$	0.62	0.57^{h}	92^{c}	13.9^{**}
Isoleucine	0.39	0.23^{h}	60^{a}	4.15	3.24	78^{b}	3.25	2.57	79^{b}	1.99	1.71	86^{c}	2.35	2.21	94^{d}	34.7^{**}
Leucine	0.59	0.32^{h}	54^{a}	6.19	4.52	73^{b}	5.78	4.51	78^{b}	3.22	2.42	75^{b}	3.51	3.30	94^{c}	38.8^{**}
Trosine	0.23	0.11^{h}	49^{a}	2.23	1.45	65^{b}	2.52	1.94	77^{c}	1.95	1.60	$82^{c,d}$	1.78	1.66	93^{d}	16.9^{**}
Phenylalanine	0.31	0.16^{h}	53^{a}	3.89	3.00	77^{b}	3.63	2.87	79^{b}	2.91	2.44	84^{b}	2.42	2.27	94^{c}	29.8^{**}
Lysine	0.34	0.19^{h}	57^{a}	1.57	1.00^{h}	64^{b}	2.81	2.16	77^{c}	4.19	3.56	85^{d}	2.88	2.74	95^{e}	47.5^{**}
Histidine	0.16	0.08^{h}	47^{a}	0.55	0.32^{h}	59^{b}	1.08	0.78	72^{c}	1.89	1.53	81^{d}	1.29	1.20	93^{e}	37.7^{**}
Arginine	0.35	0.18^{h}	52^{a}	5.44	4.19	77^{b}	5.45	4.58	84^{c}	4.44	3.82	86^{c}	3.47	3.30	95^{d}	68.3^{**}
Average	0.46	0.26	57^{a}	4.55	3.09	68^{b}	4.13	3.18	76^{c}	3.39	2.82	83^{d}	2.71	2.54	93^{e}	50.1^{**}
Crude protein %		28.98			83.04			63.65			56.62			44.29		

a, b, c, d, e, Any two means in a row with the same superscript are not significantly different ($p < 0.05$) by Duncan's multiple range test.
f Average of 2 replicates per feedstuff
g Availability percentage: average of 8 birds
h Limited essental amino acids in relation to the National Research Council (NRC, 1977) requirements according to Soares and Kifer (1971)
** Highly significant ($P < 0.01$) in the analysis of variance

Advantages and disadvantages of DPW

Dried poultry waste, after proper treatment, could be used as a feedstuff because it contains undigested feed, metabolic excretory products and residues resulting from microbial synthesis. Micro-organisms in the poultry excreta convert some of the uric acid to microbial protein, which can be utilized by poultry. DPW provide an unknown growth factor (UGF). Moreover, it contains some true protein in addition to the NPN. If this feedstuff can be recycled no doubt pollution will be reduced (El Boushy and Vink, 1977).

In contrast to ruminants, poultry cannot utilize NPN as a protein source. Under specific conditions, however, some monogastrics such as poultry are able to utilize the N from NPN sources to a limited extent. Ackerson, Ham and Mussehl (1940) and Jones and Combs (1953) reported that NPN could not be utilized by young chicks. On the other hand other reporters concluded that NPN could replace protein under the special conditions of balanced essential amino acids, when some amino acids are replaced by their hydroxy analogues (Sullivan and Bird, 1957; Featherston, Bird and Harper, 1962; Blair and Waring, 1969; Trakulchang and Balloun, 1975).

Uric acid is known to cause a growth depression in growing chicks, but Bare, Wiseman and Abbott (1964) found that uric acid additions of 2% were required to cause a significant weight depression after 4 weeks, while levels of 0.5% and 1% had no effect. They suggested that this depression was due either to uric acid acting as an irritant interfering with the absorption of nutrients from the intestinal tract, or to the inhibition of microbial biosynthesis of vitamins or other nutrients being essential to the host.

2.2.2 Processing of Poultry Waste

Owing to the high moisture content (75-80%) (Biely *et al.*, 1972) and possible contaminants, poultry waste must be processed before it can be incorporated into poultry rations.

In general processing methods are complicated and costly for use in developing countries. In some areas, the use of DPW as a feedstuff is attractive owing to the high price of conventional feedstuffs or import restrictions. Financial investments are not available for those countries to obtain the relevant processing equipment. Therefore, several alternative methods have been developed. For example, several simple methods are reported for drying cage layer manure (Kese and Donkoh, 1982; Olorede *et al.*, 1995), such as sundrying, steam heating and roasting.

(a) Simple methods

Sundrying. Drying could be carried out by spreading out a portion of partially dried waste (sun dried for 24 h at an ambient temperature of approximately 30 °C and cleaned of extraneous materials) on a flat metal sheet in the open and sun drying for two days at ambient temperature ranging from 30-35 °C. Excreta were turned over periodically and collected overnight to protect the waste from being moistened by dew. This method for drying poultry waste/involved the pressure of non-pathogenic *(Salmonella spp.)*. Olorede *et al.* (1995) reported also the air-drying procedure for poultry droppings.

Their poultry manure samples were collected from a concrete floor under cage layers in a semi-dried condition. The manure was sun-dried for 5 days till 10.36% moisture and then hammer-milled.

Steam heating. The apparatus used for steam heating consisted of two 220 l drums, one with a solid and the other with a perforated bottom. The perforated one, containing the waste, was placed on top of the other one, which contained water. The drum containing the waste was covered with burlap to conserve the heat. Water was heated to produce steam of burning logs. The process took 4 h during which the steam rose through the excreta, attaining temperature up to 80 °C. The steamed excreta were sun dried for 3 h. No microbial growth on culturing was noticed by this method (Kese and Donkoh, 1982).

Roasting. For this method the excreta were treated by direct heat for 2 h in an open 200 l capacity drum. During heating the excreta were stirred constantly. The temperature in this process ranged between 102 and 105 °C.

Sheppard *et al*, (1971) reported that temperatures from 149-385 °C were used to dry poultry manure. In addition, a reduction was observed in the nitrogen content of the dried manure at the higher temperatures. Olorede *et al.* (1995) reported also that the collected poultry droppings from concrete floor were of semi-dried condition. Then the samples were oven-dried at 105 °C for one day till a moisture content of 8.80% and then hammer milled. The authors concluded that drying the cage manure with heat would appear to be the best procedure for the production of a possible feed ingredient of more consistent nutrient composition.

(b) Drum drying

A relatively new technique in waste management has been the development of the manure dryer (Sturtevant Engineering Company Ltd, 1979). The method of operation is as follows. The layers' manure is gathered by belts under the cages from the poultry house and is transported to the dryer feed hopper. Subsequently it is fed through a variable speed auger to the dryer drum. The auger permits a constant feed at a pre-determined feed rate.

The dryer drum is heated by an oil-fired burner which raises the drum temperature up to 340-400 °C. The manure is moved by agitating reflector plates and passes to the discharge end of the drum. On input of the dryer drum, the moisture content of the wet manure is approximately 80% which after processing is reduced to 8-14%. Both main burner and afterburner are protected by the inclusion of a photo-electric cell for automatic flame monitoring.

The dried product is pneumatically conveyed from the drum to a collecting cyclone, which separates the solids from the gases. It then passes through a sifter to remove feathers and on to the bulk bin for transportation or bagging off.

The moisture-laden gases separated by the cyclone pass via a specially developed oil-fired afterburner in which the temperature of the gases is raised sufficiently to remove any unpleasant smell before being released to the atmosphere (Figure 2.1).

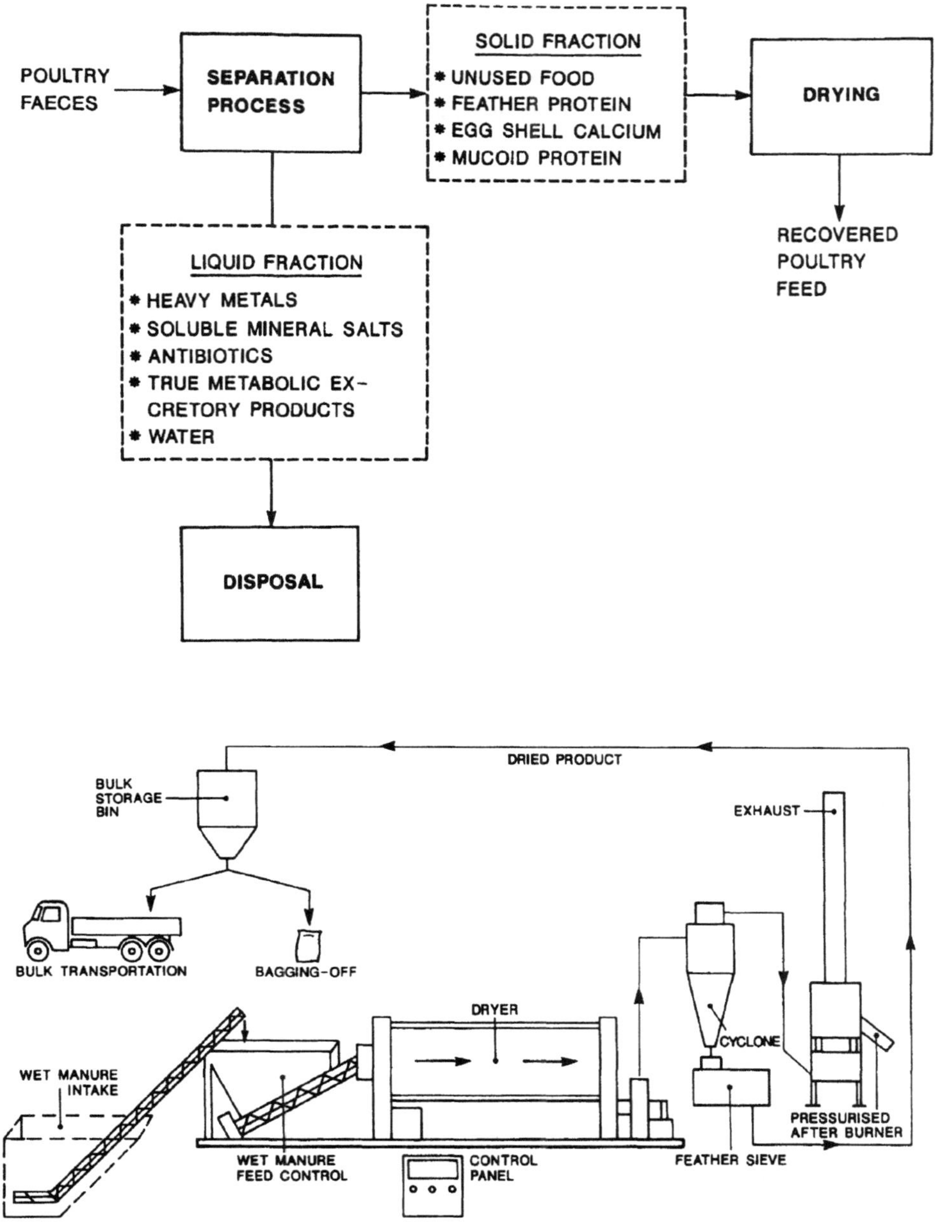

Figure 2.1. Flow diagram (a) and layout (b) of a complete processing plant for manure. Reproduced with permission from Sturtevant (1979).

2.2.3 Biohazards of DPW

Dehydrated poultry waste obtained by processing manure from caged layers contains little, if any, drug residues, since drugs are not normally included in feed formulation for layers. Poultry litter from broilers may contain various drugs including coccidiostats, antibiotics, arsenic acids, and others (Couch, 1974). On the other hand Webb and Fontenot (1975) found residues of penicillin, chlortetracycline, oxytetracycline, nicarbazin and amprolium in broiler litter; this is not of value when cage layer manure is used.

Fontenot (1981) reported some safety considerations concerning the use of untreated layer manure, broiler manure or poultry litter in the nutrition of ruminants and the effects on quality of products. The most relevant points are toxic elements such as heavy metals, aflatoxins owing to contamination with *Aspergillus flavus*, insecticides and medicinal drugs. As indicated before, cage layer manure is mostly free from the above mentioned residues and it should be kept in mind that these products are sterilized. Table 2.3 presents some results of the bacteriological examination of DPW samples.

Table 2.3. Bacteriological examination of DPW samples, without sterilization. From Shannon *et al.* (1973)

Sample no.	*Direct culture*	*Selenite broth culture*
1	Few anthracoid bacilli	Paracolon bacilli
2	Few anthracoid bacilli	Paracolon and anthracoid bacilli
3	No organisms isolated	Paracolon bacilli
4	Few paracolons and anthracoid bacilli	Anthracoid bacilli
5	No organisms isolated	No organisms isolated
6	Few anthracoid bacilli	Anthracoid bacilli
7	Several staphylococci	Staphylococci and anthracoid bacilli
8	Few *E.coli* and anthracoid bacilli	-
9	Few anthracoid bacilli	-

2.2.4 Use of DPW in Poultry Diets

Feeding trials indicate that DPW has a low energy content, contains some amino acids, and is high in ash content, mainly calcium phosphorus. In addition, it contains few trace elements. All those elements may effect the growth, feed consumption and conversion of broilers. For laying hens DPW may be an acceptable element. The high ash content is creating a problem in formulating mainly broiler rations due to the high Ca and P levels.

(a) Laying hens

More emphasis has been placed on feeding DPW to layers than to broilers or baby chicks. Inclusion of 10, 20, 30 and 40% DPW in rations for cage layers resulted in an increase in feed conversion and caused no significant differences in egg production, shell thickness or egg weight. Eggs from the DPW-fed groups were of better quality as measured by Haugh units (Flegal and Zindel, 1969, 1970[b]).

York *et al.* (1970) reported that egg production, shell thickness and egg weight were not affected by DPW in the diets, but that feed efficiency was inversely affected, proportional to the amount of DPW in the diet. Also, too high levels of DPW in the ration will decrease egg production and feed efficiency, and will increase mortality according to Trakulchang and Balloun (1975).

Castro *et al.* (1984) conducted trials in which DPW was used in layer rations calculated to be isocaloric and isonitrogenous. The inclusion levels of DPW were 10, 15 and 20%. They concluded that the highest inclusion of DPW caused reduction in feed consumption, feed conversion ratio and egg weight. Layers in the later stages of production seem to utilize this by-product more efficiently.

Biely (1975) measured the effect of several levels of DPW in diets of starter, grower and layers on production, conversion, egg weight and fertility and hatchability. He concluded that the inclusion of 10 and 20% DPW showed no deficiency in any essential nutrient nor an excess of metabolites which may affect the above mentioned productive characteristics (Table 2.4 and 2.5).

Ramteke *et al.* (1994) examined the use of sun-dried poultry droppings with the inclusions of 0, 5, 10, 15 and 20% to replace rice polishing. They concluded that shell % and shell thickness were significantly ($p < 0.01$) high on feeding 20% DPW, but albumen %, yolk %, yolk index and Haugh unit did not show any significant effect. Organoleptic tests revealed no significant difference in taste, flavour and odour of eggs from different groups.

(b) Broilers and chicks

Flegal and Zindell (1970[a]) fed several levels of DPW to white Leghorn chicks and broiler chicks at 4 weeks of age. The mean body weight of the Leghorn chicks was not influenced by 20% dietary DPW. However, the inclusion of 10 and 15% reduced the body weight of the broilers. Feed efficiency was inversely related to the DPW level in the diet. Biely *et al.*, (1972) reached similar conclusions with Leghorn chicks; at high DPW levels, their growth rate was lower and their feed efficiency was poorer. Rinehart *et al.* (1973) observed that broiler faecal volume increased in a direct relationship with increasing consumption of DPW, suggesting an atmost complete lack of nutrient utilization.

Cunningham and Lillich (1975) studied the effects of feeding 9.6, 19.1 and 38.2% DPW to broilers. The highest level group showed lower average life weight and the poorest feed conversion. They concluded that DPW may be fed to broilers at a level below 20% without serious consequences.

El Boushy and Vink (1977) studied the effects of the inclusion of DPW in levels of 5, 7.5, 10, 12.5 and 15% in broiler diets formulated to be isocaloric, isonitrogenous, equal in methionine and lysine, and constant in Ca : P ratio and fed up to 6 weeks of age (Table 2.6). They concluded that a level of 5% DPW can be used in broiler rations with no harmful effects. However, the 15% level slightly decreased body weight, growth and increase in feed consumption. All differences at a 5% inclusion level were statistically insignificant. They concluded also that DPW was found to be an acceptable feedstuff in an N-source. Its true protein content is rather low, however the level of methionine and lysine could be obtained by the addition of synthetic versions. No harmful effects were noticed due to NPN or uric acid at the 15% inclusion level of DPW, but it is not a reliable source of energy. Olorede *et al.* (1995) reported from their broiler experiment that the inclusion of DPW in a broiler standard diet of 0, 7.5 and 15% had no significant effect on body weight gain. Feed consumption was increased with an increase in faecal volume and depression of feed conversion as DPW increased.

Table 2.4. Body weight and egg production characteristics of growing and laying hens receiving 10 and 20% DPW in comparison with a basal diet. From Biely (1975)

	Body weight				*Average hen-day egg production*	*Average egg weight*	*Feed per doz.eggs*
	4 wk.[a] *(g)*	*12 wk.*[a] *(g)*	*16 wk*[a] *(g)*	*44 wk.*[b] *(g)*	*168 days (%)*	*168 days (g)*	*168 days (g)*
Basal	280.0	1528	2103	2506	60.4	55.4	2279.0
Basal + 10% DPW	250.0	1212	2086	2446	60.7	55.5	2251.7
Basal + 20% DPW	237.0	1177	2004	2484	57.5	54.8	2524.4

[a] Mixed chicks
[b] Pullets only

Table 2.5. The effect of feeding basal diets with the inclusion of DPW at 10 and 20% on fertility, hatchability and average weight of day old chicks. From Biely (1975)

	Eggs set	*Eggs Infertile*	*Dead germs pipped and dead in shell*	*Chicks hatched*	*Hatch-ability of fertile eggs (%)*	*Average weight day old chicks (g)*	*Average weight of chicks (g)*
HATCH 1							(11 days)
34 week old pullets							
Basal	136	-	10	126	92.6	39.2	76.8
Basal + 10% DPW	136	3	5	128	96.2	38.1	82.3
Basal + 20% DPW	136	2	9	125	93.3	38.3	79.5
HATCH 2							(13 days)
39 week old pullets							
Basal	136	4	9	123	93.2	35.2	99.3
Basal + 10% DPW	136	9	3	124	97.6	35.5	95.7
Basal + 20% DPW	136	8	12	124	96.9	35.2	102.5

Table 2.6. Average weights, growth, feed consumption and feed efficiency of broilers fed various levels of DPW in comparison with the basal ration. From El Boushy and Vink (1977)

	Average weight (g/chick)				Growth (g/chick)				Feed consumption (g/chick)				Feed efficiency (F/G)			
Rations	*3 weeks*		*6 weeks*		*0-3 w*		*0-6 w*		*0-3 w*		*0-6 w*		*0-3 w*		*0-6 w*	
	g	*%*	*g*	*%*	*g*	*%*	*g*	*%*	*g*	*%*	*g*	*%*	*abs.*	*%*	*abs.*	*%*
Basal	602[a]	100	1535[a]	100	561[a]	100	1493[a]	100	878[a]	100	2928[a]	100	1.57[a]	100	1.96[a]	100
5% DPW	608[a]	101	1537[a]	100	568[a]	101	1497[a]	100	902[b,c]	103	3007[a]	103	1.58[a,b]	101	2.01[a,b]	103
7.5% DPW	588[a]	98	1506[a]	98	558[a]	99	1466[a]	98	894[b]	102	2989[a]	102	1.60[a,b,c]	102	2.04[b,c]	104
10% DPW	601[a]	100	1493[a]	97	559[a]	100	1451[a]	97	916[c,d]	104	2981[a]	102	1.64[b,c,d]	104	2.06[b,c,d]	105
12.5% DPW	592[a]	98	1477[a]	96	551[a]	98	1436[a]	96	926[d]	105	3016[a]	103	1.68[d,e]	107	2.10[d]	107
15% DPW	594[a]	99	1469[a]	96	553[a]	99	1429[a]	96	948[e]	108	3094[a]	106	1.72[e]	110	2.17[e]	111

Means within a column with same superscript are not significantly different at 5% level by Duncan's multiple range test.

Effect of DPW on the flavour of eggs and broiler meat
To find out to what extent the flavour of eggs is affected by the inclusion of DPW till 30% in layer diets, the boiled eggs produced by this treatment were presented to a consumer preference panel. They did not detect a difference between control eggs 0% DPW and 30% DPW inclusion eggs (Flegal, Goan and Zindel, 1970; Zindel, 1972).

In broiler meat, flavour was studied by a triangle taste test to study flavour differences caused by feeding 0, 9.6, 19.1, or 38.2% DPW to broilers. Flavour differences could not be detected between the extreme treatments (Cunningham and Lillich, 1975). Kese and Donkoh, 1982 also concluded from their trials that the carcass flavour was not significantly different among dietary treatments influenced by the inclusion of 10% DPW in broiler diets.

(c) Turkeys

When DPW was included in turkey diets with levels of 0, 5, 10 and 30% on the basis of isocaloric and isonitrogenous formulation, body weight gain was not significantly affected at 17 weeks of age. Also, plasma uric acid was not altered while plasma phosphorus was raised in birds fed 30% DPW (Fadika, Walford and Flegal, 1973). Also, Zindel (1974) reported from his work on turkey anaphage that the liveability was not affected by using DPW in turkey diets.

2.3 BIOLOGICAL CONVERSION METHODS TO IMPROVE CAGE LAYER MANURE

Cage layer manure in its fresh form contains 75-80% moisture which is closely associated with the organic matter. This water makes fresh manure difficult to handle and to transport. Therefore, fresh manure cannot be used in poultry nutrition without dehydration and sterilization. Moreover, the high content of NPN as uric acid makes it not very promising for use with monogastrics.

There are several possibilities for changing the poor structure of layer cage manure and to result in a final product which is very rich in true protein and a useful feedstuff for poultry:

- Biodegradation of manure
- The use of some natural living organisms to break down the hen excreta, such as house-fly larvae, earthworms
- Fermentation
- Upgrading by aerobic digestion
- Oxidation ditch
- Algae culture

2.3.1 House-Fly Larvae: *Musca Domestica L.*

Biodegradation of cage layer manure by *Diptera* (flies) has been reported (El Boushy, Klaassen and Ketelaars, 1985; Boda, 1990; El Boushy, 1991). Many species of *Diptera* grow and develop naturally in animal wastes. Particularly, the house-fly (*Musca domestica*) was selected for this purpose because of its short generation interval. Under proper conditions, it will complete a life-cycle in seven days or less, depending mainly on the temperature. The idea of rearing flies on manure, however, is rather repulsive to the general public when first considered. Rearing flies is also highly undesirable because once they escape from the breeding area, they become a nuisance to man and animals. It should be realized, however, that the end stage is not the house-fly itself, but the pupae and larvae as a rich source of protein growing on poultry manure.

The biological digestion of poultry manure by *Diptera* pupae and larvae appears to be a solution for the accumulation of large amounts of manure, if a reasonable technological process can be developed. The house-fly larva is able to break up and dry out the poultry droppings. The pupae and larvae with or without manure residue, are then harvested and sterilized and can be used as a feedstuff with a high nutritive value. Finally, a group of American scientists came to a new convertor of poultry manure to a high quality feedstuff by using the black solder fly, *Hermetia illucens* L.

Sheppard *et al.* (1998) concluded that the black solder fly has a simple life cycle egg hatches within 4 days, the larvae take 2-3 weeks to reach pupal stage. The final dried larva meal contains C.P. 42.1%, Fat 34.8%, Fibre 7.0%, NFE 1.4%, Ash 14.6%, Ca 5.0%, Moisture 7.9%, Methionine 0.86% and Lysine 3.37%. They concluded that this fly is an excellent convertor of poultry manure to a rich feedstuff.

(a) Technological aspects

The stages in the life-cycle of the house-fly *(Musca domestica)*; family *(Muscidae)* are egg, larva, pupa and adult. The larva moults twice so larvae have first, second and third instars, each stage being larger than the preceding instar (Figure 2.2).

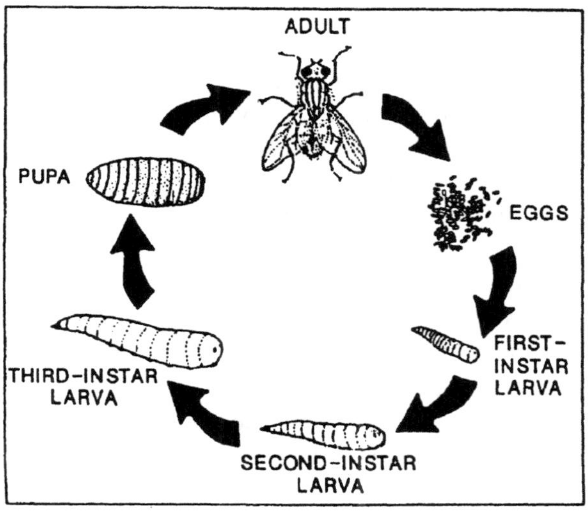

Figure 2.2. Life-cycle of the house-fly (*Musca Domestica*) showing stages: eggs, larvae of 1st, 2nd and 3rd instar, pupa and adult. Reproduced with permission from Axtell (1986).

The overall life-cycle of the house-fly (from egg to adult fly) is about 7-10 days in the summer in warm areas and in cold areas is about 40-49 days (Table 2.7). Often the temperature of the decaying and fermenting larval medium is considerably higher than the prevailing air temperature. In this way, development occurs much faster than anticipated by the climatic conditions.

Table 2.7. Time required for the total life-cycle of the house-fly at various temperatures. From Axtell (1986)

Temperature	*Time (days)*	
(C)	*average*	*range*
16	44.8	40-49
18	26.7	23-30
20	20.5	19-22
25	16.1	14-18
30	10.4	9-11
35	7.0	6- 8

Cage layer manure is highly suitable for the development of fly eggs to pupa. The idea of harvesting the pupa was first suggested by Miller (1969), who reported that the separation of pupa from the manure can be easily done by flotation on water. This, however, was difficult on a commercial scale. In addition, the prevention of pathogen proliferation during incubation requires breeder flies developed from disease-free stock.

Calvert, Morgan and Martin (1970) reported that 3 eggs per gram of manure is quite effective for an optimum production of larvae. Miller and Shaw (1969) concluded that the production of pupae is optimum when 0.5-1.0 g of house-fly eggs are inoculated per kg of fresh hen manure.

A system was patented by Calvert, Morgan and Martin (1973) in which light caused the larvae to migrate out of the manure through a screen into a lower compartment, where they pupate and subsequently can be harvested.

During the digestion of the manure the larvae convert the sticky mass of manure into an odourless loose crumbly product which can easily be dried and used as a feedstuff. The great number of flies needed to catabolize a reasonable amount of manure and to yield enough pupae requires close co-operation of several specialists. Calvert (1977) estimated the maximum yield of pupae plus larvae to be 3.2% of the fresh manure. Dry matter yield was about 4.0% on the basis of 25% DM content for manure and 32% DM content for larvae plus pupae.

In practice manure is scraped from belts under the cages or it drops into a deep pit known as deep manure accumulation. The growth of the pupae has to be introduced. The first system has to be used because it allows conveyance of the scraped manure from the house to a separate storage house. Scraping can be done once a week. The manure then can be conveyed to a storage house by means of a swivelling conveyor which can produce a manure bed of constant depth over the whole floor (Figure 2.3).

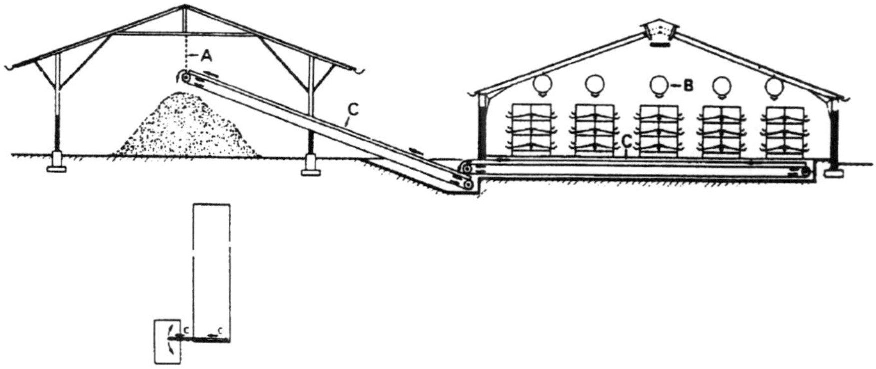

Figure 2.3. Cross-section of a layer house and a storage house equipped with swivelling conveyer. Reproduced with permission from Kroodsma, Arkenhout and Stoffers (1985).

Every week the dropped manure should be inoculated with fly eggs (mixed with water for an equal distribution), obtained from a disease-free breeder stock. The inoculation ratio as mentioned before would be 0.75 g of eggs for each kg of fresh manure.

The incubation time of the house-fly is 7 days at high temperatures (35 °C) so this means that the fresh manure added would have to be inoculated every 7 days. When the manure storage house is full, the digested manure could be transferred by conveyors to trucks and then dried and sterilized as shown in Figure 2.3. A detailed description of the whole process of poultry waste has been reported previously (El Boushy, 1991).

A heavy invasion of house-flies in the farm is possible. A hot-air cannon built into the manure storage house would be sufficient to burn these flies and prevent invasion. The final product after drying would be a mixture of fly pupae, larvae and manure residue, a combination that is richer as a feedstuff than the untreated manure but incurs the same costs as far as drying is concerned (El Boushy, 1991).

(b) Chemical analysis and nutritive value

The digestion of poultry manure by the larva reduces moisture content from 75-50% and removed about 80% of the organic matter resulting in a product that is odourless, loose and crumbly in texture (Miller, 1969; Miller and Shaw, 1969; Calvert, Martin and Morgan, 1969; Miller, Teotia and Thatcher, 1974).

The residue from the digested manure and the pupae is rich in protein and low in NPN. The ME values of the house-fly pupae and digested manure are 10.58 and 2.43 MJ/kg, respectively (Teotia and Miller, 1970[a]; Miller, Teotia and Thatcher, 1974).

Pupa meal is a good source of limiting amino acids, particularly arginine, lysine and methionine. Its amino acid content is higher than those of soybean meal and equal to that of meat and fish meal. House-fly pupa meal is a good source of protein, amino acids, fat and minerals for chick starter rations (Teotia and Miller, 1974). Tables 2.8 and 2.9 are summarizing the chemical composition of fresh, dried and digested poultry waste and of the dried house-fly pupa.

Table 2.8. Chemical composition of fresh, dried and digested poultry waste and the dried pupae on air-dry basis. From [(1)] Teotia and Miller (1974) and [(2)] El Boushy and Vink (1977)

	(1)		*(2)*	
%	*Dried pupa*	*Fresh manure*	*Digested manure*	*Dried manure*
Crude protein	61.4	46.9	17.6	24.3
Crude fat	9.3	2.6	1.6	4.1
Crude fibre	-	18.1	19.1	10.1
Ash	11.9	22.7	30.1	35.8
Minerals				
Ca	0.9	5.1	2	10.6
P	1.4	3.4	1.9	2.7
Magnesium	-	4.3	4.6	-
Potassium	0.9	1.7	3.1	2.4
Manganese (ppm)	370	276	432	-
Copper (ppm)	34	68	59	-
Zinc (ppm)	275	376	545	-
Iron (ppm)	465	560	544	-

Table 2.9. Content of amino acid percentage in dried pupa and in manure on air dry basis. From [(1)] Teotia and Miller (1974) and [(2)] El Boushy and Vink (1977)

Amino acids (%)	*(1)*	*(2)*
	Dried pupa	*Dried manure*
Arginine	3.7	0.53
Glycine	2.4	0.93
Histidine	2.0	0.19
Isoleucine	2.4	0.66
Leucine	3.4	0.94
Lysine	3.8	0.56
Methionine	1.6	0.29
Phenylalanine	3.0	0.53
Thyronine	2.1	0.60
Valine	2.7	0.83
Glutamic A	7.2	1.69
Alanine	2.5	1.07
Tyrosine	3.8	0.40
Cystine	-	0.21
Proline	2.3	-
Serine	1.9	0.72
Aspartic A	5.3	1.22

(c) Use of pupa meal in feeding

Dried, ground house-fly pupae meal (63.1% protein) was equal to soybean meal (50% protein) in supporting growth of the chick through the first two weeks of life (Calvert, Martin and Morgan, 1969).

Teotia and Miller (1970[a,b]) investigated the nutritive value of fly pupae and digested manure by means of two feeding trials. In the first trial, dry, ground pupae or manure residue were used to replace all protein supplements to a chick starter diet. White Leghorn chicks were fed the diets from one day old until 4 weeks of age. Chicks fed the diet containing fly pupae had slight but not significantly better feed conversion and weight gains compared with control birds. Replacement of soybean meal with manure residue failed to support growth at control levels. In the second trial, broiler chicks were fed a diet containing dry ground pupae as the only protein supplement from one day to seven weeks of age. These chicks received no trace minerals or B-vitamin supplements in their diet. In comparison to control chicks receiving a fully balanced diet, no significant differences in body weight or feed conversion were noticed. The results of this evaluation indicated that dried fly pupae have potential as a protein supplement in chick starter and broiler diets.

Ernst *et al.* (1984) fed dried and ground house-fly pupae to laying hens for 7 months. Pupae meal was added in diets to replace meat and bone meal, and raise the protein content from 18-19.3%. The pupae meal increased egg yield by 3.6% and hatchability of eggs by 12% in comparison with control. However, Peter *et al.* (1985) found that broilers had significantly greater weight gains when given feed with 2.5 or 5% poultry excreta dressed with house-fly (*Musca domestica*) than those given a mash equal in protein and energy but without poultry excreta.

2.3.2 Earthworms

Biodegradation of animal and bird waste and municipal sludges by earthworms has been reported by many research workers (Collier and Livingstone, 1981; Satchell, 1983; Lofs-Holmin, 1985; Edwards and Neuhauser, 1988). Earthworms are harvested for use as a high-protein feed ingredient. The waste residual dirt is called castings which can be used effectively as a fertilizer for plant growth. Faecal material from animals may serve as a sole source of feed for worms. Culturing of earthworms on the faecal wastes of farm animals represents a great potential both for the production of animal protein, and as a method for biological digestion and disposing of such wastes.

(a) Technological aspects

The stages of the life-cycle of the earthworm are adult, cocoons and young worm. The earthworm is hermaphroditic, which means that each worm has both male and female organs. Copulation occurs through the clitellum, the glandular swelling located about one-third the length of the worm from the head. The worms meet and overlap each other with their heads going in opposite directions. As the clitella of the two worms meet, large quantities of mucus are secreted which will bind the worms together.

Each worm acts as a male and gives off seminal fluid which is stored. After the worms have separated, a slime tube which is formed by the clitellum of each worm, is worked forward over the body and collect a few eggs from the oviducts, and receives sperm from the seminal receptacles where they have been stored. The slime tube is gradually slipped off over the head, closing to form a greenish-yellow cocoon.

The cocoon of the "red wigglers" is about 0.3 cm in diameter and lemon shaped. Under favorable conditions copulation and production of cocoons takes place once every three to five days. Each cocoon usually produces two to ten worms after an incubation period of approximately three to five weeks. The young worms grow rapidly and within one month from the time of hatching they may be able to reproduce, while six months' time is needed for a full- size worm. Earthworms may live to be more than ten years (Abe, Braman and Simpson, 1985).

The particular species of earthworm used is a matter of choice as far as the media (organic waste), temperature and humidity are concerned. There are nearly 2000 species of earthworms in the world but only a few species are suitable for culturing organic waste materials. Generally they grow and reproduce at high rates, but each has its special properties which make it more or less suitable for different purposes. The most attractive earthworm as far as length and weight are concerned is the *Lumbricus* species, *L. terrestris*. This is the largest *lumbricid* found in Northern Europe. However, it reproduces and grows slowly, and moreover, is difficult to cultivate. The species is attractive for its size, up to 30 cm (10 g), and is mostly used for fish bait (Lofs-Holmin, 1985).

In general, the production of earthworms is greatly influenced by temperature, the survival percentage and the total cocoons produced. Production increases sharply with increasing temperature up to 25 °C (Table 2.10). The production characteristics and the biological data of the several species fed on manure or sludge are presented in Table 2.11.

Table 2.10. Survival and cocoon production of three earthworm species at different temperatures. 5 replicates with 4 worms fed with sludge for 10 weeks. From Loehr *et al.* (1984)

Species	*Temperature °C*					
	10	15	20	25	30	35
E. foetida						
% survival	100	100	100	100	50	0
total cocoons	0	111	353	410	0	0
E. eugeniae						
% survival	40	90	100	90	15	0
total cocoons	0	4	183	867	0	0
P. hawayana						
% survival	65	75	60	80	0	0
total cocoons	0	0	1	32	0	0

Table 2.11. Biological data (maximal values) for some earthworm species fed on manure or sludge. From Lofs-Holmin (1985)

Species	*Maximum weight (g)*	*Start of cocoon production at age (days)*	*Cocoons perworm per week*	*Young per cocoon*	*Optimum temp. (C)*
E.foetida	2.4	30	6	1.6-3.8	20-25
L.rubellus	2	60	5	1	15-18
L.terrestris	10	100	0.7	1	15-18
E.eugeniae	8	30	11	1.2-2.7	25-28
P.excavatus	1.58	20	24	1	25-28
D.veneta	4	40	5	1	20-25
P.hawayana	2.5	40	10	1	25

(b) Production of earthworms

It is outside the scope of this chapter to deal with the technology of worm production in detail; the references available have been mentioned already. Systems to produce earthworms commercially in beds, boxes or continuous flow systems, as a source of protein must be controlled environmentally. Waste must be added regularly in small amounts to prevent high temperatures from developing and earthworms can colonize all of the organic matter.

Under these circumstances the worms can utilize the micro-organisms growing on it to the maximum. Wastes used as layer deep litter manure must be modified to produce a suitable moisture content. Temperature, ammonia and salt content must be reduced to acceptable levels by leaching, composting or some other methods, before the worms can be grown successfully. A rapid rate of growth and multiplication of earthworms, and an efficient conversion of organic wastes into earthworm tissue protein would mean that systems could be developed using a minimum of labour or sophisticated technology (Edwards and Niederer, 1988).

Growing worms in beds, filled with animal waste to a depth of about 50 cm, in successive shallow layers, at regular intervals, by automatically operated gantries, is an easy and non-labour-intensive system.

(c) Harvesting of earthworms from wastes

Worms reach their optimum growth at relatively high moisture levels. This raises harvesting problems, since it is not easy to separate worms mechanically from the organic matter at such a high moisture contents. Therefore, some drying before harvesting is necessary.

Machinery for separating worms from organic materials has been developed at Rothamsted and the National Institute for Agricultural Engineering, Silsoe, and has been patented (Phillips, 1988). The efficiency of this machinery in terms of percentage recovery of worms is very high. The machinery that has been developed currently, will separate worms from about 1 t of waste per hour, if necessary, this machine can be automated and scaled up to increase throughput. However, the rate of separation of worms from waste is still too slow and labour-intensive to make production of protein for animal feed and economic process on its own. Figure 2.4 shows the separation and the vermicompost facilities.

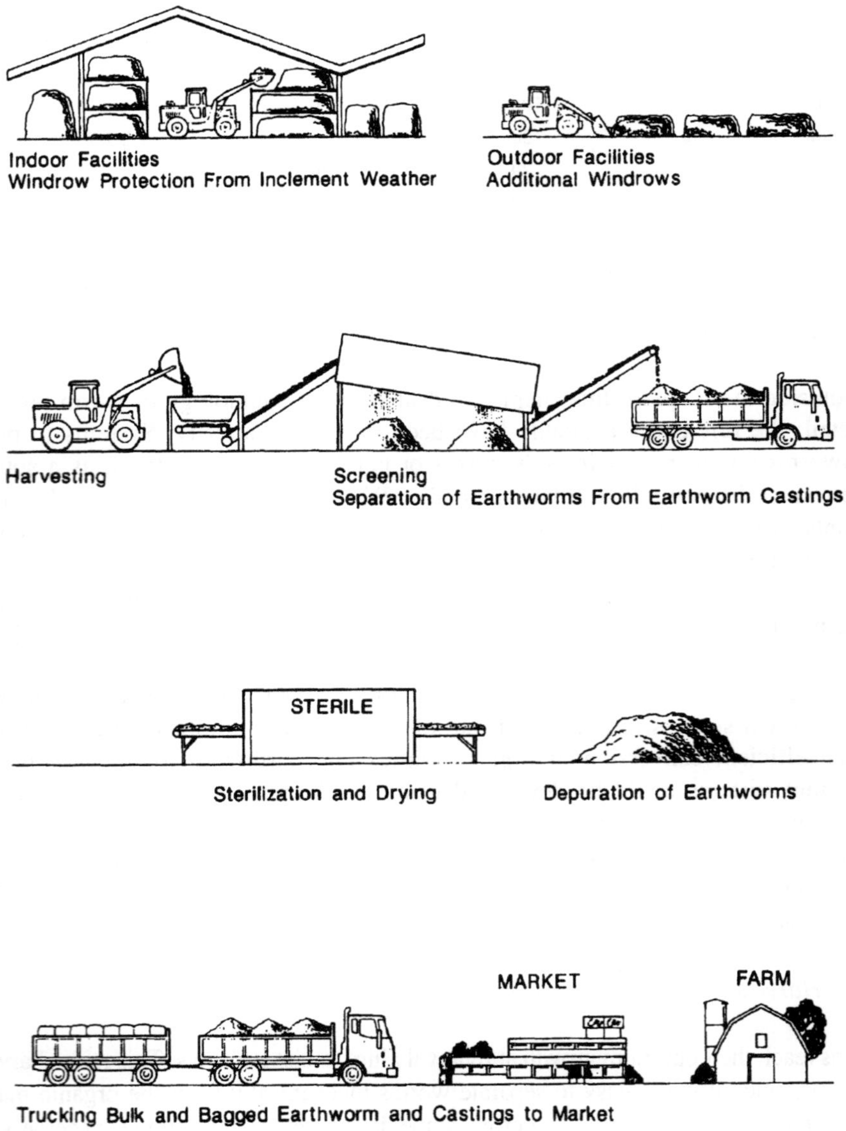

Figure 2.4. Lay-out of vermicomposting facilities. Reproduced with permission from Collier and Livingstone (1981).

The nutritive value of earthworms plus its dirt (Castings), plus the remaining waste after drying, is of great importance and needs further research. Fosgate and Babb (1972) used calf manure to produce earthworms and they reached a production level of 1 kg (fresh) earthworms-2 kg manure dry matter. On the basis of information obtained from their study, a herd of 100 cows, producing 3174 kg of manure per day would yield 42.3 kg dried earthworm protein per day or 15.4 t annually (Fosgate and Babb, 1972).

(d) Processing of earthworms for animal feed

After the mechanical separation of the worms, they are washed thoroughly and left in water for several hours to evacuate residual waste in their guts. Processing of worm meal was developed at Rothamsted Experimental Station, Harpenden, UK. Edwards and Niederer (1988) reported the several methods of processing worms as follows:

- Incorporation with molasses
- Ensiling with formic acid
- Air drying
- Freeze drying
- Aceton heat drying
- Oven drying

Two of the methods produced a mixed product and the other four a dry worm meal. The most relevant ones for poultry nutrition are air and oven drying. In air drying the worms are blanched in boiling water for 1 min., then dried in air and ground. This needs high ambient temperatures. In oven drying a dry meal was produced by killing and drying worms in trays in a large commercial oven at 80 °C for 2-4 h and then grinding them up into a dry powder. Figure 2.5 shows a scheme for worm processing to composts, wet protein and dry protein.

(e) Chemical analysis and nutritive value

Earthworm meal is a rich source of protein which contains a well balanced mixture of essential amino acids and also a reasonable percentage of minerals.

Fosgate and Babb (1972) reported a partial analysis of the earthworm meal as follows, 22.9% DM, containing 58.2% crude protein, 3.3% crude fibre, 2.8% crude fat, 0.90% P, 0.54% Ca, 0.88% K and 0.19% Mg. Taboga (1980) concluded from his complete analysis that the protein percentage is high (or average 66.5%) and the amino acid content of the product corresponded fairly well with the amino acid requirement of poultry recommended by NRC (1971). Fisher (1988) reported that the worm meal has the following composition: moisture 5%, ash 10%, crude protein 68%, Ca 0.58%, available phosphorus 0.8%, Na 0.11%, K 0.16%, Cl 0.006% and ME 11.5 MJ/kg. As far as the earthworm protein quality is concerned, only a few *in vivo* results are available. Results from rat growth assays were reported by Schulz and Graff (1977) that earthworm protein (*E. foetida*) has a biological value of 84% and a net protein utilization of 79%. Both values were excellent in comparison with fish meal protein tested at the same time. Tables 2.12 and 2.13 show the analysis of earthworm meal by several authors. It may be concluded that this feedstuff has excellent nutritional values which are equivalent to fish meal or meat meal. This feedstuff will be produced as pure earthworm meal grown on manure or mixed with castings plus manure it will be a solution to the huge accumulation of layers' cage waste.

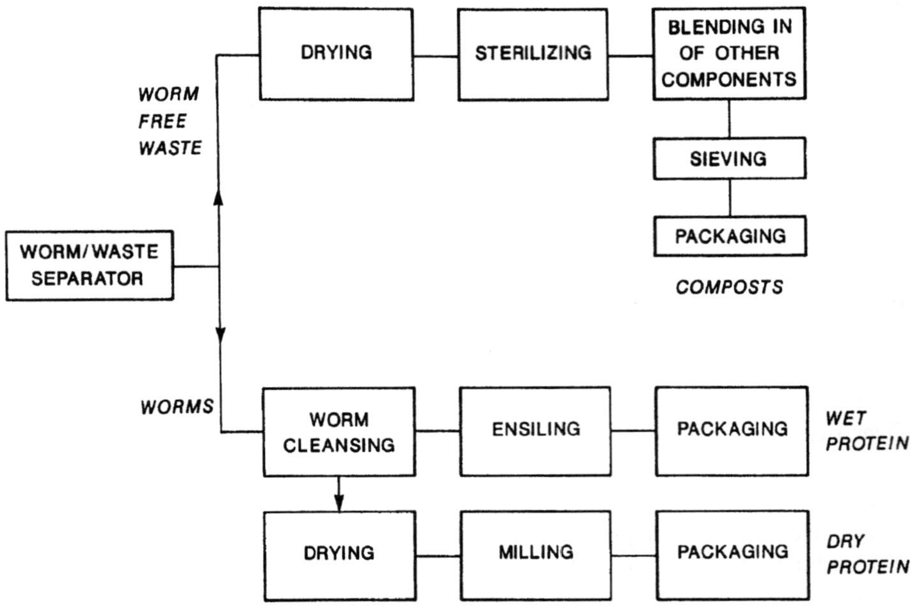

Figure 2.5. Lay-out of earthworm processing to compost, wet and dry protein feedstuffs. Reproduced with permission from Phillips (1988).

Table 2.12. Gross energy and nutrient content of earthworms *(Eisenia foetida)*. From (1) McInroy (1971); (2) Fosgate and Babb (1972); (3) Schulz and Graff (1977); (4) Sabine (1978); (5) Taboga (1980) and (6) Hartenstein (1981)

	(1)	*(2)*	*(3)*	*(4)*	*(5)*	*(6)*
Dry matter (%)	12.9	22.9		20-25	15-20	18
As % of dry matter:						
Protein (N x 6.25)	68.1	58.2	66.3	2-64	62-71	65
Fat (ether extract)	6.4	2.8	7.9	7-10	2.3-4.5	9
Fibre		3.3				
Carbohydrate			14.2			21
Ash	5.2		11.6	8-10		5-8
Calcium		0.54		0.55		0.3-0.8
Phosphorous		0.90		1.0		0.7-1.0
Gross energy (MJ/kg)				16.4-17.2		

Table 2.13. Amino acid composition of earthworm protein (g/100 g of protein). From [(1)] McInroy (1971); [(2)] Taboga (1980); [(3)] Sabine (1981); [(4)] Graff (1981) and [(5)] Morrison (1957)

	(1)[a]	*(2)*[b]	*(3)*[a]	*(4)*[a]	*(4)*[c]	*(5) FM*[d]	*(5) MM*[d]
Ala	-	5.4	-	6.0	5.2	-	-
Arg[e]	6.1	7.3	6.8	6.1	6.1	6.7	6.5
Asp	-	10.5	-	11.0	10.3	-	-
Cys	1.8	1.8	3.8	1.4	1.6	1.1	1.3
Glu	-	13.2	-	15.4	13.8	14.8	13.8
Gly	-	4.3	4.8	-	-	4.0	7.2
His[e]	2.2	3.8	2.6	2.3	2.6	2.0	2.5
Ile[e]	4.6	5.3	4.2	4.7	4.5	3.5	6.0
Leu[e]	8.1	6.2	7.9	8.2	7.9	6.4	8.4
Lys[e]	6.6	7.3	7.1	7.5	7.1	6.9	10.4
Met[e]	1.5	2.0	3.6	1.8	2.0	1.5	3.0
Phe[e]	4.0	5.1	3.7	3.5	4.1	3.5	4.2
Pro	-	5.3	-	-	-	-	-
Ser	-	5.8	4.7	4.8	4.8	-	-
Thr[e]	5.3	6.0	4.8	4.7	4.8	3.3	4.6
Try[e]	-	2.1	-	-	-	0.5	1.1
Tyr	-	4.6	2.2	3.0	3.4	1.6	3.0
Val[e]	5.1	4.4	4.9	5.2	5.0	4.7	5.7

[a] *Eisenia foetida.*
[b] Mixture of *E. foetida* and *Lumbricus rubellus.*
[c] *Eudrilus eugeniae.*
[d] FM, Fish meal; MM, Meat meal
[e] Essential amino acids.

(f) Use of earthworm meal in feeding

Earthworm meal has been tested in poultry rations by many scientists throughout the world. Harwood (1976) in Australia; Yoshida and Hoshii (1978) and Mekada *et al.* (1979) in Japan; Taboga (1980) in the United States and Fisher (1988) in the United Kingdom. Birds fed on earthworm meal as the major source of protein in the diets have all grown at rates equal to or better than those displayed by birds fed conventional protein meals. Harwood (1976) and Mekada *et al.* (1979) reported that chickens fed on earthworm meal had better feed-conversion ratios than control birds, i.e. the same weight gain could be obtained from less feed consumed.

Mekada *et al.* (1979) made an extremely complex substitution when using 5% worm meal in diets for chickens. No effect on growth rate was observed but there was a trend to reduced feed consumption. They also reported successful experiments with laying hens in which raw worms were fed, which is similar to the natural behaviour of adult poultry. Taboga (1980) mainly fed live worms with maize and vitamin supplements.

He concluded that growth rates were equivalent to controls. He used also worm meal "powder" (desiccated naturally for 24 h) which was fed at an inclusion level of 35% together with maize and vitamins. Chickens showed a normal feed intake and a slight growth depression in comparison with control diets.

Practical trials in China showed that worms (boiled and chopped) could replace fish meal and give improved growth performance (Jin-You *et al.*, 1982), while Sugimura *et al.* (1984) substituted 6% freeze-dried worm meal for fish meal with no effects on growth or feed consumption.

Finally, Fisher (1988) concluded that when worm meal formed 7.2% of the diet and provided about 25% of the dietary protein, there was virtually no effect on the growth and feed intake of the chickens nor on dietary ME or N-retention. However, at higher levels of inclusion, there is a small but significant depression in growth and feed conversion efficiency (Table 2.14). Feed consumption tended to decline but the overall effect of the treatments was not significant ($P > 0.05$). It was concluded that this reduction in feed intake may have been due to some physical or chemical characteristics which tend to affect the palatability of the meal for chickens.

Table 2.14. Performance of chicks given feeds containing different levels of worm meals. [a] From Fisher (1988)

	Level of worm meal, %				
	0	7.2	14.4	21.5	SEM[b]
Proportion of protein from worm meal	0	0.24	0.48	0.71	
Initial liveweight, g	203	201	198	201	
Final liveweight, g	735	722	677	674	
Weight gain, g	532	521	479	473	16.4*
Feed consumption, g	820	830	774	778	19.9NS
Gain/feed	0.649	0.628	0.619	0.608	0.009**
Diet GE, kJ/g	17.752	18.016	18.020	18.154	
ME/GE	0.738	0.730	0.713	0.710	0.003***
N-retention, g/g diet N	0.588	0.573	0.569	0.599	0.011NS

[a] Each feed was given to four replicate groups of six broiler chicks (Ross I) from 14-28 days of age. No birds died. Diet ME and N-retention were determined at 25-27 days of age by incorporation of an indigestible marker (TiO_2) in the feed.

[b] Standard error of mean and significance of diet effect; NS - not significant, * $p < 0.05$, ** $p < 0.01$; *** $p < 0.001$.

(g) Biohazards of earthworm meal

Before starting to use earthworm meal in poultry nutrition, some further work is needed to ascertain whether it might carry disease or contain toxic residues. Brown and Mitchell (1981) have shown that the concentration of *Salmonella spp.*, a major disease organism of poultry, can be significantly reduced in laboratory culture by the presence of *E. foetida.* Augustine and Lund (1974) showed that earthworms are not ordinarily significant in the natural distribution of an important parasitic nematode, *Ascaridia Galli* in chickens. However, studies on the persistence of disease organisms in earthworm meal that is prepared for inclusion in poultry rations will be essential, especially when the worms are cultured on poultry manure.

Toxic residues accumulated by earthworms, including metals and agrochemicals in particular, were cited by many scientists. Accumulation of metals such as lead, cadmium, chromium, copper, nickel, mercury and zinc by earthworms was cited by Gish and Christensen, (1973), Hartenstein, Neuhauser and Collier (1980) and Beyer (1981). Worms seem to tolerate large concentrations of these metals in their tissues and continuous monitoring will be necessary if earthworms are introduced in large quantities as a feedstuff.

2.3.3 Aerobic Fermentation of Poultry Waste

Upgrading poultry manure by aerobic fermentation will convert the harmful uric acid NPN to a non-toxic form. Meanwhile the N of this uric acid will be used by the aerobic bacteria to form a cell mass which is higher in true protein and N content than the original manure.

Few results have been published in this field except some popular articles with promising results concerning processing (Jackson, 1970; Harmon *et al.*, 1972, 1973; Müller, 1977; Taiganides, 1977; Vuori and Näsi, 1977; Austic *et al.*, 1978; Shuler *et al.*, 1979; Taiganides, Chou and Lee, 1979; Martin, 1980; El Boushy, Klaassen and Ketelaars, 1985 and Dafwang *et al.*, 1986). Rapid advances in physical, chemical and fermentation technology could provide better methods for the conversion of manure into a product of high nutritive value for poultry nutrition. When pollution control processes are costly, it is wise to combine waste treatment with resource recovery processes to produce feedstuffs for poultry nutrition (Figure 2.6).

(a) Technological aspects

Vuori and Näsi (1977) concluded that upgrading of poultry manure by fermentation is based on the conversion of uric acid to a non-toxic form and on the addition of energy to assure a good fermentation. They isolated 41 bacteria and 8 yeast strains effective in eliminating uric acid. The final fermented product showed a much higher concentration of amino acids than the original poultry manure.

Austic *et al.* (1978) and Shuler *et al.* (1979) developed a two stage continuous fermentation process for converting poultry manure into a high protein containing product. The protein was particularly influenced by the N-content of the fresh manure

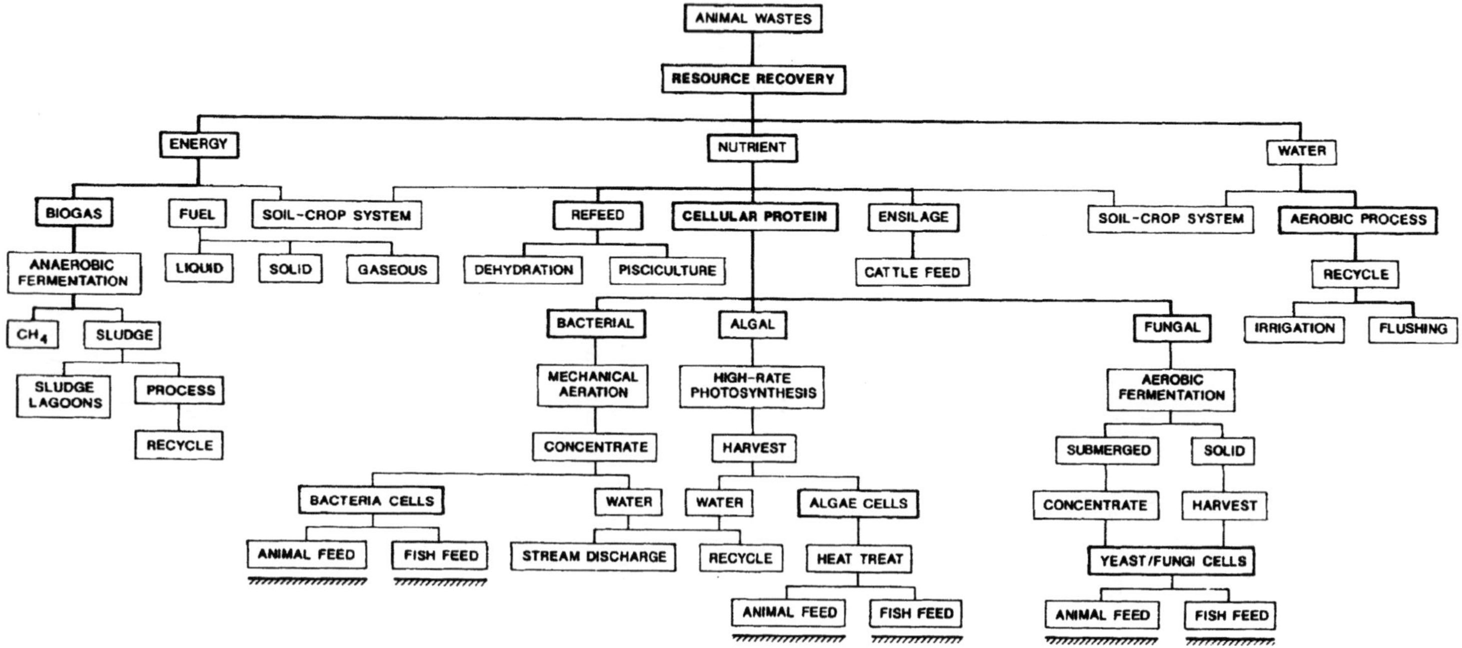

Figure 2.6. Promising pathways from animal waste treatments with resource recovery processes of feedstuffs. Reproduced with permission from Taiganides, Chou and Lee (1979).

and by the retention of N by the fermentation process. The two stages are needed because the uric acid catabolism is inhibited by the presence of many readily assimilable carbon sources.

In the first stage uric acid is converted into ammonia that can be used in the second stage for the production of an almost pure culture of *Pseudomonas fluorescens*. Less then 5% of the biomass consists of other organisms.

In the second stage a carbon source needs to be added to the nitrogen-rich, carbon-poor effluent of the first stage. Centrifugation of the material from the second stage gives:

- a clear yellowish brown supernatant;
- a light brown "creamy" layer of cellular biomass single cell protein (SCP); and
- a hard-packed dark-brown layer (thick layer), containing bacterial cells and undigested material (sand, oyster shell fragments, grain hulls *etc.*).

Figure 2.7 shows the process scheme for converting poultry manure into a high-protein feedstuff. This process shows two stages giving the chemical reaction of the uric acid and the carbon source.

Shuler *et al.* (1979) are research workers in this field in the U.S.A. who developed a process of aerobic conversion of poultry manure into high protein feedstuff. Their method was described as follows:

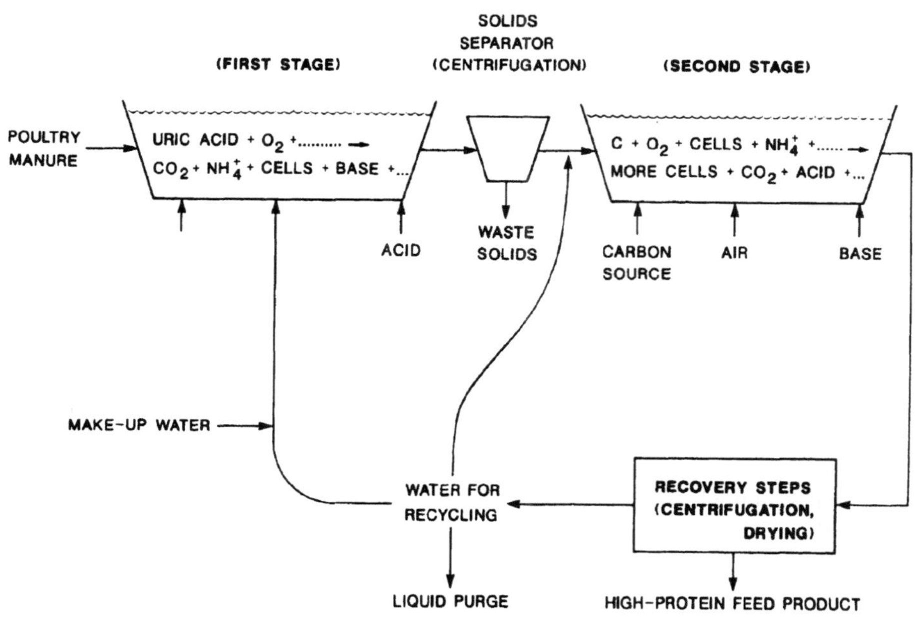

Figure 2.7. A scheme showing converting poultry manure to a high-protein feedstuff. The stages show the chemical reaction of the uric acid and the carbon source. Reproduced with permission from Shuler *et al.* (1979).

• *Manure*: manure without litter mostly from layers (25-28% solids), was collected, slurried and blended with adding chilled water, and was screened to remove feathers and shells (screening removes 15% of the total solids). Water was added to bring the slurry to the desired solids concentration (about 10 g/liter).

The supernatant liquid from slurried manure contains approximately 80% uric acid nitrogen, 10% ammonia nitrogen, 2% protein nitrogen and 8% other nitrogenous compounds (urea, nucleosides, *etc.*).

• *Fermentation*: fermentation took place in a 14 l vessel fermentor equipped with mechanical foam breakers. The pH was controlled, peristaltic pumps were used to feed the system by pumping first-stage effluent into the second stage. Chemicals were added also with the aid of the pumps. Aeration and agitation were maintained at high levels to prevent oxygen from becoming the rate-limiting nutrient. Sugar was added to the product of the second stage. The content of the reservoir, which was kept at 2-3 °C was mechanically mixed. During the fermentation process the following was measured: ammonia nitrogen, uric acid, total oxidizable carbon, turbidity and amino acids. Some of the microbes involved during the fermentation were *Micrococcus*, *Pseudomonas*, *Arthrobacter* and *Corynebacterium* and others (18 strains in total).

It may be concluded from their work that at least two stages are required to upgrade manure, since uric acid catabolism is inhibited by the presence of many readily assimilable carbon sources. The main purpose of the first stage is to convert the uric acid into ammonium ion. The second stage a carbon source is added to the nitrogen-rich, carbon-poor effluent from the fist stage to produce a medium suitable for the rapid growth of bacteria and conversion of the fixed nitrogen into predominantly protein. The product recovery has been explained before as pointed out by Shuler *et al.* (1979).

(b) Chemical analysis and nutritive value

Müller (1977) reported that the end result of the aerobic fermentation process [single cell protein (SCP) and yeast] was very promising for yield and final product quality, which is characterized by a 50% crude protein and an amino acid pattern rather similar to that of soybean meal.

Austic *et al.* (1978) concluded from their experiments that poultry manure can be upgraded by controlled fermentation to a high protein material which may have potential value in the feeding of monogastric animals and birds. They analyzed the SCP of the aerobic fermented manure and found a true protein level of 40% and an amino acid content of lysine 3.4%, methionine 1.0-1.16% and cystine 0.16%. They concluded that the thick layer of fermented product contained a similar amino acid profile in comparison with the SCP fraction except for lysine which was lower.

Shuler *et al.* (1979) noticed that a similar product with tube settlers in place, contains about 45% crude protein and has a quite attractive amino acid profile averaging about 9.3% lysine and 2.9% methionine with 59.1% as essential amino acids. Lysine and methionine are typically the limiting amino acids in many feedstuffs. The amino acid profile from this product is compared with several feedstuffs (Table 2.15).

Table 2.15. Amino Acid Content of poultry manure and its various fermentation products in comparison with some selected feedstuffs. From [(1)] Vuori and Näsi (1977); [(2)] Shuler *et al.* (1979); [(3)] El Boushy and Vink (1977) and [(4)] Scott *et al.* (1982)

	(1)			*(2)*			*(3)*	*(4)*
% *Amino Acids*	*PM*	*PML*	*PMS*	*CM*	*Ferm. Waste and product*[a]	*DPW*	*Soybean Meal*	*Herring Meal*
Aspartic acid	0.78	1.47	0.75	3.49	-	1.22	-	-
Threonine	0.36	0.85	0.43	1.04	3.48	.60	1.80	2.80
Serine	0.43	0.82	0.46	0.56	-	.72	-	-
Glutamic acid	1.04	1.88	1.03	2.41	-	1.69	-	-
Proline	0.33	0.49	0.41	0.90	-	-	-	-
Glycine	1.68	0.85	0.48	1.81	-	.93	2.10	5.90
Alanine	0.53	1.26	0.60	2.34	-	1.07	-	-
Valine	0.34	0.83	0.48	1.74	4.29	.83	2.30	3.50
Cystine	0.00	0.00	0.00	0.00	0.27	.21	0.67	1.20
Methionine	0.11	0.35	0.06	1.11	1.94	.29	0.65	2.00
Isoleucine	0.35	0.82	0.40	1.71	3.82	.66	2.50	3.70
Leucine	0.44	1.03	0.66	2.27	5.23	.94	3.40	5.10
Tyrosine	0.24	0.53	0.19	1.55	2.61	.40	0.70	2.10
Phenylalanine	0.28	0.68	0.42	0.00	3.28	.53	2.30	2.80
Lysine	0.28	0.73	0.41	1.27	6.23	.56	2.90	6.40
Histidine	0.13	0.29	0.14	0.63	2.55	.19	1.10	1.60
Arginine	0.26	0.74	0.31	1.47	5.16	.53	3.20	6.80
Tryptophan	-	-	-	-	1.01	-	0.60	0.90

PM: poultry manure; PML: poultry manure fermented (laboratory) + cell mass; PMS: poultry manure fermented + cell mass; CM: cell mass only (separated from PMS).

[a] true protein was 67%

The bacterial product of the pure cultures of the strain of *Ps. fluorescens*, yields a product with 67% true protein, although lysine (8.3%) and methionine (2.5%) values are slightly lower than with the waste-grown bacterial product with the tubesettlers. The product contains 7.5-11.3% lipids, 12-19% ash, 1.9-2.5% Ca, and 1.5-2.8% P. ME has been determined to be 15.09 MJ/kg (Shuler *et al.* 1979).

(c) Its use in feeding

As far as degradation of poultry waste is concerned, few research workers have published the effect of this product in poultry nutrition. Dafwang *et al.* (1986) used a combination of broiler house litter and offal from broiler processing plant. This combination (Fermway) is aerobically fermented. This product will not be discussed because aerobically fermented poultry manure is not similar to the well known traditional cage layer manure. For example this product possibly contains residuals as antibiotics or other medicines. The most relevant research which may be related to this point is the work of Bragg *et al.* (1975). They reported that there was no adverse effect on body weight by the addition of digested waste to the chick diet at the 5, 10, 20 and 25% level. However, final body weights for the 5 and 10% waste diets were consistently above those for the control diet. Feed conversion improved when the digested waste was increased in the chick diet using fraction one (high in crude fibre). The use of the second fraction did not affect feed utilization (Table 2.17).

2.3.4 Oxidation Ditch of Poultry Waste

The oxidation ditch is a highly technological process which is used on large farms for all livestock wastes and for the treatment of municipal sewage to control odours and to upgrade its nutrient level. The aerobic process utilizing oxidation ditch was found to be very effective in stabilizing the manure, when mixed with water and in increasing the percentage of microbial cells (Irgens and Day, 1966; Day, 1967). The efficacy of the oxidation ditch in reducing the biological oxygen demand (BOD) of the liquid manure up to 90% was reported by Jones, Converse and Day (1969). The procedure and facility of the oxydation ditch system have been described in greater detail by Day, Jones and Converse (1970).

(a) Technological aspects

The theory of the oxidation ditch system is based on supplying the liquid manure with air to reduce the BOD. The result is a clear supernatant and a residue.

Harmon *et al.*, (1972) described the technological aspects of oxidation ditch as follows: swine were housed in environmentally controlled buildings, with partially slotted floors. Mostly, hens are arranged in such a way that slotted sections are over a continuous raceway (gutter). In the case of layers kept in cages, the system is known as Indoor Lagoons (Al-Timimi, Owings and Adams, 1964). Normally the manure is propelled around a raceway at approximately 0.4 m/s with a roter forming an oxidation ditch. Oxidation ditch mixed liquor contains an average of 3.5 ppm dissolved oxygen with a pH range of 6.6-6.7. The oxidation ditch mixed liquor is maintained at a depth of 0.5m using an overflow stand pipe.

To collect solids, the rotor has to be raised to obtain a blade immersion depth of 0.08 m or less. This will reduce the velocity of the liquids in the ditch and allow solids to settle on the bottom. After the end of the process, the supernatant may be drained and used as an animal drinking water. The solids (oxidation ditch residues) may be dried by spreading them in thin layers at 40 °C.

Taiganides, Chou and Lee (1979) reported that the mixed liquor is odourless and brown in colour. Solids settle out readily, producing a clear supernatant. The effluent does not contain any solid material. Its BOD varies between 50-70 mg/l. The clear supernatant can be used for drinking water for birds and animals (Figure 2.8), and it contains rich nutrients which add a certain value to the total ration of animals and allow a lower protein level to be given in the diet (Müller, 1977). The oxidation ditch residue consists of an

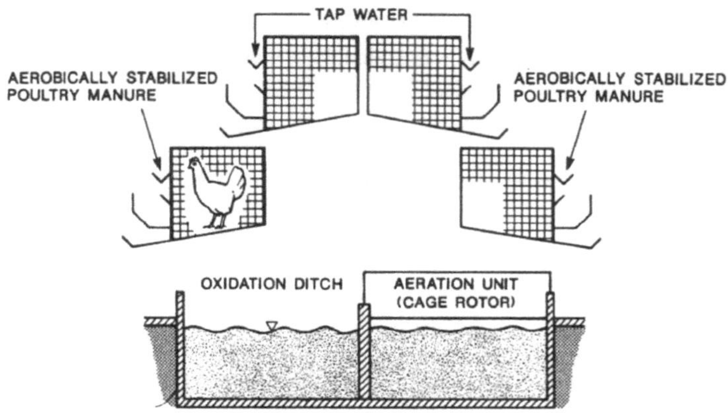

Figure 2.8. Cross-section of experimental facilities employed for utilization of aerobically stabilized oxidation ditch using poultry manure as a tap water substitute. Reproduced with permission from Martin (1980).

upper fraction, containing a slimy appearing, microbial protein which is higher in protein and amino acids than the lower "thick" layer (Harmon *et al.*,1972).

Disadvantages of the oxidation ditch are:

- Disease transmission and concentration of feed additive residues (Calvert, 1977).
- The production of nitrates/nitrites in case of a large supply of oxygen to the system Johnson *et al.*, 1977).

If the oxidation ditch system is functioning efficiently, health hazards will be of a minor importance (Müller 1980).

(b) Chemical analysis and nutritive value

The most relevant reports on the chemical analysis and nutritive value of the oxidation ditch fractions were reported by Harmon *et al.* (1972) who worked on swine manure. They found a significant increase in the amino acid fraction of dry matter in swine manure following aerobic stabilization in an oxydation ditch (Tables 2.16 and 2.17).

The nutritional value of the oxidation ditch mixed liquor (ODML) is high. Its protein content varies between 25-50% crude protein, and it has a high level of essential amino acids. The energy content is low because of the high mineral level (Müller, 1980).

Taiganides, Chou and Lee (1979) analysed the mixed liquor suspended solids from the oxidation ditch and found that the composition of the solids on dry matter basis, was 48% crude protein, 8.6% crude fibre, 31% ash, 3.6% P and 2.8% Ca.

(c) Its use in feeding

Harmon *et al.* (1972) used settled solids from swine waste collected from an oxidation ditch to substitute other protein sources in studies with weanling rats. They concluded that the oxidation ditch residue (ODR) could replace 33 or 50% of the protein of casein or soybean meal without affecting growth.

Table 2.16. Chemical analysis (air dry basis) of ODML, upper and lower fraction from swine and poultry in comparison with dried poultry waste. From [1] Harmon *et al.* (1973); [2] Harmon *et al.* (1972); [3] Müller (1980) and [4] El Boushy and Vink (1977)

	(1)	*(2)*	*(2)*	*(3)*	*(4)*
%	*ODML swine*	*Lower fraction of ODR swine*	*Upper fraction of ODR swine*	*ODML poultry*	*DPW*
Crude protein	49.00	27.70	31.80	-	24.3
Ca	3.33	1.60	1.60	-	10.6
P	3.83	1.50	1.50	-	2.7
Mg	1.49	-	-	-	-
Na	2.75	-	-	-	-
K	4.14	-	-	-	2.40
Lysine	1.42	0.32	0.72	-	0.56
Histidine	0.47	0.10	0.28	0.80	0.19
Arginine	1.28	0.33	0.74	1.43	0.53
Threonine	1.96	0.39	0.67	-	0.60
Valine	2.06	0.43	0.75	-	0.83
Isoleucine	1.49	0.29	0.51	1.40	0.66
Leucine	2.79	0.51	0.92	2.48	0.94
Methionine	0.77	0.13	0.22	0.69	0.29
Aspartic A	3.73	0.74	1.28	-	1.22
Serine	2.55	0.33	0.57	-	0.72
Glutamic A	5.06	0.89	1.59	-	1.69
Proline	1.29	0.33	0.61	-	-
Glycine	2.29	0.49	0.86	-	0.93
Phenylalanine	-	0.23	0.72	1.34	0.53
Alanine	2.83	0.58	0.98	-	1.07
Tyrosine	1.17	0.29	0.56	1.15	0.40
Cystine	-	-	-	0.43	0.21
Tryptophan	0.28	-	-	-	-

ODML = oxidation ditch mixture liquor; ODR = oxidation ditch residue; DPW = dried poultry waste.

Harmon et al. (1973) fed ODML to swine, which served as a nutrient solution mixed with dry feed. It improved weight gain and feed efficiency significantly ($P < 0.05$) when fattening pigs were fed a dry diet that was margined in protein. The nutrient solution contained 3% dry matter. The addition of ODML increased protein intake by 2.5% and lysine intake by 0.1%. Addition of ODML to maize alone did not significantly increase gain or feed efficiency.

Johnson *et al.* (1977) studied recycling ODML as the only source of drinking water for laying hens. When the dissolved oxygen content of the ODML rose from 4.7-6.0 ppm and the nitrate level increased from 210-1300 ppm, egg production dropped sharply. However, the laying hens were able to recover much of their production loss after being subjected to these levels, or slightly lower ones, for several weeks. The initial weight of the experimental birds was 1679 g. At the end of the study, these birds averaged 1769 g. The corresponding weights for the controls were 1695 and 1817 g, respectively.

Martin (1980) reported an experiment with laying hens in which three separate feeding trials over three years were carried out. In these studies aerobically stabilized poultry manure from an undercage oxidation ditch was used which substituted for tap water (Tables 2.18 and 2.19). For the three trials, egg production by the hens receiving the aerobically stabilized oxidation ditch poultry manure exceeded that of the control group by an average of 2.6-1.7%. No significant differences were observed in final body weights between the two groups of hens in any of the three feeding trials.

Table 2.17. Chemical analysis of poultry waste, wet and dry, in comparison with the two fractions after fermentation. From [(1)] El Boushy and Vink (1977); [(2)] Shuler *et al.* (1979); [(3)] Austic *et al.* (1978)

%	*(1)*		*(2)*		*(3)*	
	D.P.W.	*Manure*	*'Thick' Layer*	*'Creamy' Layer (SCP)*	*'Thick' Layer*	*'Creamy' Layer (SCP)*
True protein	14.73	-	-	-	28.0	40.0
Amino acids	-	13.0	29.0	40.0	-	-
Ash	35.79	24.0	11.0	14.0	13.0	15.0
Lipids (fat)	4.07	6.5	7.8	9.5	7.0	10.0
P	2.71	-	-	-	2.4	1.7
Ca	10.61	-	-	-	3.4	2.0

SCP: single cell protein

Table 2.18. Comparison of egg production levels of hens receiving tap water versus aerobically stabilized poultry manure under cage oxidation ditch. From Martin (1980)

	Average egg production hen-day %			
Trial	*Water*	*Aerobically stabilized poultry manure*	*Difference[a] (%)*	*SD*
I	66.6	68.6	+2.0	±2.0
II	76.7	79.8	+3.1	±1.9
III	77.2	80.0	+2.8	±1.3

[a] Differences in average egg production within each trial are significant ($P < .01$).

Table 2.19. Comparative final body weights, mortality, and average egg weights of hens receiving tapwater and aerobically stabilized poultry manure under cage oxidation ditch. From Martin (1980)

Trial	*I*	*II*	*III*
Final body weights (kg)			
Tap water	1.91	1.98	1.92
ASPM	1.90	1.95	1.88
Mortality (no. of birds)			
Tap water	11	4	5
ASPM	11	6	6
Average egg weights (g)			
Tap water	54.4	60.8	55.7
ASPM	54.7	60.9	57.8

ASPM = Aerobically stabilized poultry manure.

2.3.5 Algae Grown on Poultry Waste

Algae are able to convert nitrogen in poultry and animal waste to protein (Cook, 1960; Powell, Nevels and McDowel, 1961; Leveille, Sauberlich and Shockley, 1962; Cook, Lau and Bailey, 1963; Lubitz, 1963; Hintz *et al.*, 1966; Venkatarman, 1979; Lipstein and Hurwitz, 1983) and have been used in livestock and poultry feeds. Grown on organic wastes at low cost, conversion efficiency is high. The ultimate cost of animal protein produced in this way is several-fold less than that of feeding conventional crops to animals. Waste material can be human sewage as well as organic industrial waste and livestock manure. The cost of algae culture can be minimized by growing the algae in open outdoor ponds. Nutrients for the algae are provided by aerobic decomposition by bacteria of the organic matter in the sewage. *Chlorella, Micraetinum* and *Scendesmus* are the three most abundant algae types encountered.

The drawback of this system at present are the amount of space required for algal ponds to process large quantities of manure. In addition, high capital outlays and certain climatic and topographic conditions are needed to establish such a system.

(a) Technological aspects

Dugan, Golneke and Oswald (1969, 1971) described a system of algae growing on a pilotplant scale and provided a schematic diagram of this operation (Figure 2.9). They described the technological process as follows: poultry manure droppings of caged layers are flushed into a sedimentation tank, the supernatant is pumped directly to an algae pond, the sediment left is pumped into an anaerobic digester for methane generation. Methane is used as an energy source to heat the digester. The product effluent from the anaerobic digester is also pumped into the algae pond. The production of algae is estimated to be 11-15 t dry matter/ha/year.

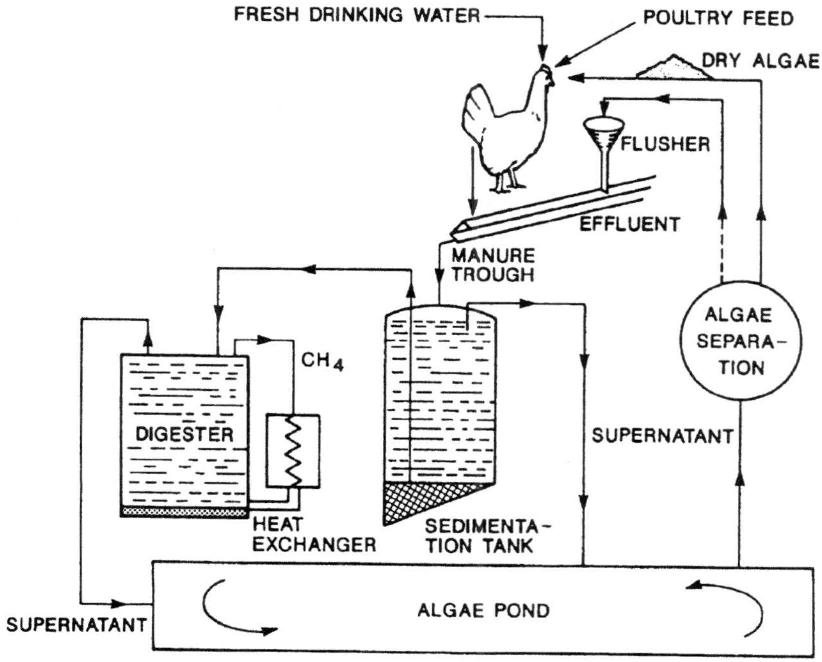

Figure 2.9. Algae production from poultry waste. Reproduced with permission from Dugan, Golneke and Oswald (1969).

The limitation of algae production was summarized by Priestley (1976) as follows:

- Separation of algae from the culture media.
- Climatic and topographic limitations on pond function.
- Large amounts of space required for algal ponds to process large quantities of waste.

There are two methods of separating the algae from the culture media. Lipstein and Hurwitz (1980, 1981 and 1983) studied a centrifugation system as well as alum flocculation; they found the latter system to be the most efficient. The final flocculated product was drum dried to a final moisture content of 8-15% of net weight.

In the operation pilot plant reported by Dugan, Golneke and Oswald (1969, 1971), the following observations were made:

- Depth: not greater than 30.5 cm.
- Pond area needed: approximately 0.18 m^2/bird.
- Amount of water needed to establish the overall system: 56.8 l/bird.
- Gas production for the system: about 0.75 m^3/kg of volatile solids introduced, methane constituting about 50-60% of the gas produced.
- Potential algal yield: 11-15 t DM/ha/year.
- Operating cost at the time: about 2 dollar cents per dozen eggs produced.

Taiganides, Chou and Lee (1979) demonstrated a complete layout of an algae production system at Singapore (Figure 2.10). This project was based on pig wastewater purification, and nutrient recovery by a high-rate, algae pond system. The ponds are operated to obtain bio-engineering data on algae productivity, degree of wastewater purification and energy inputs. They demonstrated also that at a later stage, more economical harvesting methods such as flotation and mechanical filtration will be tested. Dried algae will be used as a feedstuff for pigs and poultry.

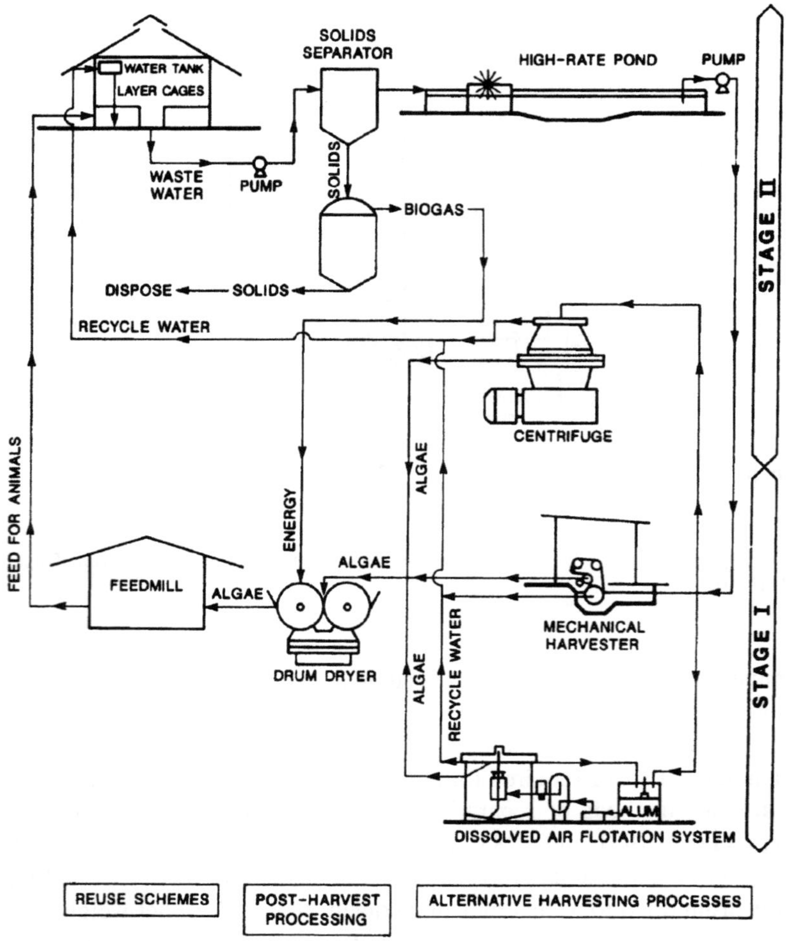

Figure 2.10. Lay-out of algae production system Reproduced with permission from Taiganides, Chou and Lee (1979).

(b) *Chemical analysis and nutritive value*

Combs (1952) and Leveille, Sauberlich and Shockley (1962) noticed a relatively low concentration of methionine, histidine and glycine in their algae samples. However, Priestley (1976) noticed that algae have an essentially similar amino acid pattern to SCP, but with a substantially lower nucleic acid content. Lipstein and Hurwitz (1980,

1981, 1983) concluded that the amino acid levels of algae samples mostly indicate lower concentrations of arginine, lysine and isoleucine but higher levels of threonine and valine in comparison with soybean meal (Table 2.20).

Table 2.20. Chemical analysis of several types of algae grown on sewage effluent with several treatments in comparison with DPW and some feedstuffs. From [(1)] Lipstein and Hurwitz (1983); [(2)] Lipstein and Hurwitz (1980); [(3)] El Boushy and Vink (1977) and [(4)] Scott *et al.* (1982)

	(1)[a]		*(2)*[b]	*(3)*[b]	*(4)*[b]	
	Alum flocculation		*Centri-fugation*	*DPW*	*Soybean meal*	*Herring meal*
%	*Chlorella*	*Micractinium*	*Chlorella*			
Crude protein	39.50	39.10	63.00	24.28	45.0	72.0
Crude fat	4.60	3.80	7.60	4.07	0.9	10.0
Crude fibre	0.50	0.70	2.40	10.11	6.0	1.0
Ash	21.80	20.70	-	35.79	-	-
Phosphorus	3.00	2.10	1.30	2.71	0.67	1.5
Calcium	0.67	0.77	0.20	10.61	0.32	2.0
Aluminium	3.90	4.90	-	-	-	-
Metabolizable Energy (MJ/Kg)	9.61	7.50	11.60	2.34	9.37	13.35
Amino acids						
Arginine	2.05	1.99	3.89	0.53	3.2	6.8
Lysine	2.08	1.91	3.69	0.56	2.9	6.4
Methionine	0.75	0.70	1.24	0.29	0.65	2.0
Cystine	0.24	0.48	0.18	0.21	0.67	1.2
Histidine	0.70	0.63	1.20	0.19	1.10	1.6
Leucine	2.73	2.68	4.74	0.94	3.40	5.1
Isoleucine	1.58	1.56	2.51	0.66	2.50	3.7
Phenylalanine	1.73	1.83	2.77	0.53	2.30	2.8
Tyrosine	1.34	1.38	2.37	0.40	2.00	2.1
Threonine	1.81	1.89	2.96	0.60	1.80	2.8
Valine	2.14	2.13	3.38	0.83	2.70	3.5
Glycine	1.89	1.82	3.17	0.93	2.10	5.9

[a] On dry matter basis
[b] on air-dry basis.

Lipstein and Hurwitz (1981, 1983) found a great variation in ME values (7.80-12.05 MJ/kg) and nitrogen absorption (from 53.8-68.2%) in various algae samples.

They also concluded that during the flocculation of algae by alum, the aluminium content had increased notably in the algae meal (Lipstein and Hurwitz, 1981, 1982). The aluminium is an inhibitor for phosphate absorption owing to aluminium phosphate binding. They calculated that 0.76 g of supplementary phosphorus were required to ameliorate the decline in bone calcification and prevent the hypophosphataemic effects of 1 g aluminium in the diet. It was also concluded that the algae meal is a good pigmenter for egg yolks owing to its high content in xanthophyll.

(c) Its use in feeding of broilers

Combs (1952), Grau and Klein (1957) and Mokady, Yannai and Berk (1977) concluded that algae meal up to 7-10% in the diet of broilers is not harmful.

Lipstein and Hurwitz (1980) reported on the use of centrifuged algae *(Chlorella)* meal as a source of protein and energy in linearly programmed diets containing 6-15% algae. They noticed a significant depression in weight gain up to 8 weeks of age with addition of 15% algae in comparison with the control groups, caused by low feed consumption in some cases.

Lipstein and Hurwitz (1983) tested the alum flocculated algae *(Microclinium* and *Chlorella)* with linearly formulated diets replacing soybean meal with 5 and 10% algae meal. These two algae meal samples, at both concentrations tested, showed no adverse effect on growth, feed efficiency or carcass fat.

Concerning the addition of alum as a flocculator, Mokady, Yamai and Berk (1976) noticed that inclusion of 30% algae meal containing 3.4% alum caused various degrees of oedema, which could be prevented by the addition of dicalcium phosphate, while Lipstein and Hurwitz (1981) concluded that the high level of aluminium in algae included in the diets up to 25% lowered the phosphorus utilization, which was reflected in a reduced plasma inorganic phosphorus level. Further addition of dicalcium phosphate to neutralize the deleterious effect of high dietary aluminium levels has to be considered in using the alum-flocculated algae.

(d) Its use in feeding of layers

Lipstein, Hurwitz and Bornstein (1980) reported that the centrifuged algae meal *Chlorella* added to diets up to 12% supplemented with Dl-methionine did not significantly affect egg output, feed conversion and egg shell quality. The 12% addition caused a deep yellow yolk colour of an acceptable appearance and increased the total linoleic acid in the diet.

Alum-flocculated algae *Micractinium* added in diets at 5, 10 and 15% significantly inhibited feed intake resulting in a decreased layer performance suggesting that the quality (algae species, location of pond, season of growth and technological processing) played an important role (Lipstein and Hurwitz 1981).

REFERENCES

Abe, R.K., Braman, W.L., and Simpson, O. (1985). Producing earthworms. *Report. Department of Animal Sci., Fort Valley State College, Fort Valley, Georgia 31030-USA.*

Ackerson, C.W., Ham, W.E. and Mussehl, F.E. (1940). The utilization of food element by growing chicks. 1) The nitrogen of urea. *Nebrasca Agr. Exp. Sta. Bull.*, p. 120.

Al-Timimi, A.A., Owings, W.J. and Adams, J.L. (1964). The effects of air and/or heat on the rate of accumulation of solids in indoor manure digestion tanks (Indoor Lagoons). *Poultry Sci.*, **43**, 1051.

Augustine, P.C. and Lund, E.E. (1974). The fate of eggs and larva of *Ascaridia galli* in earthworms. *Avian Diseases*, **18**, 394-8.

Austic, R.E., Henry, A.E., Shuler, M.L., Kargi, F., Seeley, J.R.H.W. and Vashon, R.C. (1978). Microbial processing of poultry manure for feeding to poultry. *Poultry Sci.*, **57**, (4), 1116.

Axtell, R.C. (1986). Fly control in confined livestock and poultry production. *Agricultural Division, Ciba-Geigy Corporation*, P.O. Box 18300, Greensborough, NC 27419 8300 USA. *Technical Monograph*, pp. 1-59.

Bare, L.N., Wiseman, R.F. and Abbott, O.J. (1964). Effects of dietary antibiotics and uric acid on the growth of chicks. *J. Nutr.*, **83**, 27-33.

Beyer, W.N. (1981). Metals and terrestrial earthworms (*Annelida: Oligochaeta*). In: Workshop on the role of earthworms in the stabilization of organic residues (M. Appelhof, Compiler), *Vol. 1, Proc. Beech Leaf Press, Kalamazoo, MI*, pp. 137-50.

Biely, J. (1975). The effect of dietary dehydrated poultry waste on fertility and hatchability of chicken eggs. *Arch. für Geflügelk.*, **6**, 222-4.

Biely, J., Soong, R., Seier, L. and Pope, W.H. (1972). Dehydrated poultry waste in poultry nutrition. *Poultry Sci.*, **51**, 1502-11.

Blair, R. (1973). Utilization of ammonium compounds and certain non-essential amino acids by poultry. *World Poultry Sci. J.*, **29**, 189-202.

Blair, R. and Knight, D.W. (1973). Recycling animal wastes. *Feedstuffs*, **45**, (10): 32, **45**(12): 31.

Blair, R. and Waring, J.J. (1969). Growth response of chicks following addition to the diet of various sources of nitrogen. *Br. Poultry Sci.*, **10**, 37-40.

Boda, K. (1990). *Nonconventional feedstuffs in the nutrition of farm animals.* Elsevier, Amsterdam, pp. 155-70.

Bragg, D.B, Kwok, M.C, Saben H.S and Kitts, W.D. (1975). Nutritive value of biological waste following degradation by thermophilic microorganisms. *Poultry Sci.*, **54**, 1736.

Brown, B.A. and Mitchell, M.J. (1981). Role of the earthworm, *Eisenia foetida.* In: Affecting survival of *Salmonella enteritidis* ser. *typhimurium*. *Pedobiologia*, **21**, 434-8.

Calvert, C.C. (1977). Systems for the indirect recycling by using animal and manure wastes as a substrate for protein production. *New Feed Resources* **4**, 245-64, *Proc. Techn. Consult.*, November 22-4, 1976, Rome.

Calvert, C.C., Martin, R.D. and Morgan, N.O. (1969). Dual roles for house-flies in poultry manure disposal. *Poultry Sci.*, **48**, 1793.

Calvert, C.C., Morgan, N.O. and Martin, R.D. (1970). House-fly larvae: Biodegradation of hen excreta to useful products. *Poultry Sci.*, **49**, 588-9.

Calvert, C.C., Morgan, N.O. and Martin, R.D. (1973). *Separator for negatively phototactic house-fly larvae from chicken hen excreta.* US Patent Office, Patented Feb.13. [Cited by Z.O. Müller (1980)].

Castro, L.F.V., Duarte, A.M.S., Fernandes, T.H.(1984). A note on the nutritive value of dehydrated poultry waste in layer feeding. *Agricultural Wastes*, **11**, (1), 20-30.

Collier, J.E. and Livingstone, D. (1981). Conversion of municipal wastewater treatment plant residual sludges into earthworm castings for use as topsoil. *Pub.Felton, C.A.Colliers' Earthworm Compost Systems*, Inc. Felton, CA, 95018 USA, p. 37.

Combs, C.F. (1952). Algae *Chlorella* as a source of nutrients for the chick. *Science*, **116**, 453-4.

Cook, B.B. (1960). The nutritive value of waste-grown algae. 88th *Annual meeting of American Public Health Association*, San Francisco, California.

Cook, B.B, Lau, E.W. and Bailey, B.M. (1963). The protein quality of waste-grown green algae. 1.Quality of protein in mixture of algae, non-fat powdered milk and cereals. *J. Nutr.*, **81**, 23-7.

Couch, J.R. (1974). Evaluation of poultry manure as a feed ingredient. *World Poultry Sci. J.*, **30**, 279-89.

Cunningham, F.E. and Lillich, G.A. (1975). Influence of feeding dehydrated poultry waste on broiler growth, meat flavor, and composition. *Poultry Sci.*, **54**, 860-5.

Dafwang, I.I., Cook, M.E., Pringh, D.J. and Sunde, M.L. (1986). Nutritional value of aerobically fermented poultry manure and offal (Fermway) for broiler chicks. *Poultry Sci.*, **65**, 1765-70. Day, D.L. (1967). Current status of the oxidation ditch summary of United States and European Research. *Proc. 10th National Pork Industry Conference*, Lincoln Nebrasca, USA.

Day, D.L., Jones, D.D. and Converse, J.C. (1970). *Livestock waste management studies termination report.* Agr. Engr. Res., University of Illinois, Urbana, USA.

Dugan, G.L., Golneke, C.G. and Oswald, W.J. (1969). Hydraulic handling of poultry manure integrated into an algal recovery system. Proceedings Natural Poultry Litter and Waste Management Seminar, pp. 57-78. [Also in: New Feed Resources, *FAO Animal Production and Health Paper no.4*, pp. 245-64 (1976)].

Dugan, G.L., Golenke, C.G. and Oswald, W.J. (1971). Poultry operation with an integrated sanitation waste materials recycling system. *Abstracts Excerpts and Reviews of the Solid Waste Literature*, Vol.IV, prepared by Golneke, C.G., *Serial Report* **71-2**, 284-5.

Edwards, C.A. and Neuhauser, E.F. (1988). *Earthworms in waste and environmental management.* Pub. SPB, Academic publishing B.V. The Hague, The Netherlands.

Edwards, C.A. and Niederer, A. (1988). The production and processing of earthworm protein. In: *Earthworms in waste and environmental management.* (Ed. C.A. Edwards and E.F. Neuhauser), SPB, Academic Publishing, The Hague, The Netherlands, pp. 169-79.

El Boushy, A.R. (1991). House-fly pupae as poultry manure converters for animal feed: a review. *Bioresource technology.*, **38**, 45-9.

El Boushy, A.R., Klaassen, G.J. and Ketelaars, E.H. (1985). Biological conversion of poultry and animal waste to a feedstuff for poultry. *World Poultry Sci.J.*, **41**, (2), 133-45.

El Boushy, A.R. and Roodbeen, A.E. (1984). Amino acid availability in dry poultry waste in comparison with relevant feedstuffs. *Poultry Sci.*, **63**, 583-5.

El Boushy, A.R. and Vink, F.W.A. (1977). The value of dried poultry waste as a feedstuff in broiler diets. *Feedstuffs*, **49**, (51): 24-6.

Ernst, L., Vagapov, R., Pogdeeva, E., Zhemchuzhina, A. and Zvereva, E. (1984). A high protein feed from poultry manure. *Plitsevodstono*, **1**, 30.

Fadika, G.O., Walford, J.H., and Flegal, C.J. (1973). Performance and blood analysis of growing turkeys fed poultry anaphage. *Poultry Sci.*, **53**, 2025-6.

Featherston, W.R., Bird, H.R. and Harper, A.E. (1962). Effectiveness of urea and amonium nitrogen for the synthesis of dispensible amino acids by the chick. *J. Nutr.*, **78**, 198-206.

Fisher, C. (1988). The nutritional value of earthworm meal for poultry. In: *Earthworms in waste and environmental management.*(Eds. C.A. Edwards and E.F. Neuhauser). SPB Academic Publishing, The

Hague, The Netherlands, pp. 182-92.

Flegal, C.J. and Zindel, H.C. (1969). The utilization of hydrated poultry waste by laying hens. *Poultry Sci.*, **48**, 1807.

Flegal, C.J. and Zindel, H.C. (1970[a]). The utilization of poultry waste as a feedstuff for growing chicks. *Michigan State University, East Lansing, USA, Research report* **117**, 21-8.

Flegal, C.J. and Zindel, H.C. (1970[b]). The effect of feeding dehydrated poultry waste on production, feed efficiency, body weight, egg weight, shell thickness, and haugh score. *Michigan State University, East Lancing, Mich.Research report*, **117**, 31-3.

Flegal, C.J., Goan, C.H.J. and Zindel, H.C. (1970). The effect of feeding dehydrated poultry waste to laying hens on the taste of the resulting eggs. *Res. Report* **117**, *Farm Sci. Michigan State University*, pp. 34-8.

Fontenot, J.P. (1981). *Upgrading residues and by-products for animals*, (Ed. J.T. Huber), CRC Press, Inc., Boca Raton, Florida, pp. 17-33.

Fosgate, O.T. and Babb, M.R. (1972). Biodegradation of animal waste by *Lymbricus terrestris*. *J. of Dairy Sci.*, **55**, 870-72.

Gish, C.D. and Christensen, R.E. (1973). Cadmium, Nickel, Lead and Zinc in earthworms from roadside soil. *Environ. Sci. Technol.*, **7**, 1060-2.

Graff, O. (1981). Preliminary experiments of vermicomposting of different waste material using *Eudrilus eugenia* King. In: Workshop on the role of earthworms in the stabilization of organic residues (M. Appelhof, Compiler), *Vol. 1, Proc. Beech Leaf Press, Kalamazoo, MI*, pp. 179-91.

Grau, C.R. and Klein, N.W. (1957). Sewage-grown algae as a feedstuff for chicks. *Poultry Sci.*, **36**, 1046-51.

Harmon, B.G., Day, D.L., Jensen, A.H. and Baker, D.H. (1972). Nutritive value of aerobically sustained swine excrement. *J. of Anim. Sci.*, **34**, 403-7.

Harmon, B.G., Day, D.L., Baker, D.H. and Jensen, A.H. (1973). Nutritive value of aerobically or anaerobically processed swine waste. *J. of Anim. Sci.*, **37**, 510-3.

Hartenstein, R. (1981). Use of *Eisenia foetida* in organic recycling based on laboratory experiments. In: Workshop on the role of earthworms in the stabilization of organic residues (M. Appelhof, Compiler). *Vol. 1, Proc. Beech Leaf Press, Kalamazoo, MI*, pp. 158-65.

Hartenstein, R., Neuhauser, E.F. and Collier, J. (1980). Accumulation of heavy metals in the earthworm *Eisenia foetida*. *J. Environ. Qual.*, **9**, 23-6.

Harwood, M. (1976). Recovery of protein from poultry waste by earthworms. *Proc.1st Australian Poultry Slookfeed Conf. Melbourne*, pp. 138-43.

Hintz, H.F., Heitman, H., Weir, W.C., Torrel, D.T. and Meyer, J.H. (1966). Nutritive value of algae grown on sewage. *J. of Anim. Sci.*, **25**, 675-81.

Irgens, R.L. and Day, D.L. (1966). Aerobic treatment of swine waste. Amer.Soc.Agr.Engr. *Symposium on Management of Farm Animal Waste. Amer. Soc. Agr. Engr. Pub.* **SP 0366:59**.

Jackson, S.W. (1970). Growth of micro-organisms in fresh chicken manure under aerobic and anaerobic conditions. *Poultry Sci.*, **49**, 1749-50.

Jin-You, X., Xian-Kuan, Z., Zhi-Ren, P., Zhen-Yong, H., Yan-Hua, G., Hong-Bo, T., Xye-Yan, H. and Qiao-Ping, X. (1982). An experimental research on the substitution of earthworm for fish meal in feeding meatchickens. *J. South China Norm. Coll.*, **1**, 88-94.

Johnson, H.S., Day, D.L., Byerly, C.S. and Prawirokusumo, S. (1977). Recycling oxidation ditch mixed liquor to laying hens. *Poultry Sci.*, **56**, 1339-41.

Jones, D.D., Converse, J.C. and Day, D.L. (1969). Aerobic digestion of swine waste. *Proc. International Conf. Agr. Engr. CIGR, Baden Baden, Germany,* **Sec 2**, p. 204.

Jones, H.L. and Combs, G.F. (1953). Effect of aurcomycine HCL on the utilization of inorganic nitrogen by the chick. *Poultry Sci.*, **32**, 873-5.

Kese, A.G. and Donkoh, A. (1982). Evaluation of methods of processing dried poultry waste in terms of performance and carcass quality of broiler chickens. *Poultry Sci.*, **61**, 2500-2.

Kroodsma, W., Arkenhout, J. and Stoffers, J.A. (1985). New system for drying poultry manure in belt batteries.*Institute of Agricultural Engineering, Wageningen, The Netherlands, Research report* **85**, (1), 1-27.

Leveille, G.A., Sauberlich, H.E. and Shockley, J.W. (1962). Protein value and the amino acid deficiencies of various algae for growth of rats and chicks. *J. Nutr.*, **76**, 423-8.

Lipstein, B. and Hurwitz, S. (1980). The nutritional value of algae for poultry. Dried *Chorella* in broiler diets. *Br. Poultry Sci.*, **21**, 9-21.

Lipstein, B. and Hurwitz, S. (1981). The nutritional value of sewage-grown, alum-flocculated *Micractinium* algae in broiler and layer diets. *Poultry Sci.*, **60**, 2628-38.

Lipstein, B. and Hurwitz, S. (1982). The effect of aluminium on the phosphorus availability in algae-containing diets. *Poultry Sci.*, **61**, 951-4.

Lipstein, B. and Hurwitz, S. (1983). The nutritional value of sewage-grown samples of *Chlorella* and *Micractinium* in broiler diets. *Poultry Sci.*, **62**, 1254-60.

Lipstein, B., Hurwitz, S. and Bornstein, S. (1980). The nutritional value of algae for poultry. Dried *Chlorella* in layer diets. *Br. Poultry Sci.*, **21**, 23-7.

Loehr, R.C., Martin, J.H., Neuhauser, E.F. and Malecki, M.R. (1984). Waste management using earthworms. Engineering and scientific relationships. *Final project report. Department of Agricultural Engineering, Cornell University. Ithaca, New York.*

Lofs-Holmin, A. (1985). Vermiculture: Present knowledge of the art of earthworm farming. *Swedish University of Agricultural Sciences, Department of Ecology and Environment Research. Uppsala, Sweden*, Report **No. 20**.

Lubitz, J.A. (1963). The protein quality digestibility and composition of algae, *Chlorella 71105*. J. Food Sci., **28**, 229-32.

Martin, J.H., Jr. (1980). Performance of caged White Leghorn laying hens fed aerobically stabilized poultry manure. *Poultry Sci.*, **59**, 1178-82.

McInory, D.M. (1971). Evaluation of earthworm *Eisenia foetida* as a food for man and domestic animals. *Feedstuffs*, **43**, (8): 37-46.

Mekada, H., Hayashi, N., Yokota, H. and Okumura, J. (1979). Performance of growing and laying chickens fed diets containing earthworms (*Eisenia foetida*). *Jap. Poultry Sci.*, **16**, 293-7.

Miller, B.F. (1969). Biological digestion of manure by Diptera. *Feedstuffs*, **41**, (51), 32-3.

Miller, B.F. and Shaw, J.H. (1969). Digestion of poultry manure by Diptera. *Poultry Sci.*, **48**, 1844-5.

Miller, B.F., Teotia, J.S. and Thatcher, T.O. (1974). Digestion of poultry manure by *Musca domestica*. *Br. Poultry Sci.*, **15**, 231-34.

Mokady, S., Yannai, S. and Berk, Z. (1976). Combined systems for algae wastewater treatment reclamation and protein production. *Nutrition Toxicology Group, Annual Report, No. 2 Technion-Israel Institute Technology, Haifa, Israel.*

Mokady, S., Yannai, S. and Berk, Z. (1977). Combined systems for algae wastewater treatment reclamation and protein production. *Nutrition Toxicology Group, Annual Report, No. 3 - Technion-Israel Institute Technology, Haifa, Israel.*

Morrison, F.B. (1957). *Feeds and feeding*. 22nd edn. Morrison Publishing Co., Ithaca, NY.

Müller, Z.O. (1977). New feed resources. In: *FAO Animal Production and Health Paper.*, **4**, 265-94.

Müller, Z.O. (1980). Feed from animal wastes: state of knowledge. In: *FAO Animal Production and Health Paper.*, **18**, 132-3.

NRC (1971) *Nutrient requirements of poultry*, 6th edn., National Academy of Science, National Research Council, Washington, DC.

NRC (1977) *Nutrient requirements of poultry*, 7th edn., National Academy of Science, National Research Council, Washington, DC.

Olorede, B.R., Ajagbonna, O.P. and Babatunde, G.M. (1995). Comparison of air-dried and oven-dried poultry droppings in broiler rations: Effect of performance, organ weights and haematological parameters. *Int. J. Anim. Sci.*, **10**, 289-93.

Peter, V., Chrappa, V., Boda, K., Zajonc, I. and Vanco, M. (1985). Effect of poultry excreta dressed with house-fly (*Musca domestica L.*) pupae in the fattening of broilers and rearing of Japanese quail. *Pol'nohospodarstvo*, **31**, (11), 1019-25.

Phillips, V.R. (1988). Engineering problems in the breakdown of animal wastes by earthworms. In: *Earthworms in Waste and Environmental Management*, (Eds. C.A. Edwards and E.F. Neuhauser), SPB Academic Publishing, The Hague, The Netherlands, pp. 111-8.

Powel, C.R., Nevels, E.M. and McDowel, M.E. (1961). Algae feeding in humans. *J. Nutr.*, **75**, 7. Price, F., (1972). Dried poultry waste as feed. *Poultry Dig..*, **31**, 248-9.

Priestley, G. (1976). Algal proteins. In: *Food from waste*, (Eds. G.G. Birch, K.J. Parker and J.T. Worgan) Applied Science Publishers, Ltd, London, pp. 114.

Ramteke, B.N., Jadhao, S.B. and Kunde, R.J. (1994). Effect of feeding dried poultry droppings on composition and quality of eggs. *Indian J. Anim. Nutr.*, **11**, (4), 259-61.

Rinehart, K.E., Snetsinger, D.C., Ragland, W.W. and Zimmerman, R.A. (1973). Feeding value of dehydrated poultry waste. *Poultry Sci.*, **52**, 2078.

Sabine, J.R. (1978). The nutritive value of earthworm meal. In: *Utilization of Soil Organism in Sludge Management* (Ed. R. Hartenstein), State University - N.Y., Syracuse, pp. 122-30.

Sabine, J.R. (1981). Vermiculture as an option for resource recovery in the intensive animal industries. In: Workshop on the role of earthworms in the stabilization of organic residues (M. Appelhof, Compiler). *Vol. 1, Proc. Beech Leaf Press Kalamazoo*, MI, pp. 241-52.

Satchell, J.E. (1983). *Earthworm Ecology*. Pub. Chapman and Hall, London, New York., 495 p.

Schulz, E. and Graff, O.G. (1977). Zur Bewertung von Regenwurmmehl aus *Eisenia foetida* (Savégny 1826) als Eiweissfuttermittel. *Landb.Forsch.-Völkenrode*, **27**, 216-18.

Scott, M.L., Nesheim, M.C. and Young, R.J. (1982). *Nutrition of the chicken*. Scott, M.L. and Associates, Ithaca, N.Y., p. 562.

Shannon, P.W., Blair, R. and Lee, D.J.W. (1973). Chemical and bacteriological composition and the metabolizable energy of eight samples of dried poultry waste produced in the United Kingdom. *4th Europ. Poultry Conf. London*. pp. 487-94.

Sheppard, C.C., Flegal, C.J., Dorn, D. and Dale, J.L. (1971). The relationship of drying temperature to total crude protein and dried poultry waste. *Michigan State Univ., Agricultural Experiment Station. Research Report*, **152**, 12-16.

Sheppard, G., Tomberlin, J.K. and Newton, L.G. (1998). Use of solder fly larvae to reduce manure, control house-flies, and produce high quality feedstuff. *Proceedings National Poultry Waste Management Symposium*. (Ed. by: J.P. Blake, and P.H. Patterson). Dept. of Poultry Sci., The Pennsylvania State University, U.S.A., p. 460.

Shuler, M.L., Roberts, E.D., Mitchell, D.W., Kargi, F., Austic, R.E., Henry, A., Vashon, R. and Seeley, H.W., Jr. (1979). Process for the aerobic conversion of poultry manure into high-protein feedstuff. *Biotechnology and Bioengineering*, **21**, 19-38.

Soares, J.H., Jr., and Kifer, R.R. (1971). Evaluation of protein quality based on residual amino acids on the ileal content of chicks. *Poultry Sci.*, **50**, 41.

Sturtevant engineering company Ltd (1979). *Manure drier.* **9703**: 1, 1-4. Westergate Road, Moulsecoomb Way, Brighton BN 24 QB, UK.

Sugimura, K., Hori, E., Kurihara, Y. and Itoh, S.(1984). Nutritional value of earthworms and grasshoppers as poultry feed. *Jap. Poultry Sci.*, **21**, 1-7.

Sullivan, T.W. and Bird, H.R. (1957). Effect of quality and source of dietary nitrogen on the utilization of the hydroxy analogues of methionine and glycine by chicks. *J. Nutr.*, **62**, 143-50.

Taboga, L. (1980). The nutritive value of earthworms for chickens. *Br. Poultry Sci.*, **21**, 405-10.

Taiganides, E.P. (1977). *Animal Wastes.* Applied Science Publishers Ltd, London.

Taiganides, E.P., Chou, K.C. and Lee, B.Y. (1979). Animal waste management and utilization in Singapore. *Agricultural Wastes*, **1**, (2), 129-41.

Teotia, J.S. and Miller, B.F. (1970[a]). Factors influencing catabolism of poultry manure with *Musca domestica. Poultry Sci.*, **49**, 1443.

Teotia, J.S., and Miller, B.F. (1970[b]). Nutritional value of fly pupae and digested manure. *Poultry Sci.*, **49**, 1443.

Teotia, J.S. and Miller, B.F. (1974). Nutritive content of house-fly pupae manure residue. *Br. Poultry Sci.*, **15**, 177-82.

Trakulchang, N. and Balloun, S.L. (1975). Non protein nitrogen for growing chicks. *Poultry Sci.*, **54**, 591-4.

Venkatarman, G.S. (1979). Indian experience with algal ponds. In: Bioconversion of organic residues for rural communities. *Food and Nutrition Bulletin Supplement, 2. JPWN-1/UNUP-43. The United Nations University*, pp. 68-71.

Vuori, A.T. and Näsi, J.M. (1977). Fermentation of poultry manure for poultry diets. *Br. Poultry Sci.*, **18**, 257-64.

Webb, K.E. and Fontenat, J.P. (1975). Medicinal drug residues in broiler litter and tissue from cattle fed litter. *J. Anim. Sci.*, **41**, 1212.

York, L.R., Flegal, C.J., Zindel, H.C. and Coleman, T.H. (1970). Effect of diets containing dehydrated poultry waste on quality changes in shell eggs during storage. *Poultry Sci.*, **49**, 590-1.

Yoshida, M. and Hoshii, H. (1978). Nutritional value of earthworms for poultry feed. *Jap. Poultry Sci.*, **15**, 308-11.

Young, R.J. and Nesheim, M.C. (1972). Dehydrated poultry waste as a feed ingredient. *Cornell Nutrition Conference, Cornell University, Ithaca, N.Y.*, pp. 46-55.

Zindel, H.C. (1972). The perfect circle recycling. *Can. Poultry Rev.*, **96**, (11), 21-4.

Zindel, H.C. (1974). Turkey anaphage. *Poult. Dig.*, **33**, 73-6.

3. PROTEIN RECOVERY FROM WASTEWATER IN POULTRY PROCESSING PLANTS

Contents

3.1 INTRODUCTION

The volume of water used in poultry processing plants per unit of product input has risen rapidly over the past several years. The costs of water and wastewater treatment have not been sufficient to warrant much management attention. However, increased regulatory requirements for point source discharge and subsequent amendments have caused these costs of water and wastewater treatment to increase more rapidly than other costs in poultry processing.

For creating an economic and relevant slaughter-house, the following objectives have to be achieved:

- Reduction of water used.
- Reduction and/or prevention of product loss to sewers.
- Reduction of wastewater.
- Recovery of a marketable product directly from the waste stream, e.g. protein from slaughter-house wastewater.
- Reduction of the overall cost of waste disposal by means of a reduction in the polluting load.

Kahle and Gray (1956) estimated the use of water in poultry processing plants to be approximately 9 l/kg and offal may consequently have altered the quality and increased the quantity of wastewater. Therefore, the figures of Erdtsieck and Gerrits (1973) may still be valid as far as the volume of water per bird is concerned.

The wastewater of poultry processing plants is composed of:

- Evisceration wastewater.
- Feather flow wastewater.
- Lung vacuum pump wastewater.
- Offal truck drain wastewater.
- Belt filter press wastewater.

The increase of wastewater usage corresponds with a rise in waste loading as measured by parameters such as the Biochemical Oxygen Demand (B.O.D.) of the wastewater (Camp, 1969; Erdtsieck and Gerrits, 1973).

However, more quality criteria can be measured, and according to Metcalf and Eddy (1979); Eilbeck and Mattlock (1987) and Merka (1989), the quality of wastewater is determined by the following points:

- Biochemical Oxygen Demand (BOD)
- Chemical Oxygen Demand (COD)
- Total Suspended Solids (TSS)
- Fat, Oil and Grease (FOG)
- Total Volatile Solids (TVS)
- Total Kjeldahl Nitrogen (TKN)
- Amonia Nitrogen (NH_3-N)

3.1.1 Biochemical Oxygen Demand (BOD)

The most widely used parameter to test the organic pollution of wastewater and surface water, is a test of 5 days (BOD_5). This determination involves the measurement of the dissolved oxygen used by micro-organisms in the biochemical oxidation of organic matter.

The advantages of the test are:

- Determination of the approximate quantity of oxygen that will be required to stabilize the organic matter present biologically.
- Determination of the size of waste-treatment facilities.
- Measure the efficiency of some treatment processes.

The disadvantages of the test are:

- A high concentration of active, acclimated seed bacteria is required.
- Pretreatment is needed when dealing with toxic wastes, and the effects of nitrifying organisms must be reduced.
- Only the biodegradable organisms are measured.
- The test does have stoichiometric validity after the soluble organic matter present in solution has been used.
- An arbitrary long period of time is required to obtain results.

The determination of the BOD test is based on the suitable dilution with water so that adequate nutrients and oxygen will be available during the incubation period. The dilution water is "seeded" with a bacterial culture that has been acclimated, if necessary, to the organic matter present in the water. The seed culture that is used to prepare the dilution water for the BOD test is a mixed culture. Such cultures contain large numbers of saprophytic bacteria and other organisms that oxidize the organic matter.

In addition they contain certain autotrophic bacteria that oxidize noncarbonaceous matter. When the sample contains a large population of micro-organisms, seeding is not necessary. The incubation period is usually 5 days at 20 °C, but other lengths of time and temperatures can be used. The temperature, however, should be constant throughout the test. After incubation the dissolved oxygen of the sample is measured and the BOD is calculated.

3.1.2 Chemical Oxygen Demand (COD)

The COD test is used to measure the content of organic matter of both wastewater and natural waters. The oxygen equivalent of the organic matter that can be oxidized is measured by using a strong chemical oxidizing agent in an acidic medium. Potassium dicromate has been found to be excellent for this purpose. The test must be performed at an elevated temperature. A catalyst (silver sulphate) is required to aid the oxidation of certain classes of organic compounds. Since some inorganic compounds interfere with the test, care must be taken to eliminate them.

The COD test is also used to measure the organic matter in industrial and municipal wastes that contain compounds that are toxic to biological life. The COD of a waste is, in general, higher than the BOD because more compounds can be chemically oxidized than can be biologically oxidized. For many types of wastes, it is possible to correlate COD with BOD. This can be very useful because the COD can be determined in 3 h, compared with 5 days for the BOD. Once the correlation has been established, COD measurements can be used to good advantage for treatment, plant control and operation.

3.1.3 Total Suspended Solids (TSS)/Total Volatile Solids (TVS)

Analytically, the total solids content of a wastewater is defined as all the matter that remains as residue upon evaporation at 103-105 °C. Matter that has a significant vapor pressure at high temperature is lost during evaporation and is not defined as a solid. Total solids, or residue upon evaporation and is not defined as solid. Total solids, or residue upon evaporation, can be classified as either suspended solids or filterable solids by passing a known volume of liquid through a filter. The filter is commonly chosen so that the minimum diameter of a particle of the suspended solids is about 1 m. The suspended solids fraction includes the settleable solids that will settle to the bottom of a cone-shaped container (called an Imhoff cone) in a 60 -min period. Settleable solids are an approximate measure of the quantity of sludge that will be removed by sedimentation.

The filterable-solids fraction consists of colloidal and dissolved solids. The colloidal fraction consists of the particulate matter with an approximate diameter range of from 1 nm - 1 m. The dissolved solids consist of both organic and inorganic molecules and ions that are present in true solution in water. The colloidal fraction cannot be removed by settling. Generally, biological oxidation or coagulation, followed by sedimentation, is required to remove these particles from suspension.

Each of the categories of solids may be further classified on the basis of their volatility at 600 °C. The organic fraction will oxidize and will be driven off as gas at this temperature, and the inorganic fraction remains behind as ash. Thus the terms "volatile suspended solids" and "fixed suspended solids", refer, respectively, to the organic and inorganic (or mineral) content of the suspended solids. At 600 °C, the decomposition of inorganic salts is restricted to magnesium carbonate, which decomposes into magnesium oxide and carbon dioxide at 350 °C. Calcium carbonate, the major component of the inorganic salts, is stable up to a temperature of 325 °C. The volatile-solids analysis is applied most commonly to wastewater sludges to measure their biological stability.

3.1.4 Fat, Oils and Grease (FOG)

Fats and oils are the third major component of feedstuffs. The term "grease" as commonly used, includes the fats, oils, waxes and other related constituents found in wastewater. Grease content is determined by extraction of the waste sample with hexane (grease is soluble in hexane).

Fats and oils are compounds (esters) of alcohol or glycerol (glycerin) with fatty acids. The glycerides of fatty acids that are liquid at ordinary temperatures are called oils, and those that are solids are called fats. They are quite similar, chemically, being composed of carbon, hydrogen and oxygen in varying proportions.

Fats are among the more stable of organic compounds and are not easily decomposed by bacteria. Mineral acids attack them, however, resulting in the formation of glycerin and fatty acid. In the presence of alkalies, such as sodium hydroxide, glycerin is liberated, and alkali salts of the fatty acids are formed. These alkali salts are known as soaps, and like the fats, they are stable. Common soaps are made by saponification of fats with sodium hydroxide. They are soluble in water, but in the presence of hardness constituents, the sodium salts are changed to calcium and magnesium salts of the fatty acids, or so-called mineral soaps. They are insoluble and are precipitated.

3.1.5 Total Kjeldahl Nitrogen (TKN)/Amonia Nitrogen (NH_3 - N)

Since nitrogen is an essential building block in the synthesis of protein, nitrogen data will be required to evaluate the treatability of wastewater by biological processes. Insufficient nitrogen can necessitate the addition of nitrogen to make the waste treatable. The nitrogen present in fresh wastewater is primarily combined in proteinaceous matter and urea. Decomposition by bacteria readily changes the form of ammonia. The age of wastewater is indicated by the relative amount of ammonia that is present. The predominance of nitrate nitrogen in wastewater indicates that the waste has been stabilized with respect to oxygen demand.

Ammonia nitrogen exists in aqueous solution as either the ammonium ion or ammonia, depending on the pH of the solution, in accordance with the following equilibrium reaction:

$$NH3 + H2O \leftrightarrow NH_4^+ + OH^-$$

At pH levels above 7, the equilibrium is displaced to the left; at levels below pH 7, the ammonium ion is predominant. Ammonia is determined by raising the pH, distilling of the ammonia with the steam produced when the sample is boiled, and condensing the steam that absorbs the gaseous ammonia. The measurement is made colorimetrically.

Organic nitrogen is determined by the Kjeldahl method. The aqueous sample is first boiled to drive off the ammonia, and then it is digested. During the digestion the organic nitrogen is converted to ammonia. Total Kjeldahl nitrogen is determined in the same manner as organic nitrogen, except that the ammonia is not driven off before the digestion step. Kjeldahl nitrogen is therefore the total of the organic and ammonia nitrogen.

Wastewater is treated by means of flotation and flocculation resulting in an effluent. The effluent from poultry slaughter-house wastewater contains large amounts of protein, fat and usually has a much higher BOD level than town sewage. It has been estimated that 2-5% of total carcass protein is lost in the effluent. Such effluents can impose heavy loads on public sewage-treatment works.

Proteins and fats from the carcass, debris and blood are the major pollutants in the wastewater, but like other poultry products, they are of high nutritional value and should be recovered for animal feed. These materials, especially proteins and fats, are usually recovered as sludges with high water content and dehydration is generally the most difficult and costly part of the whole treatment (Ross, 1968; Grant, 1976; Beszedits, 1980; El Boushy, 1980; El Boushy, Roodbeen and Hopman, 1984; Rockey and Zaror, 1988).

3.2 PROCESSING OF THE RECOVERED PROTEIN EFFLUENT

3.2.1 Methods Used for the Recovery

(a) Ion exchange

Ion exchange is a very effective method for recovering proteins from slaughter-house wastewater. Ion-exchange resins consist of cross-linked regenerated cellulose modified by the introduction of anion or cation exchange groups. The resins are registered (under the name "Protion") and available in a granular form in a wide range of grain sizes, and have markedly superior hydraulic properties and physical stability compared with the fibrous cellulose ion exchangers usually employed for protein adsorption in laboratories. Conventional ion-exchange resins based on synthetic organic materials such as polystyrene have negligible capacities for protein molecules. The granular regenerated cellulose ion exchangers, however, have protein capacities comparable with fibrous or micro-crystalline cellulose ion exchangers of the order of 0.5 g protein/g resin. The new resins have been evaluated over a considerable period of time in applications ranging from laboratory-scale enzyme production to effluent and water treatment. Owing to their excellent hydraulic properties, "Protion" resins are suitable for use in conventional types of ion-exchange plants as well as in continuous and fluidized bed systems. By varying the grain size and degree of cross-linking, it is possible to "tailor" the resin for specific applications in water and effluent treatment and biochemical processing. The new resins are available in four main types - weakly and strongly basic or acid - corresponding to the main types of conventional resins used for water treatment and covering a working pH range of 2-12. In general, adsorbed protein is readily desorbed by washing the resin bed with salt or buffer solutions; alkaline brine has been found particularly suitable as a regenerant solution where the resins are used for effluent treatment (Ross, 1968; Grant, 1976; Metcalf and Eddy, 1979; Beszedits, 1980; Eilbeck and Mattlock, 1987).

(b) Membrane processing

Ultrafiltration and reverse osmosis are powerful processing tools for separating sugars, starches, proteins, *etc.*, from food processing wastewater. Both ultrafiltration and reverse osmosis employ pressure permeation through membranes, however, while reverse osmosis requires pressures of 2760-13800 kPa, ultrafiltration operates in a range of 173-690 kPa. This method is very useful in the extraction of whey protein, soy whey protein and protein from blood plasma (Metcalf and Eddy, 1979; Beszedits, 1980).

(c) Electrolytic process

Proteins can also be recovered by electrolytic methods. For example, the Lectro-Clear process has been successfully demonstrated on meat, poultry and fish processing wastewater. This process uses a continuous upflow electro-coagulation apparatus having an iron plate as a cathode and a carbon plate as an anode for the purification of effluents from fish paste manufacturing plants. At optimum operating conditions, COD was reduced by 62% while protein recovery was 96% (Beszedits, 1980; Eilbeck and Mattlock, 1987).

(d) Coagulation with chitosan

The shells of crustaceans such as shrimps and crabs are a source of a number of valuable products, including chitin, a cellulose-like polysaccharide. De-acetylation of chitine yields chitosan, a cationic polymer.

Chitosan is an excellent coagulant for the treatment of food processing effluent, particularly for discharges containing proteins. It is competitive with and often more effective than synthetic polyelectorolytes. Moreover, chitosan is completely biodegradable. By-products recovered from food processing wastewater with the aid of chitosan on average contain 30-70% protein.

All types of food processing wastewater are amenable to treatment with chitosan and reductions in suspended solids reductions as high as 90% can be attained. Gravity settling or dissolved air flotation are usually employed for recovering the coagulated solids. For example, only 5 mg/l chitosan was very effective in increasing the suspended solids removal from poultry wastewater effluents (Beszedits, 1980; Eilbeck and Mattlock, 1987).

(e) Precipitation with Lignosulfonic acid

In an aqueous medium, lignosulphonic acid (LSA), a by-product of the sulphite wood pulping industry, reacts almost instantaneously with proteins to yield an insoluble precipitate.

Adjusting the pH of a proteinaceous waste stream to below the isoelectric point imparts a net positive charge on the proteins. Acidification to pH less than 3 ensures a pH below the isoelectric point for most proteins. At these low pH values, the LSA carries a net negative charge. Consequently, the optimum pH range for protein precipitation with LSA is between 2 and 3. Applied dosages vary according to the nature of the waste-waters treated and typical dosages are 250-550 mg/l. During protein precipitation, fatty materials are also removed. Precipitated proteins are usually recovered by dissolved air flotation. After dewatering and drying, the protein-rich sludge can be used for livestock feeding.

Facilities based on this technique have been installed in numerous countries for the treatment of meat, poultry and fish processing wastewaters. From trials it was proved that adding an average dose of 422 mg/l LSA to the effluents of abattoirs gave a BOD reduction of 84% while total solids were reduced by 96%. Protein content of the dry sludge was 42%, fat 24% and crude fibre 4.3% (Beszedits, 1980; Eilbeck and Mattlock, 1987).

3.2.2 Industrial Applications

Methods for industrial applications are based on adding various chemicals such as iron salts and synthetic polyelectrolytes to the wastewater. This results in the following treatment of the effluent:

- Screenings to remove coarse materials such as bones, skin, lungs and feathers
- Flotating of the sludge by means of micro air-bubbles
- Addition of chemical additives for flocculation
- Sludge thickening, partial dehydration
- Sludge drying

These points have been discussed previously by Grant (1976); Beszedits (1980); Eilbeck and Mattlock (1987) and Meyn, (1990). The processing lay-out is shown in Figure 3.1.

(a) Coarse screening

Screening to remove coarse materials such as bones, skin, lungs and feathers by static and vibrating screens is a very important step for water treatment.
The use of stainless steel wedge wire in the form of static inclined screens provides a simple effective solution, achieving maximum liquid extraction and solids recovery.

(b) Micro-flotation

The micro-flotation technology, also known as dissolved air flotation (DAF), can be used for most industrial effluents where the waste products have a specific gravity

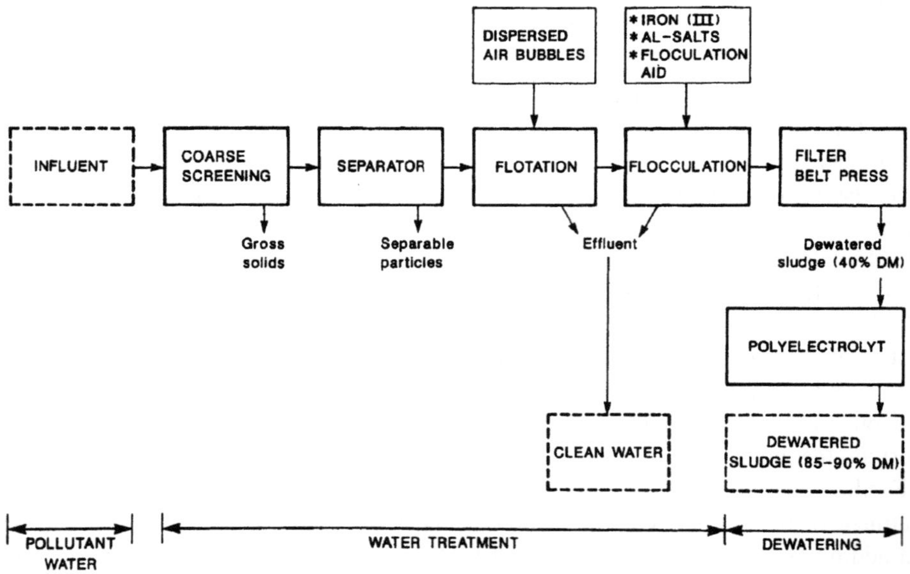

Figure 3.1. Process flow diagram of the several effluent treatments to form sludge and clean water.

less than or equal to that of water. Using very finely dispressed air bubbles, a higher density difference between the pollutants and water is produced by the adsorption of the wastes particles on to the surfaces of the bubbles. This results in a rapid separation of the pollutants by flotation.

(c) Flocculation

Effluents can contain both organic and/or inorganic pollutants which can be flocculated by iron or aluminium salts followed by a flocculant aid. The flocculated effluents can be floated to the surface of a flotation tank and finally scraped away by an electrically driven scraper. This double-action device ensures that the flotation sludge is low in humidity as it is being removed. This process of effluent treatment is leading to a high reduction in: (BOD), (COD), (TKN), (TSS), (FOG).

Flocculation of wastewater by mechanical or air agitation leads to: a) increase the removal of suspended solids and BOD in primary settling facilities; b) conditioning the wastewater containing the slaughter-house wastes; and c) improving the performance of secondary settling tanks, following the biological treatment process, especially the activated-sludge process. More detailed specifications of this process are discussed in details (Metcalf and Eddy, 1979; Eilbeck and Mattlock, 1987 and Merka, 1989).

(d) Sludge dewatering

The most expensive problem of the sludge disposal is its high water content. The high humidity content present in the treated sludge (90-93%) and the excess activated sludge

(98-99%) has major financial consequences for firms practising effluent treatment. Meyn (1990) reported a new method for thickening and drying sludge to a water content of 40%. This drying takes place by means of using heat coagulation and pressing with a simple compact stainless steel filter belt press. The advantage of this treatment method is that direct processing of fresh flotation sludges prevents them turning rancid owing to their high content of free fatty acids. Flotation sludge treated in this way has reduced bacterial content (approximately 90% reduction) and can be a valuable animal feedstuff. Finally, the end product can be dried and sterilized in a dryer drum with a jacket wall, kept at 180-190 °C (El Boushy, Roodbeen and Hopman, 1984).

3.3 CHEMICAL ANALYSIS AND NUTRITIVE VALUE

Effluent protein recovered from the flocculation and ion-exchange stages showed a high protein percentage varying from 61.3-73.9% and ash 5.1-2.3%. Essential amino acids such as methionine and lysine are considered to be high and equal to those in eggs and casein (Grant, 1976; Tables 3.1 and 3.2).

Table 3.1. Amino acid composition of recovered solids fractions A and B of effluent in comparison with reference proteins. From Grant (1976)

	Recovered solids		*Reference proteins*			
	Fraction					
Amino acid[a]	*A*	*B*	*Fibrin*	*Hemoglobins*	*Serum proteins*	*Casein*
Lysine	8.8	8.5	9.1	9.1	10.0	8.5
Histidine	3.9	5.9	2.9	8.0	3.3	3.2
Arginine	4.4	4.7	7.8	3.9	5.8	4.2
Aspartic acid	14.3	9.4	11.9	9.8	10.3	7.0
Threonine	7.9	4.5	7.3	5.0-6.0	12.6	4.5
Serine	7.7	5.7	12.5	5.5	18.2	6.8
Glutamic acid	19.3	10.0	15.0	8.1	14.2	23.0
Proline	6.6	3.2	5.3	4.7	5.5	13.1
Glycine	5.5	3.5	5.4	5.3	2.0	2.1
Alanine	8.8	6.8	4.0	9.8	-	3.3
Half-cystine	Trace	Trace	3.8	1.0-2.2	7.0	0.8
Valine	11.0	8.2	5.6	9.0	7.5	7.7
Methionine	2.8	3.2	2.6	1.0-3.0	4.0	3.5
Isoleucine	5.5	4.1	5.6	0-2.0	3.4	7.5
Leucine	17.1	15.0	7.1	14.4	10.1	10.0
Tyrosine	5.5	3.2	6.0	2.9	5.5	6.4
Phenylalanine	9.9	7.9	4.5	7.0-8.0	5.2	6.3
Ammonia	-	1.1	-	-	-	-

[a] g Amino acid/16 g nitrogen.

Table 3.2. Essential amino acids of recovered solids fractions A and B of effluent in comparison with other relevant parameters. From Grant (1976)

Amino acid [a]	*FAO* [b]	*Egg*	*A*	*B*	*Casein*
Isoleucine	4.2	6.8	5.5	4.1	7.5
Leucine	4.8	9.0	17.1	15.0	10.0
Lysine	4.2	6.3	8.8	8.5	8.5
Phenylalanine	2.8	6.0	9.9	7.9	6.3
Tyrosine	2.8	4.4	5.5	3.2	6.4
Threonine	2.8	5.0	7.9	4.5	4.5
Tryptophan	1.4	1.7	-	-	-
Valine	4.2	7.4	11.0	8.2	7.7
Sulphur containing:					
Total	4.2	5.4	2.8	3.2	4.3
Methionine	2.2	3.1	2.8	3.2	3.5

[a] g Amino acid/16 g nitrogen.

[b] Food and Agricultural Organization, "Provisional Pattern" of Essential Amino Acids for human Nutrition, Rome, 1957.

Grant reported also that the nutritional value of the dried effluent was approximately equal to that of meat meal and casein, slightly inferior to that of fish meal and superior to that of meat and bone meal and to that of grass extract. The results of Grant were based on effluent protein recovered from the flocculation and ion-exchange processes, however, iron as flocculant aid was not used.

El Boushy, Roodbeen and Hopman, (1984) analysed the effluent from poultry processing wastewater in which iron chloride was used as a flocculant aid (Table 3.3). They found levels of: crude protein 37.5%; fat 28.6%; ME (MJ Kg) 12.77%; ash 13.2%; iron 4.10%; lysine 2.11% and methionine 0.67%. They concluded that herring meal was significantly superior to that of dried effluent protein, fed on equivalent protein levels. Blood and meat meal were comparable to effluent protein. The only great disadvantage of this feedstuff is its high content in iron (4.10%) and chlorid (0.42%). The iron content may cause a colloidal suspension of insoluble iron phosphate which may absorb vitamins or trace inorganic elements. Consequently, this may cause growth depression for broilers consuming the sterilized effluent added to broiler diets. To avoid this disadvantage, another flocculant should be used.

The newest idea in coagulation with chitosan to recover protein effluent was practised by Lyons and Vandepopuliere (1997). They used an alternate coagulant-flocculent system using volclay, polysolisic acid, and chitosan to remove secondary poultry nutrients (SPN) from the wastewater stream of a turkey processing plant. The sludge (SPN) was treated by several methods with the following composition as a feedstuff by EXT (extrusion) or Dry (drying) (Table 3.4).

Table 3.3. Chemical analysis and amino acid composition of the effluent from poultry processing wastewater (on air-dry basis). From El Boushy, Roodbeen and Hopman (1984)

Crude protein	37.15	Cl	0.42
Ether extract	28.62	Fe	4.10
ME (MJ kg)	12.77[a]	Na	0.21
Ash	13.22	K	0.24
Ca	1.52	Mg	0.04
P	1.67	Zn	0.04
Amino acids			
Lysine	2.11	Methionine	0.67
Histidine	0.87	Isoleucine	1.90
Arginine	1.62	Leucine	2.88
Aspartic acid	3.00	Tyrosine	1.19
Threonine	1.50	Phenylalanine	1.67
Serine	1.69	Cystine	1.32
Glutamic acid	3.47	Tryptophan	-
Glycine	1.58	Proline	1.66
Alanine	2.11	Valine	2.42

[a] The ME was analysed by Spelderholt Institute for Poultry Research, Beekbergen, The Netherlands on a sample from the same slaughterhouse.
All values (%) on air-dry basis; dry matter of original effluent, 4.95%.

Table 3.4. Chemical analysis and amino acids methionine and lysine composition of corn soybean meal, and secondary poultry nutrients (SPN) on (air-dry basis). From Lyons and Vandepopuliere (1977)

	Moisture	*C-protein*	*C-fat*	*Methionine*	*Lysine*
SPN-Ext [a]	11.36	14.70	6.55	0.28	0.80
SPN-Dry EXT [b]	11.36	20.88	19.12	0.40	1.36
SPN-Dry [c]	11.20	35.91	37.76	0.69	2.66
Corn	14.23	8.14	4.18	0.18	0.32
Soybean meal (43%)	11.83	46.92	1.92	0.69	3.14

[a] A mixture of 67% corn + 33% SPN (centrifuged, containing 47% moisture).
[b] A mixture of 50% corn + 50% SPN (centrifuged and partially dried containing 27.5% moisture).
[c] Dried SPN contains 11.2% moisture.

3.4 ITS USE IN FEEDING

There is only few literature on the effect of feeding dried effluent mixed into diets for poultry. Grant (1976) carried out a standard feeding trial, with protein recovered from effluent without iron as a flocculant, on crossed cockerels and on young rats.

Table 3.5. Composition of the standard broiler rations: six concentrations of effluent and meat, blood and herring meal providing protein levels equivalent to 4 or 7% effluent. From El Boushy, Roodbeen and Hopman (1984)

Ingredients (%)	*Control*	*Effluent (%)*						*Control meat meal equivalent to*[a]		*Control blood meal equivalent to*[a]		*Control herring meal equivalent to*[a]	
		2	*3*	*4*	*5*	*6*	*7*	*4%*	*7%*	*4%*	*7%*	*4%*	*7%*
Corn, Yellow	54.52	55.21	55.63	56.07	56.49	56.97	57.47	58.14	60.90	57.98	60.67	58.17	60.69
Soybean meal 44%	30.65	28.67	27.58	26.49	25.40	24.23	23.04	26.30	22.96	26.30	22.92	26.27	22.92
Fish waste	7.00	7.00	7.00	7.00	7.00	7.00	7.00	7.00	7.00	7.00	7.00	7.00	7.00
Effluent	-	2.00	3.00	4.00	5.00	6.00	7.00	-	-	-	-	-	-
Meat meal 58%	-	-	-	-	-	-	-	2.60	4.60	-	-	-	-
Blood meal 8.5%	-	-	-	-	-	-	-	-	-	1.80	3.20	-	-
Herring meal 73%	-	-	-	-	-	-	-	-	-	-	-	2.10	3.70
Animal fat	5.22	4.72	4.46	4.19	3.92	3.64	3.35	3.86	2.82	4.28	3.56	3.99	3.06
Di Ca Phosphate	0.97	0.80	0.72	0.63	0.55	0.47	0.38	0.57	0.27	0.99	0.99	0.78	0.64
Limestone	0.27	0.30	0.31	0.32	0.34	0.35	0.37	0.16	0.08	0.28	0.29	0.34	0.39
Salt	0.30	0.30	0.30	0.30	0.30	0.30	0.30	0.30	0.30	0.30	0.30	0.30	0.30
Premix[b]	1.00	1.00	1.00	1.00	1.00	1.00	1.00	1.00	1.00	1.00	1.00	1.00	1.00
Methionine 99%	0.07	-	-	-	-	-	-	0.07	0.07	0.07	0.07	0.05	0.03
Lysine HCl 78%	-	-	-	-	-	0.04	0.09	-	-	-	-	-	-
Total %	100.00	100.00	100.00	100.00	100.00	100.00	100.00	100.00	100.00	100.00	100.00	100.00	100.00

[a] Effluent (%).

[b] Vit. mineral mixture (Farmix 10 PS) 1 kg contains: vit.A, 1 200 000 IU; vit.D_3, 240 000 IU; vit.E, 1 000 mg; vit.k_3, 150 mg; vit.B_1, 100 mg; vit.B_2, 500 mg; nicotinic A, 3 000 mg;
D.Panth.acid, 750 mg; vit.B_6, 100 mg; vit.B_{12}, 1.5 mg; folic acid, 100 mg; Choline chloride 35 000 mg; Fe, 2 000 mg; Mn, 7 000 mg; Cu, 1 000 mg; Zn, 5 000 mg; I, 100 mg;
Co, 50 mg; Se, 10 mg; Ethoxyquin, 5 000 mg.

He measured the relative growth rate, feed consumption and feed efficiency until the cockerels were 4 weeks old. He concluded that this protein could be used as a concentrate for feeding poultry, and that there was no evidence of toxic side effects that could be attributed to the recovery process in either the chick or the young rats.

El Boushy, Roodbeen and Hopman (1984) reported on the effect of six levels of slaughter-house wastewater effluent mixture in diets in comparison with a control diet (2, 3, 4, 5, 6, 7% pure effluent) and the effects of 4 and 7% of effluent in comparison with equivalent protein levels supplied by meat, blood and herring meal (Table 3.5). All diets were calculated on a linear programming basis isonitrogenously and isoenergetically and contained the same calcium and phosphorus levels. Methionine and lysine were added when needed. Their results showed that the effluent (sludge) was found to be a reasonable animal protein feedstuff as far as methionine (0.67%), cystine (1.32%) and energy (ME 12.77 MJ/kg) are concerned. Its disadvantages are the high cost of drying and its high iron content (4.1%). Increasing the proportion of effluent in the diet from 2-7% leads to a clear growth depression at 4 and 7 weeks of age because of the high iron concentration. Herring meal was significantly superior to that of dried effluent protein, supplied on equivalent levels where blood and meat meal were comparable to effluent protein.

Lyons and Vandepopuliere (1997) reported from their trials, by using the sludge chitosan - produced secondary poultry nutrients (SPN) as a feedstuff for broilers by either extrusion or drying. They concluded that feed conversion was not significantly increased by the SPN from products which had been partially dried prior to extrusion. SPN can be used at levels up to 4% in broiler diets without affecting body weight or feed conversion.

3.5 BIOHAZARDS OF RECOVERED SLUDGE

From the toxicological point of view and contents of trace elements, there is no objection to using recovered sludge from wastewater in certain percentages in diets of poultry. From the microbiological point of view, direct feeding of recovered sludge to poultry is not safe due to *Campylobacter* (Havelaar, Notermans and Oosterom, 1983). However Bürger (1978, 1982) reported that, eggs of parasites are also present in high numbers, especially *Ascaris*. Heating during drying and sterilizing the product will kill these eggs (Knapen, Franchimont and Otter, 1979). A recovered sludge from wastewater, dried and sterilized will be free from viruses and spore-forming bacteria (Edel, 1983).

The safest method for the utilization of the recovered sludge from wastewater as a feedstuff added in the diets of poultry is to process it in a rendering plant, where temperatures go up to 130 °C for at least half an hour. The use of the recovered sludge without sterilization may be a hazard in spreading pathogens and might intensify infection transmission cycles. These cycles have been reported mainly for *Salmonella* (Edel, 1972, 1973, 1978; Edel, Van Schothorst and Kampelmacher, 1976; Hobbs and Cristian, 1973). The cycle of this species may be considered as a model for those of other pathogens and precautions should be taken to avoid infections and risks of those pathogens.

The direct feeding of the recovered sludge from wastewater to poultry is not recommended on the ground of hygiene and maintaining human and animal health. Untreated sludge as a feedstuff is prohibited by the law (Weiers and Fisher, 1978). Treatment of the sludge by sterilization will guarantee a feedstuff free from pathogens (Havelaar, Notermans and Oosterom, 1983).

REFERENCES

Beszedits, S. (1980). Recovery of animal feedstuffs from food processing wastewaters. *Feedstuffs*, **52**, (47), 26-8.

Bürger, H.J. (1978). Parasitologische Risiken bei der Verfütterung getrockneter oder silierter Schlachtabfälle. *†bers. Tierernährg.*, **6**, 184-8.

Bürger, H.J. (1982). Large-scale management systems and parasite populations. Prevalence and resistance of parasitic agents in animal effluents and their potential hygienic hazard. *Veterinary Parasitol.*, **11**, 49-60.

Camp, W.J. (1969). Waste treatment and control at livestock poultry processing plant. *Eighteenth Southern Water Resources and Pollution Control Conference*, North Carolina State University, Raleigh, North Carolina, USA.

Edel, W. (1983). Health aspects of slaughter-house waste converting by animals. In: *Animals as waste converters. Proceedings of an International Symposium*, Wageningen 30 September - 2 October, The Netherlands, pp. 87-91.

Edel, W., Guinée, P.A.M., Van Schothorst, M. and Kampelmacher, E.H. (1973). Salmonella cycles in foods with special references to the effects of environmental factors, including feeds. *Proceedings Symposium. Microbial food borne infections and intoxications*, Ottawa, USA, pp. 17-20.

Edel, W., Van Schothorst, M., Guinée, P.A.M. and Kampelmacher, E.H. (1972). Mechanism and prevention of Salmonella infections in animals. In: The microbiological safety of food. (Eds. B.C. Hobbs and J.H.B. Cristian, 1973), Academic Press, London. New York, pp. 247-55.

Edel, W., Van Schothorst, M. and Kampelmacher, E.H. (1976). Epidemiological studies on Salmonella in a certain area (Walcheren projects). I. The presence of *Salmonella* in man, pigs, insects, seagulls and in foods and effluents. *Zentralblatt für Bakteriologie und Hygiene*, 1.Abt. Originale A, **325**, 476-84.

Edel, W., Van Schothorst, M., Van Leusden, F.M. and Kampelmacher, E.H. (1978). Epidemiological studies on Salmonella in a certain area (Walcheren project III). *Zentralblatt für Bakteriologie und Hygiene*, 1.Abt. Originale A, **242**, 468-80.

Eilbeck, W.J. and Mattlock, G. (1987). *Chemical processes in wastewater treatment*. Ellis Horwood Limited, Publishers, Chichester, England, p. 331.

El Boushy, A.R. (1980). Aspects of waste recycling as feedstuffs for poultry. *Feedstuffs*, **52**, (42), 16, 24, 28, 32.

El Boushy, A.R., Roodbeen, A.E. and Hopman, L.C.C. (1984). A preliminary study of the suitability of dehydrated poultry slaughter-house wastewater as a constituent of broiler feeds. *Agricultural wastes*, **10**, 313-8.

Erdtsieck, B. and Gerrits, A.R. (1973). Characteristics and volume of liquid waste from poultry processing plants. *4th Europ. Poultry Conf.*, London, 529-33.

Grant, R.A. (1976). Protein recovery from meat, poultry and fish processing plants. In: *Food from waste*. (Eds. G.G. Birch, K.J. Parker and J.T. Worgan). Applied Science Publishers LTD, London, pp. 205-20.

Havelaar, A.H., Notermans, S.H.W. and Oosterom, J. (1983). Pilot study into the microbiological nature of the wastewater from a slaughter-house for poultry and the effect of chemical-biological treatment. *Report of the National Institute of Public Health*, The Netherlands, no. 831001.

Hobbs, B.C. and Cristian, J.H.B. (1973). The microbiological safety of food. *Proceedings of the 8th International Symposium on Food Microbiology*, Reading, England, 1972. Academic Press, London. New York.

Kahle, H. and Gray, L. (1956). Utilization and disposal of poultry by-products and wastes. *Marketing Research Report*, No. 143, *U.S. Dept.of Agriculture*, Washington, D.C.

Knapen, F. van, Franchimont, J.H. and Otter, G.M. (1979). Steam sterilization of sand pits. *Br. Med. J.*, **1**, 1320.

Lyons, J.J. and Vandepopuliere, J.M. (1997). Dehydrated and extruded chitosan-produced secondary poultry nutrients (SPN) as broiler feedstuffs. *J. Appl. Poultry Res.*, **6**, 100-6.

Merka, B. (1989). Characteristics of wastewater. *Broiler Industry*, **17**, 20-7.

Metcalf & Eddy, Inc. (1979). *Wastewater engineering: Treatment disposal reuse*. McGraw-Hill Book Company, New York. 920 p.

Meyn, J. (1990). Water treatment. *Information Bulletin. Meyn Water Treatment B.V.*, Zaandam, The Netherlands.

Rockey, J. and Zaror, C. (1988). Alternative strategies for the treatment of food processing waste waters. In: *Developments in food microbiology*. 4th Ed., (Ed. R.K. Robinson). pp. 187-221.

Ross, R.D. (1968). Waste liquid treatment. In: *Industrial Waste Disposal*. Reinhold Book Corporation, New York. Amsterdam. London. pp. 99-189.

Weiers, W. and Fischer, R. (1978). The disposal and utilization of abattoir waste in the European communities. *Commission of the European Communities*, EUR 5610 E. p. 148.

4. POULTRY BY-PRODUCTS

Contents

4.1 INTRODUCTION

The tremendous growth of the poultry industry is creating a large amount of offal and waste. If sufficient care is taken, this offal may contribute to animal feed as essential diet ingredients, thus replacing parts of other expensive feed ingredients. Poultry industries now exist in many countries and generate large amounts of residues.

The centralization of poultry processing has intensified the problem of disposal of poultry waste, the volume of which may be large enough to develop techniques of processing this offal or present possibilities for future use. The processing of poultry by-products will have a role in countries where rendering plants are developing and animal proteins are not abundant.

The wastes from poultry slaughter are blood, feathers and offal (viscera, heads, and feet), and, if collected separately, can be processed into blood meal, hydrolized feather meal, poultry offal meal, and fat respectively.

A large proportion consists of poultry feathers which are keratinous proteins, as well as offal. Since feather protein in its natural state is very poorly digested by monogastric animals, various methods have been developed for processing these keratinous proteins into a more digestible form.

The various methods of processing raw feathers may cause differences in the nutritive value of the treated feathers, related to protein and amino acid digestibility (Sullivan and Stephenson, 1957; Moran, Summers and Slinger, 1966; Morris and Balloun, 1973[b]; Papadopoulos, 1984; El Boushy *et al.*, 1990[a]).

Specific feed-ingredients, including additives to enrich the feather meal quality have been proposed by many workers, who have suggested dried brewers' yeast (Balloun, Miller and Spears, 1968), dried whey (Balloun and Khajareren, 1974), Menhaden fish meal (Potter and Shelton, 1978[b]), antibiotics (Potter and Shelton, 1978[a]) and synthetic amino acids (Baker *et al.*, 1981).

The production of separate hydrolized feather meal has been a point of discussion. Methods have been devised whereby poultry offal and feathers can be processed together in natural proportions, yielding a product having the approximate composition represented by a mixture of poultry by-product meal (PBPM) (45%), hydrolized poultry feathers (HPF) (40%) and poultry fat (15%). This formula poultry offal meal (POM) has the physical and nutritional characteristics which make it an excellent feedstuff for poultry (Potter and Fuller, 1967). However, it was also concluded that a mixture of poultry by-product and hydrolized feather meal (PBHFM) is also an excellent animal protein feedstuff which may be used for broilers and layers (Davis, Mecchi and Lineweaver, 1961; Naber, 1961; Naber *et al.*, 1961; Potter and Fuller, 1967; Bhargava and O'Neil, 1975).

Hatchery by-products consisting of infertile eggs, dead embryos, shells of hatched eggs and unsaleable chicks, attract attention of many research workers. Its use as a feedstuff in poultry rations, its holding techniques and its processing have been reported (Wisman, 1964; Wisman and Beane, 1965; Hamm and Whitehead, 1982; Miller, 1984).

Shell waste from egg-breaking plants includes egg shells, shell membranes and adhering albumen. The evaluation of this feedstuff in the nutrition of layers and their performance was performed many years ago, the study of Wilcke, (1940) being notable in this respect. Its chemical composition was determined by Walton *et al.* (1973) and its complete survey was carried out by Vandepopuliere *et al.* (1974).

4.2 FEATHER MEAL

4.2.1 Processing Methods

Various methods of processing raw feathers (Draper, 1944; Binkley and Vasak, 1951; Sullivan and Stephenson, 1957; Naber *et al.*, 1961) have been reported and reviewed by El Boushy, van der Poel and Walraven, 1990). These reports indicated that the different processing methods caused differences in nutritive value of feather meal.

The development of an effective method of processing feathers into a friable, high density meal (Binkley and Vasak, 1951) stimulated new investigations on the nutritive value of feather keratin. A wet cooking process is essential for this method, in which the feathers are treated with saturated steam at pressures of 275-415 kPa for 30-60 minutes with constant agitation. The feathers are then dried and ground to produce a free-flowing meal of relatively high density. With a steam pressure above 415 kPa and constant agitation the feathers tended to "gum", leading to a non free-flowing meal.

The proper processing of feathers is dependent on temperature (steam), pressure, processing time and the humidity percentages, followed by drying and grinding.

The definition of the Standard Feather Meal has been reported by the Association of American Feed Control Officials (1960) as follows: "Hydrolyzed poultry feathers is the product resulting from the treatment under pressure of clean undecomposed feathers from slaughtered poultry, free of additives and/or accelerators. Not less than 70% of its crude protein content shall consist of digestible protein."

In the known processes for hydrolyzing poultry feathers, the feathers are normally plucked from freshly slaughtered birds and washed into a sluiceway, employing water as a carrier to transport the feathers from the slaughter-house directly to the processing plant or to transport by trucks. The feathers are then tumbled through a screen to remove the bulk of the sluiceway water. In addition, the feathers are processed by either one of two methods, referred to as a "batch cooker" method and the "continuous method", respectively. Both methods employ similar process conditions in that the feathers, with a moisture content of about 60-70%, are pressure cooked in a steam-heated batch cooker at about 207-690 kPa for about 30-60 minutes. This permits hydrolysis and causes the feathers to break up into a final hydrolized and sterilized product.

(a) Batch cooker

Feathers are usually processed in dry-rendering cookers of 1,000-15,000 l content. These cookers are steam jacketed, with horizontal shaft agitators, and have a dome for loading and a bottom-front opening for discharge (Figure 4.1). They are generally designed for steam pressures of 414-850 kPa and charged with an internal pressure of 207-350 kPa. Internal pressures as low as 138 kPa and as high as 414 kPa have also been used. The processing time is 4-7 h from the time the cooker is closed until it is opened to discharge. The critical point appears to be that the entire charge should be maintained at the equivalent of 207 kPa for at least 15 minutes (El Boushy, van der Poel and Walraven, 1990). As soon as the hydrolysis has taken place, pressure is released and the contents will be dried. Drying of the hydrolized, sterilized feathers is done in the same batch cookers, or in other driers such as disc- or flash-driers.

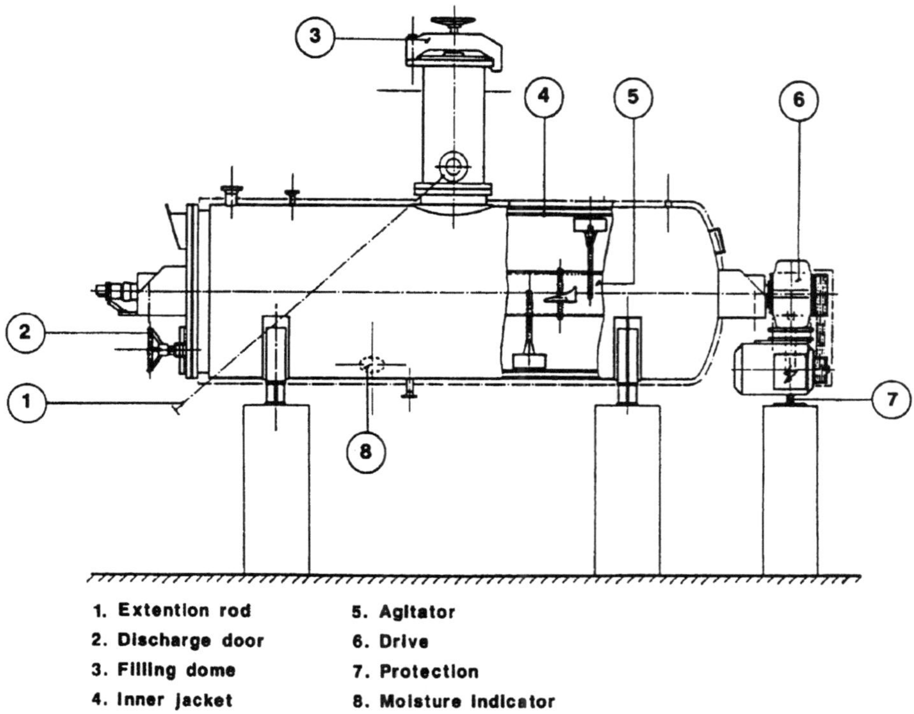

Figure 4.1. A typical batch cooker-drier, steam jacketed with horizontal shaft agitators. Reproduced with permission from Van der Poel and El Boushy (1990).

Steam-heated disc driers

These systems are used in most of the modern installations with a disc type agitator (Figure 4.2). The dried product is known to be consistent in quality and is light in colour (Davis, Mecchi and Lineweaver, 1961). The agitation to which the heated feathers are subjected during processing and drying breaks up the feathers until little remains that is recognizable as feather. This point will be explained in details later in this chapter.

Direct-fired type driers

These are driers, such as flash-driers, which are less expensive than the steam-heated drier but their control is more critical with respect to reach the desired moisture content and to prevent darkening of the product (Figure 4.2). This system will be explained in details later in this chapter.

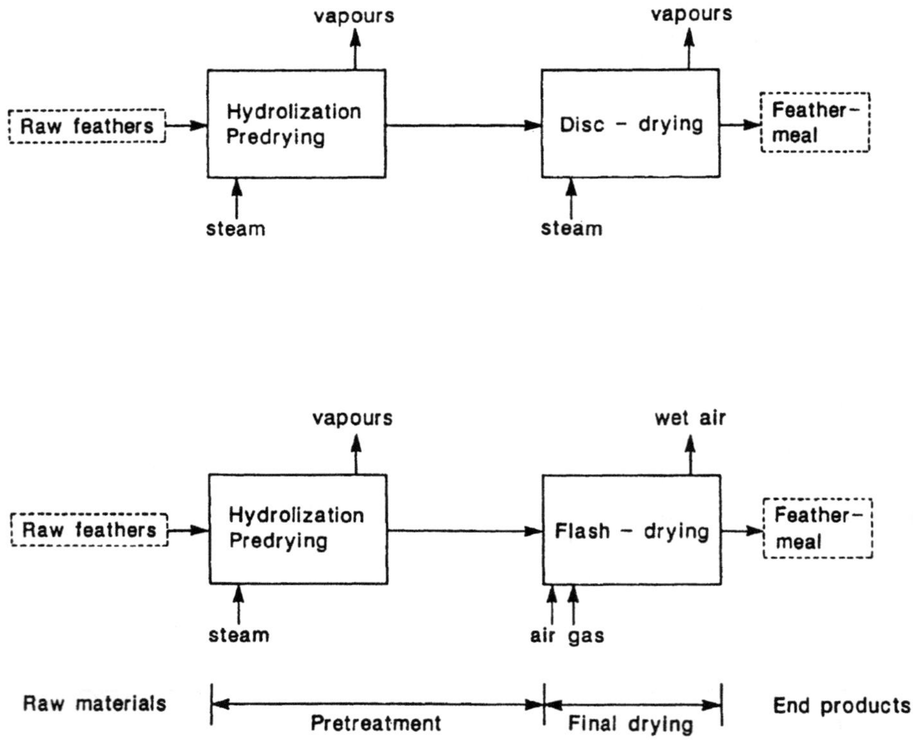

Figure 4.2. A diagram of the batch cooker and its driers in the form of a disc-drier and flash-drier. Reproduced with permission from El Boushy, van der Poel and Walraven (1990).

(b) Continuous processing

In this method the following steps are employed: feathers are fluffed to facilitate their transport through the intake lock of a pressurized vessel; and the feathers are hydrolized at a high moisture content (60-80%) inside the pressurized vessel, which is continuously operated through intake and outlet locks. The average residence time is 6-15 minutes and the pressure is about 483-690 kPa. The feathers are subsequently subjected to a first drying stage of the liquid slurry discharged from the hydrolized feathers to a moisture content of about 40%. Finally the feathers are dried in a disc drier to a moisture content of about 8%. The continuous process is merely an extension of the batch process and has some more practical advantages as far as capacity is concerned. These two processes involve hydrolyzing feathers with steam at high moisture levels and employs thermal drying to reduce the moisture content of the final product (Williams, Horn and Bronikowski, 1979; Figure 4.3).

(c) High shear extrusion

In this procedure raw feathers are firstly passed through a screw press (fluidizer) where the moisture level of the raw feathers is reduced to approximately 30-40%. The water

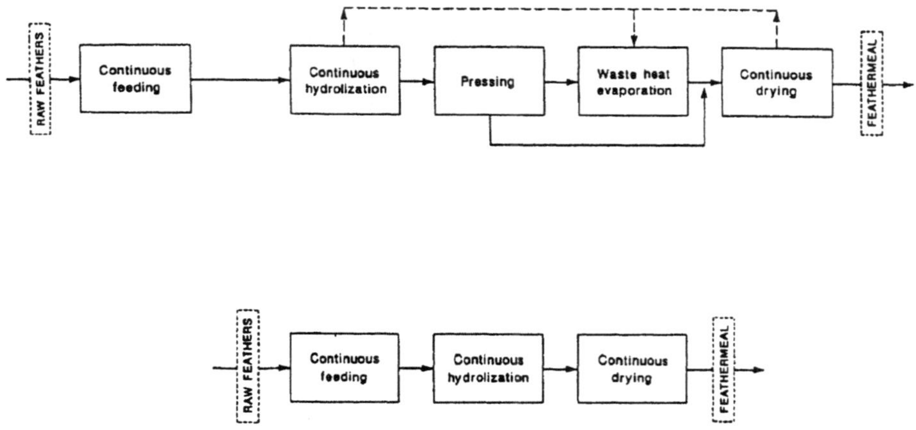

Figure 4.3. A diagram of the continuous processing from raw feathers to feather meal. Reproduced with permission from El Boushy, van der Poel and Walraven (1990).

squeezed from the screw press is returned to the sluiceway in which the fresh feathers are transported from the slaughter-house. The mechanically dewatered feathers are next passed through an extruder (used as a hydrolizer) wherein the keratinous material is subjected to high shear, causing a temperature and pressure build-up in the material, while maintaining the water in its liquid phase to hydrolize the keratin. Typical processing conditions in the extruder are 30 s residence time, 5518 kPa pressure, about 150 °C and 35% moisture content in the treated material. The hydrolized feathers exit the extruder and are conveyed to a steam-heated drier where the moisture is reduced to approximately 8% (Williams, Horn and Bronikowski, 1979; Vandepopuliere, 1988; Davis, 1989; Figure 4.4).

Extrusion. It is well known that rendering, more efficiently utilizes feathers or poultry by-products to increase its nutritional value, but it cannot always be economically justified. Most poultry processors choose not to operate their own rendering plants because such a plant has a high capital cost and a capacity far in excess of the processors' needs. Thus the poultry industry often gives away its valuable by-products to conventional renderers at a negative cost. Extrusion may be a cheap solution for processing feathers or poultry by-products, their technical problem is the high moisture content. Therefore they cannot be effectively dried or dehydrated without adversely affecting their nutritional value. Diluting the by-product with an ingredient such as soybean meal, corn, or wheat middlings, however, reduces its moisture to a level that the extruder can process. Extrusion then can achieve full cooking expansion, sterilization, and partial dehydration on a mixture of 50-60% feathers or poultry by-products with the chosen dry ingredient (Said, 1996).

There are several different types of extruders. Some have a single screw and others have double screws. The most up to date extruders are equipped with a feeder to adjust the required input followed by a conditioner to adjust the steam input of the extruder, which may be equipped with a jacket for extra heating (Figure 4.5).

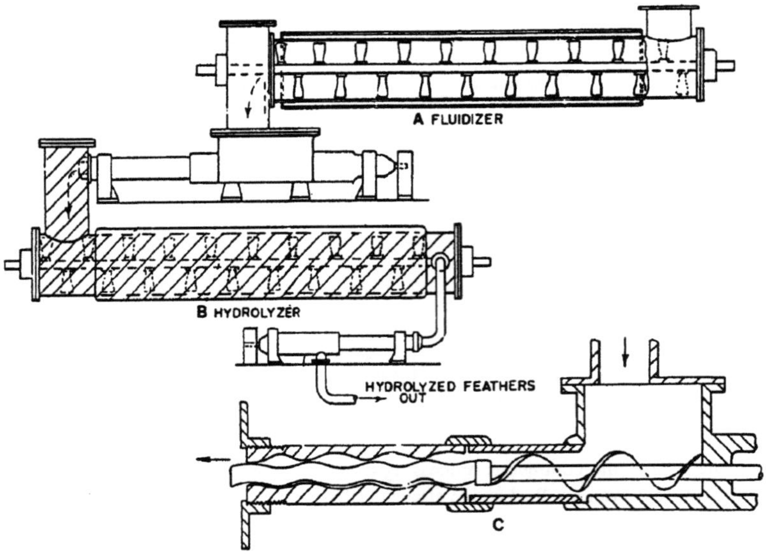

Figure 4.4. Showing the high shear extrusion system: (A) fluidizer, (B) hydrolizer and (C) complete extruder. Reproduced with permission from Williams, Horn and Bronikowski (1979).

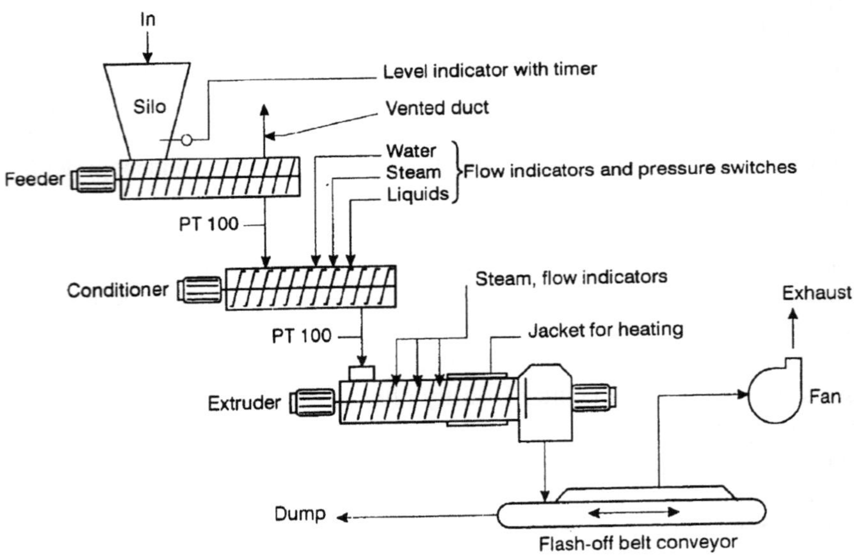

Figure 4.5. A flow sheet of a single screw extruder. Reproduced with permission from Almex extrusion techniques (1995).

(d) Low moisture hydrolysis

"To reduce the moisture content to a level sufficient to support a subsequent hydrolizing step" means reducing the moisture content of keratinous materials being treated to a level approaching the optimum moisture level required to hydrolize all of the keratinous material. The recommended moisture percentage for hydrolysis is 30-40% (Williams, Horn and Bronikowski, 1979).

Low-moisture hydrolysis is mostly applied by means of a screw press; subsequently, hydrolysis takes place as described in the previous paragraphs.

If the low-moisture feathers could be used in the batch-or continuous processing it would bring a remarkable saving in the cost of processing, e.g. thermal drying of the hydrolized meal and it would establish a preferred moisture level for the hydrolysis step. This application needs further investigation as reported previously (El Boushy, van der Poel and Walraven, 1990).

Stork rendering (1995) concluded that the most effective technique to lower the moisture content of the raw feathers is by means of the continuous feather press. Through their investigations they concluded that the continuous feather press effectively reduces the moisture content of the wet feathers from 70% to 45%. They added also that in absolute terms the feather press removes at least 450 l of water from 1.000 kg of wet feathers. Accordingly, the transport costs will drop to 45%, as will the energy required for processing the pressed feather mass.

Figure 4.6 and 4.7 show the feather press and its technical data: slaughtering capacity of broilers, and the production, kg of wet feathers per hour, Stork rendering (1995).

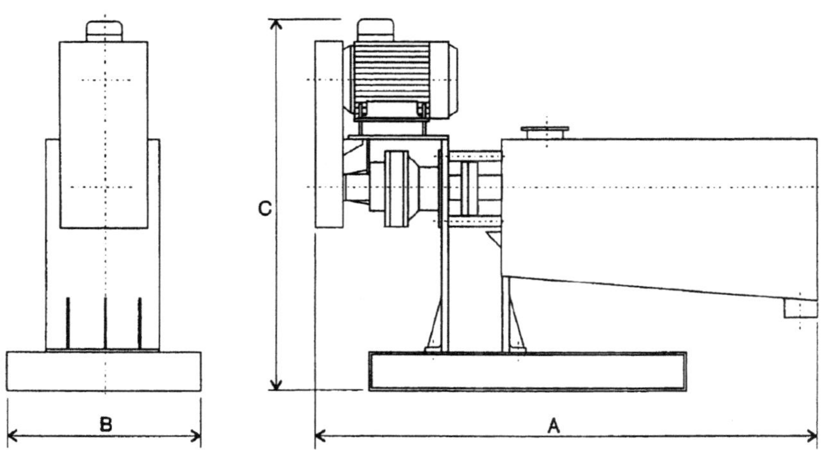

Capacity (Broiler feathers)		**Technical data**				
Type	Wet feathers kg/h	*Dimensions in mm*			Weight (kg)	Istalled power (kW)
		A	B	C		
WFP - 650	650	2165	820	1495	1000	17
WFP - 8000	8000	4750	1500	2800	10000	110

The capacity for other types of poultry depends on weight

Figure 4.6. The continuous feather press showing the capacity and the technical data. Reproduced with permission from Stork rendering (1995).

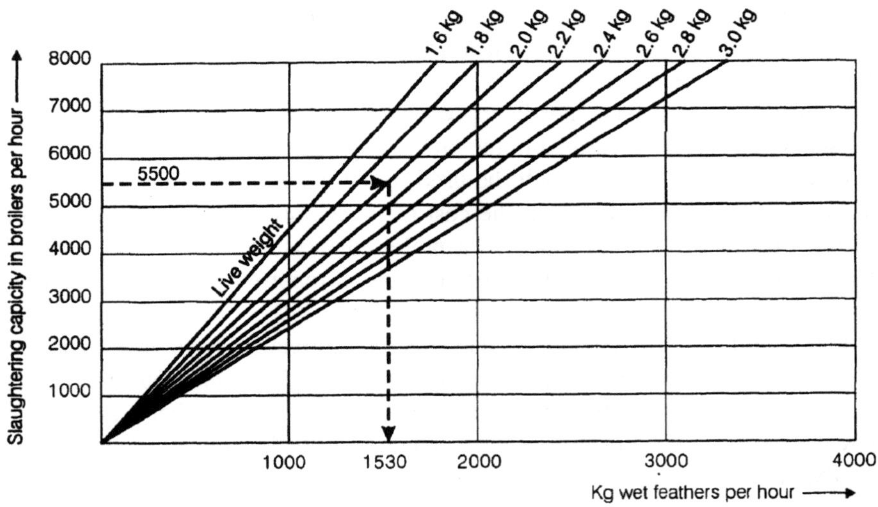

Figure 4.7. Slaughtering capacity in broilers in relation to the quantity of wet feathers per hour. Reproduced with permission from Stork rendering (1995).

4.2.2 Chemical, Enzymatical and Physical Evaluation of Feather Meal Quality According to the Method of Processing

The effect of processing methods on feather meal quality has been measured in many ways, such as using *in vitro* methods and *in vivo* methods and by routine chemical analysis. The *in vitro* methods are based on biological digestion of the treated material by means of enzymes. The best known method is the pepsin-HCl test for protein digestibility, which was applied for processed feather meal evaluation by Naber *et al.* (1961), Morris and Balloun (1973[b]), Aderibigbe and Church (1983) and Papadopoulos (1984). Other methods deal with the digestibility of amino acids such as lysine by digestion with pronase (Payner and Fox, 1976) and methionine and cystine by pancreatin (Pienazek, Grabarck and Rakowska, 1975).

On the other hand, *in vivo* methods are based on determining the balance between the tested feedstuff consumed by poultry and the faecal output. The difference between the feed intake and faecal excretion determines the apparent digestibility of amino acids (Bragg, Ivy and Stephenson, 1969; Burgos, Floyd and Stephenson, 1974; El Boushy and Roodbeen, 1980; 1984; Papadopoulos, 1984; Liu, Waibel and Noll, 1989).

(a) *Effects of time, temperature, pressure and moisture*

The effects of different processing variables, such as time, temperature, pressure and moisture level on digestibility of proteins and amino acids in feather meals have been studied by many research workers. Davis, Mecchi and Lineweaver (1961) demonstrated that different processing times of 90, 20 and 6 min. were required at pressures of 207, 414 and 621 kPa respectively to obtain approximately 70% pepsin-HCl digestibility

(Figure 4.8). They concluded that the important feature of this relationship is the rapid increase in time, required as pressure is reduced below 207 kPa and the rapid decrease in time as pressure is raised above that point. Also, they observed a relation between pepsin-HCl digestibility and hydrolysis time at a pressure of 207 kPa. Apparently, *in vitro* digestibility of protein increases rapidly until it reaches 65-70% but increases only slowly with prolonged heating (Figure 4.9).

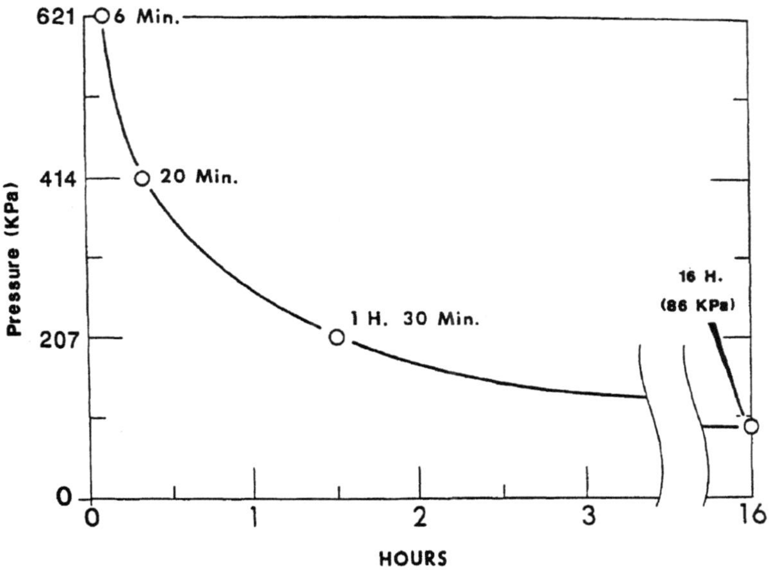

Figure 4.8. Pressure and period of processing required to produce feather meal that is 70% digestible in pepsin-hydrochloric acid. Reproduced with permission from Davis, Mecchi and Lineweaver (1961).

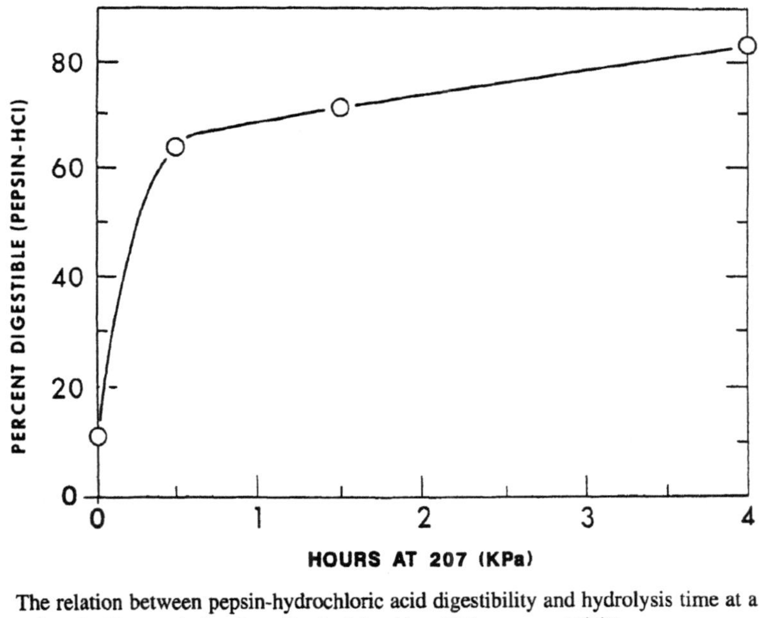

Figure 4.9. The relation between pepsin-hydrochloric acid digestibility and hydrolysis time at a pressure of 207 kPa. Reproduced with permission from Davis, Mecchi and Lineweaver (1961).

Retrum (1981) came to similar conclusions as Davis, Mecchi and Lineweaver (1961) but he added that the new systems for feather protein hydrolysis contain a heating jacket and use agitation, both of which will play a role in the pepsin-HCl digestibility figures. Table 4.1 shows the effect of time and pressure on the chemical analysis and the most limiting amino acids in the treated feathers as reported by several workers.

The crude protein level is not a basis for evaluating a feedstuff protein quality but the percentage digestibility of the most limiting -and the essential- amino acids in this protein is. Therefore, it is advisable to compare the most limiting amino acids mentioned above in Table 4.1. There is no agreement on the optimum temperature-time profile necessary to obtain the highest percentage of amino acids. It was reported by Davis, Mecchi and Lineweaver (1961) that at 30 minutes and 207 kPA pressure the highest lysine percentage was obtained, while Morris and Balloun (1973*b*) showed that at 60 minutes and 345 kPa the highest lysine and methionine percentage could be reached. On the other hand, Papadopoulos (1984) showed the highest percentages of lysine and methionine at 30 minutes and 436 kPa pressure.

Latshaw (1990) carried out an experiment on the effect of pH and of pressure on feather processing efficiency. The pH used was 5, 7 or 9, and the steam pressures used were 207, 276 or 345 KPa for 30 min. respectively. Increasing the pH or pressure increased the percentage of the protein that was pepsin digestible. As the pepsin digestibility increased, the cystine content of the feather meal decreased and that of lanthionine increased. From this comparison it is difficult to draw a clear conclusion of the optimum time and pressure to obtain the maximum limiting amino acids and the digestibility of the amino acids of the several treated samples.

Table 4.1 shows that the highest digestible lysine value was obtained in the hydrolized feathers at 60 min. and 345 kPa pressure (Morris and Balloun, 1973*b*) while Papadopoulos (1984) showed the highest digestibility of lysine, methionine and cystine in the feathers hydrolized at 30 minutes and 436 kPa pressure. Lanthionine is an indicator of the conversion by heating of cystine. The discovery of this unusual amino acid has been reported by various authors (Wheeler and Latshaw, 1980; Baker *et al.*, 1981; Latshaw, 1990). The progressive changes of cystine and lanthionine owing to the time of processing are shown in Figure 4.10. The loss of cystine, rather than the level found, is plotted to show the approximate parallel with the appearance of lanthionine; with increasing time of processing, the conversion of cystine to lanthionine is increased (Davis *et al.*, 1961). However, Latshaw (1990) concluded that increasing the pH or pressure increased the percentage of protein that was pepsin digestible. As the pepsin digestibility increased, the cystine content of the feather meal decreased and that of the lanthionine increased. Loss of cystine probably occurs through desulphurization reactions that may lead to unstable residues of dehydroalanine. These condense with cystine to form lanthionine or with the -amino group of lysine to form lysinoalanine, as a result of thermal degradation (Bjarnason and Carpenter, 1970). Loss of lysine could be explained by the addition of an -amino group of lysine residues to dehydroalanine (Ziegler, Melchert and Lürken, 1967; Bjarnason and Carpenter, 1970).

The optimum digestibility of the most limiting amino acids, lysine and methionine, in processed feathers needs further research. On a laboratory scale as well as at a practical scale several processing conditions in large autoclaves have to be studied and evaluated in practical diets for both broilers and layers.

Table 4.1. The effects of processing conditions on the raw feathers and the chemical analysis of the treated feathers (Laboratory processing). From [(1)] Davis, Mecchi and Lineweaver (1961); [(2)] Morris and Balloun (1973b); [(3)] Papadopoulos (1984); [(4)] Latshaw (1990) and [(5)] El Boushy *et al.* (1990a)

Processing conditions		Crude protein	Pepsin-HCl Digestibility	*Lysine*		*Methionine*		*Cystine*		Lanthionine	References
Time	*Pressure*			*Total*	*Digestible*	*Total*	*Digestible*	*Total*	*Digestible*		
Min.	*kPa*	%	%	%	%	%	%	%	%	%	
0	0	94.1	16.0	1.80	-	0.40	-	8.20	-	0.00	(1)
6	621[a]	95.0	74.0	1.80	-	0.40	-	5.50	-	1.80	
20	414[a]	95.6	72.0	2.00	-	0.40	-	5.80	-	1.70	
30	207[a]	95.0	64.0	2.10	-	0.40	-	7.10	-	1.30	
40	207[a]	95.0	71.0	1.90	-	0.40	-	5.10	-	2.10	
30	276[b]	84.7	71.8	1.95	1.43	0.43	-	3.99	-	-	(2)
30	345[b]	83.1	73.8	2.14	1.68	0.41	-	4.43	-	-	
60	276[b]	83.4	74.6	1.41	0.95	0.41	-	2.51	-	-	
60	345[b]	82.9	74.2	2.85	1.92	0.66	-	1.87	-	-	
30	241[c]	83.1	72.9	Lost	1.71	0.53	-	2.59	-	-	
0	0	96.9	-	1.97	-	0.57	-	6.87	-	-	(3)
30	436[b]	96.9	93.6	1.96	1.13	0.66	0.45	4.63	2.58	1.96	
50	436[b]	97.5	90.3	1.93	0.92	0.64	0.38	4.04	1.96	2.33	
70	436[b]	96.9	93.5	1.94	0.88	0.67	0.42	3.69	2.07	2.28	
30	207[a]	88.5	52.0	-	-	0.51	-	6.14[d]	-	1.07	(4)
30	276[a]	88.8	66.0	-	-	0.36	-	5.83[d]	-	1.51	
30	345[a]	88.4	71.0	-	-	0.24	-	4.42[d]	-	1.63	
90	307[b]	96.0	73.0	1.80	1.00	0.60	0.40	5.60	2.40	-	(5)
60	376[b]	96.6	75.4	1.80	1.00	0.60	0.40	4.70	1.90	-	
20	514[b]	96.8	75.0	1.80	0.90	0.60	0.40	4.70	1.20	-	
6	700[b]	96.8	74.0	1.70	0.90	0.50	0.30	6.20	2.20	-	

[a] No agitation was used; [b] Continuous agitation was used; [c] Intermittent agitation; [d] Half cystine

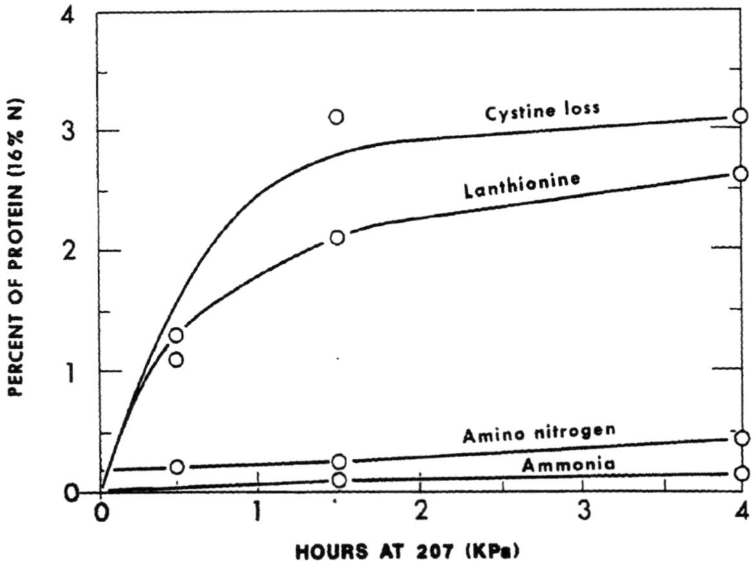

Figure 4.10. Loss of the amino acid cystine and formation of the amino acid lanthionine in feather meal owing to increased processing time. Reproduced with permission from Davis, Mecchi and Lineweaver (1961).

Moisture effect

There is little information available about changes caused by different moisture levels of the processed feather meals. Papadopoulos *et al.*, (1986) reported that with the levels of moisture of 50, 55, 60, 65 and 70% and processing times of 30, 40, 50, 60 and 70 minutes, at a constant pressure of 436 kPa and temperature of 146 °C, there was a positive linear effect of moisture on the sulphur amino acids, cystine and methionine. He also found that most of the essential and all the non-essential amino acids were lower in more highly moistened meals than in those processed with a lower moisture content.

Figure 4.11 and Table 4.2 show that the moisture content affected pepsin digestible protein, lysine and methionine. Myklestad, Biørnstad and Njaa (1972) showed the same trends for autoclaved fish meal with moisture percentages of 7,5 and 27%. This point of moisture content needs more research on a practical basis in large autoclaves.

(b) *Effect of chemical and enzyme additives*

The low content of the most limiting amino acids, methionine and lysine in feather meal have created an interest in developing alternative methods that will reduce autoclaving requirements, such as adding chemicals or enzymes, to establish improvements of the low percentage of these amino acids.

Table 4.2. Amino acid composition and protein quality characteristics of hydrolyzed feather meals as affected by processing time and moisture content. From Papadopoulos *et al.* (1986)

Treatment No.	*1*	*2*	*3*	*4*	*5*	*6*	*7*	*8*	*9*
Time (min)	*40*	*40*	*60*	*60*	*50*	*30*	*70*	*50*	*50*
Moisture (%)	*55*	*65*	*55*	*65*	*60*	*60*	*60*	*50*	*70*
Thr.[a]	4.75	4.72	4.86	4.70	4.77	4.73	4.65	4.87	4.60
Cys.	4.12	4.32	3.91	3.80	3.91	4.54	3.42	4.00	3.99
Val.	7.64	7.68	7.93	7.78	7.77	7.70	7.59	7.99	7.66
Met.	0.58	0.60	0.59	0.59	0.59	0.61	0.61	0.56	0.59
Ile.	4.95	4.97	5.17	5.03	4.97	4.85	4.99	5.07	4.92
Leu.	7.52	7.55	7.63	7.54	7.51	7.53	7.43	7.63	7.56
Tyr.	2.33	2.42	2.31	2.47	2.30	2.46	2.35	2.36	2.54
Phe.	4.15	4.26	4.39	4.48	4.21	4.31	4.20	4.66	4.71
Lys.	1.92	1.93	1.96	1.97	1.95	1.94	1.86	1.91	1.96
His.	0.69	0.66	0.70	0.70	0.67	0.67	0.68	0.66	0.67
Arg.	6.52	6.46	6.71	6.51	6.53	6.57	6.34	6.46	6.47
Essential AA	45.18	45.57	46.16	45.57	45.18	45.91	44.12	46.17	45.67
Asp.	6.50	6.40	6.45	6.29	6.56	6.36	6.47	6.41	6.24
Ser.	11.57	11.29	11.27	11.29	11.61	11.46	11.51	11.41	11.03
Glu.	10.86	10.71	11.12	10.90	10.93	10.82	10.91	11.19	10.42
Pro.	9.00	8.92	8.87	8.64	9.17	8.82	9.22	9.11	8.97
Gly.	6.91	6.90	6.99	6.94	7.13	6.96	7.05	6.89	6.82
Ala.	4.24	4.26	4.29	4.27	4.34	4.29	4.30	4.41	4.27
Non-essential AA	49.08	48.48	48.99	48.33	49.74	48.71	49.46	49.42	47.75
Total AA	94.26	94.05	95.15	93.90	94.92	94.62	93.58	95.59	93.42
NH_3	1.61	1.82	1.59	1.68	1.77	1.57	1.78	1.67	1.71
Cp[b]	96.48	96.27	97.40	95.85	97.32	96.63	95.75	98.07	96.85
Ash	1.24	1.24	1.40	1.22	1.19	1.21	1.26	1.38	1.17
PDP[c]	94.23	92.06	94.23	93.87	91.80	94.02	93.94	94.08	93.56
NSS[d]	64.77	62.88	78.53	76.03	74.29	56.12	82.42	72.47	66.21
NSH[e]	31.02	32.52	42.49	41.82	37.20	26.95	49.01	34.81	37.60

[a] Treatment means; [b] Crude protein; [c] Pepsin digestible protein; [d] Nitrogen solubility in 0.02 N NaOH;

[e] Nitrogen solubility in 6 N HCl

AA, amino acid

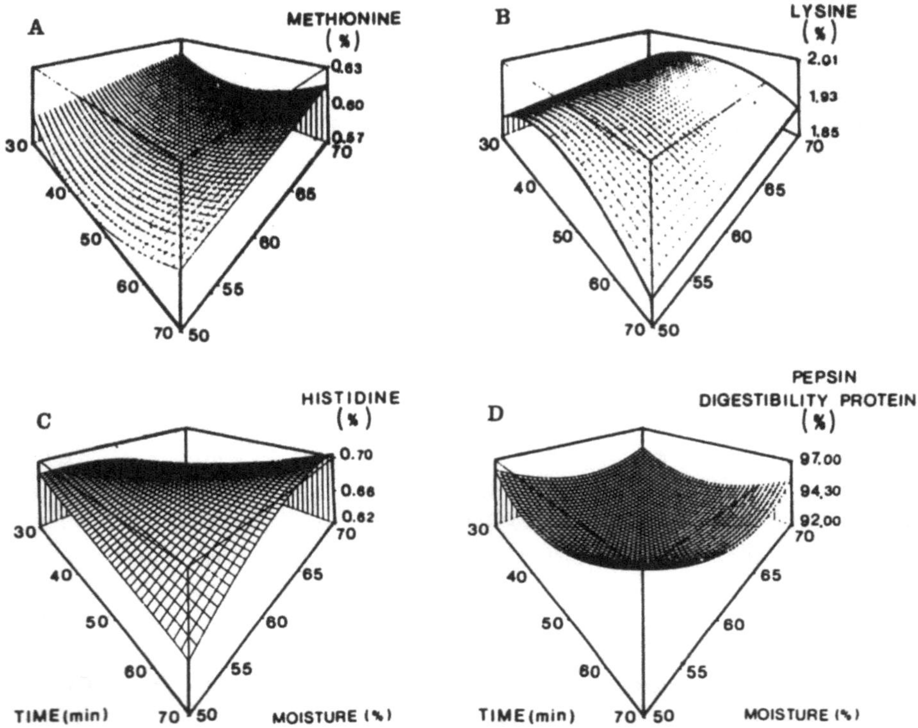

Figure 4.11. Three-dimensional response surfaces of methionine (A), lysine (B), histidine (C) and pepsin-digestible protein (D) for feather meals treated under varying processing times and moisture contents. Reproduced with permission from Papadopoulos *et al.* (1986).

Chemical additives

Davis, Mecchi and Lineweaver (1961) reported from his trials that odours were reduced by adding hydrated lime to feathers after picking. Because odour reduction has been a serious problem in urban areas, some meal has been manufactured from feathers to which lime has been added. Feathers processed after addition of 5% of hydrated lime, showed some changes in the amino acid pattern in comparison with control samples. The conversion of cystine to lanthionine was high owing to the addition of hydrated lime. However, there were losses of 20-25% of isoleucine, threonine, and serine. Addition of lime creates an alkaline reaction environment (pH 8.5-9) in which arginine may become unstable and a loss of arginine is to be expected. They added also that the natural L-form of amino acid in feathers will be changed to the D-form due to the high pH, a form which is not always available to animals. This conversion, which is known as racemization, may represent a loss of nutritional quality.

The amino acid composition of chemically treated feathers has been reported by Eggum (1970). In his trials the addition of 1% HCl solution to hydrolized feathers reduced the fall of cystine compared with feather meals processed under the same conditions of heat and pressure without HCl. It was also shown that the addition of

HCl reduced the contents of all amino acids, except for lysine, tyrosine, arginine and tryptophan, when compared with feather meals treated without HCl. Wolski, Klinek and Rompala (1980) observed that in feathers treated with dimethylsulphoxide the content of all the amino acids increased in comparison with the non-modified feathers, except for cystine, methionine, lysine and histidine. Further studies by Papadopoulos (1984) showed that in chemically-treated feather meals there was a significant variation in the amino acid responses to sodium hydroxide addition. All the essential amino acids, with the exception of valine, were reduced, while the non-essential ones with the exception of serine, were increased as sodium hydroxide concentration increased (Tables 4.3 and 4.4). However, Latshaw (1990) concluded that increasing the pH from 5-9 lowered cystine and methionine, and increased lanthionine.

Enzyme additives

The use of proteolytic enzymes as a digestive aid has been applied practically in processing keratinous materials. The reports show that enzymes from *Streptomyces Fradia*, isolated from soil (Noval and Nickerson, 1959), *S. microflavus* (Kuchaeva *et al.*, 1963) and from *Trichophyton granulosium*, a fungus of human and mammalian dermatophytes (Day *et al.*, 1968; Yu, Harmon and Blank, 1968) have been shown to be effective proteases with keratinolytic activity. It was reported that the unusual ability of the enzymes to decompose keratin rapidly and completely is due to their ability to reduce disulphide bonds in keratin. Elmayergi and Smith (1971) compared commercial feather meal, fermented by *Streptomyces Fradia*, with unfermented meal, in feeding trials with chicks. They found no significant difference in nutritional value between the two products, although the fermented meal was 90% digestible by pepsin-HCl solution as compared with 65-70% for unfermented meal. In their experiments, the levels of methionine, tyrosine, lysine and histidine, usually present in small quantities in feather meal, were increased considerably during fermentation with *Streptomyces Fradia*. However, Papadopoulos (1984) reported that using a commercial proteolyic enzyme, "Maxatase", did not improve the levels of amino acids except for leucine, tyrosine and phenylalanine, although his experiment was rather a preliminary one (Tables 4.3 and 4.4). It may be concluded from Table 4.4 that the true digestibilities of the limiting amino acids, methionine and lysine, were not affected by either chemicals or enzymes under investigation.

Dalev (1990) reported from his trials with alkaline-treated feathers with a pH of 8-8.3 mixed with an enzyme (alkaline protease B 72 from *Bacillus subtiles* with an activity of 50 000 U/mg produced at the Plant for Enzyme Preparations, Botevgrad, Bulgaria) with a concentration of 0.5g enzyme per 100g of feather pretreated material. The content of feather protein concentrate obtained by the alkaline-enzyme - disintegration treatment, as percentage of dry matter is: protein 85.4; fats 1.22; ash 8.60; fibre 0.68; Ca 0.55; P 0.16. The amino acid content of feather protein concentrate as percentage per 100g of protein is lysine 2.17; histidine 0.58; arginine 6.98; aspartic acid 7.79; threonine 3.49; serine 10.24; glutamic acid 12.74; alanine 7.63; valine 8.02; methionine 1.20; isoleucine 4.90; leucine 8.11; tyrosine 2.85; phenylalanine 3.01 and proline 8.90. He concluded that the feather meal protein concentrate significantly exceeds cereal feedstuffs in lysine and methionine content but is inferior to fish meal and soybean meal.

Table 4.3. Comparison between the chemical composition of feather meals either untreated or treated with chemicals and enzymes. From Papadopoulos (1984)

	Feather meals[a]								
Item	*FM30*	*FM50*	*FM70*	*FM30CH*	*FM50CH*	*FM70CH*	*FM30EN*	*FM50EN*	*FM70EN*
Proximate analysis %									
Dry matter	93.42	93.21	92.81	93.05	93.08	92.44	91.65	92.90	92.14
Crude protein	96.56	97.63	96.88	94.94	95.81	95.38	96.81	97.69	97.00
Ash	1.21	1.19	1.26	2.47	2.42	2.46	1.28	1.19	1.31
Essential amino acids %									
Threonine	4.88	4.81	4.66	4.72	4.67	4.52	4.80	4.79	4.54
Cystine[b]	4.63	4.04	3.69	3.44	3.04	2.67	4.65	3.51	2.94
Valine	7.51	7.41	7.36	7.41	7.36	7.20	7.44	7.36	7.14
Methionine	0.66	0.64	0.67	0.66	0.67	0.64	0.64	0.69	0.65
Isoleucine	5.14	5.09	5.08	5.08	5.04	4.95	5.07	5.08	4.93
Leucine	7.68	7.64	7.71	7.66	7.58	7.52	7.62	7.57	7.48
Tyrosine[b]	2.88	2.86	2.87	2.86	2.91	2.72	2.87	2.89	2.77
Phenylalanine	4.77	4.79	4.78	4.79	4.79	4.66	4.72	4.74	4.61
Lysine	1.96	1.93	1.94	1.94	1.90	1.87	1.93	1.94	1.90
Histidine	0.70	0.74	0.74	0.67	0.72	0.71	0.70	0.75	0.70
Arginine	6.76	6.68	6.62	6.59	6.55	6.37	6.61	6.59	6.66
Non-essential amino acids %									
Aspartic acid	6.41	6.26	6.26	6.35	6.32	6.28	6.25	6.22	6.14
Serine	11.37	11.24	11.13	11.24	11.51	11.77	11.17	10.77	11.07
Glutamic acid	10.72	10.63	11.03	11.03	10.89	10.93	10.47	10.54	10.69
Glycine	6.81	7.05	6.89	6.85	6.77	6.98	6.68	6.61	6.63
Proline	8.62	8.40	8.80	8.67	8.60	8.38	8.87	8.55	8.19
Alanine	4.36	4.33	4.44	4.34	4.34	4.37	4.27	4.27	4.32
Lanthionine	1.96	2.33	2.28	3.28	3.31	3.05	2.08	2.25	2.27

[a] FM 30, 50 and 70 = hydrolizing time in minutes, CH = chemical treatment with NaOH (0.4%) and EN = enzymatic hydrolyses (0.4% enzyme, Maxatase)
[b] Semi-essential amino acids (Scott, Nesheim and Young, 1982)

Table 4.4. Comparison between average true digestibilities of amino acids in differently processed feather meals either untreated or treated with chemicals and enzymes. From Papadopoulos (1984)

	Feather meals [e]									
Amino Acid	*FM30*	*FM50*	*FM70*	*FM30CH*	*FM50CH*	*FM70CH*	*FM30EN*	*FM50EN*	*FM70EN*	*Pooled SE*
Thr.	$69.1^{b,c}$	$67.0^{b,c}$	$64.8^{a,b,c}$	$66.5^{b,c}$	$64.9^{a,b,c}$	58.9^{a}	71.6^{c}	$63.3^{a,b}$	$62.3^{a,b}$	2.26
Cys.	55.7^{c}	$48.5^{a,b,c}$	56.1^{c}	$52.7^{b,c}$	$51.2^{b,c}$	$43.9^{a,b}$	55.8^{c}	39.1^{a}	40.8^{a}	3.18
Val.	$77.7^{b,c}$	$76.9^{a,b,c}$	$76.7^{a,b,c}$	$78.1^{b,c}$	$75.2^{a,b,c}$	70.8^{a}	80.6^{c}	$73.3^{a,b}$	$71.6^{a,b}$	2.02
Met.	$67.6^{b,c}$	59.6^{a}	$63.4^{a,b}$	$69.0^{b,c}$	$63.3^{a,b}$	58.2^{a}	71.8^{c}	59.8^{a}	58.2^{a}	2.32
Ile.	$83.8^{b,c}$	$83.2^{b,c}$	$82.6^{b,c}$	$84.0^{b,c}$	$81.8^{a,b,c}$	77.3^{a}	86.5^{c}	$79.2^{a,b}$	$80.4^{a,b}$	1.54
Leu.	$75.4^{a,b}$	$75.7^{a,b}$	$77.8^{a,b}$	$75.9^{a,b}$	$75.7^{a,b}$	72.3^{a}	79.0^{b}	$74.4^{a,b}$	$74.7^{a,b}$	1.86
Tyr.	74.0^{c}	$71.1^{b,c}$	$70.9^{b,c}$	$69.8^{b,c}$	$71.0^{b,c}$	64.0^{a}	74.4^{c}	$67.9^{a,b}$	$68.0^{a,b}$	1.84
Phe.	$80.3^{b,c}$	$80.5^{b,c}$	$79.1^{b,c}$	$78.5^{b,c}$	77.5^{b}	71.9^{a}	83.3^{c}	77.1^{b}	$75.6^{a,b}$	1.52
Lys.	57.9^{d}	$47.9^{b,c}$	$45.6^{a,b}$	$55.0^{c,d}$	$45.9^{a,b}$	38.3^{a}	59.5^{d}	$44.3^{a,b}$	$42.2^{a,b}$	2.54
His.	$56.0^{b,c,d}$	$52.8^{b,c}$	$49.8^{b,c}$	$57.6^{c,d}$	$54.8^{b,c,d}$	41.5^{a}	62.4^{d}	$52.3^{b,c}$	$47.9^{a,b}$	2.72
Arg.	$79.2^{b,c}$	$78.7^{b,c}$	$76.2^{a,b}$	77.1^{b}	$75.9^{a,b}$	71.2^{a}	83.0^{c}	$75.5^{a,b}$	$74.2^{a,b}$	1.78
Essential (mean)	$70.6^{c,d}$	$67.4^{b,c,d}$	$67.5^{b,c,d}$	$69.5^{b,c,d}$	$67.0^{b,c}$	60.7^{a}	73.5^{d}	$64.2^{a,b}$	$63.3^{a,b}$	1.98
Asp.	$47.2^{a,b,c}$	$42.9^{a,b,c}$	$39.6^{a,b}$	$48.5^{b,e}$	$45.8^{a,b,c}$	36.3^{a}	51.2^{c}	$40.3^{a,b}$	36.7^{a}	3.37
Ser.	$72.5^{b,c}$	$70.9^{b,c}$	$71.2^{b,c}$	$68.3^{a,b}$	$68.5^{a,b}$	63.9^{a}	76.2^{c}	$69.6^{a,b}$	$69.2^{a,b}$	1.94
Glu.	$68.3^{c,d}$	$63.4^{a,b,c,d}$	$65.2^{a,b,c,d}$	$66.8^{b,c,d}$	$64.3^{a,b,c,d}$	58.9^{a}	70.1^{d}	$61.2^{a,b}$	$61.9^{a,b,c}$	3.55
Pro.	$66.7^{b,c}$	$61.7^{a,b}$	$61.6^{a,b}$	$65.2^{b,c}$	$59.7^{a,b}$	53.8^{a}	70.4^{c}	56.9^{a}	53.5^{a}	2.57
Ala.	$76.1^{b,c}$	$74.4^{b,c}$	$74.3^{b,c}$	$74.3^{b,c}$	71.8^{b}	66.3^{a}	79.8^{c}	73.1^{b}	$71.1^{a,b}$	1.86
Noness. (mean)	$66.2^{c,d}$	$62.7^{a,b,c,d}$	$62.4^{a,b,c,d}$	$64.6^{b,c,d}$	$62.0^{a,b,c}$	55.9^{a}	69.5^{d}	$60.2^{a,b,c}$	$58.5^{a,b}$	2.29
Total (mean)	$69.2^{c,d}$	$66.0^{b,c,d}$	$65.9^{b,c,d}$	$68.0^{b,c,d}$	$65.4^{a,b,c}$	59.2^{a}	72.2^{d}	$63.0^{a,b,c}$	$61.8^{a,b}$	2.07

[a,b,c,d] mean values for individual treatments followed by different superscripts horizontally are significantly different ($P < 0.05$)

[e] see footnote to ***Table 4.3***

(c) Effect of physical characteristics, X-rays, bulk density and optical appearance

X-rays

Some experiments on the effect of X-rays on feather meal took place without any practical application. The objective of these studies was to test the changes in the chemical structure. X-ray diffraction data led Schor and Krimm (1961) to postulate a *β*-helix as the structural unit of feather keratin, i.e. an extended chain which coils slowly to form a helix with a relatively large pitch. Such helices tend to aggregate by hydrogen bonding to form cylindrical units which in turn associate into cable-like structures. The result of the reaction due to X-ray showed that reagents which oxidize or reduce disulphide bonds as well as solubilize keratin would tend to support a further slowly increase in susceptibility to enzymate digestion (Schroeder and Kay, 1955).

Bulk density

Davis, Mecchi and Lineweaver (1961) concluded that bulk density increases as processing time increases with a constant pressure level (307 kPa). The correspondence between bulk density and pepsin-HCl digestibility proved to be very good (Figure 4.12), however, no correlations were calculated. The method of determining the bulk density was briefly described and not very accurate as far as the unit of estimate is concerned. El Boushy *et al.* (1990), failed to find a clear significant trend between bulk density and pepsin-HCl digestibility of feather meal protein. Apparently, bulk density gives a rough estimation of the *in vivo* pepsin digestibility but is not suitable for detecting minor quality differences.

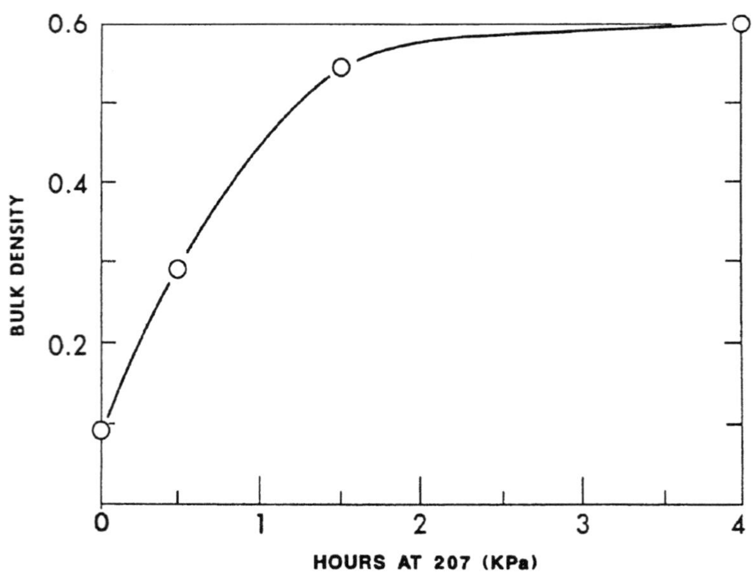

Figure 4.12. Effect of processing on bulk density of feather meal. Reproduced with permission from Davis, Mecchi and Lineweaver (1961).

Optical appearance

Colour: The effect of the processing treatment on the colour of the hydrolized feathers was discussed by Davis, Mecchi and Lineweaver (1961). Differences were found to be more pronounced in the laboratory processed samples than in the commercial ones, owing to the low iron content in the first. In the commercial processing, a small amount of iron is picked up from the interior of the cooker and is converted to black iron sulphide in combination with a part of the sulphur released from the feathers during the processing reaction. This darkening reaction (iron related) tends to mask the darkening (heating) produced by processing. Slightly less darkening may be caused by processing for a short time at high pressure in comparison with processing at low pressure for a long time.

El Boushy *et al.* (1990) reported from their trials that there is a very slight colour difference between their experimental treatments. The only clear difference was between the commercial sample which was darker and processed at 480 kPa for 1.5-2 h (Figure 4.13). This dark sample may be affected also by the iron sulphur complex.

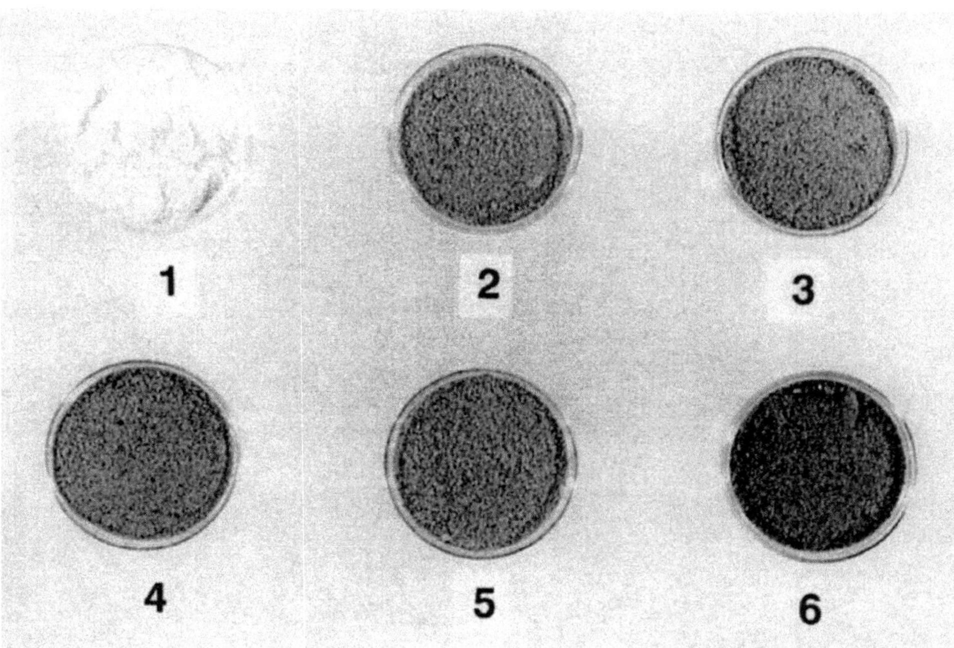

Figure 4.13. Effect of processing treatments on the colour of feather meal: (1) Raw feathers; (2) 307 kPa/90 min; (3) 376 kPa/60 min; (4) 514 kPa/20 min; (5) 700 kPa/6 min; (6) Commercial sample. Reproduced with permission from El Boushy *et al.* (1990).

Microscopic appearance: Davis, Mecchi and Lineweaver (1961) concluded that the original characteristic structure disappears progressively with increasing treatment until very little remains after normal processing. With severe processing for a long time, the original structure is completely gone, so that, after grinding, the material has the appearance of dark sand. Experienced feed microscopists can always recognize the material as overprocessed feather meal in feed.

El Boushy *et al.* (1990) observed the changes in microscopic appearance in relation to the processing conditions (Figures 4.13 and 4.14) in comparison with untreated and commercial samples. The original characteristic structure disappears with processing. However, no clear relation between pepsin-HCl digestibility and microscopic structure was noticed.

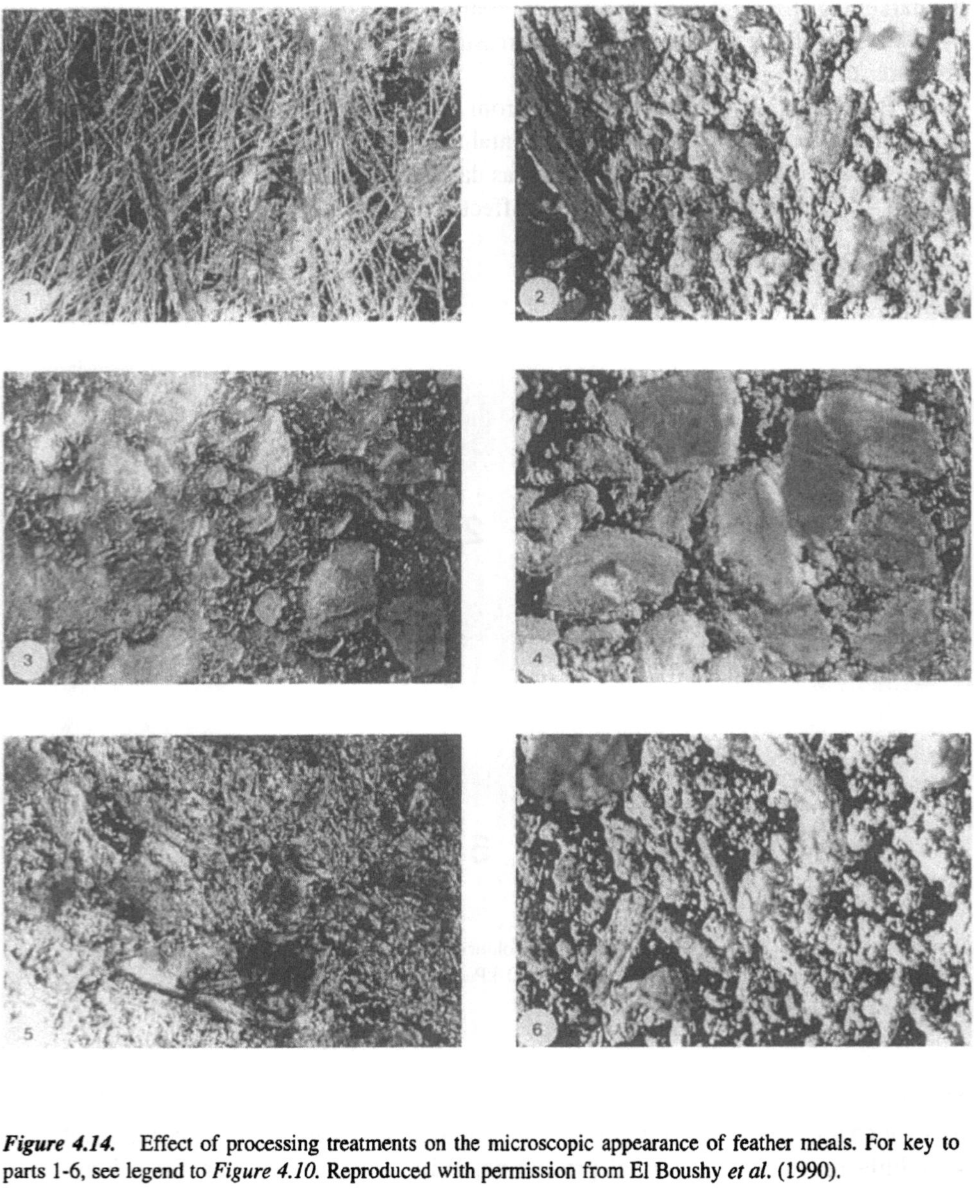

Figure 4.14. Effect of processing treatments on the microscopic appearance of feather meals. For key to parts 1-6, see legend to *Figure 4.10.* Reproduced with permission from El Boushy *et al.* (1990).

4.2.3 Improving the hydrolized Feather Meal Quality

When feather meal contains more than 85% crude protein, it is rich in cystine, threonine and arginine, but deficient in methionine, lysine, histidine, and tryptophan (Routh, 1942; McCasland and Richardson, 1966; Moran, Summers and Slingers, 1966; Wessels, 1972; El Boushy and Roodbeen, 1980; Papadopoulos, 1984; El Boushy, van der Poel and Walraven, 1990; El Boushy *et al.*, 1990). The addition of synthetic amino acids, fish meal, dried whey powder and/or antibiotics may improve its quality as a feedstuff in the diet of broilers and laying hens.

(a) Synthetic amino acids

A number of investigators have demonstrated that broiler chicks and laying hens performed normally when up to 4% of their diet was supplied by hydrolized feather meal, while 5-8% dietary feather meal induced methionine, lysine, histidine and tryptophan deficiencies and worsened chicken performance (Naber *et al.*, 1961; Moran, Summers and Slinger, 1966; Morris and Balloun, 1973[a]; Vogt and Stute, 1975; Luong and Payne, 1977; MacAlpine and Payne, 1977). Also, supplementation of the diet containing the high levels of feather meal with the synthetic form of the deficient amino acids resulted in improvement in performance.

Baker *et al.* (1981) showed that in diets with feather meal as the sole source of protein, methionine and lysine were the first- and second-limiting amino acids, while histidine and tryptophan were equally third limiting. With methionine-fortified corn-soybean meal diets it was shown that at least 10% of the dietary crude protein (24%) could be provided as feather meal. With methionine and lysine supplementation, up to 40% of the dietary crude protein could be supplied by feather meal with little depression in chick growth rate and efficiency of weight gain.

(b) Fish meal

It was noted before that feather meal is very rich in crude protein (85%), cystine, threonine and arginine, and poor in methionine, lysine, histidine and tryptophan. Any combination of feedstuff which is rich in the deficient amino acids will improve the total patterns of the most-limiting amino acids in feather meal.

Fish meal is sometimes available at reasonable prices and in several qualities as a fish meal or as a waste. Fish meals such as anchovy, herring or menhaden, as well as fish waste, are very rich in the most-limiting amino acids, lysine and methionine (Soares *et al.*, 1971), and the other essential ones, besides having high energy and rich minerals and an unknown growth factor (Tables 4.5 and 4.6). Fish meal added at 1-5% to feather meal will improve the essential amino acids pattern of a ration which contains mostly soybean meal and corn.

Gerry (1956) reported that when feather meal replaced fish-meal protein in broiler diets, growth rates were significantly poorer than the controls. If some part of the fish meal and/or soybean oil meal was replaced by feather meal in all-mash broiler rations, good results were obtained. Naber and Morgan (1956) reported similar results to Gerry (1956) by combining feather meal, fish meal and whey products.

Table 4.5 Chemical analysis (air-dry basis) of hydrolized feather meal in comparison with several fish meals. From Scott, Nesheim and Young (1982)

Feedstuff	*Protein*	*Metabolizable energy*	*Fat*	*Crude fibre*	*Calcium*	*Phosphorus*		*Sodium*	*Potassium*	*Chloride*	*Manganese*	*Zinc*
						Total	*Available*					
	(%)	*(MJ/kg)*	*(%)*	*(%)*	*(%)*	*(%)*	*(%)*	*(%)*	*(%)*	*(%)*	*(ppm)*	*(ppm)*
Feathers, hydrolyzed poultry	85	9.66	2.5	1.0	0.5	0.32	0.32	-	-	-	-	-
Fish meals:												
Anchovetta	65	11.84	4	1	4.0	2.6	2.6	0.8	0.7	0.3	22	110
Herring	72	13.35	10	1	2.0	1.5	1.5	0.5	1.1	1.0	10	-
Menhaden	61	12.89	9	1	5.5	2.8	2.8	0.3	0.7	1.2	36	150
Pilchard	64	11.05	5	1	3.8	2.5	2.5	0.2	0.3	0.8	9	-
Whitefish, waste	60	11.05	5	1	7.9	3.6	3.6	0.6	0.5	-	14	-
Redfish, waste	58	12.43	9	1	7.7	3.9	3.9	0.6	0.5	-	8	-
Fish solubles, dried	62	15.65	19	1	0.3	1.5	1.5	0.4	1.8	-	50	76

Table 4.6 Amino acid composition (air-dry basis) of hydrolized feather meal in comparison with several fish meals. From Scott, Nesheim and Young (1982)

Feedstuff	*Protein* %	*Arginine* %	*Cystine* %	*Glycine* %	*Histidine* %	*Isoleucine* %	*Leucine* %	*Lysine* %	*Methionine* %	*Phenylalanine* %	*Tryptophan* %	*Tyrosine* %	*Valine* %	*Threonine* %
Feathers, hydrolized poultry	84	5.6	3.0	-	-	-	-	1.5	0.5	-	0.5	-	-	-
Fish meals:														
Anchovetta	65	3.4	1.00	4.6	1.5	3.6	5.0	5.2	1.8	2.7	0.80	2.0	3.4	2.6
Herring	72	6.8	1.20	5.9	0.8	2.0	3.4	2.6	2.0	2.8	0.90	2.1	3.5	2.8
Menhaden	60	3.8	0.94	4.4	1.4	3.6	5.0	5.0	1.8	2.7	0.80	2.0	3.4	2.6
Pilchard	64	3.4	1.00	4.5	1.5	3.5	5.0	5.1	1.8	2.7	0.80	2.0	3.4	2.6
Whitefish, waste	60	3.5	0.90	4.2	1.5	3.1	4.5	4.9	1.7	2.5	0.70	2.0	3.2	2.5
Redfish, waste	58	4.0	0.90	4.2	1.3	3.4	4.9	6.5	1.8	2.5	0.60	1.7	3.3	2.6
Fish solubles, dried	62	3.0	0.42	6.0	2.6	2.0	3.0	3.2	1.0	1.5	0.25	0.7	2.2	1.5

(c) Dried whey powder

Some investigators concluded that whey products contain an unidentified growth factor (UGF). Atkinson, Ferguson and Couch (1955) found that dried whey and whey products, dried brewers' yeasts, and condensed fish solubles all contained UGF for turkey poults. Supplee *et al.* (1956) obtained an increase in weight gain of poults by supplementing a purified diet with a mixture of condensed fish solubles, distillers' dried solubles, distillers' dried yeast, dried whole whey or molasses distillers' dried solubles.

Balloun and Khajareren (1974) concluded that for poults between 1 and 4 weeks of age, feather meal in a diet at 5% was used efficiently in the presence of 4-5% whey. They also reported that whey, in general, increases nitrogen retention, protein digestibility and fat digestibility, but higher levels of feather meal with whey were not promising. From the above mentioned reports it is not possible to draw a conclusion that whey powder will improve the utilisation of feather meal protein in diets with more than 5% feather meal.

(d) Antibiotics

Antibiotics in small doses as a feed additive are known to be growth permittant agents. Potter, Shelton and Kelly (1971, 1974) reported that when zinc bacitracin was added to diets of young turkeys an increase of body weight of 8-10% was obtained at 2 and 4 weeks of age.

Potter and Shelton (1978[a]) studied the effect of hydrolized feather meal and zinc bacitracin. From their experiment they concluded that 5% feather meal with 27.6 ppm Zinc bacitracin increased the body weight and feed consumption against 5% feather meal alone.

4.2.4 Additions to Raw Feathers During Processing to Improve its Quality

It is a fact that hydrolized feather meal is very rich in crude protein and very poor in the essential amino acids, the most limiting ones being methionine and lysine. Any addition during cooking, hydrolizing and sterilizing the feathers to a form of a well-balanced amino acid feedstuff will improve the nutritional quality of the end product.

(a) Adding poultry manure to the raw feathers

Poultry manure, mainly cage layer manure, was discussed in details in chapter two. However a few important points will be mentioned to explain the expected nutritive value of several combined inclusion percentages. Flegal and Zindel (1970), Biely *et al.* (1972), El Boushy and Vink (1977) and El Boushy and Roodbeen (1984) concluded that poultry manure has many advantages as an additive as it contains undigested feed, metabolic excretory products, residues resulting from microbial synthesis and UGF. It contains some true protein besides the non-protein nitrogen (NPN) and important amino acids, and is very rich in minerals. Its disadvantages are a low energy value, high content of uric acid, and low content of limiting amino acids. The complete chemical analysis of dry poultry wastes (DPW) found by several workers is shown in Table 2.1.

From Tables 2.1 and 4.7 it is clear that the DPW is a very rich source of minerals. Its true protein level varies between 10 and 23%. In comparison, the feather meal true protein level varies between 81 and 87%. The levels of the most-limiting amino acids such as lysine and methionine, are not high, but hydrolized feather meal (FM) showed higher levels of lysine and methionine than the DPW. In Table 4.8 the true digestibilities or availabilities of the amino acids are shown for both DPW and FM. The total availabilities of amino acids were 57 and 68%, respectively. While methionine was 61 and 65%, and lysine 57 and 64% available, respectively, both amino acids were limited. The pattern of the essential amino acids was much better in FM than in DPW. If these two waste products could be hydrolized together perhaps a certain interaction due to processing would build another combination which would be much more valuable than both separately. This point is a practical solution for disposing this tremendous quantity of manure, and needs further investigation.

Table 4.7. Chemical analysis and amino acid composition of commercial hydrolized feather meals. From (1) Scott, Nesheim and Young (1982); (2) Ewing (1963); (3) Davis, Mecchi and Lineweaver (1961) and (4) Allen (1981)

%	*HYDROLIZED FEATHER MEAL*			
	(1)	*(2)*	*(3)*	*(4)*
Crude protein	85.00	87.4	81.8	85.00
ME (MJ/kg)	9.67	-	-	12.57
Fat	2.50	2.9	2.5	2.50
Crude fibre	1.00	0.6	1.0	1.50
Calcium	0.50	-	-	0.20
Phosphorus	0.32	-	-	0.33
Amino Acids				
Arginine	5.6	5.9	6.4	3.9
Cystine	3.0	3.0	4.9	3.0
Glycine	-	6.8	7.2	4.8
Histidine	-	-	0.4	0.3
Isoluicine	-	-	4.8	2.7
Leucine	-	-	6.7	7.8
Lysine	1.5	2.0	2.0	1.1
Methionine	0.5	0.6	0.4	0.6
Phenylalanine	-	-	4.5	2.7
Threonine	-	-	4.2	2.8
Tryptophan	0.5	0.5	-	0.4
Tyrosine	-	-	2.4	-
Valine	-	-	7.4	4.6

- = not determined

Table 4.8. Amino acids (AA), expressed as total (TAA), available content (AAA), availability %, percentage of total and their limitation in relation to the National Research Council (1977) requirements in dried poultry waste (DPW), feather meal (FM), poultry by-products (PBM), meat meal (MM) and soybean meal (SBM). From El Boushy and Roodbeen (1984)

	DPW			*FM*			*PBM*			*MM*			*SBM*			
AA	TAA[f]	AAA	%[g]	TAA	AAA	%	TAA	AAA	%	TAA	AAA	%	TAA	AAA	%	F-value
Aspartic acid	0.89	0.59	66[b]	5.02	2.81	56[a]	5.46	3.66	67[b]	5.77	4.62	80[c]	5.15	4.84	94[d]	32.5**
Threonine	0.38	0.20[h]	52[a]	3.36	2.08	62[b]	3.22	2.45	76[c]	2.52	2.07	82[c]	1.70	1.53	90[d]	38.0**
Serine	0.37	0.22	59[a]	6.73	4.31	64[a]	6.09	4.93	81[b]	2.83	2.29	81[b]	2.04	1.86	91[c]	16.6**
Glutamic acid	1.10	0.65	59[a]	7.96	4.94	62[a]	8.00	6.16	77[b]	7.93	6.66	84[b]	8.05	7.65	95[c]	31.0**
Proline	0.59	0.37	62[a]	9.39	6.67	71[b]	6.13	4.72	77[c]	4.45	3.83	86[d]	2.27	2.11	93[e]	38.0**
Glycine	2.91	-	-	4.47	-	-	6.59	-	-	6.55	-	-	1.90	-	-	-
Alanine	0.65	0.40	61[a]	4.85	3.78	78[b]	4.35	3.39	78[b]	4.74	4.03	85[b,c]	2.42	2.25	93[e]	14.7**
Cystine	0.35	0.19	55[a]	4.26	2.77	65[a]	2.43	1.15	62[a]	0.79	0.63	80[b]	0.88	0.80	91[b]	15.5**
Valine	0.48	0.27[h]	57[a]	6.41	4.81	75[b]	4.81	3.70	77[b]	3.53	2.93	83[b]	2.55	2.42	95[c]	24.7**
Methionine	0.16	0.10[h]	61[a]	0.79	0.51[h]	65[a]	1.14	0.88	77[b]	1.08	0.93	86[b,c]	0.62	0.57[h]	92[c]	13.9**
Isoleucine	0.39	0.23[h]	60[a]	4.15	3.24	78[b]	3.25	2.57	79[b]	1.99	1.71	86[c]	2.35	2.21	94[d]	34.7**
Leucine	0.59	0.32[h]	54[a]	6.19	4.52	73[b]	5.78	4.51	78[b]	3.22	2.42	75[b]	3.51	3.30	94[c]	38.8**
Trosine	0.23	0.11[h]	49[a]	2.23	1.45	65[b]	2.52	1.94	77[c]	1.95	1.60	82[c,d]	1.78	1.66	93[d]	16.9**
Phenylalanine	0.31	0.16[h]	53[a]	3.89	3.00	77[b]	3.63	2.87	79[b]	2.91	2.44	84[b]	2.42	2.27	94[c]	29.8**
Lysine	0.34	0.19[h]	57[a]	1.57	1.00[h]	64[b]	2.81	2.16	77[c]	4.19	3.56	85[d]	2.88	2.74	95[e]	47.5**
Histidine	0.16	0.08[h]	47[a]	0.55	0.32[h]	59[b]	1.08	0.78	72[c]	1.89	1.53	81[d]	1.29	1.20	93[e]	37.7**
Arginine	0.35	0.18[h]	52[a]	5.44	4.19	77[b]	5.45	4.58	84[c]	4.44	3.82	86[c]	3.47	3.30	95[d]	68.3**
Average	0.46	0.26	57[a]	4.55	3.09	68[b]	4.13	3.18	76[c]	3.39	2.82	83[d]	2.71	2.54	93[e]	50.1**
Crude protein %		28.98			83.04			63.65			56.62			44.29		

a, b, c, d, e, Any two means in a row with the same superscript are not significantly different ($P < 0.05$) by Duncan's multiple range test.

f Average of 2 replicates per feedstuff

g Availability percentage: average of 8 birds

h Limited essental amino acids in relation to the National Research Council (NRC, 1977) requirements according to Soares and Kifer (1971)

** Highly significant ($P < 0.01$) in the analysis of variance

(b) Adding poultry offal to the raw feathers

The combination of feather meal plus poultry offal (viscera, heads, feet and blood) (PBHFM) has been studied by some investigators (Gregory, Wilder and Ostby, 1956; Wisman, Holmes and Engel, 1957; Bahargava and O'Neil, 1975; El Boushy and Roodbeen, 1984). They concluded that the amino acid analysis showed that protein of poultry offal contains much more histidine, isoleucine, lysine and methionine than does feather-meal protein. Since feather meal is low in lysine and methionine, poultry offal and feather meal together would have a better balanced assortment of amino acids than either of the products alone (Table 4.9). El Boushy and Roodbeen (1984) reported (Table 4.8) that the total amino acid availabilities of feather meal and poultry by-product meal were 68 and 76%, and the availabilities of methionine and lysine for feather meal were 65 and 64% and for Poultry by-product meal (PBM) were 77 and 77%, respectively. There was no limitation in available amino acids in the poultry by-product meal. It seems to be a good process to combine feather meal and poultry offal meal together in the raw phase to produce a poultry by-product meal (PBHFM) which is comparable with meat meal and better than soybean meal (Table 4.9) in relation to the chemical composition.

Table 4.9. Chemical analysis and amino acid composition of commercial poultry by-product meal plus hydrolized feather meal (PBHFM) and poultry by-product meal (PBM). From [(1)] Scott, Nesheim and Young (1982); [(2)] Bhargava and O'Neil (1975); [(3)] North (1972) and [(4)] Allen (1981)

	(PBHFM) % [a]		*(PBM) %*		
	(1)	*(2)*	*(1)*	*(3)*	*(4)*
Crude protein	70.00	70.30	60.00	55.00	58.00
ME (MJ/kg)	11.30	11.33	12.18	11.62	12.13
Fat	12.00	12.30	13.00	12.00	14.00
Crude fibre	2.70	2.70	2.50	2.50	2.50
Calcium	2.30	2.30	3.60	3.00	4.00
Phosphorus (total)	1.40	1.40	2.20	1.40	2.40
Sodium	0.30	-	0.40	-	0.30
Potassium	0.40	-	0.60	-	0.60
Chloride	0.40	-	0.60	-	-
Amino acids					
Arginine	3.90	3.90	3.80	3.80	3.80
Cystine	2.20	2.20	1.00	1.00	1.00
Glycine	5.90	5.90	2.90	-	2.90
Histidine	0.70	8.70	0.80	-	1.60
Isoleucine	2.90	2.90	2.30	-	2.30
Leucine	5.20	5.20	4.20	-	4.40
Lysine	2.40	2.40	2.60	3.70	2.60
Methionine	1.20	1.20	1.10	1.00	1.00
Phenylalanine	3.20	3.20	1.80	-	1.80
Threonine	2.90	2.90	2.00	-	2.00
Tryptophane	0.50	-	0.50	0.50	0.60
Tyrosine	2.20	2.20	0.50	-	-
Valine	3.80	3.80	2.90	-	2.60

[a] A mixture of slaughtery waste is composed of 3.5% blood meal plus 15.8% offals (heads, feet and vicera) and 13.9% wet feathers with 60-65% water (El Boushy, 1986). All values as % (dry matter).

Poultry by-product meal alone or combined with hydrolized feather meal is very rich in protein, energy, fat, calcium, phosphorus, with methionine and lysine as the most important essential amino acids. Burgos, Floyd and Stephenson (1974) studied the amino acid bio-availability in samples of poultry by-product meal and feather meal which had been processed in different ways. They concluded that the amino acid availabilities in meals separately processed and then blended were higher than in those where the feathers and offal were processed together. The lower amino acid availability when the products were cooked together suggests that optimum processing conditions for poultry offal differ from those for feathers (Tables 4.10 and 4.11). Poultry meal and poultry by-product meal (poultry offal plus feathers) will be discussed in details in section 4.3.

Table 4.10. Comparison of amino acid content of feather meal, poultry by-product meal, and poultry by-product meal and feather meal blended (cooked separately). From Burgos, Floyd and Stephenson (1974)

	(1)				*(2)*	
	Feather meal batch process (%)	*Feather meal, cont. process* (%)	*P. by-product meal, batch process* (%)	*P. by-product meal, continuous process* (%)	*Blended product meal, batch process* (%)	*Blended product meal, cont. process* (%)
Asp.	$5.90^{a,c}$	5.75^{a}	5.20^{a}	5.17^{a}	5.57^{a}	5.40^{a}
Thr.	$4.05^{a,b}$	4.35^{a}	2.40^{e}	2.33^{e}	3.23^{c}	$3.13^{c,d}$
Ser.	7.50^{b}	9.25^{a}	2.70^{e}	2.70^{e}	5.70^{c}	4.57^{d}
Glu.	10.10^{a}	10.35^{a}	9.83^{a}	9.70^{a}	10.03^{a}	10.13^{a}
Pro.	9.55^{a}	8.85^{a}	6.43^{a}	6.50^{a}	8.70^{a}	6.93^{a}
Gly.	6.75^{a}	6.85^{a}	7.87^{a}	7.40^{a}	6.90^{a}	6.97^{a}
Ala.	5.35^{a}	$4.75^{b,c,d}$	$4.43^{b,c,d,e}$	$4.93^{a,b}$	$4.77^{b,c}$	$4.27^{c,d,e}$
Val.	$5.40^{a,b}$	5.80^{a}	2.87^{e}	3.03^{e}	$4.87^{a,b,c}$	$4.47^{a,b,c,d}$
Cys.-Cys.	$2.60^{a,b}$	3.00^{a}	0.63^{e}	0.60^{e}	1.80^{c}	$1.57^{c,d}$
Met.	0.50^{e}	0.40^{e}	1.07^{b}	1.43^{a}	$0.83^{b,c,d}$	$1.03^{b,c}$
Ile.	4.15	4.25^{a}	2.23^{e}	2.30^{e}	3.13^{c}	$2.83^{c,d}$
Leu.	$7.00^{a,b}$	7.25^{a}	4.20^{e}	4.37^{e}	5.40^{c}	$5.13^{c,d}$
Tyr.	$2.35^{a,b}$	2.40^{a}	$1.80^{c,d,e}$	$2.00^{c,d,e}$	$2.07^{b,c}$	$2.03^{b,c,d}$
Phe.	4.30^{a}	$4.10^{a,b}$	$2.40^{d,e}$	$2.53^{d,e}$	3.10^{c}	$2.80^{c,d}$
Lys.	1.80^{e}	1.90^{e}	$3.70^{a,b}$	3.80^{a}	2.90^{d}	3.17^{c}
His.	0.60^{a}	0.55^{a}	1.10^{a}	1.20^{a}	0.83^{a}	0.97^{a}
Arg.	6.65^{a}	$6.60^{a,b}$	$4.77^{c,d,e}$	$4.77^{c,d}$	$5.13^{b,c}$	$4.20^{c,d,e}$

(1) Comparison of individual amino acids between treatments: values represent the average of three replicates

(2) Blended poultry by-product meal and feather meal (see footnote to ***Table 4.9***)

a, b, c, d, e Any two means with the same superscript are not significantly different ($P < 0.05$) of probability by Duncan's multiple range test

Table 4.11. Amino acid availability [a] of feather meal, poultry by-product meal and poultry by-product meal and feather meal blended (cooked separately). From Burgos, Floyd and Stephenson (1974)

	Feather meal batch process (%)	*Feather meal,cont. process (%)*	*P. by-prod. meal,batch process (%)*	*P. by-prod. meal,cont. process (%)*	*Blended prod. meal,batch process (%)*	*Blended prod. meal,cont. process (%)*
Asp.	96.01	93.85	97.01	97.36	95.44	96.17
Thr.	96.56	95.50	97.14	97.68	95.60	97.42
Ser.	97.70	96.87	97.43	98.06	96.09	97.85
Glu. Ac.	96.04	94.77	97.30	98.31	96.45	97.19
Pro.	99.33	98.37	98.64	99.18	96.92	100.00
Gly.	95.44	92.72	94.84	95.77	95.44	93.79
Ala.	98.47	96.63	97.52	98.44	97.25	97.12
Val.	96.76	96.55	96.82	97.69	97.28	97.72
Cys.[b]-Cys.	97.32	94.52	94.03	96.57	95.44	97.96
Met.	95.42	93.73	98.05	98.76	97.54	97.83
Ile.	97.58	96.82	96.43	97.57	97.19	97.71
Leu.	97.48	96.24	96.31	98.14	96.86	97.81
Tyr.	96.66	95.50	96.53	98.34	96.52	97.18
Phe.	97.59	95.43	96.56	98.03	97.19	98.13
Lys.	94.92	94.10	97.04	98.39	97.78	96.46
His.	94.54	92.42	96.15	98.00	96.28	96.71
Arg.	98.25	97.39	98.01	98.76	97.69	97.46
Average	96.83	95.44	96.81	97.94	96.59	97.32

[a] Values represent the average of five birds. [b] ½ Cys.
Note: No significant difference between treatment for any amino acid ($P < 0.05$) of probability.

(c) Adding kitchen waste to the raw feathers

The combination of feathers and kitchen waste has never been studied before. The only studies reported concerned the nutritive value of kitchen waste and its use in poultry and swine nutrition (Soldevila, 1977; Yoshida and Hoshii, 1979; Hoshii and Yoshida, 1981; Lipstein, 1985). They all concluded that treated kitchen waste is an acceptable feedstuff for broilers and layers. This point will be discussed in details later in Chapter 7.

Hoshii and Yoshida (1981) reported that garbage from restaurants and large hotels contained 31.6% crude protein; 24.3% nitrogen-free extracts (NFE) and 18.3 MJ/kg of gross energy. Lipstein (1985) added that kitchen waste contained 0.47% phosphorus, 3.26% calcium, 2.2% sodium chloride and 11.9 MJ/kg metabolizable energy. The chemical analysis of feather meal was discussed earlier (Table 4.7). If these two ingredients are mixed together, hydrolized (by a batch cooker or extruder), sterilized and dried, the product will have a better nutritive value compared to the ingredients when considered separate.

This point needs additional technological, chemical and bioassay research to resolve the theory and the optimum levels of mixing the two ingredients, kitchen waste and raw feathers.

4.2.5 The Effect of Processed Feather meal level in rations on Poultry Performance

Several trials have been performed to evaluate the effects of protein and amino acids of hydrolized feather meal in diets for poultry.

(a) Broilers

Moran, Summers and Slinger (1966) reported that substitution of 5% protein in a practical 20% protein, corn-soybean ration with feather meal protein resulted in equally good chick growth. Replacement of all the soybean protein with corn and feather meal severely depressed 3-week chick weight; supplementation with methionine, lysine and tryptophan completely overcame the depression. Portsmouth, Cherry and Sharman (1970) obtained satisfactory growth of chickens fed on diets containing 2.5% feather meal but failed with 5%, while Thomas *et al.* (1972) obtained satisfactory performance with 7% feather meal.

Morris and Balloun (1973*a*) found that methionine and lysine supplementation to broiler chick rations, was not necessary when feather meal protein provided 2.5% of the diet. These two amino acids should be added to the diet when feather meals provide 5% or more of the protein in the diet (which is about 6% feather meal). Compared with "standard" processing, feather meals processed with intermittent agitation (agitate 1 min, stop one min) at higher temperatures for longer periods tended to increase the chicks' biological response (Table 4.12). Moreover a "new" feather meal (345 kPa for 60 min with intermittent agitation) included in a basal diet showed an increased availability of its amino acids for utilization compared to the "standard" feather meal (241 kPa for 30 min with constant agitation).

Daghir (1975) reported from his trials with feather meal in comparison with fish meal for broilers that in a good high protein broiler ration, 5% of feather meal and offal meal could be used instead of fish meal if Dl-methionine was added to the ration. Feather meal alone at 5% could be improved by methionine and lysine supplementation but this did not make it equivalent to fish meal (Table 4.13 and 4.14).

Bhargava and O'Neil (1975) indicated that poultry by-product and hydrolized feather meal (PBHFM) may be used up to 10% to replace a protein equivalent amount of soybean meal without deleterious effects in starter diets for broilers which have an adequate amount of methionine and lysine (Table 4.15).

MacAlpine and Payne (1977) reported from their experiments on feather meal protein and its utilization for broilers that 2.5% (for birds kept on floor) and 6% (for birds kept in cages) feather meal in the diet did not depress performance. In their trials they used a very well balanced ration containing fish meal, lysine and methionine. Therefore it was possible to include 6% feather meal as indicated previously without any depressing action on performance.

Table 4.12. The effect of feather meal processing on mean weight gains (wg) and feed conversion efficiency (fce) for male broilers of 1-5 weeks of age. From Morris and Balloun (1973[a])

	Pressure (kPa)	*Time (min)*	*Agitation*	*Feather meal protein* [a]				*Mean*	
				2.5%		*5%*			
				wg	*fce*	*wg*	*fce*	*wg*	*fce*
Control [b]				-	-	-	-	759	(2.03)
FM	276	30	Intermittent	751	(1.85)	751	(2.08)	751	(1.97)
FM	276	60	Intermittent	765	(1.95)	723	(2.09)	744	(2.02)
FM	345	30	Intermittent	761	(1.90)	734	(1.95)	747	(1.92)
FM	345	60	Intermittent	748	(2.04)	735	(1.86)	742	(1.95)
FM	241	30	Constant (standard process)	724	(2.03)	700	(1.96)	712	(2.00)
Mean				750	(1.95)	729	(1.9)	-	

[a] Feather meal protein was added to replace soybean meal in the control diet; 2.5 and 5% feather meal protein is equivalent to 2.6 and 5.2% feather meal in the diet.
[b] A basal corn-soybean meal diet of 21.8% crude protein, no feather meal.

Table 4.13. Growth and feed utilization of broilers fed feather and offal meal and feather meal supplemented with methionine in the diet. From Daghir (1975)

Treatment	*Body Weight (g/bird)*	*Feed Consumption (g/bird/day)*	*Feed Utilization (g feed/g wt)*
5% Fish meal	456	29.1	1.79
5% Feather and offal meal [a]	392	26.0	1.91
5% Feather and offal meal + 0.1%	433	27.4	1.79
Dl-methionine 5% Feather meal	407	27.4	1.94
5% Feather meal + 0.1% Dl-methionine	412	27.5	1.88

[a] See footnote to ***Table 4.9.***

Table 4.14. Growth and feed utilization of broilers fed feather meal supplemented with both lysine and methionine. From Daghir (1975)

Treatment	*Body Weight (g/bird)*	*Feed Consumption (g/bird/day)*	*Feed Utilization (g feed/g wt)*
5% Fish meal	384	29.4	2.11
5% Feather meal	330	27.3	2.23
5% Feather meal + 0.2% Dl-lysine HCl	344	28.1	2.11
5% Feather meal + 0.2% Dl-methionine + 0.1% L-lysine HCl	353	27.7	2.14

Table 4.15. Effect of feeding various levels of PBHFM on growth, feed efficiency and performance index (P.I.) of chicks (0-4 wk.). From Bhargava and O'Neil (1975)

(%)	*Experiment 1*			*Experiment 2*			*Experiment 3*		
PBHFM	*Body weight (g)*	*Feed/gain*	*P.I.*	*Body weight (g)*	*Feed/gain*	*P.I.*	*Body weight (g)*	*Feed/gain*	*P.I.*
0.0	615^{a}	1.54^{a}	370^{a}	496	1.69	270	587^{a}	$1.57^{a,b}$	349^{a}
2.5	599^{a}	$1.64^{a,b}$	$340^{a,b}$	498	1.72	264	601^{a}	1.55^{a}	363^{a}
5.0	611^{a}	$1.61^{a,b}$	354^{a}	523	1.68	285	624^{a}	1.53^{a}	383^{a}
7.5	613^{a}	$1.69^{a,b}$	$339^{a,b}$	523	1.77	271	591^{a}	1.55^{a}	357^{a}
10.0	611^{a}	$1.61^{a,b}$	351^{a}	477	1.79	242	$558^{a,b}$	1.62^{b}	$321^{a,b}$
12.5	-	-	-	534	1.70	289	494^{b}	1.63^{b}	281^{b}
15.0	511^{b}	1.75^{b}	273^{b}	467	1.80	239	372^{c}	1.76^{c}	190^{c}
20.0	314^{c}	1.99^{c}	135^{c}	-	-	-	-	-	-

Poultry By-product Hydrolized Feather Meal (PBHFM)

P.I. = [(Gain in body weight)^2 / feed consumption] x 100

a, b, c Numbers with different superscripts are significantly different ($P < 0.05$) from other numbers in that column

Feeding feather meal during the finishing period proved to be an effective method of reducing abdominal fat in broilers, which represents a loss in efficiency of production (Cabel, Goodwing and Waldroup, 1987, 1988). However, this theory is based on altering the calorie: protein ratio and its effect on carcass fat content. It appears that increasing the calorie: protein ratio enhances carcass fat deposition whereas decreasing the ratio has the opposite effect (Kubena *et al.*, 1974; Griffiths, Leeson and Summers, 1977; Summers and Leeson, 1979; Jackson, Summers and Leeson, 1982). The depressing effect of high protein diets on hepatic lipogenesis (Yeh and Leveille, 1969) and the high energy cost of uric acid synthesis (Buttery and Boorman, 1976) may explain the observed reduction in fat deposition with high protein diets.

In an experiment of Cabel, Goodwing and Waldroup (1988) feather meal was added at 4, 6 and 8% and glycine at levels of 0.125, 0.25 and 0.5%, which is similar to amounts contributed by corresponding levels of feather meal. Corn-soybean diets were also formulated at protein levels corresponding to those of feather meal diets. All experimental diets were fed from 35-49 or from 42-49 days of age. The results showed that there were no significant differences in weight gain and feed efficiency between the treatment and control groups during the study. The addition of glycine resulted in a significant ($P < 0.05$) reduction in abdominal fat content and appeared to be partially responsible for the observed reduction in fat from feather meal fed birds (Tables 4.16 and 4.17). Increasing the dietary protein level also significantly ($P < 0.05$) reduced abdominal fat deposition regardless of protein source. The study indicates that lower quality protein sources such as feather meal can be effectively used as a non-specific nitrogen source for reducing abdominal fat deposition during the finishing period.

(b) Layers

Daghir (1975) reported that laying hens receiving feather meal instead of fish meal, without amino acid supplements, had the lowest rate of egg production and the poorest feed efficiency on a corn-soybean diet. The combinations of feather and offal meal were a better protein supplement than feather meal alone. Supplementation of lysine and methionine to feather meal showed a significant improvement in performance, although egg weight was not affected by any of the treatments (Table 4.18). With additions of methionine at 0.075 and 0.15%, egg production, egg weight and feed efficiency were improved but not to the same level as those of birds receiving the fish meal diet (Table 4.19).

Luong and Payne (1977) tested the effect of hydrolized feather meal protein up to 7% in a wheat-based diet for laying hens. They concluded that egg production was reduced due to the deficiency in lysine, methionine and tryptophan. They added also that supplementation of the feather meal diet with 3% lysine (as L-lysine HCl), 0.07% Dl-methionine and 0.05% L-tryptophan resulted in a daily egg mass output similar to that achieved with a conventional layer's diet.

Table 4.16. Effect of inclusion of feather meal and glycine on body weight gain, feed efficiency, live weight, dressing percentage and abdominal fat content. From Cabel, Goodwing and Waldroup (1988)

Feather meal (%)	*Glycine (%)*	*Calculated protein content (%)*	*35-49 day weight gain (g)*	*35-49 day feed efficiency (g/g)*	*49 day live weight (g)*	*Dressing (%)*	*Abdominal fat pad weights (g)*	*Abdominal fat[d] (%)*
0	0	16.33	675	0.360	1927	60.67	40.35[a]	3.27[a]
4	0	18.05	700	0.363	2024	61.07	36.94[a,b,c]	3.10[a,b]
6	0	19.09	708	0.366	2032	61.58	37.43[a,b,c]	3.13[a,b]
8	0	20.42	682	0.344	1964	60.77	36.62[a,b,c]	2.97[a,b]
6e	0	19.09	742	0.380	1914	61.46	39.12[a,b]	3.21[a,b]
8e	0	20.42	722	0.374	1967	61.63	37.50[a,b,c]	3.11[a,b]
0	0.125	16.48	738	0.366	2017	61.61	39.68[a]	3.30[a]
0	0.250	16.62	689	0.361	1991	60.51	37.24[a,b,c]	3.08[a,b]
0	0.500	16.91	727	0.396	1981	61.12	35.06[b,c]	2.88[b]
0	0.125[e]	16.48	729	0.473	2047	61.15	36.93[a,b,c]	3.11[a,b]
0	0.250[e]	16.62	724	0.375	1940	61.55	35.14[b,c]	2.85[b]
0	0.500[e]	16.91	758	0.387	2002	60.99	34.23[c]	2.84[b]
Pooled SEM	-	-	30.4	0.025	34.7	0.35	1.69	0.15

[a,b,c] Within columns, means with no common superscripts differ significantly ($P < 0.05$).

[d] As a percentage of dressed weight

[e] Diets were fed from 42 to 49 days of age; all other diets were fed from 35 to 49 days of age

SEM = Standard error of the mean

Table 4.17. Effect of protein source of finisher diets on body weight gain, feed efficiency, live weight, dressing percentage, and abdominal fat content. From Cabel, Goodwing and Waldroup (1988)

Source	*Calculated protein content (%)*	*35-49 day weight gain (g)*	*35-49 day feed efficiency (g/g)*	*49 day live weight (g)*	*Dressing (%)*	*Abdominal fat pa^{d} weight (g)*	*Abdominal fatd (%)*
Soybean	16.33	778	0.380	2062^{b}	62.22	45.25^{a}	3.54^{a}
	18.05	774	0.393	2065^{b}	61.83	$42.08^{a,b}$	$3.27^{a,b}$
	19.09	801	0.419	1984^{c}	62.33	46.08^{a}	3.59^{a}
	20.42	712	0.377	$2006^{b,c}$	61.86	40.70^{b}	3.13^{b}
Feather meal	18.05	774	0.384	2075^{b}	62.00	41.23^{b}	3.25^{b}
	19.09	771	0.388	2151^{a}	62.49	$42.00^{a,b}$	$3.34^{a,b}$
	20.42	747	0.396	2068^{b}	62.28	40.469^{b}	3.18^{b}
Pooled SEM	-	34.5	0.014	26.0	0.26	1.58	0.12

^{a,b,c}Within columns, means with no common superscripts differ significantly ($P < 0{\cdot}05$)

^{d} As a percentage of dressed weight

SEM = Standard error of the mean

Table 4.18. Effect of inclusion of feather meal and feather plus offal meal on performance of laying hens. From Daghir (1975)

Treatment	*Egg Production (%)*	*Egg Weight (g)*	*Feed consumption (g/bird/day)*	*Feed efficiency (kg feed/doz eggs)*
5% Fish meal protein	71.6	57.8	118	1.98
5% Feather meal protein	67.7	57.5	120	2.14
5% Feather plus offal meals protein	71.2	57.4	122	2.05
5% Feather meal protein + 0.15% methionine + 0.19% lysine	71.6	57.6	120	2.01
5% Feather plus offal meals protein + 0.08% methionine	71.6	57.9	121	2.05

See footnote to *Table 4.9.*

Table 4.19. Effect of methionine supplementation on laying hen diets containing feather meal. From Daghir (1975)

Treatment	*Egg production (%)*	*Egg Weight (g/egg)*	*Feed Consumption (g/bird/day)*	*Feed efficiency (kg feed/doz eggs)*
5% Fish meal protein	76.9	56.9	111	1.74
5% Feather meal protein	66.6	55.9	114	2.11
5% Feather meal protein + 0.075% methionine	71.8	56.5	114	1.92
5% Feather meal protein + 0.15% methionine	72.5	57.1	113	1.88

(c) Turkeys

Some research workers have investigated the level of hydrolized feather meal in the diets of poult turkeys and its effect on growth, feed consumption and conversion. The effect of adding whey and yeast to diets containing feather meals was studied by Balloun and Khajareren (1974). They found that for poults between 1 and 4 weeks of age, feather meal in the diet at a 5% level was used efficiently in the presence of 4-5% whey. Potter and Shelton (1978*a*) reported that by adding 27.6 ppm Zinc bacitracin to 5% feather meal in the diet an increased body weight and feed consumption was noticed against 5% feather meal without Zinc bacitracin.

It was also concluded from the work of Balloun and Khajareren (1974) that an increased inclusion level of feather meal level up to 15% led to a remarkable depression in body-weight gain from 1-4 weeks which was slightly improved from 4-8 weeks. Feed conversion at the two ages was not significantly affected in comparison with the control (Table 4.20).

Table 4.20. Effect of feather meal on weight gain and feed gain in corn-soybean meal diets for turkey pullets. From Balloun and Khajareren (1974)

Feather meal	*Average weight gain (g)*			*Feed conversion g feed/g gain*		
(%)	1-4 wks	4-8 wks	1-8 wks	1-4 wks	4-8 wks	1-8 wks
0	593^{b}	1711	2304	155^{b}	1.21	1.81
5	582^{c}	1704	2286	145^{a}	1.92	1.80
10	561^{e}	1704	2265	152^{b}	1.95	1.84
15	558^{e}	1647	2204	154^{b}	1.94	1.84
Mean	574	1692	2266	152	1.93	1.83

All values represent means of two replicate groups values; means within a column followed by different superscript letters are significantly different ($P < 0.05$).

Potter and Shelton (1978*a*) showed from their two experiments with a corn and soybean diet that without additions of methionine, 5% addition of feather meal negatively affected body weight at 8 weeks of age, feed consumption was significantly negatively affected and feed efficiency was significantly positively affected (Table 4.21).

Table 4.21. Effects of 5% feather meal on body weight, feed consumption and feed efficiency of turkeys at eight weeks of age experiments. From Potter and Shelton (1978a)

	Feather meal %	
	0	*5*
Body weights g (8 weeks)		
Trial 1	1802	1775
Trial 2	1816	1823
Average	1809	1799
Percent change		- 6
Feed consumption; g (0-8 weeks)		
Trial 1	3306	3172
Trial 2	3335	3317
Average	3321	3245^{a}
Percent change		-2.3^{a}
Feed efficiency (0-8 weeks)		
Trial 1	5279	5419
Trial 2	5279	5339
Average	5279	5380^{a}
Percent change		1.9^{a}

[a] Significant at 5% level

It may be concluded from the above mentioned work that the addition of synthetic methionine and lysine will improve the feather-meal quality for turkey pullets from 1-4 weeks of age with less effect for ages from 4-8 weeks.

4.3 POULTRY BY-PRODUCT MEAL

Poultry by-products processed by rendering can be divided into five general categories depending upon the composition of the raw materials. They are:

- Blood meal which consists of ground dried blood.
- Poultry by-product meal which contains only clean, dry-rendered, wholesome parts such as the heads, feet, undeveloped eggs, gizzards, and intestines, but not feathers, except the few that might be included in the normal processing and collection practices.
- Hydrolyzed feathers which are in the product resulting from the treatment of clean, undecomposed feathers, free of additives.
- Mixed poultry by-product meal which contains blood, offal, and feathers, generally in their natural proportions. Mixed poultry by-products produce a more balanced product nutritionally than the other formulations, but it takes longer to process mixed by-products because feathers are harder to decompose or hydrolize than offal.
- In some cases fat is extracted from the resulting product (Lortscher *et al.*, 1957). The term poultry by-product meal may be also called poultry offal meal (POM) or poultry by-product plus hydrolized feather meal (PBHFM).

4.3.1 Processing Methods (Rendering)

The following practices are generally used in processing poultry by-products. A large proportion of the water is evaporated in a dry-rendering cooker to hydrolize, sterilize and reduce the material to about 8% moisture either in the same cooker or in a separate dryer. The dehydrated material is then pressed to remove excess fat so the rendered material has a fat content of about 10%. Finally, the product is ground to a size small enough to pass through 8-12 mesh screens. By transferring the by-products of poultry slaughtering to a mixed poultry meal (soft offal, blood, feathers and dead birds in their natural proportion), rendering is taking place.

The rendering system includes five phases as follows:

- Storage of raw material
- Cooking and drying or drying separately
- Condensation
- Meal handling
- Fat extraction

(a) Storage of raw material

Soft offal and feathers which are a continuous output of the slaughterhouse have to be stored prior to processing. Storage containers may vary from simple ones with a capacity of 500 l, mostly handled manually, to large full automatic ones. The automatic system is based on receiving feather/water and soft offal and transport to the cooker by means of feed screw conveyor. Blood is stored in the tank and transferred in the cooker by means of a flexible hose in specific amounts corresponding with the exact ratio of feathers: offal: blood, this point will be discussed under the yields of offal.

(b) Cooking and drying

Hydrolizing/sterilizing takes place in a cooker which may also work as a dryer. When the cooker is filled, the dome is closed, the offal is heated through the steam in the double wall of the cooker, as discussed before with feather meal (Figure 4.1). During heating the offal to 142 °C with a pressure of 380 KPa for 30-40 min, the hydrolysis and sterilization is taking place (Hamm, Childs and Mercury, 1973; Filmich, 1987; El Boushy, van der Poel and Walraven, 1990; Van der Poel and El Boushy, 1990).

At the end of the hydrolization/sterilization process, the vapor line to the condensor is opened to start drying. During the rest of the process the cooker contents are dried (or moved to a separate dryer) to form a mixed poultry by-product meal at the lowest temperature to preserve quality of the protein and amino acids.

Separate drying

This system is mostly used in the case of continuous processing, the batch type (Figures 4.2 and 4.3). In general, two types of driers are used, the disc-drier and the flash-drier.

Disk-drier

This is a very simple system, and its benefits are to allow drying by:

- Positive condensate removal from each section of the disk.
- Use of more effective heating/drying surface thus providing a greater evaporative capacity per m^2.

Through the contact of the wet product with the heated surface and the movement of the conveyors and scrapers fixed between the disks, a constant intensive mixing is achieved.

Drying is achieved by heating, the source being either saturated steam, or a heat carrier, mostly oil or hot water. The product is cooled using cold water through the rotor shaft.

The vapor escape is supported by generating a slight underpressure. Inspection and cleaning takes place through covers in the vapor dome.

By measuring the moisture and temperature in the drier, it can be regulated either manually or fully automatically. An electronic steering processor regulates the charge and discharge units. Lumpy, granular, fluid, pasty or liquid products can be dried and cooled. Examples of animal waste recovery are poultry by-products, feathers, meat scraps, bone, blood, fish and manure as well as industrial and municipal sludge. Examples of plant and vegetable waste recovery are brewers' grains, potato residues and citrus pulp. Figure 4.15 shows the disk dryer with its inlets and outlets.

Flash drier

This is mostly used in a complete continuous integrated process unit that delivers the final offal product ready for packaging. The main benefits of these driers are:

- The heater is suitable for both oil and gasfiring
- Exceptionally high in end product quality
- It is suitable for all rendering materials having a low fat content, such as blood, feathers, hair, stomach content *etc.*
- It is ecologically sound
- It uses minimal floor space

The main components of the flash drier are shown in (Figure 4.16) with the main components being the air heater; the disintegrator; the feed bin; the drying duct; the separator and the cyclones.

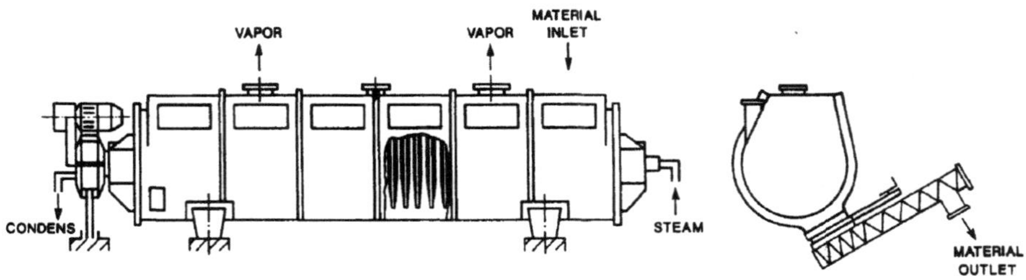

Figure 4.15. Disk dryer showing its inlets and outlets.

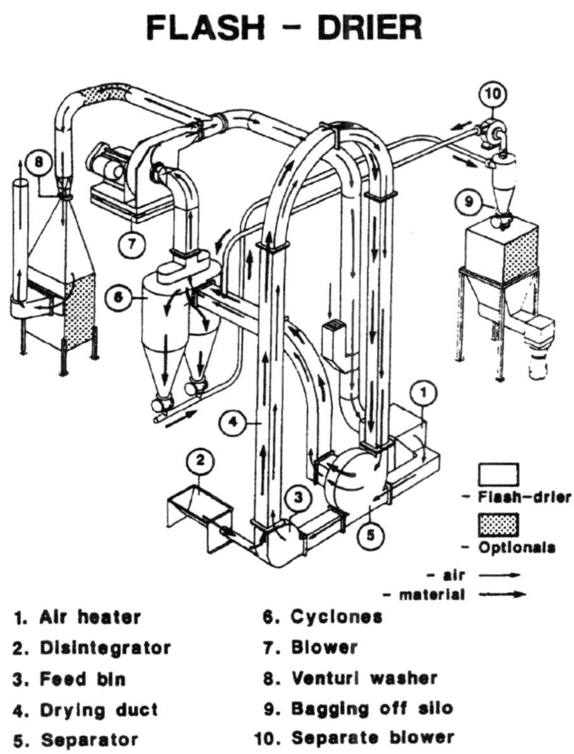

Figure 4.16. Flash dryer with its main components: (1) air heater; (2) disintegrator; (3) feed bin; (4) drying duct; (5) separator; (6) cyclones; (7) blower; (8) Venturi washer; (9) bagging off silo; (10) separate blower. Reproduced with permission from El Boushy *et al.* (1990).

(c) Condensation

The raw materials have a water content of approximately 70%. During the condensation process, this water is vaporized and evacuated from the cooker/drier. It may be released into the air but, according to the environmental regulations, this practise is prohibited in some countries. Therefore condensors are used to condense the fetid vapours to water. The most useful condensing system works by air cooling; it requires no water and is very safe.

(d) Meal handling

The processed poultry offal meal can be delivered into the feeding industry in a bulk or bag system. In some cases the end product may be ground and sieved.

(e) Fat extraction

The normal poultry offal meal prepared from broilers with a live weight of 1600 g has a fat content of approximately 16%. This meal needs no handling and can easily be stored. In case of heavier broilers and using parent stock, or old layers, or using offal without feathers, the fat content will be high and care should be taken. This processed high fat offal should be defattened to assure that the product has a low fat content of about 10-12%. The extracted fat may be used as an addition to feed or as fat for other purposes. The output capacity of a small defattening unit is 400-500 kg/h of defattened meal which is sufficient for a rendering system with a raw material input of approximately 2 t/h.

(f) The poultry offal rendering system

This system is shown in details in Figure 4.17. Feathers and offal reach the mixed-raw material bin via a feather/water separator and a vacuum transport system where they are temporarily stored until the cooker is ready to be filled again. Blood is pumped from the slaughter-house into the blood storage tank. The feathers and offal collected in the mixed-raw-material bin are transported to the cooker by the feed screw conveyor. The conveyor is designed to transport the total quantity of raw materials for one batch in only a few minutes. Blood is drained into the cooker, directly from the storage tank. Hatchery waste and dead birds may simply be added to the contents of the raw material bin and processed with the other offal.

A new development in the processing of a poultry by-product with or without feathers was described by Vandepopuliere (1988). He used a high temperature extrusion reaching 132-138 °C assuring that the extrusion employed effectively sterilized all the diets. The principal method of extrusion was based on mixing the ingredients of the diets including the (raw and ground) poultry by-product meal, with or without feathers prior to extrusion.

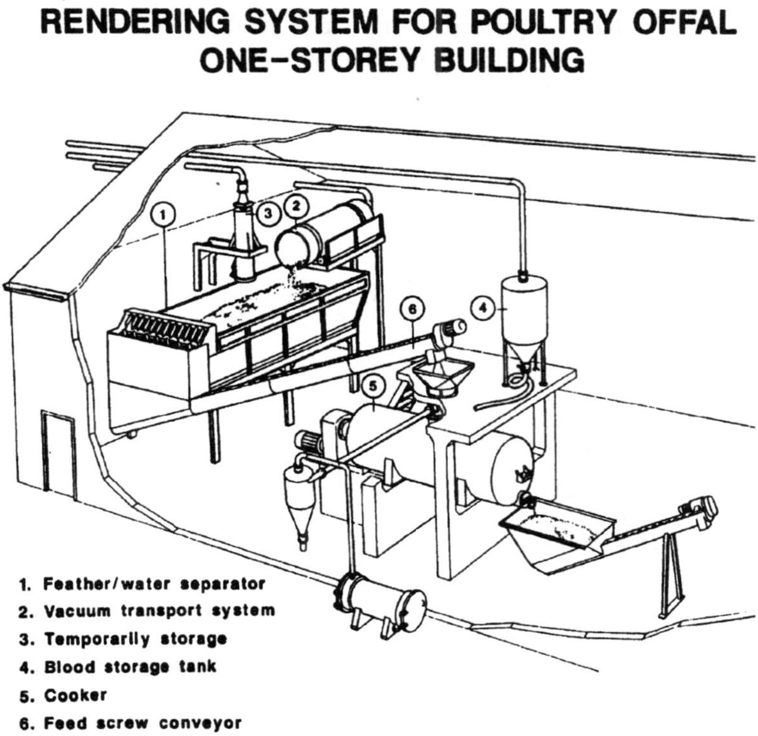

Figure 4.17. A rendering system for poultry offal: (1) feather/water separator; (2) vacuum transport system; (3) temporary storage; (4) blood storage tank; (5) cooker; (6) feed screw conveyor. Reproduced with permission from Fielmich (1987).

4.3.2 The Yield of Offal

The slaughtering capacity/production yield ratio for mixed poultry offal is shown in Figure 4.18. From this figure the following data can be calculated: for example, 24 000 broilers per day, each with a live weight of 1.5 kg, yields the following:

- Blood 1.25 t
- Offal including heads and feet, 5.7 t
- Wet feathers, 5.0 t
- Total slaughtering offal, 11.95 t
- This offal will produce 3.6 t mixed poultry meal

The several components of the raw offal can be summarized as follows (El Boushy, 1986):

- Total offal (heads, feet, intestines), 15.8%
- Blood, 3.5%
- Feathers, 6.0%
- Moisture, 9.0%

  Total raw material for rendering, 34.3%.

Collectable by-products, expressed as the mean yield per 1,000 kg live bird put into slaughter-house and the typical plant-handling procedures, respectively, were as follows Hamm, Childs and Mercury (1973):

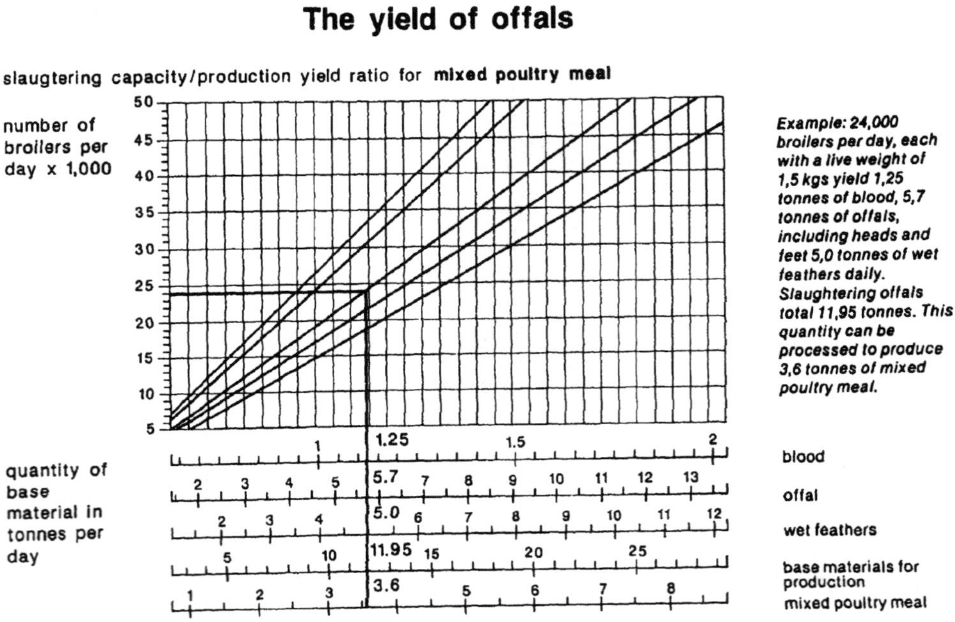

Figure 4.18. Slaughtering capacity/production yield ratio for mixed poultry meal. Reproduced with permission from El Boushy (1986).

(a) *Feathers*

Feathers 54 kg (30-80 kg). Feathers are removed from the water flowaway systems by screening, and are usually move immediately into trucks for delivery to the renderer. Since feathers are exposed to water during scalding and picking, and then to water plumed before separation, the product leaving the processing plant will contain on average 70% water, whereas prior to processing on a live weight basis, feathers do not exceed 30% water (Kahle and Gray, 1956).

(b) *Blood*

Blood 37 kg. This value is much higher than the actual amount collected by the industry, as it is based on the amount obtained experimentally in the laboratory using a 90-s. bleeding time. Other investigators, however, have reported similar levels for 1.3-1.8 kg birds (Newell and Shaffner, 1950; Kotula and Helbacka, 1966). Blood is collected, kept in tanks and may be dried separately in a flash drier or may be mixed with the other offal and processed in the main cooker.

(c) Offal or viscera

Offal or viscera 118 kg. Usually all the non-edible materials removed from the carcass during evisceration are water plumed away and recollected by screening. These include the viscera, heads, and trimmed tissues (gizzard lining and unusable parts).

(d) Feet

Feed, 30 kg. Feet are usually removed before evisceration, but the plume carrying them usually joins the evisceration plume just up stream from the screen. After collection by screening, the offal, feet included, is moved to the renderer. Feet are sometimes collected separately for a special market in some countries.

(e) Lungs

Lungs, 9 kg. Lungs are removed almost exclusively by vacuum aspiration and are maintained separately only for convenience of using the vacuum system. They are usually dumped in with the offal before leaving the processing plant. The vacuum system is also used to remove the pancreas and spleen.

(f) Unwholesome carcasses

Unwholesome carcasses, 10 kg. A small number of the carcasses processed are unfit for human consumption, these must be decharacterized, usually with dyes, so that they can not later re-enter the human food chain. They are kept separate, and either sold to petfeed processors or added to the offal for rendering.

(g) Fat and grease

Fat and grease, 6 kg. A small part of the fat and grease, attached to the carcass, will be separated from it. Thereafter, few fat and grease is found in the wastewater. Final plant effluents may run as high as 1,000 ppm fat, but usually average 300 ppm. Air flotation can remove 60-70% of the fat plus some suspended materials. When collected, this sludge is usually held separately for industrial use or is added to the offal for processing.

(h) Settling or sludge

Settling or sludge, 10 kg. Plants that apply some primary treatments to their effluents have installed solids settling tanks along with fat flotation systems. These settlings are primarily proteinaceous in nature and are usually added to the offal.

These eight categories account for a collectable by-product yield of 34.3% of liveweight (El Boushy, 1986). The yield of by-products from broilers, fowl and turkeys is shown in Table 4.22.

Table 4.22. Yield of by-products from broilers, fowl and turkeys [a]. From Lortscher *et al.* (1957)

Material	*Percentage of live weight*		
	Broilers	*Fowl*	*Turkey*
Waste yield from pantry			
Offal	17.5 (15-20)	17.0 (17-18)	12.5
Blood	3.5 (3.2-4.2)	3.0	3.5
Feathers	7.0 (4.8-7.5)	7.0	7.0
Feathers, wet	22.0	20.0	14.0
Mixed (dry feathers)	28.0	b	23.0
Water pick up			
Offal	1.0	1.0	-
Blood	-	-	-
Feathers	15.0	13.0	7.0
Mixed	16.0	b	7.0
Total waste yield			
Offal	18.5	18.0	12.5
Blood	3.5	3.0	3.5
Feathers	22.0	20.0	14.0
Mixed	44.0	b	30.0
Water evaporated			
Offal	12.7	10.6	7.5
Blood	2.7	2.3	2.7
Feathers	16.5	14.5	8.1
Mixed	31.9	b	18.3
Dry product (8% Moisture)			
Offal	5.8	7.4	5.0
Blood meal	0.8	0.7	0.8
Feather meal	5.5	5.5	5.9
Mixed	12.8	b	11.7
Pressed product (1% fat)			
By-product meal	5.2	4.3	4.2
Grease	0.6	3.2	0.8

[a] Range is given in parenthesis

[b] No advantage to mixing prior to cooking, since fat level will require pressing prior to grinding

4.3.3 Chemical Analysis and Nutritive Value

The value of poultry by-product meal plus feather meal combined (PBHFM) has been studied by some investigators (Gregory, Wilder and Ostby, 1956; Wisman, Holmes and Engel, 1957; Burgos, Floyd and Stevenson, 1974; Bahargava and O'Neil, 1975; Daghir, 1975; McNaughton, May and Strickland, 1977; El Boushy and Roodbeen, 1984; El Boushy, van der Poel and Walraven, 1990). They all concluded that PBHFM is a feedstuff very rich in protein, energy, fat, calcium, phosphorus and essential amino acids. This may be also concluded from Table 4.23, which summarizes the results of several workers.

Table 4.23. Chemical analysis of the mixed poultry by-product meal. From (1) Scott, Nesheim and Young (1982); (2) Bhargava and O'Neil (1975); (3) McNaughton, May and Strickland (1977) and (4) El Boushy and Roodbeen (1984)

%	*(1)*	*(2)*	*(3)*	*(4)*
Crude protein	70.00	71.10	53.7[a]	63.7
Metabolizable energy (MJ/kg)	11.30	11.33	-	-
Fat	12.00	13.30	25.1[a]	-
Crude fibre	2.70	2.70	4.7	-
Calcium2.30	2.30	1.60	-	
Phosphorus (total)	1.40	1.40	1.1	-
Sodium 0.30	-	-	-	
Potassium	0.40	-	-	-
Chloride	0.40	-	-	-
Amino acids %				
Aspartic acid	-	4.60	4.1	5.5
Threonine	2.90	2.90	2.4	3.2
Serine -	5.50	3.70	6.1	
Glutamic acid	-	7.30	7.1	8.0
Proline -	6.20	-	6.1	
Glycine	5.90	5.90	5.3	6.6
Alanine	-	3.60	3.3	4.4
Cystine 2.20	2.20	2.20	2.4	
Valine 3.80	3.80	3.30	4.8	
Methionine	1.20	1.20	0.8	0.9
Isoleucine	2.90	2.90	2.3	2.6
Leucine	5.20	5.20	4.1	5.8
Tyrosine	2.20	2.20	1.4	2.5
Phenylalanine	3.20	3.20	2.4	3.6
Lysine 2.40	2.40	2.10	2.8	
Histidine	0.71	0.70	0.6	1.1
Arginine	3.90	3.90	3.6	5.5

[a] Crude protein and crude fat are expressed as percentage of wet weight

El Boushy and Roodbeen (1984) reported from their comparison between poultry by-product meal and feather meal (Table 4.8) that the total amino acid bioavailabilities of feather meal and poultry by-product meal were 68 and 76%, and the availabilities of methionine and lysine for feather meal were 65 and 64%, and for poultry by-product meal were 77 and 77% respectively. There was no limitation in available amino acids in the poultry by-product meal, making it comparable with meat meal and better than soybean meal.

Wang and Parsons (1998) concluded from their work on amino acid limitation in poultry by-product meal that the order of the limitation was: 1) cystine; 2) tryptophan; 3) threonine and lysine; 4) valine, and 5) isoleucine and histidine with a degree of deficiency of valine, isoleucine and histidine being similar. The results also suggested that methionine may be slightly limiting in poultry by-product protein. Their results were based on amino acids determination which are not applied in practice with the calculation of rations based on linear programming.

4.3.4 Its Use in Feeding

The use of PBHFM as an animal protein feedstuff was studied by few research workers, however, it is an excellent feedstuff for broilers and layers, and it may be compared with meat meal as far as its levels in protein, energy, minerals and amino acids are concerned (Acker *et al.*, 1959).

Bhargava and O'Neil (1975) indicated that the PBHFM may be used up to 10% to replace a protein equivalent amount of soybean meal without deleterious effects on broiler starter diets adequate in methionine and lysine (Table 4.15).

As far as PBHFM is concerned in the nutrition of layers, Daghir (1975) concluded from his trials that combinations of feather and offal meal were a better protein supplement than feather meal alone based on the performance of laying hens.

4.4 HATCHERY BY-PRODUCTS

The usual disposal of hatchery waste is through landfill or rendering plants. Landfill operators are reluctant to receive hatchery wastes, and some operators refuse to accept it. A critical factor influencing the supply of hatchery waste to renderers is the cost of accumulation and delivery to the rendering plants. Only very large hatcheries produce sufficient waste daily to supply a full truckload to the render. Small hatcheries, therefore, have sought out local means of disposal, but they are now encountering greater resistance from authorities for access to landfills.

Hamm and Whitehead (1982) defined hatchery waste as all the collectable material remaining in commercial hatchery trays after saleable chicks have been removed. This normally includes shells from hatched chicks, infertile eggs, dead embryos in the shell and dead chicks. The nutritive value of the hatchery by-products (or processed hatchery waste, PHW) was studied by few research workers, who concluded that it is a relevant feedstuff for layers owing to its high Ca content (Wisman, 1964; Wisman and Beane, 1965; Vandepopuliere, Walton and Cotterill, 1974; Hamm and Whitehead, 1982; Miller, 1984; Ilian and Salman, 1986).

4.4.1 Processing Methods

As mentioned before, the cost of accumulation and the quantity delivered to the rendering plants are limiting this industry. Moreover, only the very large hatcheries produce sufficient waste daily. Small hatcheries, therefore have sought out local means of disposal and this has lead to finding reasonable holding techniques for hatchery wastes.

(a) Holding technique

To solve the problem of low production of waste in small hatcheries with their limited output, several holding techniques were studied. Hamm (1974) reported that inedible egg meats from egg grading and breaking plants could be slightly acidified and held at 13 °C for up to 10 days to allow economical accumulation. The hatchery waste could then be transported to processing plants, neutralized and rendered. Hamm and Whitehead (1982) studied several methods of treatment to retard spoilage during holding of hatchery wastes. They reported that passing wastes through a disposer to reduce its volume coupled with an ensiling system, showed satisfactory results. Blending in 10% W/W of feed grade cane molasses during grinding of the collected wastes retained adequate freshness for 1 week of storage at about 22 °C. The second possibility they tried was adding 10% of cane molasses plus 10% of ground corn which resulted in a safe storage up to 14 days.

(b) Processing technique

There are very few reports about the processing of the hatchery wastes, however, the best known method is through rendering mixed with other offals as feathers, blood, intestines, heads and legs, *etc.* Ilian and Salman (1986) reported a cooking method of the hatchery waste in water, with a ratio of 2 waste-1 water (w/w). After the cooking, a drying process took place for 24 h with temperatures of 60, 80 and 105 °C, respectively. The dried end product was finally ground to a fine texture. This method seems rather impractical for a large scale production.

Miller (1984) extruded hatchery waste to produce a fine ground sterilized product, this product was a mixture of 75% corn and 25% hatchery waste. His method of preparing this product was as follows: the hatchery waste was treated with chloroform to kill surviving embryos and then immediately ground with a large meat grinder with 12 mm holes in the grinder disc to allow a complete mix and a reduced particle size. A mixture of twenty-five kilograms of hatchery waste was blended with 75 kg of ground yellow corn for 5 min. in a horizontal mixer. This blended material was fed into a Brady Model 206 extruder driven through the power take-off from a 60 kW (80 hp) farm tractor. Extruding temperature was 140 °C with an estimated retention time of 10 s. Feeding rate of the extruder was approximately 5.3 kg/min. Finally the end product was analysed chemically and was found to contain 13.6% crude protein, 7.1% ash, 2.2% fat, and 1.6% calcium while no *salmonella* organisms were found. This method seems to be cost efficient by using large-scale extruders driven with electrical motors.

The combination of hatchery waste with feathers or manure or other feestuffs needs further research.

Vandepopuliere, Vors and Jones (1977[b]) suggested that a drying manure unit as discussed in Chapter 2 will be very useful in processing hatchery waste.

Hatchery by-product meal is officially designated as "Poultry hatchery by-products" (Association of American Feed Control Officials, 1973), including a mixture of egg shell, infertile and unhatched eggs and cull chicks which have been cooked, dried, and ground with or without removal of part of the fat. It is considered as an approved feed ingredient and no special restrictions are placed on use of poultry by-products.

Deshmukh and Patterson (1997[a]) preserved hatchery waste by lactic acid fermentation and evaluated the nutritive value of this waste. Their waste product was composed of dead cockerel chicks, shell waste, infertile eggs, and dead embryos. The waste product was separately ground in a meat grinder with a ratio of 60 : 40 (cockerel chicks: shell waste) and then mixed with a by-product carbohydrate at 15 and 16.7% wt:wt, respectively and a liquid bacterial culture (0.2%). After sealing, the final products were fermented in special barrels for a period of 21 days at 20 °C. Finally the product was extruded, dried and hammer milled.

4.4.2 Chemical Analysis and Nutritive Value

Hatchery by-product meal or poultry hatchery waste (PHW) is rich in crude protein (22.2-26%). The crude fat ranged from 11.4-18% and metabolizable energy (MJ/kg) is 7.09-11.3. In addition, Ca levels ranged from 17.2-24.6% and P levels from 0.3-0.6%. Levels of the essential most limiting amino acids were as follows: methionine 0.5-3% and lysine 1.1-7.5%. Table 4.24 shows the chemical analysis of the PHW adapted from several research workers. This by-product was shown to be an excellent source of macro minerals such as Ca and P, of reasonable protein and energy levels, and of reasonable methionine and lysine in comparison with several non-traditional and traditional feedstuffs as shown in Table 4.25.

4.4.3 Its Use in Feeding

Few research workers evaluated the effect of processed hatchery by-product meal in the diets of layers and broilers. The most important objective was replacing the calcium from the limestone with the calcium of the test material.

(a) Layers

Wisman and Beane (1965) concluded that 15% hatchery by-product meal may be incorporated in laying hen rations without affecting performance. At the 15% level, the by-product supplied approximately 4% of the protein and 3% of the calcium in the diet. They concluded that there was no evidence that any of the animal by-products supplied unidentified growth factors.

Table 4.24. Chemical analysis of the poultry hatchery waste PHW. From (1) Ilian and Salman (1986); (2) Vandepopuliere *et al.* (1977a) and (3) Wisman (1964)

%	*(1)*	*(2)[a]*	*(2)[b]*	*(3)*
Moisture	3.90	65.0[c]	71.00[c]	-
Crude protein	22.80	22.2	32.30	26.00
True protein	21.50	-	-	-
Crude fat	14.40	9.9	18.00	11.40
Metabolizable energy (MJ/kg)	11.32	-	-	7.09
Ash	60.40	-	-	33.70
Calcium	22.60	24.60	17.2	20.60
Phosphorus	0.40	0.3	0.60	0.50
Amino acids %				
Aspartic acid	1.70	1.9	2.80	-
Threonine	0.80	0.9	1.30	0.90
Serine	0.90	1.00	1.7	-
Glutamic acid	2.40	3.3	5.20	-
Proline	1.10	1.90	2.8	-
Glycine	1.00	1.5	2.50	1.40
Alanine	1.30	1.0	2.10	-
Cystine	0.80	0.60	0.7	0.30
Valine	1.70	1.40	2.1	1.30
Methionine	0.70	0.6	0.80	0.50
Isoleucine	1.00	1.2	1.70	0.90
Leucine	1.70	2.0	3.00	1.60
Tyrosine	-	0.4	1.00	0.60
Phenylalanine	1.80	0.9	1.50	0.90
Lysine	1.70	1.2	1.8	1.10
Histidine	0.70	0.4	0.60	0.50
Arginine	1.60	1.6	2.40	1.20

a Broiler hatchery by-product

b Egg-type chick hatchery by-product

c Original moisture

Table 4.25. Chemical analysis of poultry hatchery waste in comparison with relevant feedstuffs. From [(1)] Ilian and Salman (1986); [(2)] Bhargava and O'Neil (1975); [(3)] Allen (1981) and [(4)] Scott, Nesheim and Young (1982)

%	*(1)* *Poultry hatchery waste*	*(2)* *Poultry by-product and feather meal*	*(3)* *Feather meal*	*(4)* *Meat and bone meal*	*(4)* *Herring meal*
Crude protein	22.80	70.30	85.00	50.00	72.00
Crude fat	14.40	12.30	2.50	10.00	10.00
Metabolizable energy (MJ/kg)	11.32	11.33	12.57	8.88	13.35
Calcium	22.60	2.30	0.20	10.60	2.00
Phosphorus	0.40	1.40	0.30	5.10	1.50
Amino acids %					
Arginine	1.00	3.90	3.90	3.50	6.80
Cystine	0.80	2.20	3.00	0.60	1.20
Glycine	1.00	5.90	4.80	7.50	5.90
Histidine	0.70	8.70	0.30	0.90	1.60
Isoleucine	1.00	2.90	2.70	1.70	3.70
Leucine	1.70	5.20	7.80	3.20	5.10
Lysine	1.70	2.40	1.10	2.50	6.40
Methionine	0.70	1.20	0.60	0.70	2.00
Phenylalanine	1.80	3.20	2.70	1.80	2.80
Thyreonine	0.80	2.90	2.80	1.70	2.80
Thryptophan	-	-	0.40	0.30	0.90
Tyrosine	-	2.20	-	0.80	2.10
Valine	1.70	3.80	4.60	2.50	3.50

Vandepopuliere *et al.* (1977[a]) reported from their trials using processed waste from broiler and from egg-type chick hatcheries as a feedstuff for laying hens. The inclusion of this waste at levels of 8 and 16% replaced soybean meal, meat and bone meal, wheat middlings and ground limestone. Egg production, feed conversion, egg shell quality (specific gravity and breaking strength) and interior quality (Haugh units) at both levels of each hatchery by-product meal were comparable to or better than with control diet (Table 4.26).

Table 4.26. Performance of laying hens fed hatchery by-product meal. From Vandepopuliere *et al.* (1977^{a})

						Egg					
Diet	Egg production (%)	Feed consumed (g/day)	Feed conversion (g/g egg)	Egg weight (g)	Body weight (g)	Haugh units	Specific gravity	Breaking strength (kg)	Shell thickness (mm)	Outer membrane (mm)	Inner membrane (mm)
1 Control	70.5	106.2^{a}	2.56	60.2	1753^{a}	70.6^{b,c}	1.084^{c}	3.14^{c}	0.346	0.069	0.015
2 HMB (8)	67.6	104.5^{a,b}	2.67	61.2	1713^{b}	70.5^{c}	1.086^{a}	3.28^{a}	0.351	0.069	0.015
3 HMB (16)	69.7	106.0^{a}	2.61	60.0	1713^{b}	69.4^{c}	1.086^{a}	3.24^{a,b}	0.351	0.071	0.015
4 HMC (8)	69.4	103.5^{a,b}	2.55	59.6	1715^{b}	72.7^{a}	1.085^{b}	3.20^{a,b,c}	0.345	0.070	0.015
5 HMC (16)	69.7	102.8^{b}	2.49	60.5	1769^{a}	72.2^{a,b}	1.084^{c}	3.16^{b,c}	0.344	0.067	0.014

$^{a, b, c}$ Column means with different letter superscripts are significantly different at the ($P < 0.05$)

HMB = Hatchery by-product meal from broiler hatcheries

HMC = Hatchery by-product meal from egg type

(b) Broilers

Wisman (1964) found that when 4.6% processed hatchery by-product meal was used to provide 1.2% protein, growth was superior to that from the basal diet, indicating the presence of unidentified growth factors activity. Their levels of inclusions were precluded by the high calcium content.

He also found that combinations of hatchery by-product meal, poultry by-product meal and poultry blood meal or hydrolized feather meal, and similar combinations where meat and bone meal replaced poultry by-products meal, proved to be satisfactory supplements when used to replace up to 50% of soybean meal protein.

However, Ilian and Salman (1986) concluded that the optimal inclusion level of processed hatchery waste in broiler rations is around 2.5%. In their broiler trials, body weight gains, feed consumption and feed efficiency were comparable for birds on all diets, namely control diet, and the experimental diets with 2.5 and 5% processed hatchery waste.

On the other hand Deshmukh and Patterson (1997) reported that nutrient dense hatchery by-products can be preserved with fermentation up to 21 days and supported broiler live performance and carcass yield as dietary ingredients equal to or better than a corn-soybean meal control diet.

4.5 SHELL WASTE FROM EGG-BREAKING PLANTS

The egg shell waste cannot be practically stored without developing bad odours or attracting insects and rodents. It is presently collected and transported to public or private dumps or is spread on agricultural lands.

In a survey Walton, Cotterill and Vandepopuliere (1973) observed that, for approximately 150 egg-breaking plants in the U.S.A., the total weight of egg shell waste would be approximately 46 000 t annually. This figure is based on an estimation for the USA and was made in 1973. However, egg-breaking plants have since increased enormously in number. This field has been described by some research workers (Walton and Cotterill, 1972; Arvat and Hinners, 1973; Vandepopuliere, 1973; Walton, Cotterill and Vandepopuliere, 1973; Vandepopuliere *et al.*, 1974; Vandepopuliere, 1983).

4.5.1 Processing Methods

Walton, Cotterill and Vandepopuliere (1973) classified the egg shell waste from the egg-breaking plants (frozen whole eggs, or albumin or yolks; fresh yolks, for processing mayonnaise and baking processing *etc.*) to be with adhering albumin or centrifuged. The shells were dried in a forced-draft oven at a temperature of 80 °C for approximately 48 hours and then ground. From their results it was evident that no microbial evaluation took place. However, 80 °C is not a very suitable temperature to assure a complete sterilization of the product.

Vandepopuliere, Walton and Cotterill (1974) conducted trials at the University of Missouri U.S.A. on egg shell waste which included egg shells, shell membranes and adhering albumin. The treatment included heating and dehydrating to convert the shell waste to a potential feedstuff. Two processing methods were used:

- A triple pass rotary dehydrator, producing egg shell meal A (EM-A).
- A single pass tube dehydrator, producing egg shell meal B (EM-B).

They reported that, EM-A included at 7.1% in the diet for laying hens showed a higher egg production ($P < 0.05$) but a lower egg weight ($P < 0.05$) in comparison with EM-B. It was concluded that the drying system of egg shell meal affects its nutritional value.

Vandepopuliere *et al.* (1977[b]) reported that processing and dehydration of egg-breaking plant waste can be accomplished by heating and removing moisture. There are several types of cookers and dehydrators available on the market. Some employ a batch cooker while others have a continuous flow. These cookers are heated with steam, gas or oil, these cookers were described in detail in the beginning of this chapter.

4.5.2 Chemical Analysis and Nutritive Value

Dried sterilized egg shell waste from egg-breaking plants in its complete form (including egg shells, shell membranes and adhering albumin) is an excellent mineral feedstuff. It is not only rich in calcium but also contains magnesium (0.40%), potassium (0.10%), iron (0.002%), sulphur (0.09%) and phosphorus (0.12%) (Walton, Cotterill and Vandepopuliere, 1973). Shell waste containing 38% calcium may be relatively compared with bone meal (24% Ca), limestone (35% Ca) and oyster shell (38% Ca) (Scott *et al.*, 1982). Shell waste contains 5.2-8.1% crude protein with 0.16-0.29% methionine and 0.20-0.37% lysine (Walton and Cotteril, 1972; Walton, Cotterill and Vandepopuliere, 1973). These essential amino acids are not found in the other mineral additives mentioned before. Table 4.27 shows the complete chemical analysis of egg shell waste from egg-breaking plants.

4.5.3 Its Use in Feeding

Evaluation of egg shell waste as a source of calcium in comparison with ground limestone and oyster shells in the diet of laying hens was reported by Arvat and Hinners (1973). They tested three calcium sources (ground limestone, oyster shells, and dried egg shells) at two calcium levels (3.7 and 5.7%) in a factorial design. They found that calcium sources failed to show significant differences in shell thickness and internal quality parameters such as, candle grade, albumin height, Haugh units, blood and meat spots or chalazae scors. Diet with a high level of calcium (5.7%) showed a significant greater shell thickness compared to diets with 3.7%. However, Vandepopuliere (1973) showed that substitution of egg shell meal for ground limestone supported comparable egg production, egg weight, feed efficiency and breaking strength. Similar conclusions were reported by Vandepopuliere *et al.* (1974).

Table 4.27. Chemical analysis of egg shell waste from egg breaking plants (dry basis). From [1] Walton *et al.* (1973) and [2] Walton and Cotterill (1972)

	(1)			*(2)*
%	*With adhering albumen*	*Centrifuged*	*Washed*	*Complete dried*
Original moisture	29.100	16.200	-	35.000
Crude protein	7.600	5.300	5.200	8.100
Ash	91.100	94.200	95.400	-
Calcium	36.400	36.700	37.300	35.200
Phosphorus	0.120	0.100	0.120	0.120
Magnesium	0.400	0.400	0.410	0.370
Potassium	0.100	0.070	0.060	0.130
Sodium	0.150	0.130	0.120	0.170
Iron	0.002	0.002	0.002	0.002
Sulphur	0.090	0.090	0.040	0.190
Amino acids				
Aspartic acid	0.870	0.520	0.450	0.830
Threonine	0.470	0.300	0.290	0.450
Serine	0.650	0.380	0.340	0.640
Glutamic acid	1.260	0.760	0.670	1.220
Proline	0.620	0.450	0.450	0.540
Glycine	0.510	0.380	0.350	0.480
Alanine	0.450	0.260	0.200	0.450
Cystine	0.410	0.200	0.350	0.370
Valine	0.540	0.320	0.290	0.550
Methionine	0.280	0.190	0.160	0.290
Isoleucine	0.340	0.190	0.150	0.340
Leucine	0.570	0.320	0.250	0.570
Tyrosine	0.250	0.150	0.120	0.260
Phenylalanine	0.380	0.180	0.100	0.460
Histidine	0.300	0.240	0.200	0.250
Lysine	0.370	0.200	0.200	0.370
Arginine	0.570	0.380	0.370	0.560

4.6 SPENT LAYING HENS

During rearing laying hens mostly Leghorn, a lot of consideration is taking place such as the length of laying cycle, moulting programs, and down time between flocks, a hudge number of hens are available for processing yearly. The majority of the hens have been utilized in hot pies and soups. The animal by-product processing industry has been faced with processing hens, that have been selected for smaller body size. The industry also has a problem with brittle bones during processing. Due to the above mentioned problems, there has been a shift away from the use of spent hens and the definite choice is to use broiler and roaster type birds to provide meat and soup stock. This situation created a vacuum for the market of spent hens and draw the attention to find an alternative for the left spent hens.

The proper processing of the spent hens to produce a new feedstuff could provide a positive economic profit and may eliminate a potential disposal problem.

Several investigators draw the attention for the processing of spent hens (Hague, Lyons and Vandepopuliere, 1991; Lyons and Vandepopuliere, 1996; 1997). They came to the conclusion that processed spent hen meal when fed at levels of up to 6% in broiler starter diets supported satisfactory performance.

4.6.1 Processing Methods

The most common processing technique for the spent hen is based on rendering in a cooker or by extrusion with a carrier or a diluter.

Patterson *et al.* (1994) used one part of broiler offal and condemned birds (75 and 25% by weight, respectively) from a local processing plant. The combined products were ground through a 6 mm screen, mixed, and combined with three parts of dry feedstuffs (cassava meal, ground barley, or wheat middlings) to reach a mixture with 75% dry matter. Then the products were extruded in a single barrel extruder with a temperature ranging between 135 and 163 °C. The extruded products were cut by a special cutter attachment to a length of 1.0 cm. The final products were cooled and dried at an ambient temperature (24 °C) then ground with a hammer mill to a dry matter of 89.0%.

Lyons and Vandepopuliere (1997) produced several products from the spent Leghorn hens. Hens were removed from their cages and sacrificed using carbon dioxide (CO_2). The birds pass through a particle-reducing device which yielded small chunks of whole hen. These chunks were treated in a pressurized rendering cooker (138 kPa at 126C for 90 min). The material passed through a screw press to remove a portion of the oil. The pressed material was then ground in a hammer mill, cooled, screened, and stored in a bulk bin for use in the experimental diet. Finally the tested ingredients were produced by the following processes:

- The rendering process as mentioned before.
- The sacrificed spent hens were mechanically deboned to produce meat (MDM), by a deboner. The raw meat fraction produced was 70% with 36% dry matter. The product was stored frozen until it was autoclaved, dried, ground, and stored in a freezer.
- The spent hen was introduced into a mechanical deboner producing raw bones and feathers, yield was 30% with 50% dry matter, then the product was stored frozen until it was autoclaved, dried at 138 kPa at 126C for 90 min, ground, hammer milled, and stored in a freezer.

4.6.2 Chemical Analysis and Nutritive Value

Processed spent hen meal is an excellent animal protein feedstuff and depends on the way of processing. Its crude protein varies from 46 to 62%, metabolizable energy varies from 12 to 20 (MJ/Kg) and is rich in Ca, P and the essential most limiting amino acids methionine and lysine, it may be slightly compared to Manhaden fish meal. Table 4.28 shows the complete chemical analysis of spent hens ingredients (Lyons and Vandepopuliere, 1997) in comparison with corn, soybean meal and Menhaden fish meal (Scott, Nesheim and Young, 1982).

Table 4.28. Chemical analysis (air-dry basis) of spent hen ingredients in comparison with corn, soybean meal and Menhaden fish meal. From [(1)] Scott, Nesheim and Young (1982) and [(2)] Lyons and Vandepopuliere (1997)

	Traditional feedstuffs (1)			*Spent hens ingredients (2)*		
Ingredients (%)	*Maize*	*SBM*	*MFM*	*A*	*B*	*C*
Dry matter	89.00[a]	90.00[a]	90.00[a]	95.70	93.00	93.00
Crude protein	8.60	48.00	61.00	62.00	45.98	55.40
Ether extract	3.90	0.90	9.00	13.20	43.28	16.95
ME (MJ/kg)	14.10	9.37	12.89	12.13	20.24	12.79
Ash	-	-	17.17[a]	15.69	3.51	20.19
Ca	0.02	0.32	5.50	5.15	0.38	7.92
P total	0.30	0.67	2.80	2.47	0.59	3.19
P available	0.10	0.29	2.80	2.47	0.59	3.19
Pepsin digestibility	-	-	95.86[a]	91.34	96.30	94.74
Amino acids %						
Arginine	0.50	3.20	3.80	3.75	2.76	3.46
Cystine	0.18	0.67	0.94	0.86	0.61	1.41
Glycine	0.40	2.10	4.40	4.78	2.34	6.29
Histidine	0.20	1.10	1.40	1.40	1.46	0.66
Isoleucine	0.40	2.50	3.60	2.40	2.10	1.74
Leucine	1.10	3.40	5.00	4.23	3.60	3.27
Lysine	0.20	2.90	5.00	3.57	3.57	1.92
Methionine	0.18	0.65	1.80	1.11	1.12	0.63
Phenylalanine	0.50	2.30	2.70	2.29	1.89	1.90
Tryptophan	0.10	0.60	0.80	0.48	0.31	0.24
Tyrosine	0.41	0.70	2.00	0.28	1.46	1.08
Valine	0.40	2.30	3.40	2.93	2.51	2.58
Threonine	0.40	1.80	2.60	2.23	1.92	1.83

SBM = Soybean meal; MFM = Menhaden fish meal
A treatment; B Mechanical deboner; C Bone + feathers
[a] Determinations are from Lyons and Vandepopuliere (1997)

4.6.3 Its Use in Feeding

Evaluation of spent hen meal compared by several feedstuffs was demonstrated by Lyons and Vandepopuliere (1997). They concluded that spent hen meal could be used in broiler starter diets up to 6% supporting optimum performance. They added also that separating the bones and feathers from the meat by using a mechanical deboner produces two fractions. Autoclaved samples of these fractions, when fed to day-old broilers, supported good production performance. They concluded that the dried spent hens meal can be used to replace Menhaden fish meal in broiler starter diets up to a level of 3%.

4.7 BIOHAZARDS OF POULTRY BY-PRODUCTS

Poultry by-products such as feather meal, poultry by-product meal, hatchery by-products and shell waste used as animal protein feedstuffs are of a high nutritive value and safe for use in the formulation of poultry diets.

The processing of all those feedstuffs takes place by using high temperatures and pressure through batch cookers, by continuous cooking procedures or by extrusion at high temperature through friction. This processing results in a sterilized, completely germ-free product.

The identification of the neurological disease of domestic cattle, bovin spongiform encephalopathy (BSE) (Wilesmith, Ryan and Atkinson, 1991) cannot be attributed to the use of poultry by-products in animal feeds. This disease is mainly derived from feedstuffs containing ruminant-derived protein in the form of meat and bone meal (Wilesmith *et al.*, 1988). Therefore, there is no chance of contamination during the processing of poultry by-products. No indications therefore have been found in literature.

Concerning the processed spent hens, Lyons and Vandepopuliere (1996) concluded that the extruded product produced *Salmonella*-negative ingredients with a marked reduction in the number of aerobic plate colonies, coliforms, yeast, and moulds.

REFERENCES

Acker, R.F., Hartman, P.A., Pemberton, J.R. and Quinn, L.Y. (1959). The nutritional potential of poultry offal. *Poultry Sci.*, **38**, 706-11.

Aderibigbe, A.0. and Church, D.C. (1983). Feather and hair meal for ruminants. 1-Effect of degree of processing on utilization of feather meal. *J. Anim. Sci.*, **56**, 1198-1207.

Allen, R.D. (1981). Feedstuffs ingredient analysis table. *Feedstuffs*, **35**, (30), 25.

Almex (1995). *Extrusion Techniques Information Bulletin.* Almex, 7207 BJ Zutphen, The Netherlands.

Arvat, V., and Hinners, S.W. (1973). Evaluation of egg shells as a low cost calcium source for laying hens. *Poultry Sci.*, **52**, 1996.

Association of American Feed Control Officials, Inc., (1960). *Official publication*, Lexington, KY, p. 189.

Association of American Feed Control Officials, Inc., (1973). *Official publication*, Lexington, KY, p. 71.

Atkinson, R.L., Ferguson, T.M. and Couch, J.R. (1955). Further studies on unidentified growth factor sources for Broad Breasted Bronze turkey poults. *Poultry Sci.*, **34**, 855-61. Baker, D.H., Blitenthal, R.C., Boebel, K.P., Czarnecki, G.L., Southern, L.L. and Willis, G.M. (1981). Protein-amino acid evaluation of steam processed feather meal. *Poultry Sc.*, **60**, 1865-72.

Balloun, S.L., and Khajareren (1974). The effects of whey and yeast on digestibility of nutrients in feather meal. *Poultry Sci.*, **53**, 1084-95.

Balloun, S.L., Miller,D.L. and Speers, G.M. (1968). Biotin, yeasts and distillers' solubles in poultry rations. *Poultry Sci.*, **47**, 1653.

Bhargava K.K. and O'Neil, J.B. (1975). Composition and utilization of poultry by-product and hydrolized feather meal in broiler diets. *Poultry sci.*, **54**, 1511-18.

Biely, J., Soong, R., Seier, L. and Pope, W.H. (1972). Dehydrated poultry waste in poultry rations. *Poultry Sci.*, **51**, 1502-11.

Binkley, C.H. and Vasak, O.R. (1951). Friable meal produced from feathers. *American egg poultry review*, **12**, 68.

Bjarnason, J., and Carpenter, K.J. (1970). Mechanism of heat damage in proteins. 2. Chemical changes in pure proteins. *Brit. J. Nutr.*, **24**, 313-29.

Bragg, D.B., Ivy, C.A. and Stephenson, E.L. (1969). Methods of determining amino acid availability of foods. *Poultry Sci.*, **48**, 2135-7.

Burgos, A., Floyd, J.I. and Stephenson, E.L. (1974). The amino acid content and availability of different samples of poultry by-product meal, and feather meal. *Poultry Sci.*, **53**, 198-203.

Buttery, P.J., Boorman, N.K. (1976). The energetic efficiency of amino acid metabolism, In: *Protein metabolism and nutrition*. (Ed. D.J.A. Cole,), Butterworth, London, UK, pp. 197-204.

Cabel, M.C., Goodwing, T.L. and Waldroup, P.W. (1987). Reduction in abdominal fat content of broiler chickens by the addition of feather meal to finisher diets. *Poultry Sci.*, **66**, 1644-51.

Cabel, M.C., Goodwing, T.L. and Waldroup, P.W. (1988). Feather meal as a nonspecific nitrogen source for abdominal fat reduction in broilers during the finishing period. *Poultry Sci.*, **67**, 300-6.

Daghir, N.J. (1975). Studies on poultry by-product meals in broiler and layer rations. *World Poultry Sci. J.*, **31**, (3), 200-11.

Dalev, P. (1990). An enzyme-alkaline hydrolysis of feather keratin for obtaining a protein concentrate for fodder. *Biotechnology letters*, **12**, (1), 71-2.

Davis, J. (1989). Low cost extrusion solves wet waste problems. *Poultry-Misset*, **5**, (5), 27-9.

Davis, J.G., Mecchi, E.P. and Lineweaver, H. (1961). Processing of poultry by-products and their utilization in feeds. 1. Processing of poultry by-products. *USDA, ARS. Utilization Research Report*, No. **3**, 1-21.

Day, W.C., Toncic,P., Stratman, S.L., Leeman, U. and Harmon, S.R. (1968). Isolation and properties of an extracellular protese of Trichophyton Granulosum. *Biochemica et Biophysica Acta*, **167**, 597.

Deshmukh, A.C. and Patterson, P.A. (1997). Preservation of hatchery waste by lactic acid fermentation. 2. Large-scale fermentation and feeding trial to evaluate feeding value. *Poultry Sci.*, **76**, 1220-26.

Draper, C.I. (1944). The nutritive value of corn oil meal and feather proteins. *Iowa Agr. Exp. Sta. Res. Bul.*, **326**, 163.

Eggum, B.O. (1970). Evaluation of protein quality of feather meal under different treatments. *Acta Agricultura Scandinavica*, **20**, 230-4.

El Boushy, A.R. (1986). Local processing industries offer food and beneficial by-products to developing countries. *Feedstuffs*, **58**, (3), 36-37.

El Boushy, A.R. and Roodbeen, A.E. (1980). Amino acid availability in Lavera yeast compared with soybean and herring meal. *Poultry Sci.*, **59**, 115-18.

El Boushy, A.R. and Roodbeen, A.E. (1984). Amino acid availability in dry poultry waste in comparison with relevant feedstuffs. *Poultry Sci.*, **63**, 583-5.

El Boushy, A.R. and Vink, F.W.A. (1977). The value of dried poultry waste as a feedstuff in broiler diets. *Feedstuffs*, **45**, (51), 24-6.

El Boushy, A.R., Poel, A.F.B. van der, Boer, H. and Gerrits, W.J.J. (1990). Effect of processing conditions on quality characteristics of feather meal. *Internal Report, 1990/7 9.32. Department of Animal Nutrition*, Agricultural University, Wageningen, The Netherlands.

El Boushy, A.R., Poel, A.F.B. van der and Walraven, O.E.D. (1990). Feather meal - a biological waste: Its processing and utilization as a feedstuff for poultry. *Biological Wastes*, **32**, 39-74.

Elmayergi, H.H. and Smith, R.E. (1971). Influence of growth of *Streptomyces fradia* on pepsin-HCl digestibility and methionine content of feather meal. *Canadian Journal of Microbiology*, **17**, 1067-72.

Ewing, W.R. (1963). *Poultry nutrition.* The Ray Ewing Company, Publisher 2690 E. Foothill Blvd., Pasadena, California.

Fielmich, E. (1987). Rendering: The neglected industry. *Poultry International*, **26**, (13), 44-52. Flegal, C.J. and Zindel, H.C. (1970). The utilization of poultry waste as a feedstuff for growing chicks. *Research Report*, **117**, *Michigan State University*, East Lansing, pp. 21-8.

Gerry, R.W. (1956). The use of poultry by-products in poultry rations. *Poultry Sci.*, **35**, 1144.

Gregory, B.R., Wilder, O.H.M. and Ostby, P.C. (1956). Studies on the amino acid and vitamin composition of feather meal. *Poultry Sci.*, **35**, 234-5.

Griffiths, L., Leeson, S. and Summers, J.D. (1977). Fat deposition in broilers: Effect of dietary to protein balance and early life caloric restriction on productive performance and abdominal fat pad size. *Poultry Sci.*, **56**, 638-46.

Hague, A.K.M.A., Lyons, J.J. and Vandepopuliere, J.M. (1991). Extrusion processing of broiler starter diets containing ground whole hens, poultry by-product meal, feather meal, or ground feathers. *Poultry Sci.*, **70**, 234-40.

Hamm, D. (1974). Storage and utilization of waste egg meats from grading plants. *Poultry Sci.*, **53**, 1548-54.

Hamm, D., Childs, R.E. and Mercuri, A.J. (1973). Management and utilization of poultry processing wastes. In: Symposium: *Processing Agricultural and Municipal Wastes.* (Ed. G.E. Inglett). The AVI Publishing Company, Inc.- Westport, Connecticut, pp. 101-18.

Hamm, D. and Whitehead, W.K. (1982). Holding techniques for hatchery wastes. *Poultry Sci.*, **61**, 1025-8.

Hoshii, H. and Yoshida, M. (1981). Variation of chemical composition and nutritive value of dried samples of garbage. *Japan. Poultry Sci.*, **18**, 145-50.

Ilian, M.A. and Salman, A.J. (1986). Feeding processed hatchery wastes to poultry. *Agricultural Wastes*, **15**, 179-86.

Jackson, S., Summers, J.D. and Leeson, S. (1982). Effect of dietary protein and energy on broiler carcass composition and efficience of nutrient utilization. *Poultry Sci.*, **61**, 2224-31.

Kahle, H.S. and Gray, L.R. (1956). Utilization and disposal of poultry by-products and wastes. *USDA, AMS Miscellaneous Research Reports*, p. 143.

Kotula, A.W. and Helbacka, N.V. (1966). Blood retained by chicken carcasses and cut-up parts as influenced by slaughter method. *Poultry Sci.*, **45**, 404-10.

Kubena, L.F., Chen, T.C., Deaton, J.W. and Reece, F.N. (1974). Factors influencing the quantity of abdominal fat in broilers. 3. Dietary energy levels. *Poultry Sci.*, **53**, 974-8.

Kuchaeva, A.G., Taplykova, S.D., Gesheva, R.L. and Kpasilnikov, V.A. (1963). Keratinase activity of actinomycetes of the "Fradia" group. *Doklady Abademii Nauk SSR*, **148**, 1400-2.

Latshaw, J.D. (1990). Quality of feather meal as affected by feather processing conditions. *Poultry Sci.*, **69**, 953-8.

Lipstein, B. (1985). The nutritional value of treated chicken waste in layer diets. *Nutr. Rep. Int.*, **32**, (3), 693-8.

Liu, J.K., Waibel, P.E. and Noll, S.L. (1989). Nutritional evaluation of blood meal and feather meal for turkey. *Poultry Sci.*, **68**, 1513-8.

Lortscher, L.L., Sachsel, G.F., Wilhelmy, O.J.R. and Filbert, R.B.J.R. (1957). Processing poultry by-products in poultry slaughtering plants. *U.S. Dept. Agri. Mktg. Res. Rept.*, no. 181.

Luong, V.B. and Payne, C.G. (1977). Hydrolized feather protein as a source of amino acids for laying hens. *Br. Poultry Sci.*, **18**, 523-7.

Lyons, J.J. and Vandepopuliere, J.M. (1997). Alternate procedures used to process spent leghorn hens. *J. Appl. Poultry Res.*, **6**, 74-80.

Lyons, J.J. and Vandepopuliere, J.M. (1996). Spent leghorn hens converted into a feedstuff. *J. Appl. Poultry Res.*, **5**, 18-25.

MacAlpine, R. and Payne, C.G. (1977). Hydrolized feather protein as a source of amino acids for broilers. *Br. Poultry Sci.*, **18**, 265-73.

McCasland, W.E. and Richardson, L.R. (1966). Methods for determining the nutritive value of feather meals. *Poultry Sci.*, **45**, 1231-6.

McNaughton, J.L., May, J.D. and Strickland, A.C. (1977). Composition of poultry offal meals from various processing plants. *Poultry Sci.*, **56**, 1659-61.

Miller, B.F. (1984). Extruding hatchery waste. *Poultry Sci.*, **63**, 1284-6.

Moran, E.T., Summers, J.D.,Jr. and Slinger, S.J. (1966). Keratin as a source of protein for the growing chicks. 1. Amino acid unbalance as the cause for inferior performance of feather meal and the implication of disulfide bonding in raw feathers as the reason for poor digestibility. *Poultry Sci.*, **45**, 1257-66.

Morris, W.C. and Balloun, S.L. (1973*a*). Effect of processing methods on util and nutritive value of herring meal. *Fiskeridreklarates Skrifter Ser. Teknd. Undersk.*, **5**, (10) 1-15.

Naber, E.C. (1961). Processing of poultry by-products and their utilization in feeds. Part II. Utilization of poultry by-products in feeds. *U.S.D.A. Utilization Research Report*, Washington, D.C., no. 3, 22-33.

Naber, E.C. and Morgan, C.L. (1956). Feather meal and poultry h ed., Natl. Acad. Sci., Washington DC.

Newel, G.W. and Shaffner, C.S. (1950). Blood loss by chickens during killing. *Poultry Sci.*, **29**, 271-5.

North, M.O. (1972). *Commercial chicken production manual*. The AVI Publishing Company Inc., Westport, Connecticut.

Noval, J.J. and Nickerson, W.J. (1959). Decomposition of native keratin by *Streptomyces fradiae*. *Journal of Bacteriology*, **77**, 251-63.

Papadopoulos, M.C. (1984). *Feather meal: Evaluation of the effect of processing conditions by chemical and chick assays*. Ph.D Thesis, Agricultural University, Wageningen, The Netherlands.

Papadopoulos, M.C., El Boushy, A.R., Roodbeen, A.E. and Ketelaars, E.H. (1986). Effect of processing time and moisture content on amino acid composition and nitrogen
characteristics of feather meal. *Anim. Feed Sci. Technol.*, **14**, 279-90.

Patterson, P.H., Ascar, N. and Coleman, W.C. (1994). Feeding value of poultry by-products extruded with cassava, barley and wheat middlings for broiler chicks; the effect of ensiling poultry by-products as a preservation method prior to extrusion. *Poultry Sci.*, **73**, 1107-15.

Payner, C.J. and Fox, M. (1976). Antino acid digestibility studies of autoclaved rapeseed meals using in vitro enzymatic procedure. *J. Sci. Food Agric.*, **27**, 643-8.

Pieniazek, D., Grabarck, Z. and Rakowska, M. (1975). Quantitative determination of the content of available methionine and cystine in food proteins. *Nutr. Metab.*, **18**, 16-22.

Portsmouth, J.I., Cherry, P. and Sharman, P. (1970). The use of feather meal in practical broiler diets. *14th*

World's Poultry Congr. Madrid, Spain, Section **3**, 631.

Potter, D.K. and Fuller, H.L. (1967). The nutritional value of poultry offal meal in chick diets. *Poultry Sci.*, **46**, 255-7.

Potter, L.M. and Shelton, J.R. (1978*a*). Evaluation of hydrolized feather meal and zinc bacitracin supplements of various purities in diets of young turkeys. *Poultry Sci.*, **57**, 947-53.

Potter, L.M. and Shelton, J.R. (1978*b*). Evaluation of corn fermentation soluables, menhaden fish meal, methionine and hydrolized feather meal in diets of young turkeys. *Poultry Sci.*, **57**, 1586-93.

Potter, L.M., Shelton, J.R. and Kelly, M. (1971). Effects of zinc bacitracin, dried bakery product, different fish meal in diets of young turkeys. *Poultry Sci.*, **50**, 1109-15.

Potter, L.M., Shelton, J.R. and Melton, L.G. (1974). Zinc bacitracin and added fat in diets of growing turkeys. *Poultry Sci.*, **53**, 2072-81.

Retrum, R. (1981). Apparatus for hydrolizing keratinaceous material. *United States Patent* Nr.4, 286, 884, Sep. 1, p. 14.

Routh, J.J. (1942). Nutritional studies on powdered chicken feathers. *J. Nutr.*, **24**, 399-404.

Said, N.W (1996). Extrusion of alternative ingredients; an environmental and a nutritional solution. *J. Appl. Poultry Res.*, **5**, 395-407.

Schor, R., and Krimm, S. (1961). ß-helix model for the structure of feather keratin. *Biophys. J.*, **1**, 489-515.

Schroeder, W.A. and Kay, L.M. (1955). The amino acid composition of keratin morphologically distinct parts of white turkey feathers and of goose feather barbs and goose down. *J. Amer. Chem. Soc.*, **77**, 3901.

Scott, M.L, Nesheim, M.C. and Young, R.J. (1982). *Nutrition of the chicken*. M.L. Scott and Associates, Ithaca, New York.

Soares, J.H., Jr. and Kifer, R.R. (1971). Evaluation of protein quality based on residual amino acids on the ileal content of chicks. *Poultry Sci.*, **50**, 41-6.

Soares, J.H., Jr., Miller, D., Fritz, N. and Sanders, M. (1971). Some factors affecting the biological availability of amino acids in fish protein. *Poultry Sci.*, **50**, 1134-43.

Soldevila, M. (1977). Garbage as feed for swine. *J. Agric. Univ. Puerto Rico*, **61**, (4), 513-15.

Stork rendering (1995). Stork Poultry Processing Information. *Technical bulletin, 9.1000-E. Stork rendering*, 3830 AC Boxmeer, The Netherlands.

Sullivan, T.W. and Stephenson, E.L. (1957). Effect of processing methods on the utilization of hydrolized poultry feathers by growing chicks. *Poultry Sci.*, **36**, 361-5.

Summers, J.D. and Leeson, S. (1979). Composition of poultry meat as affected by nutritional factors. *Poultry Sci.*, **58**, 536-42.

Supplee, W.O., Keene, O.D. Combs, G.F. and Romoster, G.L. (1956). Unknown factor studies with purified diets. *Poultry Sci.*, **35**, 1175.

Thomas, O.P., Bossard, E.H., Nicholson, J.L. and Twining, P.J. (1972). The use of feather meal in poultry diets. *Proc. Maryland Nutr. Conf.*, March 16-17, pp. 86.

Vandepopuliere, J.M. (1973). Value of egg shell meal as a poultry feedstuff. *Poultry Sci.*, **52**, 2096.

Vandepopuliere, J.M. (1983). Animals as converters of by-products from animal processing. In: *Animals as waste converters*. Proceedings of an international symposium. Wageningen, The Netherlands. (Ed. E.H. Ketelaars and S. Boer Iwema). Session **3**, pp. 92-8.

Vandepopuliere, J.M. (1988). High temperature extrusion used to process poultry by-products. *Proceedings XVIII World's Poultry Congress*, Vagoya, Japan, pp. 131-5.

Vandepopuliere, J.M., Kanungo, H.K., Walton, H.V. and Cotterill, O.J. (1977[a]). Broiler and egg type chick hatchery by-product meal as laying hen feedstuffs. *Poultry Sci.*, **56**, 1140-4.

Vandepopuliere, J.M., Vors, L.A. and Jones, H.B. (1977). The potential of hatchery waste as a feed ingredient. *Agric. Exp. Stat., Univ. Missouri-Columbia*, USA, SR 200, p. 12.

Vandepopuliere, J.M., Walton, H.V. and Cotterill, O.J. (1974). Poultry egg waste: Potential feedstuff. *XV World Poultry Congress*, SC5, pp. 227-8.

Vandepopuliere, J.M., Walton, H.V., Jaynes, W. and Cotterill, O.J. (1977[b]). Elimination of pollutants by utilization of egg breaking plant shell-waste. *University of Missouri, Columbia*, Missouri 65201 USA, Grant No. S 8033614, p. 36.

Van der Poel, A.F.B. and El Boushy, A.R. (1990). Processing methods for feather meal and aspects of quality. *Neth. J. Agric. Sci.*, **38**, 681-95.

Vogt, H. and Stute, K. (1975). Scheinbare Aminosäureverdaulichkeit des Federmehles bei Legehennen. *Arch. Geflügelk.*, **39**, 51-3.

Walton, H.V. and Cotterill, O.J. (1972). Composition of egg shell wastes from egg breaking plants. *Poultry Sci.*, **51**, 1884.

Walton, H.V., Cotterill, O.J. and Vandepopuliere, J.M. (1973). Composition of shell waste from egg breaking plants. *Poultry Sci.*, **52**, 1836-41.

Wang, X. and Parsons, C.M. (1998). Order of amino acid limitation in poultry by-product meal. *Br. Poultry Sci.*, **39**, 113-6.

Wessels, J.P.H. (1972). A study of the protein quality of different feather meals. *Poultry Sci.*, **51**, 537-54.

Wheeler, K.B. and Latshaw, J.D. (1980). Evaluation of the sulphur amino acid and lanthionine content of feather meal. *Poultry Sci.*, **59**, 1672.

Wilcke, H.L. (1940). Egg shells good poultry feed. *U.S. Egg Poultry Mag.*, **46**, 617-18.

Wilesmith, J.W., Ryan, J.B.M. and Atkinson, M.J. (1991). Bovine spongiform encephalopathy: epidemiological studies on the origin. *Veterinary Record*, **128**, 199-203.

Wilesmith, J.W., Wells, G.A.H., Cranwell, M.P. and Ryan, J.B.M. (1988). Bovine spongiform encephalopathy: Epidemiological studies. *Veterinary Record*, **123**, 638-44.

Williams, M.A., Horn, R.E. and Bronikowski, C. (1979). Process for hydrolizing proteinaceous derivatives of the skin. *United States Patent* No. 4.151.306: Apr. 24, p. 8.

Wisman, E.L. (1964). Processed hatchery by-product as an ingredient in poultry rations. *Poultry Sci.*, **43**, 871-6.

Wisman, E.L. and Beane, W.L. (1965). Utilization of hatchery by-product meal by the laying hen. *Poultry Sci.*, **44**, 1332-3.

Wisman, E.L., Holmes, C.L. and Engel, R.W. (1957). Composition of poultry offal. *Poultry Sci.*, **36**, 1099-100.

Wolski, I., Klinek, J. and Rompala, A. (1980). Characteristics of modified proteins of keratin feathers. *Acta Alimentaria Polonica*, **6**, 167-72.

Yeh, Y.Y. and Levielle, G.A. (1969). Effect of dietary protein on hepatic lipogenesis in the growing chick. *J. Nutr.*, **98**, 356-66.

Yoshida, M. and Hoshii, H. (1979). Nutritive value of garbage of supermarket for poultry feed. Japan. *Poultry Sci.*, **58**, 350-5.

Yu, R.J., Harmon, S.R. and Blank, F. (1968). Isolation and purification of an extra cellular keratinase of Trichophyton mentagrophytes. *J. Bacteriol.*, **96**, 1435-6.

Ziegler, K., Melchert, I. and Lürken, C. (1967). N (2-amino-2-carboxyethyl)-ornithine, a new amino acid from alkali heated proteins. *Nature*, London, **214**, 404-5.

5. HIDE AND TANNING WASTE BY-PRODUCTS

Contents

5.1 INTRODUCTION

During the production of leather several waste by-products are produced which may cause pollution unless they are used as a feedstuff. The yearly world production of fresh hides (cattle, buffalo, sheep and goat) is estimated to be 8-9 million t per year (FAO, 1990). During the processing of these hides, the waste production is estimated to be 1.4 million t per year.

Tannery has the unevitable reputation for being one of the filthiest and evil smelling of industries. Today, with the large increase in population and the strain being put on our world for saving our natural resources, it is becoming more apparent that the disposal of tannery wastes is a matter of the industry's responsibility to the society around it. The pollution control regulations which were relatively loose until a few decades ago are now extremely tight. Efforts are being made to limit and eliminate as much as possible the discharge of polluting materials into the atmosphere, ground and primarily, the discharge into water.

Tannery waste is formed during the processing of hides to leather in which a solid waste (from trimming, cutting, fleshing and shaving hair, and buffing material) is formed. This waste should be utilized if possible. A hide weighing 45.4 kg looses about 9.1-11.3 kg fat, flesh, hair and manure during fleshing. Usually the trade agrees upon a 16% loss owing to fleshing (Ockerman and Hansen, 1988).

During the modern processing of hides into leather, trivalent chromium salts are used as tanning agents. A part of these chromium salts is discharged in effluents and solid waste.

This waste is a polluting chemical waste owing to its high chromium content. This is approximately 3% on a dry matter basis and is known as wet-blue shavings and trimmings. For this reason, material contaminated with chromium is not allowed to be used in rendering plants. A small amount of this waste is used for artificial leather but most of it is without any value and is dumped at waste disposal sites. TNO Leather and Shoe Research Institute, The Netherlands developed a new process for extracting chromium from the wet-blue waste (Koene and Dieleman, 1987). The recovered chromium is recycled in the tanning process. The dechromed protein material can be used as a source material for glue or as an animal feedstuff with a promising nutritive value (El Boushy *et al.*, 1991).

5.2 TECHNOLOGICAL ASPECTS

Leather technology may be classified into the traditional process, which is used in many developed and developing countries, and the modern process, called chrome tanning, which is used in developed countries only. It is not the intention to explain the complete processing in detail but only to put emphasis on the waste by-products from the processing. Leather technology is described in detail in several articles and books (McLaughlin and Theis, 1945; Gustavson, 1956; Highberger, 1956; Marselli, Marselli and Burford, 1958; Koppenhoefer, 1965; Mann, 1967; ; Taylor *et al.*, 1982; Thorstensen, 1985 and Ockerman and Hansen, 1988). A few points are summarized from these sources below.

Hide composition

Before going into technical details of leather processing, one should have an idea about the composition of the hide. The thickness of the skin varies with species, age, sex and region of the body (thicker on the back and on the external surfaces of the limbs; thinner on the neutral surfaces (places of cutting) and on the flexor surfaces). The skin is composed of three major layers (Table 5.1): the surface pigmented epidermis, the underlying connective tissue corium and the deep subcutis. The thin epidermis covers the surface and extends downward as tubular invagination. It thus forms part of the surface of the hair follicles. The underlying corium is associated with the hair follicles. The upper portion of the corium contains sebaceous glands, the erectile follicular smooth muscles and elastic, reticulum and collagenous fibres. The deeper portion of the corium is interwoven bundles of collagen. In bovine animals the hair root extends about one-third the depth of the corium, but in swine the hair follicle penetrates the corium and extends down into the subcutis (Figure 5.1). The subcutis consists of a loose membrane network of collagen and elastin fibres. The subcutis portion contains fatty deposition and determines the tautness or slackness of the skin. Thorstensen (1985) reported that the hide is made up of a) fibrous proteins, keratin, collagen, and elastin, b) the soluble or soft proteins, albumins, globulins, mycoproteins, and soft keratins; and c) the fat components of the body. The most resistant materials to chemical and bacterial attack are the fibrous proteins, hair and elastin, followed by collagen. The least resistant components are the mycoproteins, the albumins and the globulins.

Table 5.1. Skin layers and two methods of splitting hides. From Moulton and Lewis (1940); Price and Schweigert (1971); Tanners' Council of America (1983)

Side	*Skin*	*Leather*	
		Five layers	*Four layers*
Hair-side	Epidermis, pigmented, thin	Buffing	
	Grain layer, papillary	Machine buff	Thop grain
	Corium, dermis, derma, cutis vera, connective tissue greatest part of hide	Deep buff Split	Split
Flesh-side	Subcutis, attachment, filled with fat	Slab	Slab

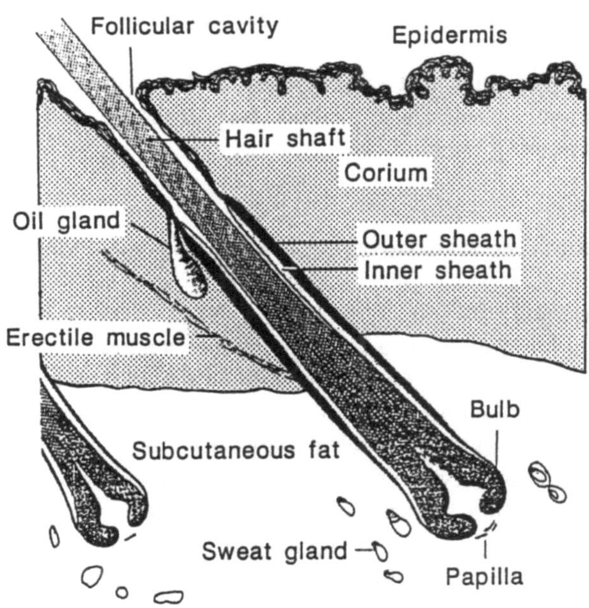

Figure 5.1. Vertical section of normal hogskin. Reproduced with permission from Price and Schweigert (1971).

5.2.1 The Traditional Hide and Skin Process

High quality leather is mostly obtained by the up-to-date techniques used for hide removal, "flaying", and the processing that takes place in the slaughter-house with its facilities.

The process can be performed as follows: hide removal, curing or preservation, fleshing and trimming. After this process, selection and grading, and subsequently storage and shipping, will take place.

(a) Hide removal

The removal of hide is mostly performed by the knife-skinning technique, using either the conventional skinning knives or the more recent developed air-driven reciprocating skinning knives, or the more modern technique for hide removal, the hide-pulling technique. Another technique used is to pump compressed air into the carcass to cause hide separation (cattle, sheep and goats). In some cases the animals' hair is removed before the hide removal. In case of cattle hair, the long high quality (winter) hair has value in mattresses, rugs *etc.* It may be also hydrolized and used as a feedstuff. In the case of hogs (pigs) the carcass is scalded to raise the hair and placed in a deharing machine (polisher).

(b) Curing and drying

After the removal of the hide from the animal, the hide should be quickly cured to avoid bacterial and enzymatic decomposition or spoilage. Drying may be used to preserve the hide or, more commonly, salt may be used as the curing ingredient. Four basic techniques are currently used. These methods are: air-drying, salt-pack curing, mixer curing and raceway curing.

Air-drying methods
These are classified as follows: ground drying, frame-drying, line-drying and parasol-drying.

Salt-pack curing method
This is based on holding the hide in a cellar, preferable at a temperature of 10-13 °C and relative humidity 85-90%. Good ventilation without draughts is of a great importance to avoid quick drying. Sodium Chloride is added in a ratio of one hide weight to one sodium chloride weight. This salt level inhibits bacterial growth and draws moisture out of the hides. Some preservatives are often used together with the salt-pack curing method such as 1% sodium fluoride (NaF, a poison; its dust should not be inhaled) or 1% naphthalene ($C_{10}H_8$) plus 1% boric acid (H_3BO_3) based on the weight of salt which was used in the beginning. The curing time needed is 20-30 days. Salt is usually not recycled, since it may become contaminated with salt-resistant bacteria. In the case of recycling, the salt should be sterilized by heating and separating the hide proteins by flocculation.

Mixer curing method
In this method a mixer similar to a cement mixer is used. The hides are washed at a temperature of 13-16 °C for 10-30 minutes with clean water. A saturated salt (brine) solution is added or a fresh salt up to 20-24% of the weight of the hides. Chlorinated lime or similar bacterial or mould deterrent is added. The hides are rotated in the mixer for

six to twelve hours in a sequence of gradually reduced rotations. The hides are then removed from the mixer and pass a squeezing machine to get rid of the excess liquid and then hanged, to drip-dry.

Raceway curing method
This is based on agitating the hides in a tank by means of two overhead paddle wheels. The hides are submerged in the brine which is circulated at 12-16 rotations per minute. A typical 250 l raceway tank can hold about 1200 fleshed cattle hides and is filled with a saturated salt brine. The brine solution is sampled frequently and maintained at 98% salinity by addition of salt if necessary. It is required to use approximately 1.8 kg of saturated brine for each 0.45 kg of green hide. Bactericides are added to the brine to prevent the growth of the proteolytic bacteria with an addition of 0.3% (of hide weight) sodium fluoride (NaF). The required time of curing is approximately 16 hours. Finally the hides are passed through the wringing machine to squeeze out the fluids and hung on hooks to drip-dry.

(c) Fleshing

In this process about 16% of the hide is removed in the form of fat, flesh, hair and manure. This process may take place by hand, using knives or by machines. The fleshing operation may be done prior to or after curing. The waste of the fleshing is added to the waste of the trimming and is processed as discussed below.

(d) Trimming

This process may take place before or after fleshing and curing. The purpose of trimming a hide is to remove parts of the hide that would have no value as leather and to make the hide conform to what is known as the standard hide trim, if the hide is unfleshed, or to conform to the modern hide trim if the hide is fleshed. The parts removed from the hide by cutting with knives include, ears, ear butts, snouts, lips, scrotal sacs, udders, tails, head skin, fat and muscle tissue from the side of the head, and ragged neutral edges (Figure 5.2).

(e) Yields of fleshing and trimming wastes

The method most commonly used is called the wet-rendering system according to the National Hide Association (1979). This method is based on an automated centrifugal system in which water and fat are removed from the tissue centrifugally. Subsequently, the fat will be separated from the water. The defatted and partially dewatered (50% moisture) tissue is then placed in a drier for further moisture removal to produce a dry-cake tankage-type feed supplement, which contains 40-60% crude protein depending on input material.

The yield varies depending on the composition of the raw material. On average, about 25% of oil and 10% other dry feed substance is obtained. The oil is somewhat different from other animal fats and is used in soaps or in industrial chemicals.

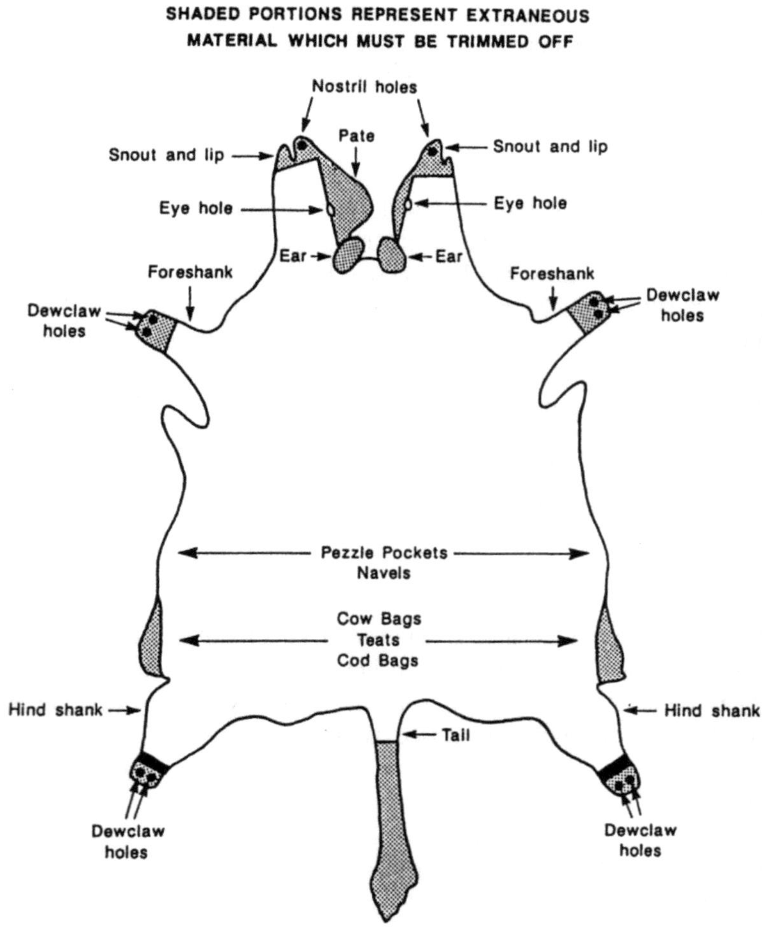

Figure 5.2. Modern trim pattern for cattle hide. Shaded areas represent extraneous material which must be trimmed off. Reproduced with permission from United States Hide, Skin and Leather Association (1985).

The steps that follow those mentioned above do not produce a significant amount of waste and are based on restoring the hide moisture by "soaking" and some "additional fleshing" and "unhairing". This waste can be added to the above-mentioned process. The "sweating" process takes place by treating the hide with temperature and humidity.

The pretanning processes are meant to prepare the hide for further processing such as "pickling" with a low pH. Tanning may be done by "chrome tanning", basically a treatment with chromic sulphate, $Cr (OH) SO_4$. "Vegetable tanning", "Zirconium tanning" ($Zr O_2$), "Alum tanning" [aluminium potassium sulphate, $ALK (SO_4)_2$], are alternative methods.

The next step after tanning is "dyeing" in which the hide is treated with chemical or vegetable dyes according to the desired colour. In these stages the dyed leather will be treated to reach the final stage, as far as the requirements are concerned, and it will pass through "drying" to make the product moisture free. "Toggling" is a process where the skins are stretched, and the "pasted" vacuum dryed and the "buffed" which involves the smoothing of the grain surface. Finally the "finishing" process is undergone by application of a film-forming material or a coating material, and then "platining" which smoothes the grain surface by means of high pressing and heat; the leather will be "gradded", "subdivided", rolled into bundles of five, rapped in paper, and placed in wooden boxes ready for sale.

5.2.2 Chrome Tanning

Modern tanning techniques lead to the use of unsalted hides which are chrome tanned [chromium sulphate, $Cr(OH)SO_4$] in the slaughter plant. A by-product is produced, known as the "wet blue" waste. This hide can be stored for months prior to continuation of leather processing. It saves the cost of salting and the fleshing material removed from the fresh hide can be rendered into a high-quality salt-free product. In general, the main advantages of this process are high speed, low cost, light colour, and excellent preservation of the hide protein. In addition, a waste with a high nutritive value is produced after dechroming.

(a) Chemical reaction of the chromium salts

Chromium tanning salts have a valency of +3. They are soluble in strong acids but will usually precipitate as chromium hydroxide, or hydrated chromium oxide, at pH above 4. They react with a number of organic materials to form coloured soluble salts at higher pH values, and will precipitate soluble proteins (McLaughlin and Theis, 1945; Thorstensen, 1985; Koene and Dieleman, 1987).

The process of chromium tanning is based on the cross-linking of chromium ions with free carboxyl groups in the collagen. The stages of the cross-linkage of the chrome tannage are as follows:

- The reaction of the chrome complexing with the protein carboxyl groups.
- As the pH of the tannage is increased, the sulphate associated with the chromium becomes displaced by the hydroxyl group. The hydroxyl groups become sharped by the chromium atoms through the complex formation (olation).

A chrome compound will be "olated" when one or more of its hydroxyl groups is held between two chromium atoms, whereby it is attached to the primary valence of one chromium atom and to the secondary valence of the second chromium atom, thus: Cr-OH-Cr.

- Upon drying, the tannage becomes more stable as the complex gives up hydrogen ions and will stabilize the complex (oxolation results).

The oxolation chrome compounds are termed "oxo". This refers to compounds in which two chrome atoms are bound by an oxygen atom: Cr-O-Cr.

In general, the main factors affecting chrome tannage are basicity, pH, temperature, time and concentration (McLaughlin and Theis, 1945; Thorstensen, 1985; Koene and Dieleman, 1987).

In Figure 5.3 a scheme of the process from raw hide to leather is given in which the by-products of the various steps are shown. Fleshings are taking place at the pretanning, shavings are produced when the thickness of the material is corrected by machine operation. Trimmings are the edges and pieces, they are too thin to be used (El Boushy *et al.*, 1991).

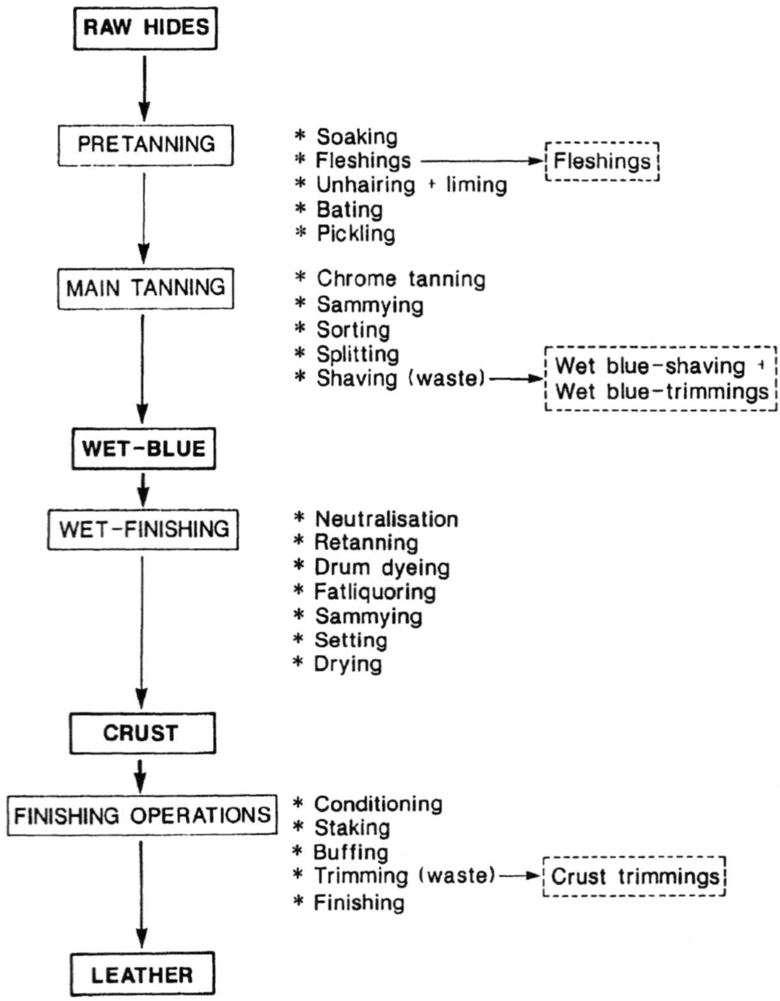

Figure 5.3. Schematic flow sheet of the process from raw hide to leather and its by-products. Reproduced with permission from El Boushy *et al.* (1991).

(b) Chromium extraction from the wet-blue waste

To use the wet-blue waste it is necessary to separate the chromium from the hide protein. The recovered chromium can eventually be recycled in the tanning process. The dechromed hide material by-product can be used as a glue material or a feedstuff (Figure 5.4) (Koene and Dieleman, 1987). This process is based on an alkaline-acid treatment where the pH is initially raised to >13.5, followed by an acid treatment where the pH is decreased to 1.0. Finally, the chromium content is decreased by means of four successive washing steps. The water used in this process contains a high percentage of NaCl. To reduce the total quantity of water and the amount of NaCl used, the process liquors are recycled. The duration of this extraction process is about 24 h. A schematic diagram of dechroming wet-blue waste is shown in Figure 5.5 (Koene and Dieleman, 1987; El Boushy *et al.*, 1991).

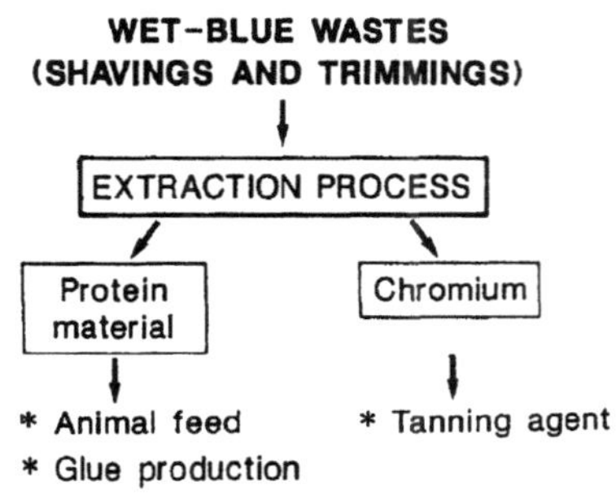

Figure 5.4. Chromium and protein separation. Reproduced with permission from Koene and Dieleman (1987).

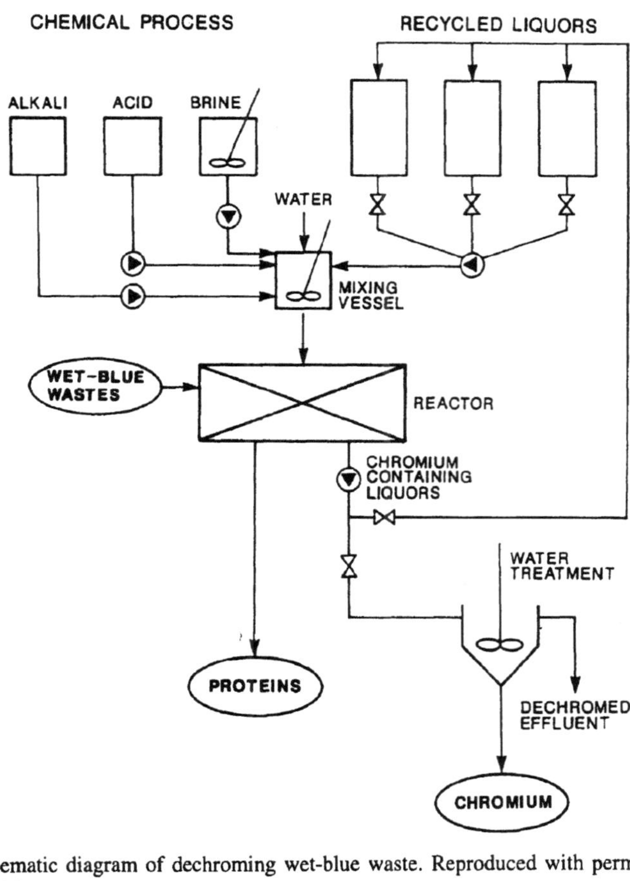

Figure 5.5. Schematic diagram of dechroming wet-blue waste. Reproduced with permission from Koene and Dieleman (1987).

The processing of tannery effluent is very important as far as the pollution control regulations are concerned. The leather industry produces tremendous waste during the total production process such as: soaking, unharing, sulphide, bat, pickle and tan operations, chrome tanning, colouring and fat liquoring and finishing. The effluent of this waste needs a special treatment to remove the various pollutants present.

The treatment of effluent from a tannery plant is generally designed to be operated in three distinct steps: screening, coagulation and settling (Peper, 1966; Thorstensen, 1985):

Screening. The removal of hair and small suspended hide particles can be accomplished by a mechanical screening system. These are usually rotating, self-cleaning screens which will screen large particles.

Coagulation. The materials of a tannery will interact mutually. For example the acid of the pickle will be neutralized by the alkalinity of the lime. The alkalinity of the lime will also bring about precipitation of the soluble chromium salts. Coagulation of tannery wastes, therefore is almost a natural process owing to the reactivity of the components from the various stages of the tannery.

Settling. By the use of lagoons, the coagulated and mutually interacted portions of the tannery effluent can settle and, subsequently, a sludge will be formed in the bottom of the lagoon. The water may be passed through several lagoons, and finally, it can be discharged directly when it has an acceptable degree of purity. Trickling filters or aeration systems may be employed to aid in the bacterial decomposition of the tannery effluent, in the clarification of the spent liquor, and to decrease biological oxygen demand (BOD).

The quantity of materials used in the tannery does not result in a completely balanced effluent (Pepper, 1966). This subject is not relevant for our point of view, because the sludge after treatment is not suitable as a feedstuff for poultry owing to their high polluting minerals.

5.3 CHEMICAL ANALYSIS AND NUTRITIVE VALUE

5.3.1 Hide and Skin

The chemical composition of the skin varies with the age of the animal, its sex and the health condition related to the fat level of the animal. In addition, it will vary according to the treatment of the hide after being removed from the carcass. In general, the hide is low in fat and minerals, and high in protein (collagen). This collagen increases dramatically and is a major component when the hide is converted into leather. Hair is composed almost entirely of the protein keratin, which normally accounts for 6-10% of the total hide protein (Table 5.2) (Moulton and Lewis, 1940; McLaughlin and Theis, 1945; Aten, Innis and Knew, 1955; Bidermann *et al.*, 1962; Henrickson, Turgut and Rao, 1984 and Ockerman and Hansen, 1988).

Table 5.2. Chemical composition of hides. From Moulton and Lewis (1940); Aten, Innis and Knew (1955); Bidermann *et al.* (1962); Henrickson, Turgut and Rao (1984)

Age of animal	*Moisture (%)*	*Protein[a] (%)*	*Fat (%)*	*Ash[b] (%)*
Average slaughter cattle	62-70	-	-	1.0
Mature cattle hide, without hair	65.0	30.0	-	-
Very fat animal	-	-	10-12	-
Wet cattle hide	83.0	15.7	0.2	0.1
Air-dried cattle hide	9.1	89.9	0.2	0.8
Newborn calf	67.9	30.8	1.0	1.0
Three-month-old calf	66.0	31.0	1.6	1.4
Two-year-old steer	61.2	35.0	3.2	1.1
Four-year-old steer	55.6	38.2	6.0	1.1
Old cow	60.2	36.0	3.1	1.1
Sheep skin	-	-	30-50	-
Goat skin	60.0	-	3-10	-
Pig skin	37.0	14.0	30-50	-
Cured cattle hide	44-48	41.0	-	14-16[c]

[a] Protein: collagen, keratin, elastin and reticulin
[b] Ash: phosphorus, potassium, sodium, arsenic, magnesium and calcium
[c] Including tanning metals

The hide and skin proteins may be classified as albumin, elastin, collagen and keratin according to McLaughlin and Theis (1945); Highberger (1956); Scott, Nesheim and Young (1982) and Thorstensen (1985). The amino acid content of the above mentioned proteins is shown in Table 5.3.

Globular proteins
These proteins may be classified as albumins, globulins, prolamines or gliadins, histones and protamines of which the albumins are of relevance in hide and skin proteins. Albumins are proteins which are soluble in water and coagulable by heat. They are characterized by having a high percentage of acid and basic amino acids.

Fibrous proteins
These proteins are classified as collagens, elastines and keratines.

Collagen is the major protein of skeletal connective tissues and represents more than one-half of the total protein in the animal body. Collagens are insoluble in water, but are readily changed into soluble, easily digestible gelatins by boiling in water or in dilute acids or bases. The collagens have a unique amino acid structure characterized by large amounts of hydroxyproline and some hydroxylysine, and a complete absence of cysteine, cystine and tryptophan. In general they contain both acid and basic amino acid groups, as well as non-polar amino acids.

Table 5.3. Amino acid percentage of hide proteins. From Thorstensen (1985)

Amino Acids/Proteins	*Collagen*	*Elastin*	*Keratin*	*Albumin*
Nonpolar				
Glycine	20	22.0	5	2
Alanine	8	15.0	3	6
Valine	3	12.0	5	6
Leucine	5	10.0	7	12
Other	4	15.0	7	9
Total	40	74.0	27	35
Acidic				
Aspartic acid	6	0.5	7	11
Glutamic acid	10	2.5	15	17
Total	16	3.0	22	28
Basic				
Arginine	8	1.0	10	6
Lysine	4	0.5	3	13
Other	2	0.5	1	4
Total	14	2.0	14	23
Others				
Serine	2	1.0	8	4
Cystine	-	-	14	-
Proline-Hydroxyproline	25	15.0	6	5
Other	3	5.0	9	5
Total	30	21.0	37	14

Elastins are the proteins of the elastic tissues such as tendons and arteries. Although similar in make-up to collagens, the elastins cannot be converted to gelatins. The fibrils of elastin are bound together by the compound desmosine, derived from the amino acid lysine. This protein has a very few acidic and basic amino acid groups.

Keratins are the proteins of hair, feathers, claws, beak, hoofs and horns. These proteins are very insoluble and indigestible. They contain as much as 14-15% cystine. Keratins have more acidic amino acids than collagen, but less than albumin. The stability of keratin is due to presence of cystine (disulphide bonds). Autoclaving of the keratinous proteins improves their digestibility and the suitability as a feedstuff as hydrolized feather meal or hog meal. This autoclaving reduces the cystine content to 5-6%. It is believed, that this may be responsible for the improvement in digestibility and that the S-S bonds within the keratins may contribute in large part to the insolubility and indigestibility of these proteins.

5.3.2 The Tannery Waste Protein

Solid tannery waste may be classified in the traditional tanning waste (hide meal) and the chrome tanning waste (wet-blue waste).

(a) Hide meal protein

Solid waste (from trimming, cutting, fleshing, shaving hair, and buffing material) and tannery offal is used as fertilizers, feed, glue, gelatin, hair, grease and leather board. The tannery solid waste contains two proteins, keratines from hair and collagenous hide fibres. These proteins are normally hydrolized and sterilized (Figure 4.14, Chapter 4) as feather meal to produce a feedstuff called hide meal. The percentage of inclusion of the hide meal in diets is normally limited due to the unbalanced amino acid composition and high ash content (Ockerman and Hansen, 1988).

Gawecki, Lipinska and Rulkowski (1981) analysed the hide meal from the tanning industry (Table 5.4). They found a crude protein of 58.5% and true protein 21.8%, fat 3.7% and ash 20.5%; minerals and trace elements, however, were not reported.

From Table 5.4 it may be concluded that the amino acids are a mixture of collagen which is high in lysine (2.30%) and proline (6.36%) and keratine (hair) with 0.62% cystine. The hide meal can serve as a source of animal protein for broilers and can replace fish meal completely in rations based on maize and soybean meal.

Table 5.4. Chemical analysis and amino acid composition of the leather meal and hide meal, on air-dry basis. From [(1)] Waldroup *et al.* (1970); [(2)] Gawecki, Lipinska and Rulkowski (1981)

Composition (%)	*(1)*[a]	*(2)*	*Composition (%)*	*(1)*	*(2)*
Dry matter	91.6	96.0			
Crude protein	65.4	58.5	Serine	1.94	1.91
True protein	-	21.8	Glutamic acid	8.71	6.79
Ether extract	6.5	3.7	Proline	9.20	6.36
N-free extract	-	12.9	Glycine	16.65	14.15
Ash	-	20.5	Alanine	7.95	5.10
ME (MJ/kg)	12.2	-	Cystine	-	0.62
			Valine	1.65	1.46
Amino acids (%)			Methionine	0.59	0.81
Lysine	2.81	2.30	Isoleucine	1.70	1.32
Histidine	0.55	0.29	Leucine	3.39	2.26
Arginine	0.96	4.11	Tyrosine	0.79	0.45
Aspartic acid	-	3.94	Phenylalanine	1.66	1.25
Threonine	1.18	1.14			

[a] Hydrolized leather meal
ME = Metabolizable energy

(b) *Cattle and hog hair protein*

Cattle hair as obtained from the commercial alkaline treatment of hides was found to be a desirable high protein meal when autoclaved in a commercial cooker for 30 minutes at 148 °C (345 KPa pressure). The metabolizable energy of the hydrolized cattle hair meal was 9.4 MJ/kg, crude protein 85.4%, fat 1.4%, ash 6.3%, Ca 2.5%, methionine 0.41%, cystine 2.9% and lysine 3.6% (Moran and Summers, 1968; Table 5.5). The substitution of 5% soybean meal by processed cattle hair caused no alteration of performance of growing chicks from that observed for the controls. A higher inclusion of cattle hair caused growth depression which could be connected by supplementary synthetic amino acids missing in the total diet. Table 5.5 shows the chemical analysis of the cattle hair meal.

Cattle hair, however, can present a possible problem from the aspect of residual calcium hydroxide and sodium sulphide from hide pre-treatment if hair will be dehaired after processing of tannery. However, this is doubtful as there was no evidence of toxicity to chicks fed with a substantial dietary level of the meal (18%) (Moran and Summers, 1968).

Processed hog hair treated at 148 °C (pressure 345 KPa) for 30 minutes has been shown to increase the metabolizable energy from 2.4-8.9 MJ/kg. Amino acid analysis of the hog hair before and after processing reveals that the cystine was greatly reduced from 10.2-3.1%, glycine increased from 4.0-5.6%. Smaller decreases were noticed for glutamic acid, lysine and phenylalanine, but serine and isoleucine would seem to have been increased (Moran, Summers and Slinger, 1967; Table 5.5).

(c) *Wet-blue waste protein*

The extraction process of the wet-blue waste according to the method mentioned above results in a recovery of hide protein, which is a fibrous protein consisting mainly of collagen. As a result of the alkaline-acid treatment during the extraction of chromium, collagen is changed into a highly digestible protein. The dechromed protein material was analysed chemically according to El Boushy *et al.*, (1991) (Table 5.6). From this table it may be concluded that this waste is very rich in crude protein (74.9%) with a very high digestibility (98%). The unique amino acid structure of collagen is shown to be high in proline, glycine and lysine, and very low in cystine and tryptophan owing to the absence of keratine (Scott, Nesheim and Young, 1982). The amino pattern of collagen is more balanced and of a better quality than feather meal, and equal to meat meal and soybean meal (El Boushy *et al.*, 1991).

As far as the minerals are concerned, the phosphorus content is very low (0.003%) while the sodium and chloride content is very high (8.8 and 12.0% respectively). These two latter elements may limit the inclusion in formulating animal rations. The requirements of poultry (NRC, 1984), swine (NRC, 1988[a]) and dairy cattle (NRC, 1988[b]) for sodium and chloride may not exceed 0.15-0.19% and 0.15-0.20%, respectively, in the diets. On the other hand, fish reared in fresh or salt water can tolerate, without any depression in production, up to 12-13% NaCl in the diet (NRC, 1983).

Table 5.5. Chemical analysis and amino acid composition of the raw processed cattle hair meals and processed hog hair. From (1) Moran and Summers (1968); (2) Moran, Bayley and Summers (1967[b])

Composition	*Cattle hair (1)*		*Hog hair (2)*	
(%)	*Raw*	*Processed*	*Raw*	*Processed*
Crude protein	87.1	85.4	92.7	88.1
Fat	2.5	1.4	7.0	6.7
Ash	6.9	6.3	-	2.2
Ca	2.6	2.5	-	-
P	0.0	0.0	-	-
Moisture	6.7	10.2	-	-
ME (MJ/kg)	7.1	9.4	2.4	8.9
Amino acids %				
Aspartic acid	4.1	4.2	7.0	6.8
Threonine	6.2	6.2	5.6	4.6
Serine	8.7	9.0	8.0	9.0
Glutamic acid	17.5	17.7	15.3	11.9
Glycine	3.6	3.9	4.0	5.6
Alanine	2.1	1.9	4.0	3.8
Valine	4.5	5.1	5.4	5.7
Methionine	0.3	0.4	0.7	0.6
Isoleucine	2.9	3.4	3.3	3.8
Leucine	7.1	8.1	7.3	7.0
Tyrosine	2.1	1.6	3.2	2.9
Phenylalanine	1.7	1.8	2.6	3.0
Lysine	4.4	3.1	3.2	2.6
Histidine	1.8	0.7	1.0	1.0
Arginine	8.6	8.3	8.6	7.0
Cystine	4.65	2.5	10.2	3.1
Proline	-	-	8.2	7.7

This waste product, therefore, could be used with some care in formulating rations of farm animals because of its low price, but with a maximum level of 1% owing to its high content of sodium chloride. In the case of fish rations the amount of this waste could be raised to higher levels as reported previously (El Boushy *et al.*, 1991). This product needs a lot of attention and improvements to lower its sodium hydrochlorid content in order to make it easy to use without restrictions.

Table 5.6. Chemical analysis and amino acid composition a of the protein recovered from wet-blue waste b, on air-dry basis. From El Boushy *et al.* (1991)

	Composition (%)		*Amino acids (%)*
Crude protein (N x 5,6)	74.900	Lysine (total)	1.91
Protein digestibility (Pepsin HCl)	98.000	Lysine (available)	1.74
Crude fat	2.200	Histidine	0.41
Crude fibre	0.300	Arginine	4.43
Ash	14.000	Aspartic acid	3.20
Ca	0.310	Threonine	1.00
P	0.003	Serine	1.91
Na	8.780	Glutamic acid	5.50
K	0.030	Glycine	11.11
Cl	12.000	Alanine	4.80
Mg	0.060	Methionine	0.48
Zn	traces	Isoleucine	0.86
Fe	0.030	Leucine	1.62
Mn	traces	Tyrosine	0.35
Cr	0.200	Phenylalanine	1.02
		Cystine	0.01
		Tryptophane	0.01
		Proline	7.01
		Valine	1.29

[a] Analyzed values are means of duplicate assays by methods of the Association of Official Agricultural Chemists (1975).

[b] Dry matter of original wet-blue waste is 13.4%.

5.4 ITS USE AS A FEEDSTUFF

There is hardly any literature about the use of hide meal or the wet-blue waste as a feedstuff in the nutrition of broilers and layers. The only literature found is from Poland, Czechoslovakia and Russia.

5.4.1 Tannery Waste (Hide Waste)

Wisman and Engel (1961) reported that tannery by-product meal could be used to replace up to one quarter of the protein supplied by soybean meal protein in practical broiler rations.

Waldroup *et al.* (1970) concluded that a level of 3% hydrolyzed leather meal in an isonitrogenous substitution for soybean meal supported adequate performance of broiler chicks from day old to eight weeks of age. Computer formulated diets, with up to 8% hydrolyzed leather meal caused no adverse effects on chick performance when fed from day old to 28 days of age (Table 5.7).

Table 5.7. Response of chicks to diets containing hydrolized leather meal. From Waldroup *et al.* (1970)

Leather meal (%)	*Body weight at 28 days (g)*[a]	*Feed-gain ratio*	*Mortality*[b]
0	486	1.98	1/40
2	503	1.92	2/40
4	497	2.00	3/40
6	481	1.97	0/40
8	504	1.85	0/40

[a] No significant differences among treatment means
[b] Indicates number of animals that died out of total number

Gawecki, Lipinska and Rulkowski (1981) carried out trials with broiler chicks till 7 weeks of age while using tannery waste (hide meal) as a source of animal protein feedstuff. They noticed that by increasing the inclusion of hide meal from 3.4-6.7% and from 2.8-5.6% in the diets, the broilers reached a higher body weight with the same feed consumption. They reported also that the body weight in the treatment groups with a high level of inclusion of hide meal was similar to the treatments used fish meal inclusion (Table 5.8). They concluded that the hide meal can serve as a source of animal protein for broilers and can replace fish meal completely in diets based on maize and soybean meal.

Table 5.8. Growth, feed consumption and conversion of broilers fed hide meal or meat and bone meal in comparison with fish meal. From Gawecki, Lipinska and Rulkowski (1981)

	Body weight gain (g)	*Feed consumption (g)*	*Feed conversion g feed/g weight*
Fish meal (control)	1564[a]	3288	2.10
Hide meal			
1) 3.4[c] / 2.8%[d]	1561[a]	3267	2.09
2) 6.7[c] / 5.6%[d]	1592[a]	3301	2.07
Meat and bone meal			
1) 3.7[c] / 3.1%[d]	1525[b]	3213	2.11
2) 7.4[c] / 6.2%[d]	1560[b]	3280	2.10

[a,b] Within columns means with no common superscripts differ significantly ($p < .01$)
[c] Inclusion level in the starter ration
[d] Inclusion level in the finisher ration

Kalous, Strádal and Volaufov· (1985) prepared a product called "leather waste meal". This is composed of 30% of leather waste (scraps) and 70% of rendering plant wastes. The analysis of this product is as follows: dry matter 93.5%, protein 43.3%, digestible protein 28.7%, ash 16.7%, fat 22.2%, nitrogen-free extract 11.3%, metabolizable energy 14.1 MJ/kg and chromium, 0.41%. The leather waste meal was used to replace meat and bone meal, fish meal and dried blood meal, the replacement percentages used were 33.3, 50, 66.6 and 100% respectively. The inclusion of the leather waste meal used in some of their experiments was 1, 2 and 3%. They concluded that those inclusions had no effect on weight gain or feed consumption of broilers in comparison, with control groups. They concluded also that with increasing the level of inclusion of the leather waste meal, increased the chromium content in the broiler organs and tissue. A level of 0.0035% was found stored in the broiler body, however, this level is permissible and not harmful according to the regulations used by the Czechoslovakian Ministry of Health on foreign matter present in food. The total chromium content in the diets was not known and it seems that with their inclusion 3% of leather waste meal in the diet, the toxic level of 300 ppm $Cr_2(SO_4)_3$ (NRC Poultry, 1984) was not reached according to the growth of the broilers tested in their trials.

Kalous and Strádal (1986) used hide meal produced from chrome tanning. Their product, however, was not dechromed and it is not the wet-blue waste. They concluded from their experiments that the replacement of 50 or 100% of meat bone meal, fish meal or dried blood with chrome-treated hide meal in a feed mixture did not effect either growth nor feed conversion or carcass yield. They noticed an increase in the Chromium content of meat, liver and bones but this was still within the limits set by the Czechoslovakian Ministry of Health.

5.4.2 Processed Tannery By-Product (Hair Meal)

Cattle hair obtained from commercial alkaline treatment of hides was found to result in a desirable high protein meal when autoclaved in a commercial cooker. Feeding chicks from one to three weeks of age on a practical 20% protein corn-soybean starting diet in which 5% of the soybean protein had been substituted for processed cattle hair caused no alteration of performance compared to that observed for the controls. Exchanging all the dietary soybeans protein for cattle hair in isocaloric rations caused a severe growth depression which could be corrected by supplementary methionine, lysine, tryptophan, histidine and glycine (Moran and Summers, 1968).

Similar results were noticed for hydrolized hog hair, substituting 5% of soybean protein in a 20% protein corn-soybean meal diet with processed hog hair protein. This resulted in a comparable chick growth and feed efficiency, although methionine appeared to be marginal.

Complete substitution of soybean protein with processed hog hair protein caused a severe growth depression which was completely overcome by supplementary lysine, methionine, tryptophan and glycine (Moran, Summers and Slinger, 1967).

5.4.3 Dechromed Hide Wastes

The dechromed hide waste which was mentioned under the name wet-blue waste is a unique product which needs further research on an *in vivo* basis to establish its real effect on the performance of the growing chicks or laying hens. The advantages and disadvantages of this protein are discussed in section 5.3.2 (b).

A Russian research took place on a dechromed hide waste but this was not the wet-blue waste. Korolev (1987) analyzed a by-product of the tanning industry called the dechromed and hydrolyzed hide waste. He reported that this waste contained 82% protein and had a biological value equal to sunflower oil meal and some cereals. It also contained 0.37% fat and calcium, zinc, chromium, iron, copper, cobalt and amino acids. The dechromed hydrolized hide waste was used as a protein supplement in diets for chickens with an inclusion of 3-5% to replace meat meal, meat and bone meal, and fish meal. It was also added in diets of layers at 2-4% resulting in an increase in egg production by 9% and body weight without any adverse effect on quality.

REFERENCES

Association of Official Agricultural Chemists (1975). *Official Methods of Analysis*, 12th edn., (Ed. Horwitz), Washington DC, USA. Aten, A., Innis, R.F. and Knew, E. (1955). *Flaying and curing of hides and skins as a rural industry.* Food and Agricultural Organization of the United Nations, Rome, Italy.

Bidermann, K., Neck, H., Neher, M.B. and Wilhelmy, V., Jr. (1962). A technical economic evaluation of four hide curing methods. *Agricultural Economic Report*, No. **16**, Marketing economics division, Economic research service, USDA, Washington.

El Boushy, A.R., Dieleman, S.H., Koene, J.I.A. and Poel, A.F.B. van der (1991). Tannery waste by-product from cattle hides, its suitability as a feedstuff. *Bioresource Technology*, **35**, 321-3.

FAO (1990). *Production yearbook*, vol. **42**. Food and Agricultural Organization of the United Nations, Rome, Italy.

Gawecki, K., Lipinska, H. and Rulkowski, A. (1981). Meal from residues of the tanning and meat industries in feeds for fattening chickens. *Roczniki Naukowi Zootechniki*, **8** (1), 185-92.

Gustavson, K.H. (1956). *The Chemistry of the Tanning Process.* Academic Press, New York.

Henrickson, R.L., Turgut, H. and Rao, B.R. (1984). Hide protein as a food additive, *J. Am. Leather Chem. Ass.*, **79**, 132-45.

Highberger, J.H. (1956). *The Chemistry and Technology of Leather.* Vol. I (Ed. F. O'Flaherty, W.T. Roddy and R.M. Lollar). Rheinhold Publishing Corporation, New York, U.S.A., pp. 65-193.

Kalous, J. and Strádal, M. (1986). Inclusion of hide meat scrap meal in mixed feeds for broiler fattening. *Sbornik Vysoke Skoly Zemedelske v. Praze*, Fakulta Agronomicka, **B 44**, 287-96.

Kalous, J., Strádal, M. and Volaufov, E. (1985). Increased content of chromium in the organs and tissues of broilers as resulting from leather waste meal supplements to mixed feeds. *Sbornik Vysoke Skoly Zemedelske v. Praze*, Fakulta Agronomicka, **B 43**, 233-45.

Koene, J.J.A. and Dieleman, S.H. (1987). Ontchroming op semi-technische schaal van wet-blue reststoffen uit leerlooierijen. *TNO, Internal report fase 1*, ILS, Waalwijk, The Netherlands, p. 45.

Koppenhoefer, R.M. (1965). Non protein constituents of skin. *The Chemistry and Technology of Leather.* Vol. 1 (Eds. F. O'Flaherty, W.T. Roddy and R.M. Lollar), Reinhold Publishing Corporation, New York, U.S.A., pp. 41-64.

Korolev, A. (1987). A feed additive for chickens. *Ptitsevodstvo*, **9**, 30-1. Mann, I. (1967). Processing and utilization of animal by-products. *Food and Agricultural Organization of the United Nations, Paper* No. **75**, pp. 80-139.

Marselli, J.W., Marselli, N.W. and Burford, M.G. (1958). Tannery wastes, pollution sources and methods of treatment. *New England Interstate Water Pollution Control Commission*, 73, Tremont Street, Boston 8, Massachusetts, USA., p. 40.

McLaughlin, G.D. and Theis, E.R. (1945). *The Chemistry of Leather Manufacture*. A.C.S. Monograph, Reinhold Publishing Corporation, New York, U.S.A.

Moran, E.T. and Summers, J.D. (1968). Keratins as sources of protein for the growing chick. 4. Processing of tannery by-product cattle hair into a nutritionally available high protein meal: Metabolizable energy, amino acid composition and utilization in practical diets by the chick. *Poultry Sci.*, **47**, 570-6.

Moran, E.T., Bayley, H.S. and Summers, J.D. (1967). Keratins as sources of protein for the growing chicks. 3. The metabolizable energy and amino acid composition of raw and processed hog hair meal with emphasis on cystine destruction with autoclaving. *Poultry Sci.*, **46**, 548-53.

Moran, E.T., Summers, J.D. and Slinger, S.J. (1967). Keratins as sources of protein for the growing chicks. 2. Hog hair, a valuable source of protein with appropriate processing and amino-acid balance. *Poultry Sci.*, **46**, 456-65.

Moulton, C.R. and Lewis, W.L. (1940). *Meat through the microscope*, revised edition. Institute of meat packing. The University of Chicago, Chicago. National Hide Association (1979). *Hides and skins*. National Hide Association. Sioux City, Iowa, USA.

NRC (1983). National Research Council. *Nutrient Requirements of Fish*, 8th edn. National Academy Press, Washington, DC, USA.

NRC (1984). National Research Council. *Nutrient Requirements of Poultry*, 8th edn. National Academy Press, Washington, DC, USA.

NRC (1988^{a}). National Research Council. *Nutrient Requirements of Swine*, 9th edn. National Academy Press, Washington, DC, USA.

NRC (1988^{b}). National Research Council. *Nutrient Requirements of Dairy Cattle*, 6th edn. National Academy Press, Washington, DC, USA.

Ockerman, H.W. and Hansen, C.L. (1988). *Animal By-Product Processing*. Hide and skin by-products. Ellis Horwood, VCH Verlagsgesellschaft, Wienheim, pp. 89-131.

Pepper, K.W. (1966). Tannery effluent problems. *J. Am. Leather Chem. Ass.*, **61**, 570-84.

Price, J.F. and Schweigert, B.S. (1971). *The Science of Meat and Meat Products*. (Ed. W.H. Freeman), San Francisco, USA. Scott, M.L., Nesheim, M.C. and Young, R.J. (1982). *Nutrition of the Chicken*, 3rd Edn. Scott and Associates, Ithaca, New York, USA.

Tanners' Council of America (1983). *Dictionary of Leather Terminology*. Tanners' Council of America, Washington D.C., USA.

Taylor, M.M., Diefendorf, E.J., Hannigan, M.V., Bailey, D.G. and Feairheller, S.H. (1982). Wet process technology research at eastern regional research center. *The leather manufacturer*, **100**, (1) 20-5.

Thorstensen, T.C. (1985). *Practical Leather Technology*. Robert E. Krieger Publishing Company, Malabar, Florida, USA.

United States Hide, Skin and Leather Association (1985). *Trade practices for proper packer cattle hide delivery*. United States Hide, Skin and Leather Association, Washington D.C.

Waldroup, P.W., Hillard, C.M., Abbott, W.W. and Luther, L.W. (1970). Hydrolized leather meal in broiler diets. *Poultry Sci.*, **49**, 1259-64.

Wisman, E.L. and Engel, R.W. (1961). Tannery by-product meal as a source of protein for chicks. *Poultry Sci.*, **40**, 1761-3.

6. FRUIT, VEGETABLE AND BREWERS' WASTE

Contents

6.1 GENERAL INTRODUCTION

Fruit and vegetable wastes as well as brewers' dried grains are produced after processing. The processing of these wastes, being sources of plant protein and energy, is dependent on economic needs. For instance, the recovery of a useful material from processing wastes seems to be an ideal solution. However, the treatment cost of the waste stream will be increased, but a credit will be introduced owing to the sale or use of the recovered waste product, with its nutritive value as a feedstuff for farm animals and especially poultry. In developing market economics, however, the need for plant protein and energy is great while the production is never adequate. Waste residue from the vegetable, fruit and brewers' industry is rarely identified or found in the developing market but as large-scale processing industries develop around the world, residues suitable for use as a feedstuff are bound to occur more often.

Processing of fruit, vegetable and brewers' wastes to a known product, can be of a great benefit to the market development and it will result in a stable economic structure towards these newly developed waste products. These developed new waste products may be used as such in feeding purposes or can be recycled in the processing of vegetable and fruit. Part of the sugar for example used in canned pineapples is recovered from pineapple wastes. However, a similar process for recovering pear sugars is not practised (Ben-Gera and Kramer, 1969). A relevant example of reducing waste by processing methods is the dewaxing of tomatoes before lye peeling (using a solution of alkaline salt), which showed a better yield of a higher quality, better-coloured product. In this process the lye quantity used was cut to one-third, and the peeling loss was reduced to one-fifth of the previous level. Thus, the method of dewaxing saved about 15% yield of the tomatoes from the waste stream as peeling loss (White, 1973).

An old example of recycling fruit process waste is apple juice processing, during which several by-products are produced and reused such as apple essence or perfume, which will fortify the taste and aroma of the apple juice. Also, vinegar from apple peels and cores are additional by-products of the mechanical apple peeling process, making recycling more attractive than it otherwise might be, as it produces a smaller quantity of peel waste (White, 1973).

In relation to the recovery of food-grade starch in the processing of potatoes to chips, Douglass (1960) reported that cutting of potatoes into slices for chips released 24.3 kg of starch per t. He noted that "the potato chip industry is badly in need of a simple, relatively inexpensive starch recovery unit". Willard, Pailthorp and Smith (1967) stressed that the processing of 200 t of potatoes per week produced 4.9 t waste of a recoverable potato starch. These recoverable solids are used as an animal feedstuff only, the solubles being pumped to municipal treatment plants. Applying a new technique for the disposal of low-grade potatoes, Shaw and Shuey (1972) used a dry milling process wherein potatoes were dehydrated, milled and subjected to air classification and screening with conventional wheat flour-processing equipment. This process resulted in two relevant products, an animal feedstuff and a starch fraction. Treating the starch fraction by washing reduced its impurities to less than 0.4% ash. In Table 6.1 this advanced dry process is compared with the conventional wet process and the composition of the products of the dry-milling process is given in Table 6.2.

Table 6.1. Recovery of products from two potato starch processes with a daily input of 250 t potatoes. From Shaw and Shuey (1972)

Conventional wet process (t)	*Dry process (t)* [a]
30.0 starch	30.0 starch
6.8 pulp	22.1 animal feed
11.9 soluble waste solids containing 3.35 soluble BOD	0 BOD

[a] Projected from pilot-scale data.

Table 6.2. Composition of the starch and feed from processed potatoes produced by the dry starch process. From Shaw and Shuey (1972)

	Composition of products	
	Starch (%)	*Feed (%)*
Nitrogen	0.044	3.9
Protein (N x 6.25)	0.370	25.7
Ash	0.370	5.0
Starch	99.200	61.8

As far as the characterization and production of the fruit waste (solid and liquid) are concerned, numerous reports have been published (Bough, 1973; Flora and Lane, 1978; Moon, 1980). In general, processing of vegetables produces a relatively larger amount of total solids compared with dairy or meat processing (Table 6.3).

Table 6.3. Solid and liquid wastes generated during the processing of food. From Moon (1980)

Processed food	*Total solid waste* [a] *(g/kg)*	*Liquid volume* *(m^3/kg)*	*BOD* [b] *(mg/kg)*
Vegetable			
Kole	16	0.004	11 000
Spinach	20	-	11 000
Mustard greens	16	-	10 000
Turnip greens	15	-	9 000
Collards	13	-	8 000
Potatoes	16	0.012	44 000
Peppers (lye peel)	65	0.020	33 000
Tomatoes (lye peel)	14	0.010	-
Dairy			
Cheese whey	-	9.000	270 000
Skim milk	-	0.070	1 500
Ice cream	-	0.080	3 000
Meat			
Red	0.440	25.000	14 000
Poultry	0.270	50.000	13 000
Eggs	0.111	-	-

[a] Waste loads calculated per unit weight of product
[b] BOD: biological oxidation demands

Woodroof and Luh (1986) estimated solid and liquid wastes derived from processed fruits in the U.S.A. Tree fruits peeled by chemical methods may be estimated to be up to one million t per year. It appeared that each ton of apricots, peaches or pears processed produces about 5.4 kg of BOD and 4.1 kg of suspended solids in rinse water, with a high total liquid waste. It was noticed that the amount of peels, cores, or trimmings and culls at processing plants varies widely with the maturity, size and grade of apples processed, as well as the operating conditions. The accumulated waste from their trial was estimated to be 227 million kg of apple peels and cores.

On the other hand Woodroof (1975) calculated from experiments that 11 770 000 t of processed vegetables (14 types) produced 118.5 m^3 wastewater with 168 million kg of BOD, 68 million kg of suspended solids, and 3 350 000 t of solid residuals.

Table 6.4 shows a part of these data indicating the yield of types of wastes and solid residuals from certain vegetables.

Table 6.4. Yield of wastes from the processing of several vegetables. From Woodroof (1975)

Vegetables	*Raw (x 1,000t)*	*Wastewater (m^3/t)*	*BOD [a] (kg/t)*	*Suspendid solid (kg/t)*	*Solids residuals (kg/t)*
Asperagus	120	37.9	4.5	3.2	327
Beans, lima	120	34.1	11.4	36.3	127
Beans, snap	630	17.0	13.6	1.8	191
Beets	270	15.1	68.1	22.7	372
Carrots	280	15.1	25.1	18.2	436
Corn	2500	6.8	11.4	4.5	608
Peas	580	19.0	22.8	4.5	118
Pumpkin squash	220	11.4	36.3	6.8	563
Sauerkraut	230	1.9	6.8	1.4	291
Spinach, green	240	34.1	11.4	4.5	145
Sweet potato	130	26.5	91.0	36.3	-
Tomato	5000	7.6	5.4	1.8	908

[a] BOD: Biological oxidation demand.

Industries in general have tried to convert wastes into profitable products or into harmless effluents (sludge). The most relevant processes that produce most of the solid and liquid wastes in fruit and vegetable processing are peeling, blanching, quick blanching, drying and water purification resulting in sludge from effluent. These solid residues are collected by belts or by screening of wash water and are usually transported by truck to disposal dumps, or are distributed in small piles as humus on adjacent fields. In this form they constitute a serious health risk in the neighbouring communities. The disposal of wash and blanch wastewaters present an even more serious problem; usually, it is flushed into streams, rivers or sewers or by spraying on urban areas. Government controls are becoming increasingly rigid, and other methods of processing this wastewater to a sludge which may be used as a feedstuff are required. The intention of this chapter is to discuss the most relevant fruit and vegetable wastes as well as brewers' dried grains, their processes, chemical analysis and nutritive value and their use in poultry nutrition.

6.2 TOMATO RESIDUES

6.2.1 Introduction

Tomato belongs to the fruits of the nightshade family (*Solanaceae*). Not only potato and tobacco, but also several fruit vegetables, notably the garden peppers and egg plants belong to this family. Tomato is known as *Solanum Lycopersicum* L. = *Lycopersicum esculentum* Mill. Several species of the genus *Lycopersicum* grow wild in Western South America, from one of which the cultivated tomato has been derived.

Figure 6.1 shows the microscopic structure of the tomato seed which is composed of spermoderm as one layer with radially arranged rods and with false hairs. Perisperm and endosperm (embryo) consist of small polygonal cells containing minute aleurone grains and fat. This microscopic structure suggests where the high percentage fat being (24.4%) is formed in the endosperm of the seed (Winton and Winton, 1949).

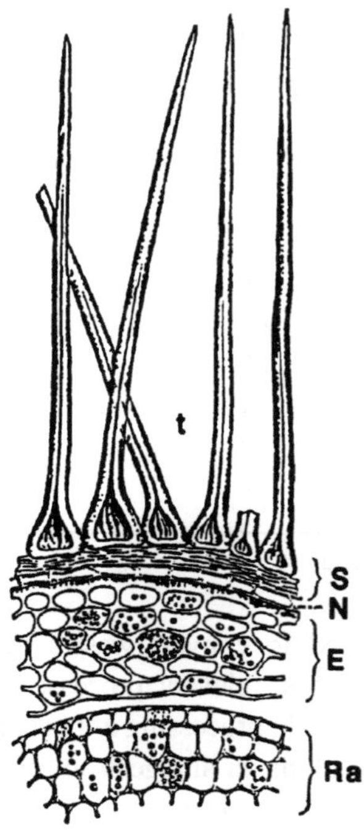

Figure 6.1. Tomato seed in a cross-section. *S*, spermoderm; *t*, false hairs; *N*, perisperm; *E*, endosperm; *Ra*, radicle. X 160. Reproduced with permission from Winton and Winton (1949).

The top producing countries in the top production of fresh and processed tomatoes are the U.S.A., Canada and Portugal and they process more than 80% of their production. A second group includes Tunisia, France and Italy and about 60, 50 and 35%, respectively of their products is processed. Other countries involved on a lower scale are Spain, Greece, Israel, Algeria, Morocco, Egypt, South Africa, and also Australia and South America.

Tomatoes serve in the human nutrition as two types of products: fresh product (in salads) or processed as juice, ketchups, paste as a whole (pulp) or without seeds, purée and in soups. The processing of tomatoes yields several waste by-products such as seeds and peel, which are mostly classified as tomato pomace, tomato seed meal, tomato seed cake and tomato seed oil.

Klimenko and Kaganskii (1969) described a pilot-scale drying plant for tomato seeds, and the possibilities of oil extraction. Their report was based on a Moldavian (Russian) tomato processing industry producing one t tomato seeds per hour. The processing of 187 000 t tomatoes resulted in a waste estimated to be 3 700 t of tomato seeds producing 940 t of oil after pressing. White (1973) measured the recoverable products from tomato processing as shown in Table 6.5.

Table 6.5. Recoverable products (tonnes) from tomato processing with an input of 1000 t per day. From White (1973)

Seeds	20.00 (at 65% H_2O)
Dry seeds	7.86 (at 11% H_2O)
Recoverable oil	1.98
Seed meal	5.02

Schultz, Graham and Hart (1976) estimated the tomato waste in California (USA) from processing to be 200 000-250 000 t per year of pomace residue and 170 000 t per year of peeling residue. Collectively, there are about 400 000 t of residue available to consider. The total waste produced from tomatoes from world production is estimated roughly to be 3.70 million t/year (FAO, 1991).

Several research workers explained the value of these types of tomato waste as a feedstuff and a fertilizer (Campbell, 1937; Edwards *et al.*, 1952; Klimenko and Kaganskii, 1969; Kalaisakis *et al.*, 1970; Wagner, 1970; White, 1973; Garcia and Gonzalez, 1984; Yannakopoulos, Tserveni-Gasi and Christaki, 1992).

6.2.2 Processing

(a) Processing of fresh tomatoes

The processing of tomatoes and its generated waste has been discussed in detail by many research workers (Edwards *et al.*, 1952; Schultz *et al.*, 1971; White, 1973; Schultz, Graham and Hart, 1976; Gould, 1983). Their reports will be summarized to fulfil the aim of this book and detailed information can be found in the above mentioned books and articles.

The preparation of tomatoes for processing is shown in Figure 6.2 and takes place as follows.

Dry sorting

After receiving the tomatoes, the first unit operation in processing is dry sorting. This step is very important to remove gross contamination and defective fruit (green, decomposed, or unfit) which would cause contamination of the washing water. Grading of size also takes place in this operation. The advantage of size grading is to standardize the whole pack portion control. All defective fruit will be used in waste processing.

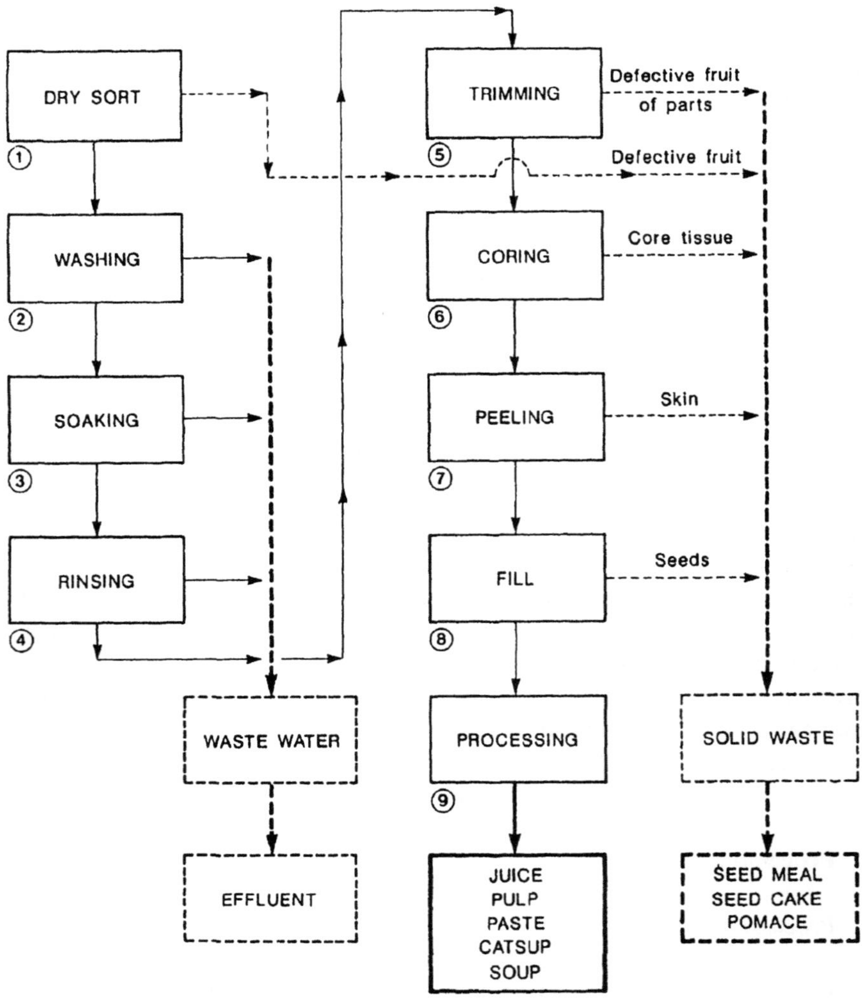

Figure 6.2. Schematic processing outline of tomatoes with the end products as food and the waste as potential feed.

Washing

This step is essential for the removal of "soil", which is a foreign substance to the fruit. Soil attached to tomatoes may include spray residues, micro-organisms, dirt, mould, *drosophila* eggs and larvae. Gould (1965) and Gould, Geisman and Sleesman (1959) concluded that soil can be removed from tomatoes by specific washing practices. The washing operation takes place in two additional steps: soaking and rinsing by spray.

Soaking

From receiving-dumping area to the factory, tomatoes are conveyed on flumes (streams of water) or by using a soak tank. A properly sized flume will allow fruits to be held for a period up to 3 min at a temperature of 54 °C. The purpose of this process is to loosen the soil and to remove *drosophila* eggs and larvae prior to final washing (Gould, Geisman and Sleesman, 1959).

The use of detergents such as alkaline solution (pH 11-12), or lye solution 0.5% and sodium hypochlorite facilitates cleaning during soaking and has been discussed intensively (El-Ashwah, 1963; Gould, Geisman and Sleesman, 1959; Gould, 1983).

Rinsing

Rinsing is the following step after soaking. In this process, the spray nozzles (at a fixed pressure) are mostly located above the fruit to generate a complete rinsing of the tomatoes. During rinsing the tomatoes travel on a belt conveyor at a particular speed that allows a certain number of turns of the fruit to assure a complete rinsing of the detergents used in the soaking process. The aim of rinsing with high pressure spray nozzles is not only to rinse away the detergents used during soaking but also to eliminate the eggs and larvae of the house-fly (*drosophila*), (Gould, Geisman and Sleesman, 1959).

Trimming and final sorting

The aim of trimming and sorting is to remove the defective fruit (rotten areas, mould portions, insect damaged, or sunburned) which were not removed by the washing, soaking and rinsing procedures or which passed through the first dry sorting. This step will deliver also waste to be processed. The number of sorted fruits depends on the accuracy of the dry sorting, the quality of the raw product, and the washing or rinsing technique. For more detailed information on trimming and rinsing, the reader is referred to the papers of Cruers (1958), Goose and Binsted (1973) and Could (1983).

Coring

The main purpose of this step is to remove the core tissue of the fruit. If excess flesh tissue is removed, a low drained weight and a lot of waste may occur. In general, tomatoes are cored by hand which is more costly and creating more fruit waste. Machine coring is much quicker, cheaper and more precise (Gould, 1965, 1983).

Peeling

The efficiency of the peeling operation influences the quality of the finished product. Tomatoes may be peeled by steam, lye or infrared, respectively. The three methods are creating different amounts of peel depending on the size of fruit and the species used (Schulte, 1965; Cruers, 1958; Lucas, 1967; Gould, 1983).

Finally the processed clean tomato flesh will be processed into chili sauce, juice, pulp, paste, ketchup or soup. The details of those points are not our subject but the most relevant point is the waste from the processing, mainly from dry sorting, final sorting and trimming, coring, peeling and, finally from seeds produced.

(b) Processing of tomato waste

The disposal of tomato processing wastes is still a great problem in many countries. If this waste could be processed, dried and sold as a feedstuff, the problem of its accumulation and pollution would be solved and a benefit would be introduced in the developing market. As mentioned previously the total solid waste is estimated to be 19% of the total solids in the original tomatoes. The processing of this waste was described by Edwards *et al.*, (1952) and by Goose and Binstead (1973) as follows:

Solid wastes

In this process the solid fraction is basically composed of:

- Broken and diseased tomatoes and debris removed during the initial washingoperation.
- Discarded materials and trimmings from the sorting lines.
- Skins, core tissue and seeds, either separated or mixed, removed by cyclones during pulp screening and refining, or by mechanical, steam or peeling operations on tomato canning lines.

The solid waste removed in the washing section of the plant normally floats from the over-flows, or sinks, and is subsequently drained from the base of the washers. Much of this waste is carried away in washing water and may be removed at any early stage by passing through grid screens. Waste and trimmings from the sorting tables is generally conveyed from the lines to a suitable collecting point either mechanically, by means of moving belts, or manually by emptying waste receiving bins. Unprocessed tomatoes, broken tomatoes and tomato waste are particularly subjected to spoilage, and may rapidly putrefy and create a considerable nuisance if not cleared away frequently.

Under the climatic conditions usually experienced during tomato seasons any accumulation of fruit wastes can soon become sour and stinking, and give rise to mould growth, and infestation with flies and maggots.

The first step in the processing of tomato residues is to obtain a high solids fraction from the culls and trimmings. This step is realized by chopping and passing them through a cyclone. The tailings from the cyclone have about the same solid content as the tailings from the juice-making operations. The combined tailings when pressed yield a press cake which can be dried, and yield liquor which is removed by pressing, and will be dried in a submerged combustion unit. Finally the dried press cake will be mixed with the concentrated liquor and conveyed to a drier to produce dried tomato waste as a feedstuff (Figure 6.3).

Liquid waste cannery effluent

The water used for tomato fluming, washing, flushing and cleaning of equipment and processing purposes, becomes quite heavily polluted by organic residues. This water cannot be directly discharged to a public sewer, and must be subjected to varying degrees of treatment according to the requirements of local authorities before being finally discharged into a water course.

Effluent treatment systems consist of the following unit operations (Goose and Binsted, 1973):

- Screening to remove gross solids.
- Settlement to remove particulate solids.
- Lime treatment and pH control.
- Biological purification treatments such as trickling filters or activated sludge.
- Final purification in shallow lagoons.

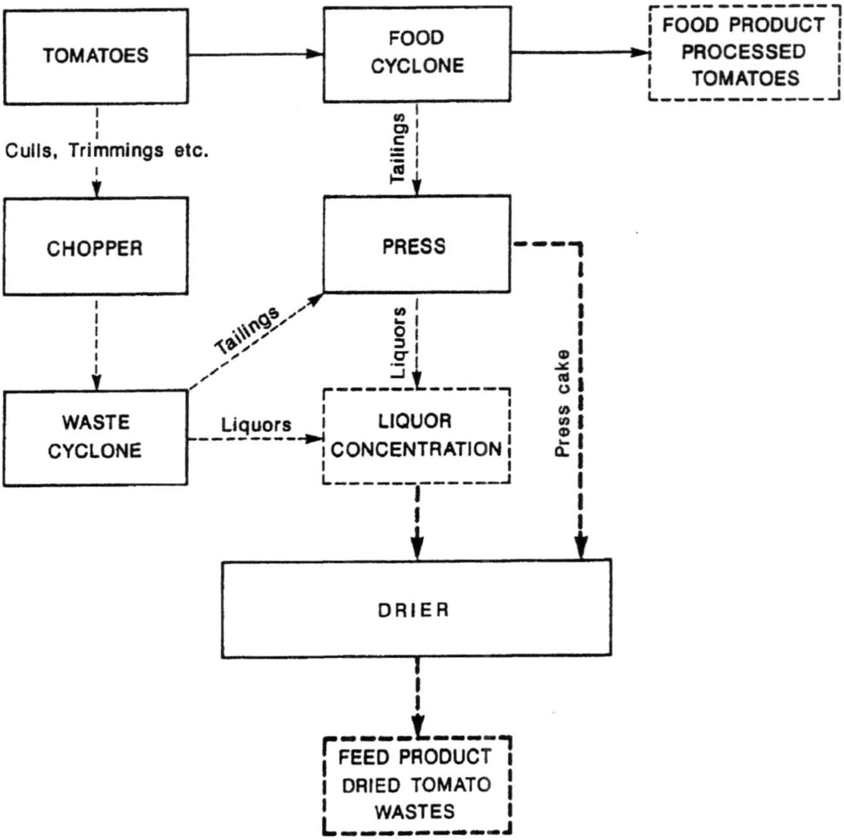

Figure 6.3. Scheme for processing tomatoes into food products and dried tomato waste.

The last point (water purification with its effluents) is still at too early a stage to deliver a suitable end product as a feedstuff for poultry.

(c) Yields of tomato waste

The yields of tomato waste are based on the percentage of unsuitable fruit defective, green, decomposed, broken or on general unfit *etc.* This waste varies according to the season, type, weather conditions and methods of transporting the tomatoes.

The second part of waste is produced in preparing the tomatoes for processing. During this stage the coring produces 3-4.8% of waste depending on the size of the fruit (Schulte, 1965). The next step is peeling and this produces about 4.8% of peel depending on the size of the tomatoes and the peeling method (steam, lye, infrared) (Schulte, 1965).

During the processing from tomato flesh to juice, seeds are separated, producing 2% of the original fresh weight (Table 6.5; White, 1973).

Edwards *et al.* (1952) reported that, for every million tons of processed tomatoes for juice and juice products, about 123 800 t of recoverable wastes resulted, having about

11 300 t of total solids. This contributes to about 19% of the total solids in the original tomatoes. The drying process of the tomato waste will recover 65% of the solids in the liquid wastes; enabling the recovery of 83% of the total waste.

6.2.3 Chemical Analysis and Nutritive Value

El Moghazy and El Boushy (1982[b]) classified tomato residues as tomato seed meal (TSM), dried tomato pomace (DTP) and tomato seed cake (TSC). Generally, these three feedstuffs are characterized by crude protein varying from 22.6-35.9%, fat 3.2-22%, crude fibre 20.8-30.5%, ash 3.1-7.4% and nitrogen-free extract (NFE) 19-32%. In case of tomato seed meal the metabolizable energy content was 6.49 MJ/kg. This waste tomato residue can contribute a reasonable protein and energy. Its contents of essential amino acids, however, is limited and its high fibre content is a limitation (up to 5%) for its use in both broiler and layer rations. From Table 6.6 it may be concluded that the tomato residue (seeds and skins), seeds and seed cake are acceptable in their vegetable protein content varying from 22.8-37.6% and their fat content varying from 11.6-24.4%. They are, however, high in their fibre content varying from 13.6-23.2%. It is also clear from the same table that seed cake is higher in fibre content than whole seeds are owing to the extracted oil, in the case of seed cake.

Table 6.6. Composition of tomato residue (seeds and skins), seed and seed cake. From Winton and Winton (1949)

	Moisture (%)	*Protein (%)*	*Fat (%)*	*N.F.E. (%)*	*Fibre (%)*	*Ash (%)*
Seeds and skin	7.0	22.8	17.0	26.7	22.4	4.1
Seeds	9.0	27.6	24.4	21.4	13.6	4.0
Seed cake [a]	8.6	37.6	11.6	29.1	22.1	4.6
Seed cake [b]	8.3	31.2	16.6	13.9	23.2	6.7

[a] Extracted
[b] Pressed
NFE: nitrogen-free extract

(a) Tomato seed meal (TSM):

Johns and Gersdorff (1922) showed that tomato seed protein is composed of α-Globulin and ß-Globulin. Table 6.7 shows the amino acid composition of the two globulines. ß-Globulin showed a higher lysine level (3.8%) in comparison with α-Globulin (1.2%).

Anwar, El Alaily and Diab (1978) showed that TSM on average contains 35.9% crude protein with a gross protein value of 73%, ether extract 3.2%, crude fibre 22.9% and NFE 23.1% (Table 6.8). They concluded that tomato seed meal showed a reasonable resemblance to cotton seed meal except for fibre which was three times higher in

TSM than in cotton seed meal. It is advisable therefore to produce a decorticated material from tomato seeds which will be very suitable as a plant protein feedstuff for poultry. The essential amino acid contents given in Table 6.8 for both CSM and TSM, showed close agreement between the two patterns except for lysine which was higher in TSM (2.2% compared with CSM 1.6%). There was also a large difference for methionine + cystine in the two feedstuffs being 1.1% in CSM and only 0.6% in TSM. Anwar, El Alaily and Diab (1978) concluded that methionine is the most limiting amino acid in TSM. The low metabolizable energy of TSM (6.49 MJ/kg) is due to its high fibre content (22.9).

Table 6.7. Amino acid composition (% of protein) of ,ß Globulin in tomato seed protein. From Johns and Gersdorff (1922)

	α- *Globulin* (%)	*ß - Globulin* (%)
Cystine	1.3	1.2
Tryptophane	1.2	1.5
Arginine	14.0	10.7
Lysine	1.2	3.8
Histidine	4.9	6.4

Table 6.8. Chemical analysis (% as such) and amino acid contents (% in product) of tomato seed meal (TSM) and cotton seed meal (CSM). From Anwar, Alaily and Diab (1978)

	TSM (%)	*CSM* (%)		*TSM* (%)	*CSM* (%)
Crude protein	35.90	41.10	*Amino acids*		
Gross protein value	75.00	65.00	Methionine +		
Metabolizable energy (MJ/kg)	6.49	11.38	Cystine	0.6	1.1
Ether extract	3.20	6.50	Threonine	1.2	1.3
Fibre	22.90	7.70	Isoleucine	1.2	1.3
Ash	7.00	7.60	Leucine	2.2	2.5
N-Free extract	23.10	29.00	Valine	2.0	1.8
Moisture	7.80	8.00	Phenylalanine	1.5	2.2
Amino acids			Histidine	0.8	1.1
Lysine	2.20	1.60	Arginine	3.1	4.3
			Tryptophan	0.3	0.5

Tomato seed oil is obtained as a by-product during the processing of tomato pulp. The refined oil contains 17.5% solid saturated fatty acids and 75.8% liquid (unsaturated) fatty acids including 45% oleic acid, 34.2% linolic acid and at least 0.4% arachidic acid (Winton and Winton, 1949).

(b) Dried tomato pomace (DTP)

This by-product in the manufacturing of tomato juice or tomato ketchup consists mainly of the skins, seeds and hard tissues of the whole tomatoes. It is sometimes dried to tomato pomace which contains 22.6-24.1% protein, 14.5-15.7% fat and 20.8-30.5% fibre (Table 6.9). This by-product is a good source of vitamin B_1 and a reasonable source of vitamins A and B_2 (Esselen and Feller, 1939; Ewing, 1963). Winton and Winton (1949) concluded that there are two colouring substances in the tomato pomace, one red and crystalline and the other yellow and amorphous, they added that they were defined as two carotenoids, *lycopersicine (lycopene)* and *carotene*, which can be of a great value for the egg yolk coloration. From Table 6.9 it may be concluded that methionine and lysine are limiting essential amino acids in this feedstuff.

Table 6.9. Chemical analysis (% as such) and amino acid contents (% in product) of dried tomato pomace. From [1] Esselen and Feller (1939); [2] Ewing (1963) and [3] Hopper (1958)

	(1) *(%)*	*(2)* *(%)*	*(3)* *(%)*
Moisture	4.3	5.3	-
Crude protein	24.1	22.6	-
Ether extract	15.7	14.5	-
Nitrogen free-extract	32.1	23.8	-
Fibre	20.8	30.5	-
Ash	3.1	3.3	-
Amino acids			
Arginine	-	-	1.3
Histidine	-	-	0.4
Isoleucine	-	-	0.8
Leucine	-	-	1.8
Lysine	-	-	1.7
Methionine	-	-	0.1
Phenylalanine	-	-	0.9
Threonine	-	-	0.8
Valine	-	-	1.1
Tryptophan	-	-	0.2
Tyrosine	-	-	1.0

(c) Tomato seed cake (TSC)

Before expelling the oil, the seeds are dried in a direct heat rotary drier at a temperature not exceeding 60 °C. The resulting press cake is ground to a meal and incorporated in the rations (Finks and Johns, 1921).

Tomato seeds contain approximately 22% of valuable oil which contains 17.5% solid acids and 75.8% liquid acids. The glycerides of this oil contains stearic acid (5.9%), palmitic acid (12.5%), oleic acid (45%) and linolic acid (34.2%) (Winton and Winton, 1949). The pressed cake which remains after expelling the oil contains about 37% of protein (Table 6.10). Tomato seed cake may also contain some fat-soluble vitamins and a significant quantity of water-soluble vitamins (Finks and Johns, 1921). Essential amino acids are well represented in the α- and ß-globulins of tomato seed cake. Arginine, cystine, histidine and lysine were present in relatively adequate amounts (Johns and Gersdorff, 1922).

Table 6.10. Chemical analysis of several dried tomato press cakes. From [(1)] Woodman (1945); [(2)] Edwards *et al.* (1952); and [(3)] Maymone and Carusi (1945)

	Dry matter *(%)*	*Crude protein* *(%)*	*Fat* *(%)*	*Fibre* *(%)*	*Ash* *(%)*	*N-free extract* *(%)*
Skin and seeds [(1)]	93.3	24.8	22.0	27.6	6.6	19.0
Press cake [(2)]	92.0	22.5	14.2	29.6	3.3	22.4
Press cake + concentrate [(2)]	92.0	21.0	9.8	21.9	5.9	33.4
Oil cake, Italy [(3)]	-	37.0	6.8	28.3	7.4	20.5

The mineral contents of the tomato fruit, skin and seeds are shown in Table 6.11. Seeds and skin are rich in potassium, calcium and phosphorus (Winton and Winton, 1949).

Table 6.11. Chemical analysis in ash of the several tomato by-products. From Winton and Winton (1949)

%	*K*	*Na*	*Ca*	*Mg*	*Fe*	*P*	*S*	*Cl*	*Si*
Tomato fruit	49.4	4.4	0.9	1.9	0.1	6.5	1.2	19.1	0.2
Seeks and skins	15.8	1.0	5.0	5.2	4.8[a]	12.4	-	-	-
Seeds	11.2	0.6	20.8	2.6	5.8[a]	20.5	-	-	-

[a] Includes aluminium oxide

6.2.4 Feeding Tomato Waste

(a) Broilers

Esselen and Fellers (1939) reported that dried tomato pomace (DTP) could be fed to chicks from two to six weeks of age at an inclusion level of 11.6% level without harmful side effects and with an acceptable palatability. Feeding DTP showed an extra gain of 33 grams over the control chicks. Kelley (1958), concluded that broilers utilized dried tomato seed cake (DTSC) more efficiently than alfalfa meal and showed a superior growth with 5% DTSC compared with wheat middlings. Anwar, El Alaily and Diab (1978) concluded that the tomato seed meal (TSM) gross protein value was increased from 75%-94% when DL-methionine was added at 0.2% indicating that this amino acid is the most limiting one in this feedstuff. From their experiments with Dokki 4 chicks (Fayomi x Barred Plumith Rock) at 4 weeks of age it was indicated that feed conversion was significantly depressed with 34% TSM in the diet owing to the relatively high fibre content. Their trials were based on a comparison between TSM and cotton seed meal, results of which are shown in Table 6.12.

Table 6.12. Body weight gain, feed consumption and efficiency of 4-week-old Dokki 4 chicks fed cotton seed meal (CSM) and tomato seed meal (TSM). From Anwar, El Alaily and Diab (1978)

Item	*Ration 1 (CSM)*	*Ration 2 (TSM)*
Initial body weight (g)	32.8	32.8
Final body weight (g)	182.0	165.1
Body weight gain (g)	149.2	132.3
Feed consumption (g)	422.1	428.2
Feed efficiency [a]	2.8	3.2[b]

[a] g feed required per g gain
[b] Significant at $P < 0.05$

(b) Layers

Abou Akkada *et al.* (1975) reported results of trials on laying hens of 3 breeds, Alexandria, Dokki and Fayomi, which were given basal diet - alone or with, 2, 4 or 6% dried tomato residue from canning. The diets were given for about 6 weeks and the effect on egg production, egg weight and yolk colour were recorded separately for each breed. The tomato residue used in these trials had a protein level varying from 12.5-22.5% and fibre and moisture levels ranged from 18.0-26.9% and from 8-10%, respectively. No differences in egg production and egg weight among diets were noticed. Yolk colour, however, was deeper in diets with tomato residue.

Tomczynski (1978) reported experiments with Sussex layers fed tomato by-products for 8 months. They replaced some ingredients in a basal diet containing soybean meal, ground millet and ground maize by 5% tomato seeds or 7.17% tomato skins. They concluded from their numerical calculations that the addition of tomato seeds or skins showed improved performance over the control diet as far as egg production, feed intake, fertility and hatchability % are concerned (Table 6.13).

Petrenko and Banina (1984) replaced grass meal in formulated layer diets with dried and ground tomato waste at a 5% level to obtain a total fibre content of 5.4%. The fibre content of the control diet was 4.75%. Layers on the tomato waste diet produced 6.5% more eggs and used slightly less feed/unit of eggs laid than those fed the control diet. Accordingly, tomato waste could be used to replace up to 5% grass meal in diets for layers.

Yannakopoulos *et al.* (1992) examined diets for layers containing 0, 8 or 15% tomato meal. The hens were 52 weeks old at the beginning of the experiment and were kept for 10 weeks of production. The diet composition was mainly based on maize, soybean meal and wheat bran while average protein and ME contents were 17.2% and 11.2 (MJ/kg), respectively. Body weight gain, egg number, feed consumption and mortality were not significantly affected by the inclusion of tomato meal. Mean egg weight tended to be increased ($P < 0.10$). Yolk colour score was significantly increased by the inclusion of tomato meal giving a deeper colour. This study indicated that replacing part of a maize-soybean-based diet with tomato meal does not affect laying performance but improves the interior egg quality late in the laying period.

Table 6.13. The effect of feeding tomato seeds or tomato skin to Sussex laying hens up to 8 months and their performance. From Tomczynski (1978)

	Control	*Tomato seeds*	*Tomato skin*
No. eggs produced	6220.0 (100)[a]	7012.0 (113)	6637.0 (107)
Egg weight (g)	55.0	55.0	55.0
Feed intake/kg eggs	6.3 (100)	6.0 (95)	5.8 (92)
Fertility (%)	88.7	91.0	87.3
Hatchability (%)	73.6	80.0	70.1

[a] Number in parenthesis indicate relative improvement over control (%)

6.3 CITRUS PULP

6.3.1 Introduction

The citrus fruits are belonging to the Rue family (*Rutaceſl*). To this family belongs the sweet orange, grapefruits, lemons, limes and mandarins. Oranges are known as *Citrus sinensis* Osbeck = *C. Aurantium* var. *sinensis* Engl. = *C. Aurantium* Risso. The common orange appears to have originated from south eastern Asia and, although formerly considered a variety of the bitter orange (*C. Aurentium L.*) is now classed as a separate species.

Figure 6.4 shows a cross-section of orange seed with its microscopic structure showing the spermoderm with its four layers; *outer epidermis*, *paranchyma*, *crystal cells* and *inner epidermis*. *Perisperm* and *endosperm* (embryo), the cells of *Cotyledon*, contain *aleuron* cells, with minute aleurone grains. This part contains a high percentage of oil (24.8%) (Winton and Winton, 1949).

The countries involved in the production of fresh and processed oranges can be divided into those with a very high production (U.S.A.), those with a high production (Japan,

Brazil, Spain), and those with moderate production (Italy, Mexico, Argentina, Israel). Countries with a low production are Morocco, Egypt, South Africa, Turkey, Algeria, Greece, Australia, Taiwan, Lebanon and those with a very low production are Tunisia, Jamaica, Chile, Belize, Trinidad, Tabogo and Surinam (Nagy, Shaw and Veldhuis, 1977).

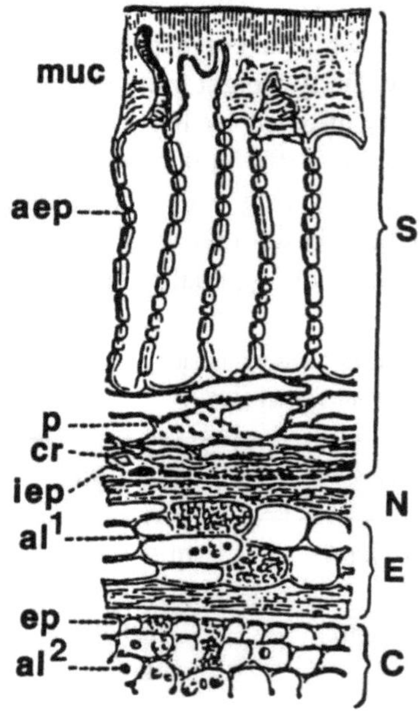

Figure 6.4. Cross-section of an orange seed. *S*, spermoderm; *aep*, outer epiderm; *muc*, mucilaginous outer wall; *p*, parenchyma; *cr*, crystal layer; *iep*, inner epiderm; *N*, perisperm; *E*, endosperm; al^1, aleurone cells; *C*, Cotyledon; *ep*, outer epiderm; al^2, aleurone grains. X 160. Reproduced with permission from Winton and Winton (1949).

In addition to the production of orange oil (essence), oranges serve in the human nutrition as a fresh product to eat or to drink, as beverage, as canned juice, as concentrated juice or syrup, frozen or dehydrated juice, or reconstituted from vacuum-dried powder.

Citrus by-products are numerous such as: pectin, citric acid, citrus-seed oils, *flavonoids* (vitamin P (rutin pro vitamin A) activity of extracts of citrus peel presumed to contain physiologically important *flavonoid* compounds), essential oils, citrus vinegar, marmalades, brined citrus and candied citrus. *Flavonoids* are organic compounds building the carbon framework of a flavone, or more broadly are C_6-C_3-C_6 compounds. The flavanone rhamnoglucosides, hesperidin and naringin, and their corresponding aglycones, hesperetin and naringenin, are the *flavonoids* associated with commercial varieties of citrus fruit.

The different types of waste encountered in citrus processing may be classified as (Von Loesecke, 1952).

Solid waste which includes:

- Cannery waste consisting of peel, rag (internal tissue) and seed.
- Screenings from citrus pulp drying effluents.
- Sludge from peel oil preparation.
- Residues from plants producing citric acid and pectins.

Liquid waste which includes:

- Cannery effluents.
- Pulp drying plant effluents.
- Distillery effluents.
- Effluents from citric acid and pectin plants.

From the citrus processing plant the aim of the feeding industry is to convert all the waste materials from the juice extraction and other operations into a saleable feedstuff suitable as an animal and poultry feed without harmful or unpalatable side effects.

The peel, pulp, rag (internal tissue) and seed are dried in rotary driers, bagged or stored in bulk and sold as citrus pulp known as a feedstuff. Some citrus processing plants are using a multiple effect steam evaporators to concentrate the liquor pressed from the peel into 72 Brix (72% soluble solids measured as sugar) citrus molasses for sale as a feedstuff. If sufficient evaporative capacity is available, other plant effluents (sludge), containing solids, are concentrated with the press liquor. The moisture content of the wet citrus pulp varies from 73-83%, and after processing, moisture in the dried citrus varies from 7-11%, therefore drying is essential to make this product suitable as a feedstuff (Rebeck and Cook, 1977).

Berry and Veldhuis (1977) reported that in the U.S.A. the citrus processing industry is concentrated in four states: Florida, California, Texas and Arizona. About 91% of the oranges grown in Florida is processed and in general, from all the total production of U.S.A., 50% or more is directed towards fresh materials. The total world production of fresh oranges in 1990 is estiminated to be 52 million t (FAO, 1991).

Rebeck and Cook (1977) calculated the percentage of wet citrus waste as peel, pulp, rag and seed after extraction of juice to be ca. 44.6% which will produce 11% dried citrus waste with 10% moisture. In some countries this waste material is still disposed off by spreading the solid waste on adjacent fields and flushing the liquid wastes into streams, lakes or sewers. If the waste material is given more attention and if new techniques can be used, then a considerable amount of citrus wastes can be converted into a useful feedstuff.

Several research workers have discussed the value of the citrus waste as a feedstuff for broilers and layers (Angalet *et al.*, 1976; Eldred, Damron and Harms, 1976; Coleman and Shaw, 1977; Karunajeewa, 1978; El Moghazy and El Boushy, 1982[a]; Velloso, 1985).

6.3.2 Processing

(a) Processing of fresh oranges

The processing of oranges and its produced waste has been described in details by many research workers (Shearon and Burdick, 1951; Veldhuis and Rushing, 1954; Agricultural Research Service, 1956; Veldhuis *et al.*, 1972; Bryan, Anderson and Norman 1974; Berry and Veldhuis, 1977; Nagy, Shaw and Veldhuis 1977). Their conclusions will be summarized to cover the aim of this book, but detailed informations can be found in the above-mentioned books and articles.

The process of preparing juice from oranges is shown in Figure 6.5 and takes place as follows:

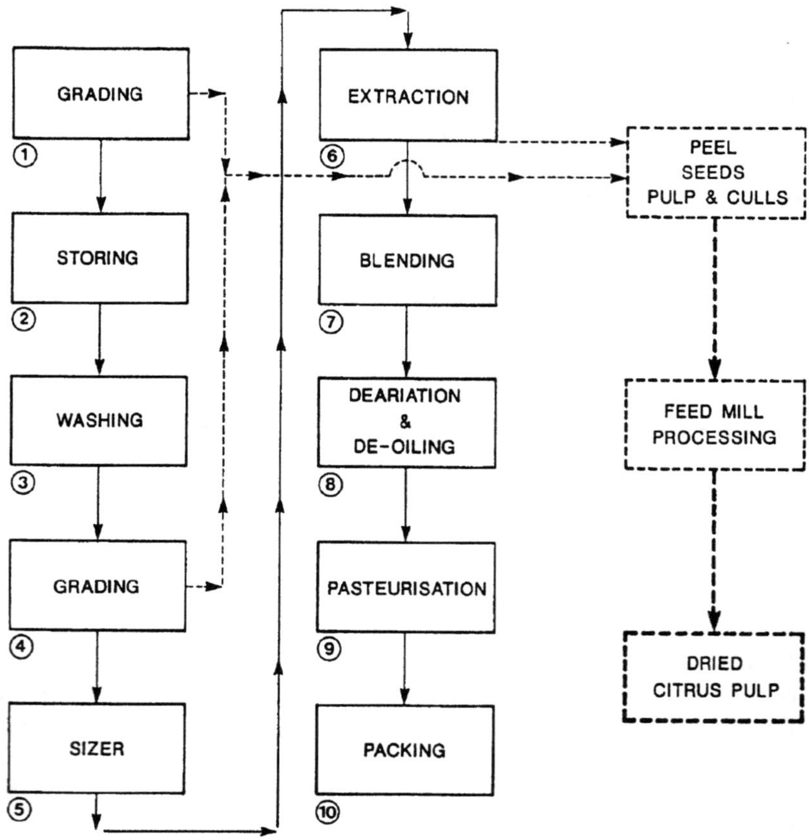

Figure 6.5. Schematic diagram of processing citrus juice and waste to dried citrus pulp.

Harvesting

There are two methods used for handling and harvesting fruit destined for processing, the traditional method (traditional ladder-climbing routine) and a mechanical one. The traditional method is based on picking by hand or pulling (or clipping) from the tree. In this method a long wooden ladder is usually laid against the tree supported by branches. The fruit is gathered in canvas bags. Finally the picked fruit is collected and transported by a tractor to a large truck. This type of labour requires a certain degree of skill and experience. The cost of hand picking is still the largest single item in the cost of producing citrus fruit. A further modification to the ladder climbing is the mechanically assisted hand picker device. The fruit picker stands on a hydraulically operated elevator platform mounted on a tractor. Fruit is picked more easily and allowed to drop on the ground and this is then gathered by a mechanical pick-up machine which also separates the fruit from sand, trash and undesirable fruit, and transports it to the truck.

The mechanical harvesting of citrus fruits is based on the use of a type of abscissions (natural separation) chemical which is usually used to loosen the fruit prior to harvesting. After a period of time varying from 3 to 5 days, fruit is gathered from the tree by means of shaking the tree parts by wind, water or mechanical attachment of a vibrating device to the tree trunk. After the fruit is shaken to the ground it may be picked up by hand, or gathered into a row by a mechanical sweeper or wind-row device, and then picked up by a mechanical pick-up machine.

Grading

Before processing, tests are performed to assure a certain minimum maturity requirements are fulfilled. Those requirements are usually based on the colour of the fruit, minimum juice content, minimum acid content, minimum percentage of total soluble solids and Brix/acid ratio.

Fruits are transported to citrus plants in long semi-trailer trucks, from which they are emptied by allowing the fruit to roll into conveyer belts by gravity. They then pass through a roller conveyer which leads to a single layer belt which passes the hand grading table. During the grading the following are discarded by hand: badly damaged, soft, over-ripe, unripe, low quality and unwholesome fruit. The load will be identified and accompanied with a complete physical and chemical determinations of the juice.

Storing

Before storing it is possible to mix several identified loads together in order to improve the juice quality as far as colour and Brix are concerned. Fruit is moved to storing bins for 12-72 h before processing. Bins are designed for an optimum ventilation to avoid mould and fungal growth. The bins are built in separate stores of a specific height to avoid gravitational pressure on the accumulated weight of fruit.

Washing

From the storing bins, the fruit is conveyed to a brush washer by adding an approved detergent. Fruit is washed and scrubbed by rotating brushes and then rinsed with sprayed water.

Grading

After washing, fruit is moved via a belt to pass the second hand-grading table. Fruit is inspected again to remove damaged fruit that may have been missed earlier or damaged through the high-speed transportation, storing and washing system.

Sizer

After secondary grading, the fruit is separated automatically according to size and directed to the juice extractors. The new mechanical system of sorting or sizing fruit is based on differences in the degree of roundness and the "index of restitution" of the fruit. This index is based on the fact that undesirable fruit is often misshapen, split or softened, and when pressed tend to return to its original shape less rapidly than round, firm, and desirable fruit. Bryan, Anderson and Norman (1974) developed a device that allows the fruit to roll down a ramp, strikes a rotating cylinder and bounces accross a projected pathway. Due to the differences in sphericity and index of restitution, better quality fruit rolls faster and bounces further, so that barriers and coveyer belts can be adjusted to separate the fruit into streams.

Juice extraction

There are many systems varying in capacity and technique that are used as of citrus juice extractors. The simplest type is the rotary juice press extractor. It consists of four drums, two top drums with cups and two bottom drums with knobs for pressing fruit halves in the cups. The fruits are fed into the cups of one top drum, each is encompassed by the mating cup in the other top drum and cut in half by a knife between the drums. Each half is held by a retainer and pressed by mating knobs on the bottom drums, then each half peel is ejected. Through an outlet the juice is collected in the bottom of the unit in pans. This system is not very satisfactory in commercial operations, but it is still used commercially in Italy, Spain, and Central and South America.

Blending

If the acidity and tartness of the juice is high it may be blended with other juices or needed to be sweetened. Quality of citrus juice products can be greatly improved and standarized through the use of sweeteners and through blending operations. Example of blending is grapefruit juice, which is often blended with orange juice and sweetened with sugar and it is mostly labelled "sugar added, orange-grapefruit juice".

Deaeration and de-oiling

Deaeration is of a great importance owing to the negative effect of the dissolved oxygen on vitamin C levels and the effect of flavour deterioration. Dissolved oxygen disappears rapidly in canned juices, particularly at high temperatures during pasteurization. Vacuum de-oilers simultaneously deaerate juice.

The main purpose of de-oiling is to control the peel oil level in citrus juices. Deoilers are small vacuum evaporators where the juice is heated to about 81 °C, creating a loss of about 3-6% by condensation. After this step the oil is separated by centrifugation or decantation and the aqueous layer is returned to the juice. With this treatment about 75% of the volatile peel oil can be removed. In the orange juice, the upper limit for oil is 0.035% by volume.

Pasteurization

Aroma and flavour in citrus juices are sensitive to heat. These delicate fresh aroma and flavour may be lost or damaged by undue exposure to heat, so they are usually pasteurized as rapidly as possible. In commercial practice the juice is rapidly heated to about 92 °C. This temperature for a very short time varying from a fraction of a second to about 40 s is sufficient to kill or deactivate most spoilage organisms. Recent trends are towards the use of high- temperature/short time pasteurization with either turbular or plate-type heat exchangers, that are heated by either steam or hot water and then cooled with cold water.

Packing or canning
The first processed citrus juices to be widely distributed are canned single-strength juices. Canned juices have been partially replaced by frozen concentrated juices because of their higher quality, lesser volume and the increasing availability of freezer space at the supermarkets and at home. The frozen concentrated orange juice is packaged in different sizes and types of containers according to the customers' taste. However, the most available and suitable forms for export are the dehydrated citrus juices. Citrus juices contain 85-90% water, and many attempts have been made to remove all the water and prepare a completely dehydrated product. The processes dealing with dehydration are freeze-drying, spray drying, vacuum pulp drying and foam-mat drying. The best known citrus juice products are the formulated citrus-flavoured beverages. These beverages do not contain high levels of citrus juices, but they may account for up to 5% of total available citrus solids.

(b) Processing of citrus waste

The value of cannery waste was early recognized by cattlebreeders located in the vicinity of citrus canneries. Because of the high water content and perishable nature of the waste, it could not be transported economically for feeding purposes. The fresh material was difficult to handle, fermented rapidly, and soured, and became a fly-breeding and odoriperous nuisance. Stabilization of the waste by drying seemed logical for preservation to permit distribution and storage.

Processing of citrus waste and citrus sludge has been described in detail by many research workers (Von Loesecke, 1952; Agricultural Research Service, 1956; Ratcliff, 1977; Rebeck and Cook, 1977). These reports will only be summarized to fulfil the aim of this book, detailed informations could be found in the above-mentioned books and articles. The process of preparing citrus pulp from peel, pulp, rag, seed and molasses is shown in Figure 6.6 and takes place as follows:

Solid waste

Peelpin storage
After extracting the juice from the citrus fruit, peel is dropped into a conveyer to be carried to the feed mill. Pulp, rag and seed are removed from the extracted juice finishing operation. The pulp will be combined with the peel and will be stored in a peel bin, usually located outside the feed mill building.

The moisture content of the combination of peel, pulp, rag and seeds varies with the season, condition and variety of the fruit. If water is not added in the extractor room, the moisture content will be about 76-78%. If oil peel will be extracted, then water has to be added for the extraction and will result in an increase of the moisture content of the peel conveyed to the feed mill.

Liming operation
From the peel bin the wet peel is conveyed by screw conveyers to a liming station where lime (calcium hydroxide or calcium oxide) is spread on the moving wet peel. Lime is

added to the wet waste at 0.3-0.5% to neutralize the fruit acids and combine with the fruit pectins to form calcium pectate which will reduce the slimy viscous pectin. The addition of lime results in the high percentage of calcium (1.6-2%) in the dried citrus pulp. The peel is then dropped into shredders where it is chopped into small pieces.

Shredding
Most of the shredders are based on a vertical or horizontal shaft hammer mills using square hole screens with 1.9-2.5 cm openings. High speed knives or hammers rotate close to the screen and cut the peel into pieces 0.6-2.0 cm in size. Following shredding, the peel, which by now is thoroughly mixed with the lime, continues to be mixed in a reaction conveyer or flug mill. A minimum reaction time of 6 min is necessary; 12 min is preferable. The high alkalinity, occuring shortly after the lime addition, brings a rapid degradation and demethylation of the pectines in the peel, and releases of the bound liquid. A mixture with lime becomes less slimy and is ready for pressing. The wet peel will be elevated by a screw and dropped through chutes into the peel press.

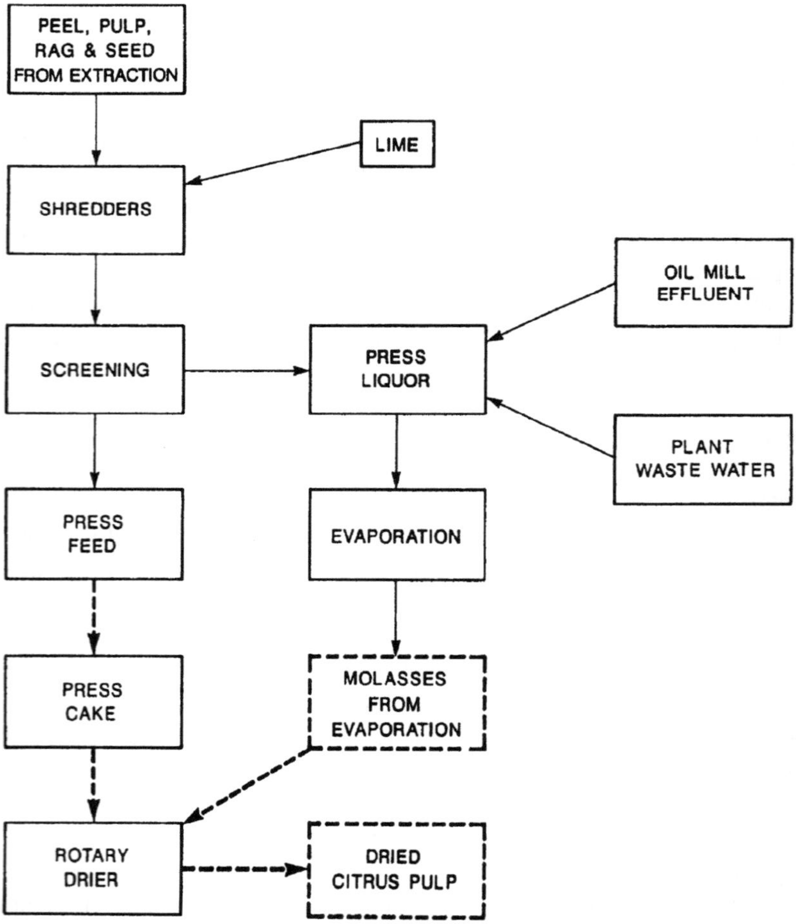

Figure 6.6. Schematic outline of the citrus processing from fruit to juice and dried citrus pulp.

Pressing
These presses generally are vertical or horizontal screw presses with a capacity starting with 32 t/h wet peel impact and reduce the moisture of the peel to 70-75%. The press cake is dropped into the dryer feed screw conveyer and the press liquor is screened and pumped to the evaporator feed tank.

Drying
This takes place mostly with a rotary direct-fired type dryer with an oil or gas furnace. Hot air enters the feed end of the rotary drum at approximately 65 °C. Hot air dries the peel and conveys it to the discharge end through a series of baffles and tumblers where the peel exits at approximately 116 °C. A single pass rotary dryer has a capacity to remove water of up to 27.2 t/h and will produce up to 12.3 t/h of citrus pulp with 10% moisture.

The peel is discharged to the discharge conveyer through an air lock. Fine particles and dust are separated from the larger pieces and will be processed into pellets. The larger pieces are first conveyed to a cooling reel, which cools the product by induced air; thereafter, they are pelleted. All pellets are cooled by air. The moisture content of the hot product leaving the drier should be about 16% to give 10% in the cooled pelletized final product.

Molasses
The press liquor mentioned in the pressing stage is fed to an evaporator where it will be concentrated and result to a product known as citrus molasses with a 72 Brix which may be pumped to a small holding tank. However, it is usually more profitable to sell the product as a concentrate citrus pulp rather than as citrus molasses. Therefore the molasses will be pumped from the holding tank and fed back to the wet peel to produce a product known as citrus pulp. Finally, the citrus pulp pellets are placed in sacks, although most of the product is handled in bulk and shipped by trucks or rail.

Peel oil
During the pressing operation peel oil is expressed and enters the evaporator with the press liquor. This oil flashes off during evaporation where it is separated in its vapour form, recondensed separately with some water vapour and separated in its liquid state by decanting. The final product is principally *d*-limonin. This is sold in bulk for use in flavouring, plastics, paints and as a substitute for terpentine.

Liquid waste
The disposal of liquid waste from citrus processing plants is of particular concern for the industry. This waste, or effluent generally consists of can-cooler overflow, fruit-washing wastewater, peeling and hand-grading table wastewater, flushing water from the floors and water dripping from waste bins. In several countries, water purification does not reach a satisfactory level and pollution takes place through disposing citrus effluents as follows: emptying into lakes, ponding, discharge into groves or into wells, disposing into tidewater rivers and city sewage system discharge into. All these systems create a lot of pollution, generating of gas, clogging and damage, spreading of flies and odours. This subject

can be found in detail in the above mentioned articles and books. The transfer the effluent produced from the citrus processing plant to a sludge of a high nutritive value as an animal and poultry feedstuff and recycling of the purified wastewater, are of a great economic value and improve the environmental sanitation and reduce pollution.

Liquid waste treatment plant

In a citrus processing plant the processing of effluent water is mostly based on conventional shaker screens to remove coarse particles. Air flotation will also remove fine particles from the juice. Addition of small amounts of lime and polyelectrolytes floc (ferrous sulphate, alum, sodium aluminate, ferric chloride, bentonite, and various combinations of these chemicals) are used. In conjunction with air flotation, they will result in a clear effluent, but in general they have a higher oxygen-consumption value. This combined process will reduce the COD (Chemical Oxygen Demand) of the effluent by 10-20%. It will give a well-dewatered float concentrate of 5-15% dry organic solids (by weight) (see Chapter 3). Finally the sludge formed may be passed for drying in a by-product plant or added to the processing of citrus pulp. Treatment by aerobic or anaerobic fermentation may take place to produce a product called activated sludge (Von Loesecke, 1952; Agricultural Research Service, 1956; Ratcliff, 1977).

Activated sludge

Liquid waste is discharged into large aeration basins into which atmospheric oxygen is diffused by releasing compressed air into the waste or by mechanical surface aerators. The created environment under these conditions will be favourable for the growth of a heavy concentration of bacteria owing to the presence of organic food supply and oxygen. The organic content of the waste is removed from the active processes of the bacteria and stored as a protoplasm bacteriae mass. The bacteriae mass, known as activated sludge, is then removed in sedimentation basins where it may be gathered and dried to produce a high quality feedstuff (Pailthorp, Filbert and Richter, 1987).

(c) Yields of citrus waste

Citrus pulp

The traditional measurement unit is a box of oranges (40.8 kg). Juice yields from extraction vary but an average is 22.7 kg per box (55.4%). The solid content of this juice also varies, but 12% soluble solids is used for calculation with a 12 Brix (soluble solids measured as sugar). Approximately 18.2 kg of peel, pulp, rag and seed remain from each box of fruit after juice extraction (44.6%). This can be assumed to contain 4.5 kg of dried citrus pulp at 10% moisture, yielding about 11%. If no water is added to the wet peel during oil extraction then the moisture content of the peel will be 77.5%. If water is added the moisture content will be 82%.

Citrus waste effluent

For a box of oranges (40.8 kg) the average water requirement is 1,385 l for processing. During the concentration process of the citrus juice another 11.4 l are needed per box. Nearly 10% of the wastewater carries the heaviest load of organic materials. An annual average for this 10% volume heavy load will be about 590 mg BOD per litre (1,850 mg COD per litre); the remaining 90% of the wastewater is very low in BOD. From a citrus waste treatment plant analysis the main results were as follows: BOD 594 mg/l, COD 1,794 mg/l; phosphorus 3.11 mg/l, pH 8.96; Turbidity 29.7 and settleable solids 3.16% (Ratcliff, 1977).

6.3.3 Chemical Analysis and Nutritive Value

El Moghazy and El Boushy (1982[b]) classified citrus residues as dried citrus pulp (DCP) and dried citrus sludge (DCS) with their sub classification as sun-dried and processed-dried sludge and aerobic and anaerobic fermented sludge.

(a) Dried citrus pulp (DCP)

Dried citrus pulp is used primarily as a carbohydrate concentrate. It is low in crude protein (6.5%), metabolizable energy (5.5 MJ/kg), fat (4.6%), phosphorus (0.12%) and in essential amino acids. However, it has a high content in crude fibre (13%), calcium (2%) and limonin (0.01%) (Scott, Nesheim and Young, 1982; Tables 6.14 and 6.15). Dried citrus pulp that has been pressed before drying is 10% lower in nitrogen-free extract. According to the processing system, the contents of ash, fibre and water are consistent, while protein and nitrogen-free extract vary according to season, the proportions of oranges and grapefruit used, and also the amount of seeds in the fruits. Owing to the addition of lime during the processing to bind pectin, the Calcium level is high in relation to phosphorus. Owing to this unbalance, care has to be taken during formulation of poultry diets to guarantee the right ratio of Ca : P (Göhl, 1975).

Table 6.14. Chemical analysis and amino acid pattern of dried citrus pulp. From [(1)] Scott, Nesheim and Young (1982); [(2)] Von Loesecke (1952) and [(3)] Kelley (1958)

%	*(1)*	*(2)[a]*	%	*(1)*	*(2)*	*(3)*
Moisture	-	7.50	Magnesium	-	0.30	
Crude protein	6.50	6.50	Sulfur	-	0.10	
M.E. (MJ/kg)	5.50	-	Iron and aluminium [c]	-	1.90	
Crude fat	4.60	5.50	Xanthophylls (mg/kg)	25.00	-	-
Crude fibre	13.00	14.00	Chlorides	-	0.05	-
Pentozans	-	14.40	AMINO ACIDS %			
Total sugars	-	11.00	Arginine	0.20	-	0.30
Pectin [b]	-	18.20	Histidine	-	-	0.10
Naringin	-	1.50	Isoleucine	-	-	0.20
Limonin	0.01	-	Leucine	-	-	0.30
Ca	2.00	1.60	Lysine	0.20	-	0.20
P	0.12	0.13	Methionine	0.08	-	0.08
K	0.04	-	Cystine	0.10	-	-
K and Na chloride	-	1.80	Phenylalanine	-	-	0.20
Manganese (mg/kg)	7.00	-	Threonine	-	-	0.20
Zinc (mg/kg)	14.00	-	Tryptophan	0.06	-	0.10
Silica	-	0.20	Valine	-	-	0.30

[a] Dried grapefruit waste; [b] Alcohol precipitate; [c] $AL_2O_3 + Fl_2O_3$
ME = metabolizable energy

Neal, Becker and Arnold (1935) reported that DCP product has a high digestibility of 83 and 81% in poultry for grapefruit and orange pulp, respectively. They reported a digestibility co'fficient for nitrogen-free extract of 88-92%. The crude protein fraction was lower in digestibility showing only 24.8 and 36.6% digestible material in dried grapefruit and orange pulps, respectively.

Table 6.15. Chemical analysis of citrus cannery waste. From (1) Von Loesecke (1952) and (2) Kelley (1958)

	Grapefruit				*Lemon*
	Florida (1)		*California (1)*	*Israel (2)*	*(1)*
%	*Peel*	*Rag*	*Peel and rag*	*Peel and rag*	*Peel and rag*
Total solids	16.7	15.6	22.0	17.9	16.2
Ash	0.7	0.8	0.7	0.7	0.8
Volatile oil	0.4	-	0.6	-	-
Citric acid	0.7	0.6	0.4	-	0.6
Crude fibre	1.7	1.4	2.0	1.9	2.7
Crude protein	7.1	6.6	8.3	7.5	9.7
Crude fat	0.3	0.2	0.2	0.3	-
Total sugar (as invert)	6.4	6.3	8.7	-	-
Pentosans	0.8	0.4	1.3	-	2.6
Pectin (calcium pectate)	3.1	3.6	3.9	-	-
Naringin	0.4	0.1	0.6	-	-

(b) Dried citrus sludge

Citrus sludge is a decomposition product of effluent wastewater discarded during the processing of citrus. Its protein percentage varies according to the treatment process, aerobic, anaerobic and sun-dried citrus sludges having 42.0, 34.3 and 20.9%, respectively (Table 6.16). The sun-dried sludge had fewer amino acids than the other two treatments which were limited in arginine, histidine and sulphur amino acids methionine and cystine for broiler chickens (Coleman and Shaw, 1977). Table 6.16 shows the effect of fermentation methods (aerobic, anaerobic) and sun-drying on citrus sludges for poultry feeds, and their percentages of crude protein and amino acids in comparison with the nutrient requirements for broilers. Microbial digestion lowers the quantity of decomposable organic matter and increases the protein content in comparison with sun-dried sludge.

Eldred, Damron and Harms (1976) analyzed the activated citrus sludge and found it had the following composition: crude protein 38.6%, crude fibre 12.6%, calcium 1.5% and phosphorus 1.6%. The determination of metabolizable energy (7.4 MJ/kg) was based on an estimation value for comparable materials. Hackler, Newmann and Johnson (1957) reported values for amino acids methionine 0.50%, cystine 0.20%, and lysine 1.3% for sewage sludge. Eldred, Damron and Harms (1976) used these for amino acids in sewage sludge for the calculation in their experiment (Table 6.17).

Table 6.16. Chemical composition of dried citrus sludges compared with amino acid requirements for broilers (%). From Coleman and Shaw (1977)

%	*Requirements* [a] *(pullets 0-6 weeks)*	*Aerobic treatment*	*Anaerobic treatment*	*Sun dried*
Arginine [b]	1.40	1.32	1.18	0.42
Glycine and / or	1.15	1.62	1.66	0.57
Serine	-	0.92	0.92	0.41
Histidine [b]	0.46	0.37	0.31	0.10
Isoleucine	0.86	1.13	1.02	0.42
Leucine	1.60	2.11	1.85	0.76
Lysine	1.25	1.25	1.30	0.38
Methionine [b] or	0.86	0.67	0.54	0.12
Cystine and	0.40	NC	NC	NC
Methionine	0.46	0.67	0.54	0.12
Phenylalanine or	1.50	1.44	1.24	0.43
Tyrosine and	0.70	0.80	0.88	0.27
Phenylalanine	0.80	1.44	1.24	0.43
Treonine	0.80	1.34	1.18	0.51
Tryptophan	0.23	0.36	0.39	0.10
Valine	1.00	1.73	1.51	0.57
Alanine	N [c]	2.61	2.34	0.92
Aspartic	N	2.74	2.65	0.99
Glutamic	N	3.61	3.10	1.08
Proline	N	1.10	1.03	0.45
Total available amino acids	25.45	23.80	8.65	
Protein [d]	23.00	42.01	34.25	20.94
Ash		0.50	12.40	11.70

NC = Not counted

[a] From National Academy of Sciences (1971).

[b] Limiting amino acids in aerobic and anaerobic sludge samples.

[c] N = not required for broilers.

[d] Kjeldahl nitrogen x 6.25.

Table 6.17. Nutrient analysis (%) of citrus sludge. From [(1)] Eldred, Damron and Harms (1976); [(2)] Hacker, Newman and Johnson (1957)

	(1)		*(2)*
Moisture	6.3	Methionine	0.5
Crude protein	38.6	Cystine	0.2
M.E. (MJ/kg)	7.4	Lysine	1.3
Crude fibre	12.6		
Calcium 1.5			
Phosphorus	1.6		

(c) *Citrus molasses*

Citrus molasses is a liquid obtained from cured citrus waste. It contains 10-15% soluble solids of which 50-70% are sugars. This material contributes to more than 50% of the total weight of cannery refuse available to the feed mill. A typical analysis of citrus molasses is as follows: concentration 72 Brix, pH 5, a viscosity of 2,000 centipoises, nitrogen-free extract 62%, total sugars 45%, moisture 29%, crude protein 4.1%. Table 6.18 shows its complete composition (Hendrickson and Kesterson, 1951).

Table 6.18. Chemical composition of citrus molasses. From Hendrickson and Kesterson (1951)

Constituent	%
Nitrogen-free extract	62.000
Total sugars	45.000
Moisture	29.000
Reducing sugars	23.500
Sucrose	20.500
Carbonate ash	4.700
Acid, as anhydrous citric	4.500
Nitrogen x 6.25	4.100
Glucoside	3.000
Pentosans	1.600
Pectin	1.000
Fat	0.200
Volatile acids	0.040
Potassium	1.100
Calcium	0.800
Sodium	0.300
Magnesium	0.100
Iron	0.080
Chlorine	0.070
Phosphorus	0.070
Silica	0.010
Manganese	0.008
Copper	0.003

Citrus molasses resembles cane molasses and is used in the feeding of dairy cattle. Its extreme bitterness makes it unsuitable for human consumption unless treated for removal of naringin, but the bitterness does not appear to affect its use in animal feeding (Burdick and Maurer, 1950). Its use in poultry nutrition is not known but it may be suggested that it could be used as a binding material during pelleting in the place of cane molasses.

6.3.4 Feeding Citrus Waste

(a) Citrus pulp

Broilers

Ewing (1963) reported that chicks fed on citrus pulp showed a decreased growth rate, compared with controls, during the first 4 weeks. The inclusion of citrus meal up to 20% in the chick diets resulted in a very high mortality, while the 10% inclusion level led to a higher feed consumption per unit of gain and no extra mortality was observed. Schaible (1970) found that citrus seed pulp caused a high mortality during the first three weeks and caused enlarged gall bladders, mottled liver and congestion of the intestinal tract owing to the presence of limonin.

Buriel, Criollo and Rivera (1976) carried out trials in which high levels of citrus pulp were used (0, 20, 30 and 40% in broiler diets for starter and finisher). The analysis of variance and orthogonal comparisons showed highly significant differences among the means of the variables studied: body weight, feed consumption and conversion (Table 6.19). They concluded that citrus pulp meal cannot be used at levels of 20% or higher in rations for broilers. Their results do not give a real picture of the effect of dried citrus pulp in levels lower than 20% which are levels that are considered to be of great importance for such a product in poultry nutrition.

Table 6.19. Average body weight, feed consumption and conversion of broilers fed various levels of dried citrus pulp in comparison with a control diet. From Buriel, Criollo and Rivera (1976)

Treatment	*Average weight (g/chick)*		*Feed consumption (g/chick)*		*Feed conversion*	
	0-4 weeks	*0-8 weeks*	*0-4 weeks*	*0-8 weeks*	*0-4 weeks*	*0-8 weeks*
Control	661[a]	1669[a]	1154[a]	3036[a]	1.6[a]	1.8[a]
20%	571[b]	1374[b]	1054[a]	3156[b]	1.8[a]	2.3[b]
30%	459[c]	1293[b]	1365[b]	3374[c]	3.0[b]	2.6[b]
40%	408[d]	1099[c]	1541[c]	3396[c]	3.8[b]	3.1[b]

[a,b,c,d] Means within a column with the same superscript are not significantly different at $P < 0.01$.

El Moghazy and El Boushy (1982[b]) investigated lower levels of dried citrus pulp in diets for broilers. Their diets were calculated on a base of 60% yellow maize and 19% soybean oil meal. Beside the control diet, the experimental diets contained citrus pulp at 2.5, 5, 7.5, 10 or 12.5% added at the expense mainly of maize. Crude fibre content was 2.03% for the basal diet and 2.28-3.72% with increasing citrus pulp inclusion. All diets were isonitrogenous (22% crude protein) and isocaloric (13.4 MJ/kg metabolizable energy). Limiting amino acids were constant by the addition of synthetic amino acids to cover the requirements. The digestible levels of methionine and lysine were 0.46% and 1.03% respectively (Tables 6.20 and 6.21). Adding citrus pulp in levels higher

Table 6.20. Composition of the experimental Mash diets. From El Moghazy and El Boushy (1982[a])

		Citrus Pulp %				
Ingredients %	*Control*	*2.5*	*5.0*	*7.5*	*10.0*	*12.5*
Yellow corn	60.30	57.04	53.70	50.44	47.20	44.44
Soybean meal (48.8%)	19.30	19.02	18.79	18.48	18.17	17.76
Rice polishing	2.75	2.75	2.75	2.75	2.75	2.75
Citrus pulp	0.00	2.50	5.00	7.50	10.00	12.50
Meat meal (58%)	7.60	8.05	8.50	9.00	9.50	10.00
Herring meal (72%)	3.00	3.00	3.00	3.00	3.00	3.00
Animal fat	3.70	4.50	5.32	6.10	6.88	7.50
Dicalcium phosphate (18.2% P; Ca 26%)	0.90	0.84	0.78	0.71	0.64	0.57
Lime stone	0.99	0.83	0.69	0.54	0.39	00.00
Salt	0.30	0.30	0.30	0.30	0.30	0.30
Vit. min. premix	1.00	1.00	1.00	1.00	1.00	1.00
Methionine (99%)	0.12	0.13	0.14	0.15	0.15	0.16
Lysine-HCl (78%)	0.04	0.04	0.03	0.03	0.02	0.02
Total	100.00	100.00	100.00	100.00	100.00	100.00

Table 6.21. Calculated contents of the diets with different levels of citrus pulp. From El Moghazy and El Boushy (1982[a])

	Control	*Citrus pulp inclusion*				
%		*2.5%*	*5.0%*	*7.5%*	*10.0%*	*12.5%*
Crude protein	22.00	22.00	22.00	22.00	22.00	22.00
M.E. M.J./kg	13.40	13.40	13.40	13.40	13.40	13.40
Ether extract	8.12	8.91	9.73	10.52	11.32	11.96
Crude fibre	2.03	2.28	2.52	2.77	3.01	3.27
Ash	6.08	6.06	6.08	6.07	6.06	5.84
Calcium	1.20	1.19	1.20	1.20	1.20	1.11
Total Phosphorus	0.79	0.78	0.78	0.77	0.77	0.76
Avail.Phosphorus	0.55	0.55	0.55	0.55	0.55	0.55
Linolic acid	1.79	1.80	1.82	1.83	1.85	1.85
Total Methionine	0.52	0.52	0.52	0.53	0.53	0.54
Diges.Methionine	0.46	0.46	0.46	0.46	0.47	0.47
Total (Meth.+ Cyst.)	0.86	0.86	0.86	0.86	0.86	0.86
Diges.(Meth.+ Cyst.)	0.72	0.71	0.71	0.71	0.71	0.70
Total Lysine	1.25	1.25	1.25	1.25	1.25	1.25
Diges.Lysine	1.03	1.03	1.03	1.02	1.02	1.01

than 7.5% depressed body weight at 4 weeks, but less at 7 weeks of age. Feed intake was decreased at 4 weeks and increased at 7 weeks of age, but not significantly. With citrus pulp at a 12.5% level of inclusion, feed conversion was low in the control and increased with increasing citrus pulp levels (Table 6.22). This effect was paralleled by an increasing crude fibre and limonin content in the diet (Table 6.23). It was suggested that fibre content of the pulp allowed a rapid passage of ingesta and poorer digestion, and limonin may cause intestinal irritation and causing poor absorption of the nutrients. The researchers concluded that dried citrus pulp was a reasonable feedstuff for broilers at an inclusion level of 7.5% of the diet.

Table 6.22. Average body weight, feed consumption and conversion of male broilers fed on various levels of dried citrus pulp in comparison with a basal ration. From El Moghazy and El Boushy (1982[a])

Treatments	*Body weight (g/chick)*		*Feed consumption (g/chick)*		*Feed conversion*	
	0-4 weeks	*0-7 weeks*	*0-4 weeks*	*0-7 weeks*	*0-4 weeks*	*0-7 weeks*
Control	851[b]	2060	1320	3910	1.63[a]	1.94[a]
Citrus pulp %						
2.5	830[a,b]	1990	1320	3820	1.67[a,b]	1.96[a]
5	856[b]	2040	1350	3910	1.66[a]	1.96[a]
7.5	817[a,b]	2040	1300	3940	1.67[a,b]	1.97[a]
10	814[a,b]	2000	1330	3907	1.71[b,c]	1.99[a,b]
12.5	792[a]	2000	1310	4000	1.74[c]	2.04[b]

[a,b,c] Means within a column with another superscript are significantly different ($P < 0.05$) by Duncan's multiple range test.

Table 6.23. Various calculations of crude fibre and limonin consumption for total mixture and dried citrus pulp. From El Moghazy and El Boushy (1982[a])

Item	*Age (wks)*	*Control*	*Citrus pulp inclusion %* 2.5	5	7.5	10	12.5
Crude fibre	4	0.0	4.3	8.8	12.7	16.9	21.3
consumed from dried citrus pulp (g)	7	0.0	12.4	25.4	38.4	50.8	65.0
Crude fibre	4	26.8	30.1	34.0	36.0	39.1	42.8
consumed from total mixture (g)	7	79.4	87.1	98.5	109.1	117.6	130.8
Limonin consumed	4	0.0	2.9	5.9	8.5	11.6	14.3
from dried citrus pulp (mg)	7	0.0	8.3	17.0	25.7	34.0	43.5

Layers

Karunajeewa (1978) reported from his studies on the effect of dried citrus pulp meal on egg yolk colour and performance of cross-bred layers. During his experiments diets were used without Xanthophyll pigments to produce depleted coloured egg yolks. He fed those layers a control diet and two experimental diets with 5% citrus pulp with or without inclusion of 1 mg/kg Canthaxanthin. He concluded that egg production, feed intake and conversion were not affected by the inclusion of 5% citrus pulp. Dried citrus pulp did not affect yolk pigmentation in the presence or absence of Canthaxanthin.

Velloso (1985) included up to 10% orange pulp in diets for laying pullets to replace maize. Results showed no significant differences concerning growth, sexual maturity or mean egg weight up to 50% of laying capacity. He concluded that the inclusion of 5% dehydrated and pelleted orange pulp could partly substitute maize without affecting laying or growth.

Yang and Choung (1985) used dried citrus peels at levels of 5, 10 or 15% to replace wheat bran in a basal diet containing 60% maize, 10% wheat bran, 16% soybean meal and 3.5% fish meal. The diet with 15% dried citrus peel decreased feed efficiency and gave darker egg yolks. There were no significant differences between the groups in weight gain. They concluded that the optimum inclusion of dried citrus peel in the diet is 10%.

(b) Citrus sludge

Broilers

Eldred, Damron and Harms (1976) reported that citrus sludge could be used as a supplement in broiler diets. Weight gains from diets containing 7.5% or less of this sludge were acceptable, while 10% or more decreased weight gain. The cause of the depression may be due to a deleterious factor known as limonin. They also used chick experimental diets containing 0, 2.5, 5, 10, 15 and 20% dried citrus sludge and 20% sludge plus additional methionine and lysine. Their data indicate that levels between 5 and 10% sludge could be included in diets of growing chicks up to 28 days without adversely affecting growth or other performance criteria. In another experiment broiler chicks from 1 day old were given a diet with 0, 2.5, 5 or 10% citrus sludge at the expense of maize and soybean meal. There was no significant difference in body weight between groups at 8 weeks old. Feed consumption and conversion values were also not significantly influenced by treatment (Table 6.24). These data indicate that a level of sludge between 5 and 10% can be used in the diets of broilers.

Angalet *et al.* (1976) evaluated citrus sludge as a poultry feed ingredient and its effect on the meat quality and flavour of broilers. In their trials with broilers they were fed diets containing either 0, 2.5, 5.0 or 10.0% citrus sludge. At the end of 8 weeks they noticed no significant ($P < 0.05$) differences among levels of citrus sludge in the diet and carcass characteristics such as carcass weight, percentage cooking loss, shear force of meat or sensory evaluation (palatability scores).

Table 6.24. Performance of broilers fed diets containing various levels of citrus sludge (8 weeks). From Eldred, Damron and Harms (1976)

Sludge (%)	*Body weight (g)*	*Feed consumption (g)*	*Feed conversion (g feed/g body weight)*
0.0	1750	3696[a,b]	2.16
2.5	1772	3528[a]	2.08
5.0	1800	3752[b]	2.16
10.0	1732	3808[b]	2.25

[a,b] Means without common letters are significantly different ($P < 0.05$) according to Duncan's multiple range test.

Layers

Eldred, Damron and Harms (1976) evaluated the activated citrus sludge in feeding trials with laying hens. They used diets containing 0, 2.5, 5.0, 7.5 or 20% citrus sludge (Table 6.25). Inclusion of up to 7.5% sludge in the diet did not significantly affect hen-day egg production, egg weight, daily feed intake, or feed efficiency. No significant differences were found in specific gravity of the eggs due to treatments. Haugh unit scores were numerically increased as the level of sludge in the diet increased up to 7.5%. Mortality was not affected by the inclusion of up to 20% citrus sludge in the diet.

Table 6.25. Performance of laying hens fed diets containing various levels of citrus activated sludge (6 months). From Eldred, Damron and Harms (1976)

Treatment (% Sludge)	*Av.egg prod. (%)*	*Egg wt. (g)*	*Feed/bird/ day (g)*	*Feed/doz. eggs (kg)*	*Specific Gravity*	*Haugh Units*
0.0	69.6[a]	62.9[a]	113[a]	2.00[a]	1.0792	64.3[a]
2.5	68.3[a]	63.4[a]	109[a]	1.98[a]	1.0801	67.4[a,b]
5.0	67.0[a]	63.6[a]	114[a]	2.13[a]	1.0787	69.4[b]
7.5	69.2[a]	63.3[a]	112[a]	2.01[a]	1.0780	68.9[b]
20.0	42.1[b]	59.0[b]	92[b]	2.86[b]	1.0781	82.0[c]

[a,b,c] Means without common letters are significantly different ($p < 0.05$) according to Duncan's multiple range test.

Angalet *et al.* (1976) evaluated activated citrus sludge as a feedstuff for layers and the characteristics of their eggs such as yolk colour and egg flavour. Eggs were collected from layers fed diets prepared with citrus sludge at levels of 0, 2.5, 5.0,

7.5 or 20.0% and were examined for differences in yolk colour and development of off-flavour. Colour differences ($P < 0.05$) were observed by reflectance colourimetry and the taste panel. The colour of the yolks increased (was more orange) as the dietary citrus sludge levels were increased. No significant flavour differences were detected by the taste panel for either the yolk or albumen.

6.4 POTATO RESIDUES

6.4.1 Introduction

Potato belongs to the tubers of the nightshade family and is the only important species of the family *Solanaceae* yielding edible tubers. Potato is known as *Solanum tuberosum* L. A great number of varieties has been produced by breeding and cultivation, the main points of difference being size, form, depth of eyes, colour, starch content, productiveness and resistance to disease.

The potato is a tuber, an abruptly thickened underground stem, closely resembling the aerial stem of the plant. Figure 6.7 shows the organization of the principal internal tissues of the mature potato tuber. The outer skin consists of a layer of corky periderm, and its function is retarding loss of moisture and resisting fungal attack. Underlying the periderm is the cortex, a narrow layer of parenchyma tissue. The vascular storage parenchyma, high in starch, lies within the shell of the cortex. Xylem and phloem are found in minute strands or bundles, most of which form a narrow, discontinuous ring (*Vascular ring*). A small central core is formed, radiating narrow branches to each eye and forming the fifth layer, sometimes called the *Water core*. It consists primarily of large cells containing less starch than the cells in the vascular area and the innermost part of the cortex. Each eye contains at least three buds together with protecting scals. A potato tuber contains 13 eyes to 5 turns of the helix. When the normal tuber sprouts, it is the eye at the bud end, corresponding to the growing tip of a stem, that develops first (Talburt, Schwimmer and Bur, 1987).

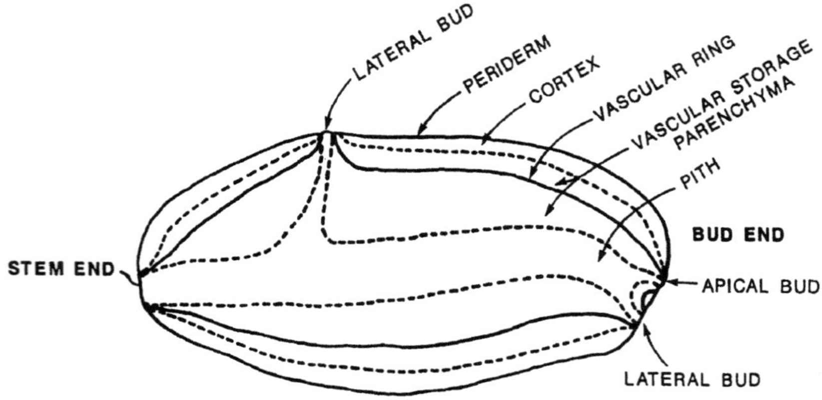

Figure 6.7. Longitudinal section of a potato tuber showing its principal structural features. Reproduced with permission from Talburt *et al.* (1987).

Figure 6.8 shows the potato tuber in cross-section. The peel is the cork layer and the very rich starch grains are combined with the outer and inner phloem, and the cambium.

The countries which are the top producers of fresh potatoes are the Commonwealth of Independent States (formerly USSR) and Poland. The second group includes the United States and what was formerly known as East Germany. The third group includes Western Germany, France, United Kingdom and The Netherlands. Other countries involved in a lower scale are Japan, Yugoslavia, Italy and Canada, in addition to other very low producers such as Austria, Denmark, Finland, Ireland, Sweden and Switzerland. The potato waste, with a total annual world production estimated to be 12.9 million t. Potato is one of the major food crops in the world. Because they yield heavily, are relatively inexpensive and can be grown in a wide variety of soils and climates, potatoes are the staple food of people in many parts of the world (FAO, 1991).

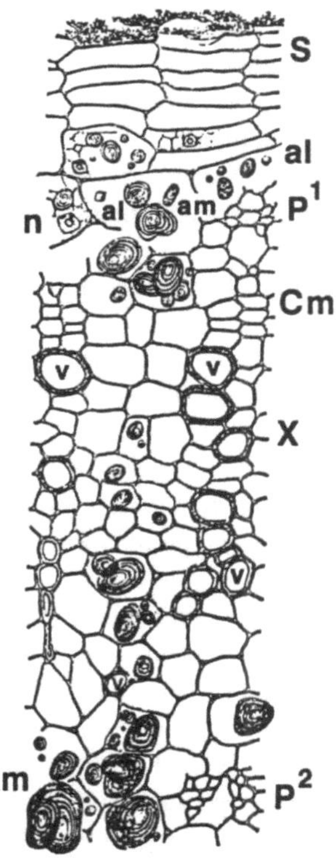

Figure 6.8. Potato tuber in cross-section. *S*, Cork; crystalloids *(al)* and cell nuclei *(n)* contained in outer cortex; *am*, starch grains; P^1, outer phloem; P^2, inner phloem; xylem *(x)* with vessels *(v)*; *Cm*, Cambium. X 160. Reproduced with permission from Winton and Winton (1949).

Potato crop is not only used as a human food, but is also used for stock feeding on the Continent of Europe (Talburt, 1987). Processing of potatoes for food has been increasing world-wide in the form of fresh (whole tuber), dried, flour, canned hash, fried slices and sticks (chips) and soups. The manufacture of starch, glucose, dextrin, and spirits from potato is also taking place. This new development in potato processing is creating tremendous quantities of waste. The methods of potato processing have also encouraged new techniques to provide an effective removal of settleable and dissolved solids from potato-processing wastes. Reduction of the total quantity of waste through selection and operation of efficient peeling systems and processing lines, as well as reduction of water flow through conservation and through water reuse systems are the main steps in a pollution control program. Pailthorp, Filbert and Richter (1987) classified the terms most often used in the effluent treatment of potato processing wastewater as follows.

- *Suspended solids.* Solids which can be mechanically filtered from the wastewater.
- *Settleable solids.* Suspended solids that will settle in sedimentation tanks in normalretention periods.
- *Total solids.* Both suspended and dissolved solids.
- *Primary treatment.* The removal of suspended and settleable solids by screening,flotation, or sedimentation.
- *Secondary treatment.* The removal of organic matter by biological decomposition (usually preceded by primary treatment).
- *Advanced wastewater treatment.* Treatment beyond secondary treatment.
- *Aerobic treatment.* Biological activity in the presence of dissolved oxygen (odours arenot produced).
- *Anaerobic treatment.* Biological activity in the absence of dissolved oxygen (causeodours).
- *BOD.* Biological oxygen demand is a measure of the oxygen necessary to satisfy therequirements for the aerobic decomposition of the waste. It indicates the organiccontent or pollution strength of the waste (see chapter 3).
- *COD.* Chemical oxygen demand is a measure of the amount of oxygen that will reactchemically with a waste. This value varies with the type of oxydant used, with thetesting method used, and with the type of waste (see chapter 3).

Figure 6.9 presents a schematic view of the relationships of the various classifications of solids in liquid waste. The settleable solids portion represents the amount of waste that can be removed by sedimentation (Pailthorp, Filbert and Richter 1987). The aim of the feeding industry is to convert all the waste materials from the potato processing (washing, peeling, trimming, shaping, second washing, separation and final processing). In addition, several heat treatment steps such as blanching, cooking and caustic and steam peeling produce an effluent containing gelatinized starch and coagulated proteins. In contrast, potato starch and chip processing produce effluents in which the components have not been heated. The disposal of all wastewater (rich in starch and protein) by means of conversion into a sludge intended as a saleable feedstuff suitable for animal and poultry feed is a matter of great importance.

Talburt (1987) calculated the total production of potatoes in the United States and estimated the part of potatoes processed for food to be 48% which will be end products in a frozen, chip and dehydrated form. The processing of potatoes to produce chips and french-fries is prosperous and is looking forward to continue expansion for the foreseeable future; the expansion is also reflected in the quantity of waste produced.

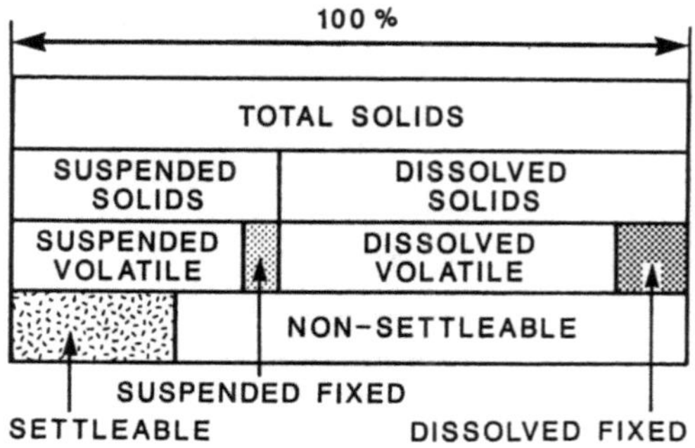

Figure 6.9. Classification of solids in wastewater produced from the potato processing plants. Reproduced with permission from Pailthorp, Filbert and Richter (1987).

Smith and Huxsoll (1987) reported from their work in potato peel losses, that the amount of product removed in peeling is of a great importance to the processor. Not only does it reduce product recovery but also such material is creating a tremendous waste. They reported also that potato size has an important effect on peel losses. As potato size decreases, the surface area per unit of weight increases exponentially. For example, one ton of 112 g weight per potato will have a 25% greater peel loss than one ton of 224 g weight potatoes, due only to the increased surface area. They concluded that the peel losses may vary from 10-50% according to their weight and the method used for peeling.

Many research workers explained the value of fresh potato and of potato waste from several processing systems as a feedstuff for broilers and layers. Potato starch as such and starch from potato waste effluents were studied (Nitsan and Bartov, 1972; Della Monica, Huhtanen and Strolle, 1975; Gerry, 1977). Some experiments have also studied cooked potato flakes (Fangauf, Vogt and Penner, 1961; Splittgerber and Gyzae, 1962; Vogt, 1969; Vogt and Stute, 1969[a,b]; Whittemore, Moffat and Mitchell-Manson, 1974; Whittemore, Moffat and Taylor, 1975).

6.4.2 Processing of Potato Tuber

Potato processing is a very specialized field which can not be described briefly. The potato processing industry produces several products and by-products. The main technological aspects are dealing with the following: peeling potatoes for processing, processing of potato chips, frozen French fries, dehydrated mashed potatoes as granules or flakes, dehydrated diced potatoes, and potato starch and flour. The processing of potato and its by-products has been described in details by many research workers (Smith, 1987; Smith and Huxsoll, 1987; Talburt and Kueneman, 1987; Talburt, Boyle and Hendel, 1987; Talburt *et al.*, 1987; Treadway, 1987; Willard and Hix, 1987; Willard, Hix and Kluge, 1987). Their conclusions are summarized, however, detailed information can be found in the above-mentioned books and articles.

The preparation of potatoes for standard preprocessing is shown in Figure 6.10 and takes place as follows:

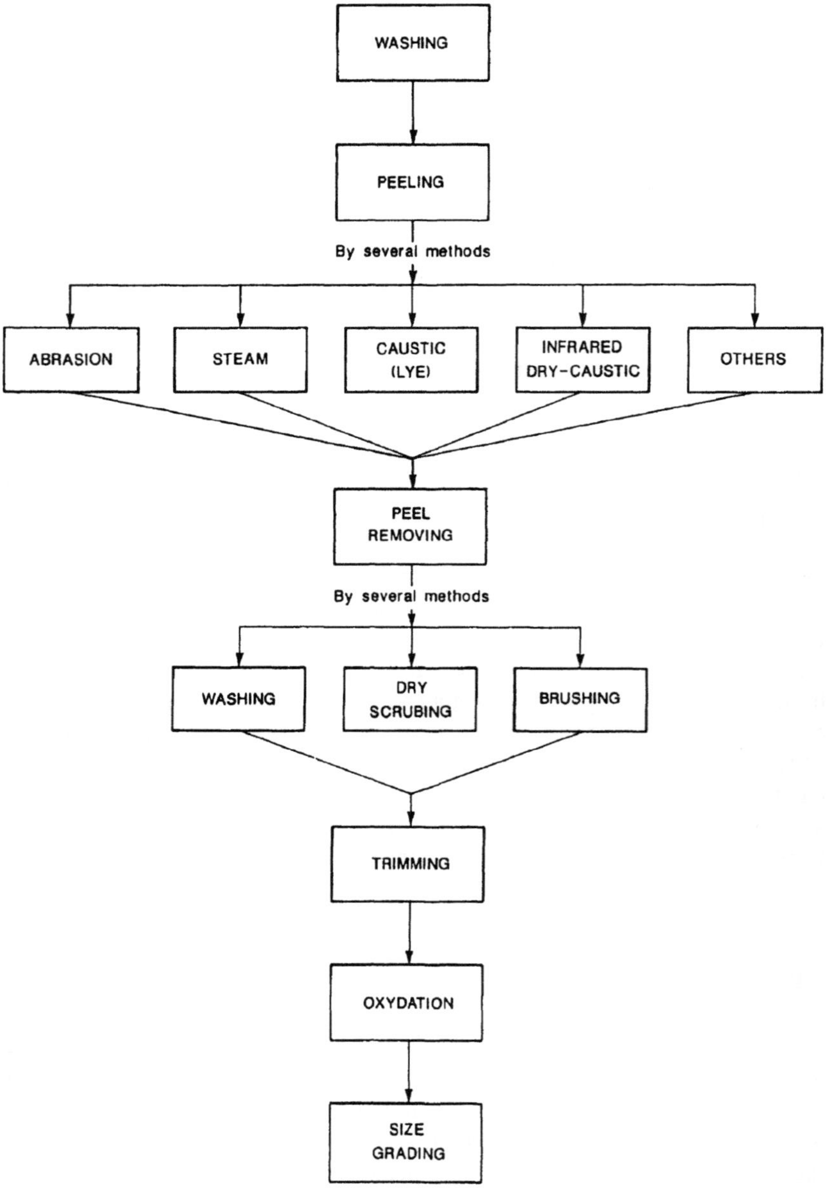

Figure 6.10. Schematic outline of the methods used for peeling potatoes.

(a) Harvesting

If harvesting is not done with care, the crop will be badly damaged and severe losses in storage may occur. Methods of harvesting vary from farm to farm and from one area of the country to another. Potatoes are normally harvested with the two-row mechanical diggers or with the mechanical harvesters. The two-row diggers drop the potatoes to the ground behind the digger and they are later picked up by hand. Potatoes may be collected into crates, wickers, wire baskets, sacks or other containers. After picking the potatoes will be loaded into trucks and transported for storage. The mechanical potato harvesters dig the potatoes, and separate them from the soil, vines and stones. Tubers will be delivered into containers or bulk-body trucks for transit to storage.

(b) Washing

By washing all mud, dirt and sand will be removed. Potatoes are washed by the water in which they are transported (fluming) from the storage room. Fluming, an economical method of conveying potatoes with a minimum amount of brushing, will encourage the dirt and dried mud to soften and wash off owing to the soaking effect. The fluming system allows recirculation of water, thus saving water and minimizing pollution.

Potatoes are often washed by machines equipped with cylinder brushes or studded rubber rolls that vigorously scrub the potatoes by vibration and water spraying. After washing, potatoes are allowed to drain for a short time, accumulated in large bins and then fed at a regulated rate to the peeling operation. Prior to peeling, potatoes will be inspected on a belt for removing trash, vines, green, rot and defects. In some cases potatoes are classified by size before peeling.

(c) Peeling

Potatoes are peeled by several methods depending on the way they will be processed (as chips, french fries, dehydrated mashed as granules or flakes or diced, or potato starch or flour). Peeling of potatoes may occur by means of heat (steam peeling), chemicals [caustic (lye) peeling] or abrasion (abrasion peeling).

Wet abrasion peeling. This type of peeling takes place by batch-type or continuous system; they are both based on a uniform contact of the potatoes surfaces being peeled with abrasive cylinders, disks or rolls in such a way as to remove as thin a peel as possible. This system is commonly used in the potato chip industry, where minimum peeling is required, and in the canned potato industry, where abrasive action is further utilized to shape the potatoes. The drawback of the abrasion peeling is the high peel loss that occurs when the potatoes are used in products which have to be intensively peeled for reasons of non-uniformity of size and shape, deep set eyes and other surface irregularities.

Dry abrasion peeling. A disadvantage of wet abrasive peelers is the discharge of a peel residue made up of fine parts of potato skins and tissue which is flushed away with water and causes pollution owing to the high suspended solids and high biological oxygen demand (BOD) in wastewater. To avoid this problem, the dry abrasive peeling system was introduced. The disposal of the dry peel residue is practized without the use of water. The potatoes are lightly rinsed after peeling, and the small amount of water used is discharged separately. The dry peel residue is mostly used as an animal feed.

Steam peeling. Potatoes are subjected to steam pressure to heat and soften the peel and the underlying surface tissue rapidly. The process depends on cooking the surface of the potatoes quickly before heat penetrates to the interior of the tubers. The steam pressure is released suddenly, causing an almost explosive vaporization of moisture in the heated surface tissue, which further loosens the peel. The peel can be removed with barrel-type washers equipped with high-pressure water sprays or by dry scrubbers. High steam pressures are more effective and produce less heat ring (partial cooking of the potato surface due to gelatinization of starch and producing a dark surface) than lower pressures. A combination of lye and steam peeling has been introduced which is known to reduce the required contact time with steam while increasing the peeling capacity (Slater, 1951; Adams, Hickey and Willard, 1960; Smith and Huxsoll, 1987).

Caustic (lye) peeling. Caustic or lye peeling of potatoes combines the effect of chemical attack and thermal shock for loosening and softening the surface skin, blemishes, and eyes. Moreover, the peels may be readily rubbed or worked off by pressure spray washers or other mechanical action.

The basic caustic peeling process involves putting the wet, washed potatoes in contact with a hot diluted lye solution followed by washing with high-pressure water sprays in a barrel-type washer to remove the softened tissue. In this method the caustic soda used has a concentration ranging from 5-20% NaOH, and a bath temperature ranging from 76-99 °C with an immersion time from 1-6 min. To avoid heat ring formation, the lye concentration and the immersion time must be adjusted in such a way that the depth of heat gelation does not exceed the depth of lye penetration. After washing, the peeled potatoes may be immersed in or sprayed with a dilute acid solution, such as citric acid, to neutralize any residual caustic on the surface (Smith and Huxsoll, 1987).

Infrared dry-caustic peeling. Since very little water is used in this process, it is called dry peeling. The surface tissue is softened by the action of lye and infrared heat, and the softened tissue is removed by the scrubbing action of rotating, rubber-tipped rolls. This method provides clear contour peeling with about one-third less peel loss than with conventional lye peeling.

The process comprises the following steps: Putting wet, washed potatoes in contact with hot dilute lye solution (6-10% NaOH), with an immersion time of about (30-80 s), and a caustic bath temperature 87-99 °C. The holding time between caustic immersion and infrared exposure is (3-5 min.), and the infrared exposure time ranges from 60-90 s. (Graham *et al.*, 1969, 1970; Smith and Huxsoll, 1987).

Other peeling methods. In addition to the commercially used peeling processes discussed before, there are other peeling techniques, which are not used frequently such as brine, flame and oil peeling.

(d) Peel removal

Removing potato peel after peeling by steam, lye and infrared dry caustic except for abrasion peeling, require separating equipment to removing the softened peel. This equipment can be classified into barrel washers, dry scrubbers and brush washers.

Barrel washers. This term refers to a horizontal rotating perforated cylinder fitted with internal water spray nozzels. Potatoes are conveyed through the cylinder and sprayed at high water pressure. Large volumes of water are usually used, as much as 1 600-2 000 l/min. for 20 t/h processing time.

Peel is removed from the potatoes by rubbing against the inner surface of the perforated cylinder, rubbing against each other, and by contact with the water jets. This system will work perfectly if the peels are very loose from potatoes. Its disadvantage is the very high water consumption and large amounts of wastewater requiring purification.

Dry scrubbers. These are machines equipped with wiping action rubber studs or brushes to wipe or scrape off softened potato peels. They are always used after steam peeling because the peel adhere rather tightly and must be mechanically wiped off to obtain clean potatoes.

In general, scrubbers are run without the addition of water. The recovered peel residue has the same percentage of solid matter as peeled potatoes. Its waste is low owing to the absence of water and it provides a feedstuff with a similar dry matter content to the whole potato and is suitable as animal feed. Dry scrubbers are used not only for conventional lye-peeling but also for infrared dry-caustic peeling. The caustic peeling waste can be fermented, after adjusting the pH level with hydrochloric acid, to convert the sodium hydroxide to sodium chloride to provide a good feedstuff for cattle.

Brush washers. Brush washers are mostly used to wash potatoes after peeling. Dry-scrubbed potatoes are mostly covered with a thin sticky layer of starch which causes the peeled potatoes to stick together. Therefore they must be washed after scrubbing. Specially designed brush washers with a strong mechanical action are used with a minimum water supply. They are fitted with cylinder brushes or rubber-studded rolls.

(e) Trimming

After peeling and washing, trimming takes place, to remove residual skin, eyes, discoloured areas, black spots, defective pieces with bacterial, fungal or insect attack and, finally, sunburned and greened material. The amount of trimmings depends on the peeling efficiency, the condition of the raw potatoes, and the product requirements.

Peeled potatoes are conveyed between two rows of inspectors who trim out the small defects and discard defected ones, where the defects are too large to trim out effectively. Trim table conveyors are equipped with troughs along each side to carry away the trimmings and rejected potatoes. The trimmed potatoes are not longer conveyed with water flumes owing to water pollution, but they are conveyed by rubber belts or vibrating conveyor troughs.

Photoelectric sorting devices are used to detect and reject discoloured potatoes automatically. These may be used also to detect potatoes which require hand trimming and transfer them to the trimming belt.

(f) Oxidation

Oxidation of potatoes takes place owing to exposure to the air after peeling, and results in dark parts. The oxidation can be prevented by using a diluted antioxidant, which can be dissolved in water in a dip tank or sprayed over the peeled or sliced potatoes. Dipping in a solution containing 100 ppm of sodium bisulphate or spraying with a similar solution is quite effective.

(g) Grading

After trimming potatoes may pass over a size grader. The grader will select a constant equal size which is needed for the processing of potatoes into special products such as chips and French fries.

6.4.3 Type of Processed Potatoes

(a) Dehydrated mashed potatoes (potato granules)

Potato granules are dehydrated single cells or aggregates of cells of the potato tuber, dried to about 6-7% moisture. The granules are made into mashed potatoes very readily by mixing with hot or boiling liquid. This product has a high bulk density, which results in minimum costs for packing and shipping. Potato granules are very useful for both homes and restaurants. The process of dehydrated mashed potatoes in the form of potato granules has been described by many research workers (Rivoche, 1948, 1950; Harrington *et al.*, 1959; Harrington, Olson and Nutting, 1960; Talburt, Boyle and Hendel, 1987).

The processing steps of potato granules are shown in Figure 6.11 and the standard commercial procedure is known as the *add-back* process. Following peeling and trimming, potatoes are usually sliced (20-22 mm thickness) to guarantee uniformity of cooking.

Cooked potatoes are partially dried by adding back enough previously dried granules to give a moist mix, which after holding can be satisfactorily granuled to a fine powder known as potato granules.

Cooking

Cooking is mostly done with steam at normal atmospheric pressure and a moving belt is used for potato transport having a depth of about 18-24 cm. Cooking time depends on the raw material and on the altitude. Usually, the time needed is 30-40 min.

Mashing and mixing

Mashing and mixing with the dry *add-back* granules of intermediate size coming from the screening process provides a material coarser about 60-80 mesh. The result will be

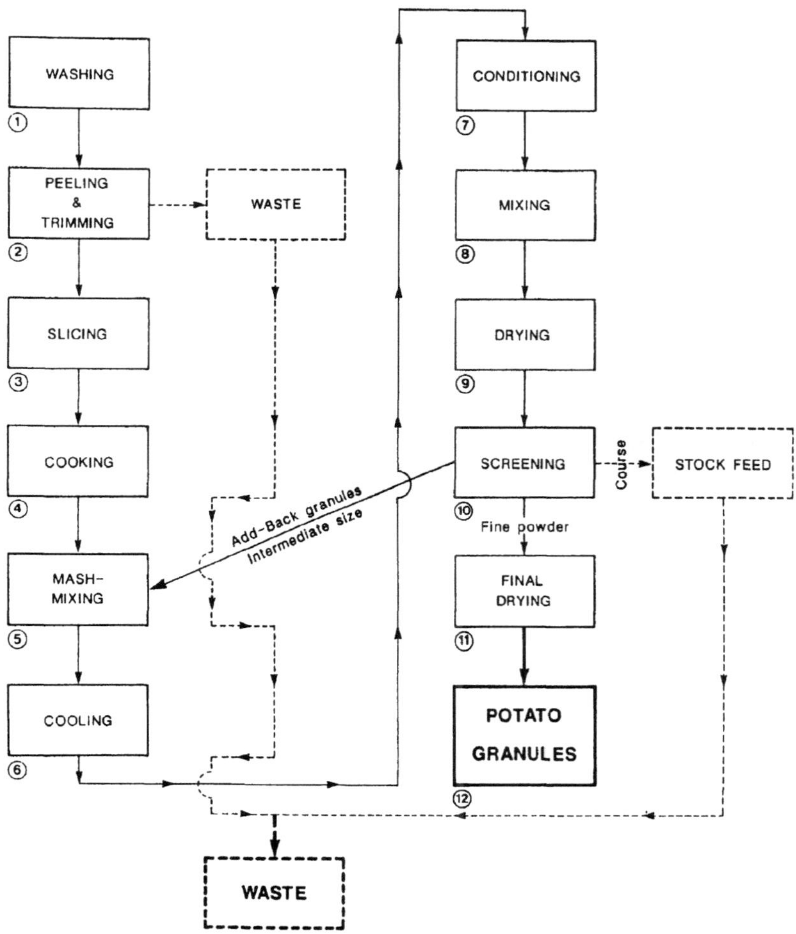

Figure 6.11. Schematic outline of the add-back process for processing potato granules. Reproduced with permission from Talburt, Boyle and Hendel (1987) with modifications.

a moist mixture which has to be cooled to a temperature of approximately 15-27 °C. After the cooling stage the conditioning phase will take place with a holding time of about 1 h at 15-27 °C. Following this stage mixing and drying will take place to reduce the moisture content to 12-13% and then the material will pass the screening stage where the add-back of granules took place, for succeeding cycles. A part of the fine material passing through the screen is also returned as add-back. The coarser part is further dried to a moisture content of about 6%. A small portion of a very coarse material, retained on the screen (16 mesh size) is discarded from the process as a waste because it does not absorb moisture fast enough from the cooked potatoes. About 12-15% is removed as packout, the remainder is used as add-back.

(b) *Dehydrated mashed potatoes (potato flakes)*

Potato flakes are dehydrated mashed potatoes made by applying the cooked, mashed product to the surface of a single-drum drier fitted with applicator rolls. The mashed potatoes are rapidly dried to a final low moisture content. Finally the dehydrated potato solid sheet is broken to a suitable size for packing. Potato flakes are the leading form of dried mashed potatoes used for retail consumption and as a food ingredient. This product is used for many by-products such as forming French fries, flake flour, soups, baby food and baked products. Owing to the easy rehydrating capacity of the product, it can be easily reconstituted with cold water and this has led to their widespread use as an ingredient. The process of producing potato flakes has been described by many research workers (Olson and Harrington, 1951; Willard *et al.*, 1956; Pader, 1962, 1964; Pendlington, Williams and Lester, 1967; Willard, Hix and Kluge, 1987). Their conclusions are summarized below.

The processing steps of potato flakes are shown in Figure 6.12. Characteristics of the raw potato tubers affect the end product as potato flakes and can be explained by potato variety or other characteristics including reducing sugar content, specific gravity, and storage temperature.

Peeling. Potatoes are washed to remove adhering soil and sorted to remove defective potatoes and debris before being peeled. All types of peeling may be used such as steam, abrasion, lye or infrared dry-caustic. It is not of great importance during peeling and trimming to remove all defects, if the operation of the drum drier applicator rolls is properly controlled to remove defective particles.

Slicing. To obtain an optimum processing, potatoes are sliced to 18 mm thickness to obtain more uniform treatment during precooking (heat) and cooling.

Precooking. During precooking, raw potato slices are heated in water at a temperature high enough to gelatinize the starch in the potato cells, but below the temperature where softening of intercellular bonds takes place. Temperature used for immersed potato slices is 71-74 °C for a period of 20 min. This process is similar to the short-time blanching of vegetables which is based on heating and then cooling.

Cooling. Cooling of precooked potatoes reduces stickiness for convenience in later processing. Free starch is also washed from the surface of blanched potato pieces to avoid sticking or scorching during dehydration. Cooling with water for 20 min. with a 24 °C temperature was suggested, but those two factors time and temperature are varying according to the total solids in the tuber. Cooling with cold air was reported to have the advantage of reducing water pollution and the disadvantage of bacteriological innations to the potatoes.

Cooking. The function of cooking is to separate the potato cells to the desired degree of agregation with a minimum rupture. Cooking potatoes may be practised by direct steam injection in a screw conveyor cooker for a time varying from 15-60 min.

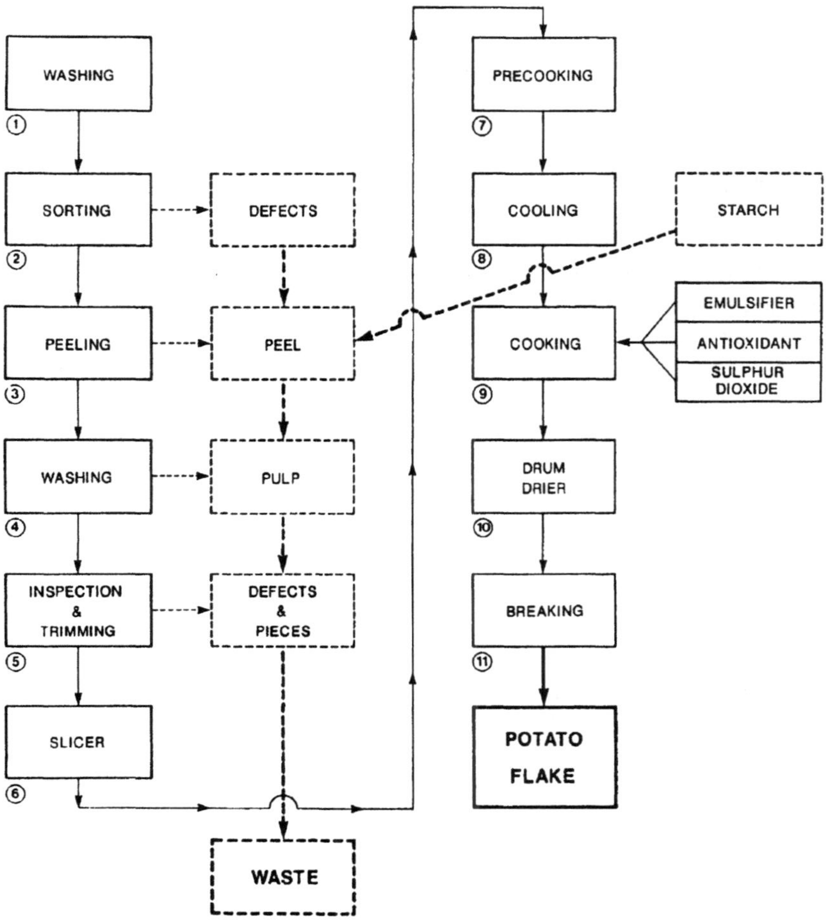

Figure 6.12. **Schematic** outline of the processing of potato flake.

depending on the solids content of the potato and many other factors; an average of 30 min. cooking time would be common.

Mashing. Cooked potatoes should be mashed immediately following cooking to obtain good mixing of additive solutions and to avoid cell rupture, which occurs if the slices are allowed to cool first. To avoid clogging of the twin screw cooker and masher, the masher is normally made from stainless steel round rods with an opening of approximately 15 cm.

Additives

Additives are incorporated into the mash before drying to improve the texture and extent the expiry date of the product.

Emulsifier. A level of about 0.5% monoglyceride emulsifier is mostly used during processing potato flakes to improve texture and reducing graininess.

Antioxidant. Antioxidants used are, for example BHT (butylated hydroxytoluene) are added to mash before drying to a level of 0,50 mg/kg to extend storage life to about 6 months in air. If the addition of antioxidants takes place during drying, 75% of the antioxidant will be lost.

Sulphur Dioxide. The use of sodium bisulphite as a source of sulphur dioxide to prevent non-enzymatic browning in dehydrated potatoes is very essential. It is mostly added in a level varying 1.0-4.0 mg/kg depending on the country.

Drying

Drying of potato flakes may take place in a single-drum drier equipped with 4-6 applicator rolls. The mash is fed to the top of the drum either at a center point or at two points equidistant from the center and the ends of the drum, and conveyed outward by a variable-speed two-way conveyor, with either solid flights or a standard open helicoid spiral rotating in the opposite direction from the drum. The effect of drum speed and sheet density on the drying and producing of potato flake is a special study which is beyond the scope of this book.

Grinding

Potato flakes made by precooking and cooling and supplemented with 0.5% emulsifier can be ground in a conventional hammermill to yield a product with acceptable texture and practical bulk density. There are several definitions for the texture grading system of the potato flake which are, however, not related to the scope of our book.

(c) Potato starch

Potato starch processing is strongly related to the season of production. It would be difficult to operate a plant economically unless raw material was available over a period of several months each year, besides having enough storage capacity of raw potato.

Nearly 10% of the potato crop is not suitable for normal table market owing to their misshapen form, small or large size or damage. Nearly all the culled and surplus potatoes not fed to farm animals are used in the starch processing. The potato starch processing has been described in detail by many research workers (Muller, 1941; Howerton and Treadway, 1948; Brautlecht, 1953; Strolle, Cording and Aceto, 1973; Treadway, 1987).

The processing steps of potato starch are shown in Figure 6.13. There are several types of processing equipments used for starch production and a simple method is selected to explain the basic principals of this procedure.

Potato transport and washing. Potatoes are removed from the storage bin by means of a flume, which carries them to a conveyor. In the mean time, stones, mud and dirt will be removed. Conveyors lift the potatoes up to the washer, where the remaining dirt will be removed by continuous water spraying.

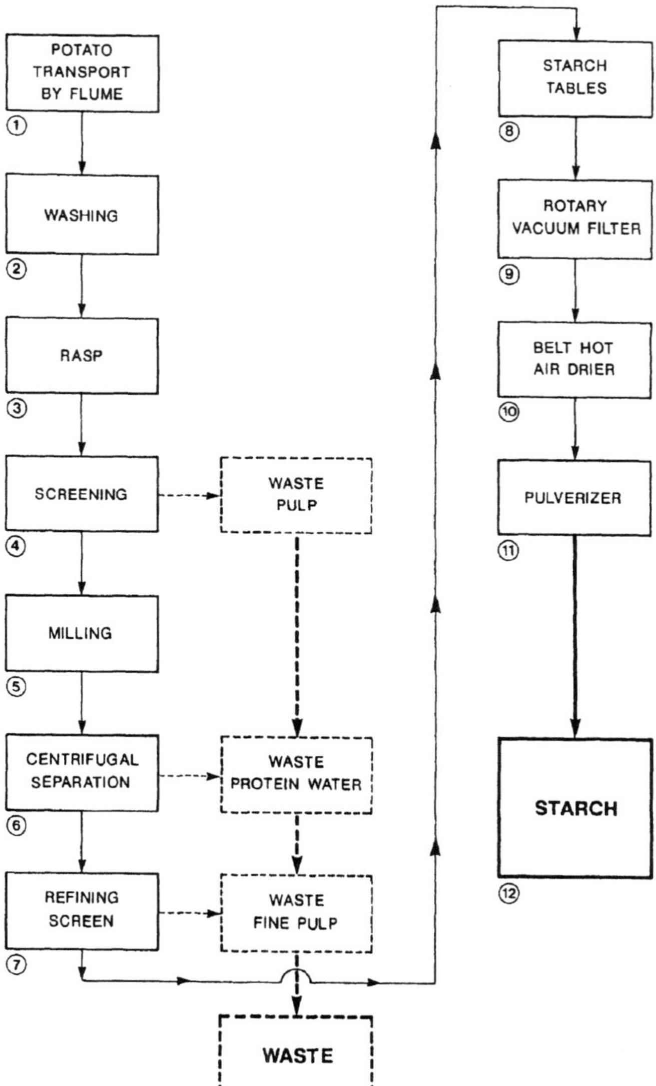

Figure 6.13. Processing scheme of potatoes into starch using screening battery and tables, and production of waste by-products.

Rasp. Washed potatoes will be elevated to a hopper from which they fall to a screwed conveyor that regulates the raw material flow to the rasp. In this process potatoes will be transformed and reduced to a slurry which is diluted with water to facilitate the following steps of processing. Sulphur dioxide is added at this stage to inhibit the action of oxidative enzymes and aiding the production of a white starch.

Screening battery
The diluted slurry is pumped to a battery of screens on which most of the cellulosic material is retained while the starch will pass through. The screening battery is composed of several screens and mounted vertically. The different diameter screens are attached to shakers and rotary brush sieves. In the screening operation, the starch is first pumped on to the bottom sieve. This step will let the starch and water pass through and the pulp is separated off, at the end of the sieve.

Attrition milling
The diluted pulp with water will drop into the attrition mill to be ground for the second time in order to release more starch. The starch suspension, with the fine pulp which passed through the lower sieve, will fall on to the lower shaker screen. The starch granules pass through the lower shaker screen and most of the fine pulp is discharged at the end of the screen later to be mixed with the reground pulp from the attrition mill. The combined pulp is pumped to the upper sieve where it is washed with a water spray. Then, pulp and starch suspension are separated.

Pomace or waste pulp. The fine pulp from the upper shaker screen and the coarse pulp from the upper sieve are combined to constitute or to form a waste product called the pomace, or waste pulp, which is discharged to the sewer.

Centrifugal separator. The starch suspension from the screening battery is pumped to a centrifugal separator where another discharged waste called protein water (washing water which contains soluble materials) is separated.

Refining screen. The starch formed from the above centrifugal separator will subsequently be diluted with water and pumped to a refining screen, which removes additional fine pulp as a third waste product.

Starch tables. The starch suspension is then pumped to tables where the starch will settle and the remaining traces of fibre and soluble substances will flow off at the end.

Rotary vacuum filter. The starch cake is scraped from the tables and is diluted to the proper density for pumping to the continuous rotary vacuum filter.

Belt hot-air drier. After dewatering to about 40% moisture, the cake is dried in a continuous hot-air belt drier at a temperature of about 35-40 °C to about 17% moisture. Finally the product will be pulverized and packed as potato starch.

Use of potato starch
Potato starch is used in a large extent in the manufacture of food such as bread and crackers, thickener in soups, instant puddings, sweets such as gum drops, orange slices, caramels and synthetic jellies. However paper manufacture requires potato starch for finishing high-grade paper, providing a surface coating for smooth finish, sealing paper board and in the fabrication of folding, corrugated, and laminated solid-fibre boxes.

In addition to these uses potato starch is used in dextrins, glucose, citric acid, adhesives, soap, dry cell batteries, *etc.*

Potato starch waste by-products
There are two relevant wastes produced during the processing of potato starch which may be converted into useful by-products: the extracted pulp (pomace) and the soluble constituents of protein water.

Extracted pulp (pomace). Dewatering and drying this product will make it suitable as a feedstuff, the complete analysis of this by-product will be discussed in detail in section 6.4.7 (c) (Chemical analysis and nutritive value). The production of the pomace is estimated to be 180-200 kg per ton starch produced (Howerton and Treadway, 1948).

Protein water. This product typically contains 1-3% solids, and nitrogenous compounds constitute about 60% of the total solids. Protein recovery from potato starch protein water is realized by lowering the pH to 5.5 or lower, heating it to at least 99 °C, pressure filtering it in a plate frame filter press and then drying it in a double-drum drier. Another method is to concentrate the entire protein water effluent by heating. The detailed chemical analysis and nutritive value will be discussed later. About 200-300 kg of protein water solids are obtained per ton of starch produced (Howerton and Treadway, 1948).

6.4.4 Processing of Potato Waste

Potato processing plants produce a tremendous waste effluent which contains many valuable nutrients suitable as feedstuffs for animal and poultry nutrition. Several processing methods have been developed to provide effective removal of screenable, settleable and dissolved solids from potato processing wastes. Moreover, reduction of waste through modern methods with efficient peeling systems and processing lines and lowering the water flow and applying water reuse systems, is a matter of efficient pollution control. Water purification and settling sludge from effluent produced from the potato industry has been reviewed and described by many research workers (Ambrose and Reiser, 1954; Pailthorp, 1965; Douglass, 1966; Echenfelder and Mulligan, 1966; Pailthorpe and Filbert, 1966; Grames and Kueneman, 1969; Culp and Culp, 1971; Willard, 1971; Landine and Dean, 1973; Voetberg, 1986; Pailthorp, Filbert and Richter, 1987).

The ingredients of potato processing waste are determined by the processes used. In general the potato processing waste can be classified according to the several processing steps: washing the raw potatoes to get rid of dirt and fixed solids; peeling including second washing to remove softened tissue; trimming to remove defective parts; shaping; washing; separation; heat treatment (optional); final processing or preservation.

Processing with heat treatment such as blanching, cooking and caustic and steam peeling, will produce an effluent containing gelatinized starch and coagulated proteins. In contrast, starch processing and potato chip processing produce effluents in which the components have not been heated.

(a) Separation of waste from effluent wastewater

Separation of waste from effluent wastewater takes place by means of water purification systems, they are as follows: Pretreatment (screening), primary treatment, secondary treatment and advanced waste treatment (AWT). The various treatments and their by-products are shown in Figure 6.14.

Pretreatment (screening)
In this stage, effluent dirt, raw pieces, and raw or cooked pulp are screened.

Dirt. Dirt as silt, sand and mud (formerly adhering to the surface of the potatoes) is removed from the effluent either by initial washing step or in the normal peeling waste stream. This line is mostly treated separately because it contains no valuable waste, but it should be purified before discharging in to open streams or recycle the water for profitable use.

Raw pieces. Unsuitable whole potatoes, small fragments, raw pieces and diseased potatoes are mostly firm pieces and easy to gather. They represent little problem in removal by screening or settling. This waste is known as Potato Waste (PW) and is commonly used as cattle feed.

Raw pulp. Raw pulp is produced from abrasion peeling, cutting and trimming waste and is combined with pulp from the starch separation process. The raw pulp may be removed from the effluent by screening or settling. This waste is known as Potato Pulp (PP) and is commonly used as cattle feed.

Cooked pulp. During peeling and processing by heat, a softening and weakening process of the intercellular bonds of the potato takes place resulting in losses of cells which will gather into balls or mass (agglomeration) during washing and handling steps, resulting in a spreading in the wastewater. Many agglomerates are removed by screening, but a large portion passes through the normal 20-mesh screen opening. These solids settle rapidly by means of a clarifier (a continuous flow of wastewater through a tank) and represents a major portion of the settleable solids. This step belongs to the primary treatment. Separated solids are used as a feedstuff for cattle feeding.

Primary treatment
The removal of the settleable solids mostly takes place by a clarifier while the buoyant solids are removed by blowing gas bubbles.

Clarifiers. In the stage of the primary treatment of potato wastewater by clarification, most of the settleable solids are removed from the effluent resulting in a decrease of 40-70% of the COD (Chemical Oxygen Demand). To reduce the volume of the settled matter which normally contains about 6% solids, a vacuum filter or a centrifuge will be attached after the clarifier which will result in a total solid content of 15-20%. The system efficiency is mostly affected by several factors such as the total solids produced during potato processing systems of peeling, (abrasive or caustic *etc.*) and the pH during caustic peeling.

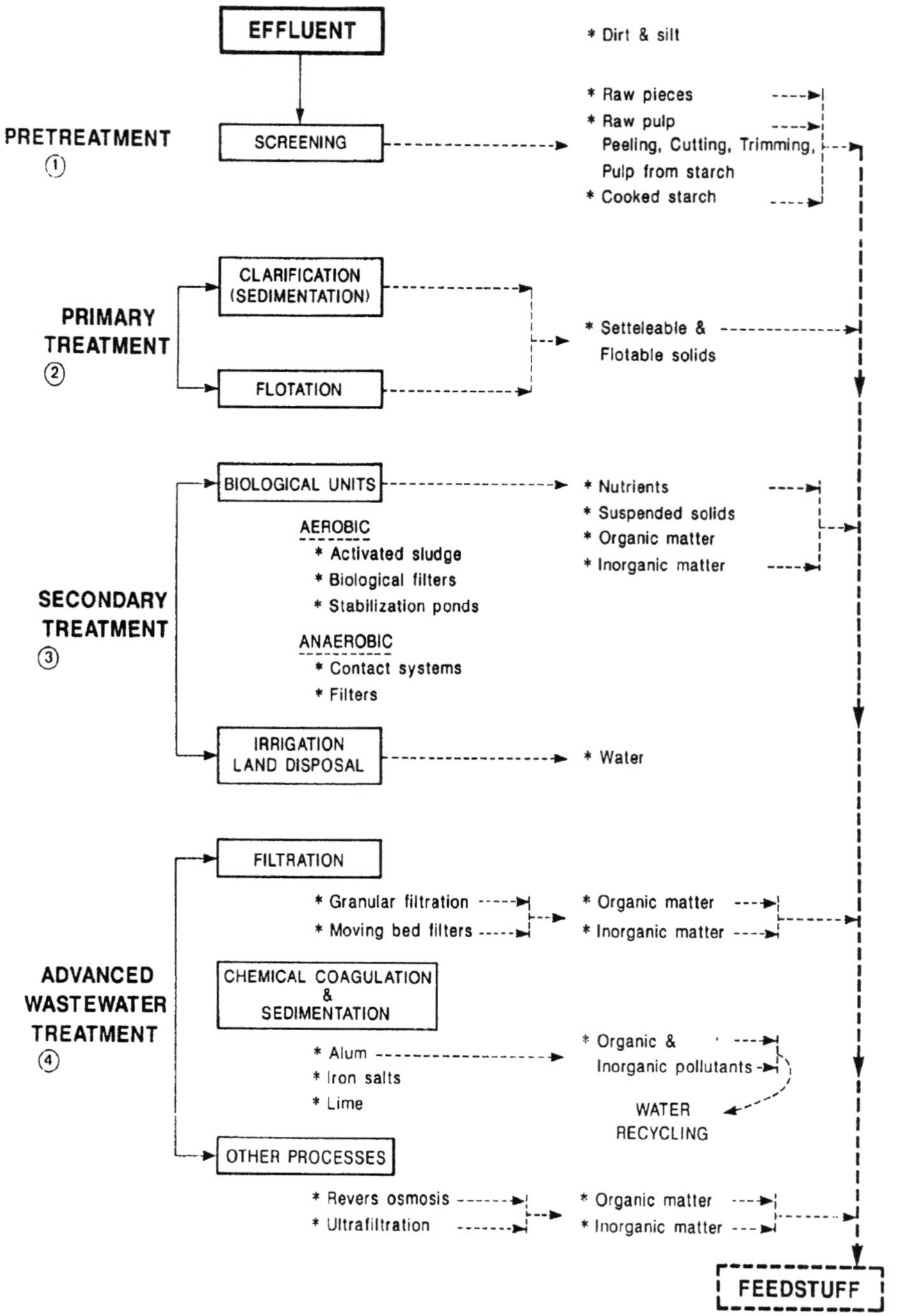

Figure 6.14. Schematic outline of the various processing treatments of potato wastewater and its by-products.

Flotation. Flotation is a further method of solids removal based on blowing air bubbles which are creating a flotation surface of potato solids which will be skimmed off by mechanical collections.

Both settleable or sedimented and flotation by-products may be used as a feedstuff called Potato Waste (PW).

Secondary treatment

Secondary treatment is based on biological treatment by growing biological organisms which can convert the potato waste from the effluent to a bacterial mass of protoplasm which is termed activated sludge. This process upgrades the waste to a better quality as far as the nutrients are concerned. The biological units are classified as aerobic and anaerobic decomposition, respectively. The result of these treatments is a sludge rich in nutrients, suspended solids, and organic and inorganic matter. In addition, sludge can be activated to provide upgraded nutrients. Besides those two systems the irrigation land disposal is also known as a method used in the secondary treatment.

(b) Aerobic biological decomposition

Aerobic biological decomposition is based on a sufficient supply of oxygen, is practically odourless, and has very high removing capacities for organic matter from the wastewater effluent of potato.

Activated sludge process

In this method, compressed air is blown by means of a mechanical surface aerators in a large aeration business. Owing to the organic feed supply and oxygen a heavy concentration of bacteria will grow. The organic content of the waste sludge is transferred to a bacterial mass of protoplasm. Part of the organic matter will be oxydized to improve the energy required for this conversion process. The mass may be used to re-inoculate again sedimentation basins to provide a highly treated sludge. This may be dried, sterilized and used as a promising feedstuff in feeding animals.

Biological filters. The idea of the biological filters is that the effluent passes through filter beds that are well supplied with air (oxygen). Biological slimes of bacterial colonies are created. The slimes grow, remove organic matter and are transferred to an upgraded product. Finally, this bacterial protoplasm is washed and removed by sedimentation as in the activated sludge process mentioned before.

Stabilization pond. Effluent of potato wastewater is poured into large ponds of standing water. Oxygen, which is essential for this process is formed from the contact of the atmosphere with the surface pond through the movement of the wind. Due to this contact the effluent waste will be oxidized biologically. The climatic temperature affects the efficiency of this treatment. If oxygen is injected by compressed air (aerators), the system is called an aerated stabilization pond. Algae may grow on the system with a benefit of upgrading the final quality. The end product is similar to that of the activated sludge plant. The collection of the end product is also the same.

(c) Anaerobic biological process

The anaerobic biological process is belonging to secondary treatment and differs from aerobic biological decomposition by the absence of oxygen. The anaerobic bacteria converts the organic pollutants to a mass of bacterial protoplasm, carbon dioxide, water and methane. The composition of this culture of anaerobic bacteria depends on pH, oxygen, temperature and heavy metals. A good control of the above-mentioned factors is necessary. A dramatic change in these factors may stop the process. With this system, a fuel (methane) can be produced and the sludge will be completely digested. The system has some drawbacks such as its sensitivity to changes in temperature and the other factors mentioned above. Moreover, the process time is long and undesirable odours are produced.

Anaerobic filters. The wastewater enters biological filters from below and exits from the top. The anaerobic bacteria culture will grow and build up a mass of protoplasm.
Anaerobic contact system. This system is similar to that for the aerobic activated sludge but without aeration. It consists of a covered basin and a mixed reactor.

(d) Land disposal (irrigation)

Unpolluted potato wastewater is a reasonable soil fertilizer and can be used as a source of irrigation. The irrigated soil will increase crop yield owing to the organic substances in the wastewater. The required irrigated land area depends on the type of crop grown, type of soil, groundwater level, weather conditions, *etc.* The most successful crops for this type of irrigation disposal are water-tolerant grasses (Pailthorp, Filbert and Richter, 1987). Still the problem of transferring the wastewater to the suitable land incurs cost; when the distances are too big, the transfer may be uneconomic.

(e) Advanced wastewater treatment

The Advanced Wastewater Treatment (AWT) takes place after the second water treatment where organic and inorganic pollutant may be removed. The system of AWT is based on physical and chemical processes such as granular filtration, sedimentation, coagulation and chlorination.

Granular filtration. After the second treatment most of the effluent from the potato wastewater contains a BOD with suspended solids from the bacteria which was grown on the activated sludge. Employing granular filters will lower the BOD in the effluent resulting in an acceptable quality of discharged water and gathering of the bacterial suspended solids. Moving bed filters can also be used as a method of filtration. If the concentration of particles of the suspended solids is raised above 250 mg/l, the effluent should be treated with the secondary treatments as mentioned before or chemical coagulation and sedimentation have to be used.

Chemical coagulation and sedimentation. Potato wastewater effluent is rich in minerals owing to the caustic peeling and formation of ammonia and to the high phosphorus content of the potato. To remove those chemicals, coagulation and sedimentation is practised with alum, iron salts or lime. The use of those coagulants will cause the sedimentation of the effluent minerals and ammonia and will create water free from these components. This water may be recycled again during the various potato processing steps.

Other advanced wastewater treatments. New methods are being developed for the water purification of the potato wastewater effluent such as reverse osmosis and ultrafiltration.

6.4.5 Processing of the Sludge Produced from Potato Wastewater

Dickey *et al.* (1965) described the various types of presses for reducing the moisture content of potato starch pomace (wet potato pulp) containing about 94% moisture. The system of drying is based on a wet lap paper dewatering machine, a hydraulic press, and a vacuum drum drier. It was possible to reduce the moisture content of the raw pomace from 94 to below 85%. Incorporation of various additives to coagulate pectins and pectin-like substances prior to pressing were tried. $Ca(OH)_2$ at a level of 0.3% was added to the pomace for a period of 7-15 min. in a continuous screw-type mixer and showed excellent results. For pomace treated in this way, the Davenport Type 3-17 press (Maine Agricultural Experimental Station, University of Maine, Orono, Maine, U.S.A.) proved to be most efficient, producing presscake of 63% moisture content. This cake was easily broken up in a hammer-mill prior to drying in a Louisville pilot-plant size direct fired drier. The resulting dried potato pulp showed the following composition: moisture 11.8%, crude protein 5.7%, fat 0.4%, fibre 13.2%, N.Free extract 66.6% and minerals 2.6%.

Grames and Kueneman (1969) studied the dewatering of the sludge produced from the potato processing wastes. They noticed that the settled sludge contained about 3% dry solids. From an average potato processing plant the amount of sludge produced would be 567 m^3/day or 75 truck loads of 7,470 l tanker. A sludge-drying bed 61 m^2 by 0.30 m deep will be filled in 2 days. Sludge concentration could be raised from 3-7% by a clarifier. Vacuum filtration will increase the dry solids from 7% to 12-16%. Sludge collected from steam peeling would not separate from the solids and would create a gelatinous slurry which would block the vacuum filter. Chemical treatment of the sludge with ferric sulphate and lime would break down the gelatinous consistency, release the water and flocculate the solids.

Finally it was discovered that the sludge becomes filterable if it is aged in the bottom of the clarifier. The holding of the sludge allows bacterial action to get started, and this action breaks down its gelatinous consistency. If the sludge is held too long, problems will occur, for example, the bacterial action breaks down the solid particles and filterability will be seriously reduced. A design rate of 5.1-6.1 kg/h/m^2 vacuum filtration can be used and the solid content of the sludge will be about 65% which reduces the volume of the sludge from a fluid to a damp paste which improves handling, disposing or drying considerably.

6.4.6 Yields of Potato Waste

The yields of potato waste vary according to the processing method used and the end product. Several waste products will be discussed under the name of the process producing the waste, such as peeling, dehydrated mashed potatoes as granules and flakes, potato starch and waste produced from the purification of wastewater effluent.

Waste from peeling. Potato peel losses mostly depend on the condition of the raw tuber, the quality standard required during processing, the type of peeling (steam, caustic or abrasion), the equipment used and the skill of the operator. However, the non-uniformity of size and shape, depth of the eyes and other surface irregularities result in a high peel losses with abrasion peeling (Wright and Whiteman, 1949). Smith and Huxoll (1987) estimated the peeling losses of the potato chip industry, which uses abrasion peeling extensively, to be about 10%. In Table 6.26 the percentage of peeling losses from several potato products is shown to vary between 2-50% according to size, type of product, processing method and season (Smith and Huxol, 1987).

Table 6.26. Percentage peel losses following several processing methods used for potato products. From Smith and Huxol (1987)

Product	*Peel loss (%)*
Canned small potatoes	40-50
Prepeeled potatoes for restaurants	20-30
French fries	10-20
Dehydrated mashed potatoes	5-10
Potato chips (early season)	2- 5
Potato chips (late season)	8-12

Potato granules. During the processing of dehydrated potato granules, a very coarse material is retained on the screen. This material is removed from the process because it does not absorb moisture during the cooking process. This waste material is estimated to be 12-15% of the total (Talburt, Boyle and Hendel, 1987).

Potato flakes. The following losses of waste would be expected in the processing of dehydrated potato flakes (Willard, Hix and Kluge, 1987): peeling, about 7%, trimming 3%, slicing 1.5%, precooking 3%, cooling 1.5%, cooking 1%, finally waste formed from the applicator rolls 3-5% (based on the total solids from each operation).

Potato starch. The processing of potatoes to potato starch produces extracted pulp (pomace) and the soluble constituents of protein water. Howerton and Treadway (1948) and Treadway (1987) reported that for every t of starch produced, about 180-200 kg of pulp solids and 200-290 kg protein water solids are obtained as waste.

Waste from wastewater purification. Potato waste solids collected from the wastewater of the primary treatment processes contain 4-7% solids which may be concentrated to 15-18% by using belt-type vacuum filters or centrifuges. These solids are difficult and bulky to handle, they may be used however as a feedstuff after proper processing (Pailthorp, Filbert and Richter, 1987).

Waste biological solids (bacteria) grown as part of the secondary treatment process may be used as a feedstuff owing to their high protein content of 35%. The total quantity of solids in this waste varies from 0.5-0.75%. This product, however, is very difficult to concentrate or to de-water (Pailthorp, Filbert and Richter, 1987).

Moon (1980) calculated the solid and liquid wastes during processing potatoes as follows: total solid waste 6.6%, liquid volume 0.012 m^3/kg and BOD 44 000 mg/kg.

6.4.7 Chemical Analysis and Nutritive Value

Chemical analysis and nutritive value will be classified according to the waste by-product produced by the various processing methods, such as potato waste meal (potatoes, potato pulp and peeling); potato flake and waste from the potato starch industry.

(a) Potato waste meal

Potato waste meal is a product produced by drying and grinding of culls of potatoes, potato trimming, pulp, peeling, and off-colour parts of French fries and potato chips. In comparison with corn potato waste analysis showed slightly less crude protein (8%), more lipid (6%), crude fibre (4%), ash (4%) and high in potassium (1.4%). Potato waste meal is not considered to be a rich source of essential amino acids. It contains considerably less of each amino acid than corn and it is limiting with respect to methionine and cystine, arginine and the aromatic amino acids. True metabolizable energy is 13.4 MJ/kg for potato waste and 14.1 MJ/kg for corn. Table 6.27 shows the chemical analysis and the amino acid composition of potato waste in comparison with potato flakes and corn (Whittemore, Moffat and Mitchell-Manson, 1974; Hubbell, 1981; Hulan, Proudfoot and Zarkadas, 1982[a,b]).

Hulan, Proudfoot and Zarkadas, (1982[a,b]) suggested that potato waste meal (methionine supplemented) might be considered as a substitute ingredient for a proportion of the ground corn in practical diets for poultry up to 20%. They concluded also that the inclusion of potato waste in the diet up to 30%, will increase the potassium content of the diet. These inclusion levels did not increase wetness of the litter nor hardness of feed pellets, two factors which may limit the use of potato waste meal in poultry diets as reported earlier (Whittemore, Moffat and Taylor, 1975).

(b) Potato flakes

Whittemore, Moffat and Mitchell-Manson (1974) and Whittemore, Moffat and Taylor (1975) indicated on the basis of the chemical analysis of the dehydrated mashed cooked potato flakes (Table 6.27) that they could be used as an alternative for corn in poultry rations. The fibre content is low (2.1%), crude protein is nearly similar to corn and the flakes have a high (in vitro) lysine availability of about 93%. Ether extract from potato flakes is lower than with corn but metabolizable energy is nearly the same. The amino acid pattern of the protein is acceptable for a carbohydrate feedstuff and suitable for growing birds. Whittemore (1977) noted in his review the high digestibility coefficients for cooked potato flakes, being 0.81 for nitrogen, 0.91 for organic matter, 0.95 for nitrogen free extract and 15 ME value (MJ/kg DM) with hens. Vogt and Stute (1969[a,b]) used dried potato shreds (not cooked) in the nutrition of broilers and reported low digestibility coefficients of 0.36-0.40 for crude protein. From these experiments it was concluded that cooked potato flakes are considered appropriate carbohydrate feedstuff for commercial use in poultry diets up to 20% with no drawbacks in the performance of broilers (Whittemore, Moffat and Mitchell-Manson, 1974). D'Mello and Whittemore (1975) studied the nutritive value of cooked potato flakes in comparison with potato starch and ground maize. They concluded that potato flakes were enhancing performance of birds over that obtained with potato starch.

Table 6.27. Chemical analysis and amino acid composition of potato waste meal and potato flake in comparison with corn. From (1) Hulan, Proudfoot and Zarkadas (1982[a]); (2) Whittemore, Moffat and Mitchell-Manson (1974); Whittemore, Moffat and Taylor (1975) and (3) Hubbel (1981)

Components %	*Potato*			*Amino acids %*	*Potato*		
	Waste (1)	*Flake (2)*	*Corn (3)*		*Waste (1)*	*Flake (2)*	*Corn (3)*
Moisture	6.000	-	14.300	Phenylalanine	0.27	0.34	0.43
Crude protein	7.900	9.50	8.800	Tyrosine	0.14	0.32	0.24
Crude fat	-	0.20	3.900	Histidine	0.13	0.21	0.28
Total lipid	6.100	0.20	2.200	Isoleucine	0.28	0.30	0.44
Crud fibre	3.900	2.10	1.500	Leucine	0.43	0.46	1.15
TME (MJ/kg) [a]	13.400	-	14.100	Methionine	0.12	0.11	0.21
ME (MJ/kg)	-	13.60	13.000	Cystine	0.09	0.09	0.19
Ash	4.200	4.30	0.010	Valine	0.39	0.41	0.44
Calcium	.080	0.05	0.300	Arginine	0.24	0.53	0.38
Phosphorus	.200	0.18	0.110	Lysine	0.41	0.50	0.28
Magnesium	.070	0.10	0.002	Threonine	0.22	0.37	0.35
Iron	.060	-	0.001	Tryptophan	0.07	-	0.05
Zinc	.001	-	0.330	Aspartic acid	1.40	1.97	-
Potassium	1.400	2.20	0.010	Glutamic acid	1.07	1.73	1.7
Sodium	.160	0.02	-	Serine	0.29	0.34	0.43
Chlorine	.240	-	-	Glycine	0.24	0.28	0.36
Digestible				Alanine	0.34	0.37	0.69
crude protein	-	8.60	-	Proline	0.23	-	0.81
Digestible lysine	-	0.47	-	4-Hydroxiproline	0.04	-	0.03

[a] TME = True metabolizable energy

(c) Potato starch

The processed potatoes for starch, produce potato starch as a human food and waste by-products as starch waste protein water and extracted pulp (pomace).

Starch. Halnan (1944) found that raw potatoes were poorly utilized because the starch escaped digestion. Boiled potatoes were readily digestible. Energy derived from one part of cooked potatoes by the bird was equivalent to that derived from 5 parts of raw potatoes.

Raw starch granules are very resistant to the action of some digestive enzymes. Sandstedt and Gates (1954) show that alpha-amylase is 165-7000 times as active on boiled starch as it is on raw starch. However, Jones (1940) showed that crushing of starch granules during milling increased their susceptibility to enzymatic digestion. He concluded that dietary utilization and digestibility of some starches can be improved through hydration, milling or mechanical damage.

Hakansson and Lindgren (1974) found the ME value for raw potato starch varied between 0 and 17 MJ/kg. The poor utilization of raw potato starch and satisfactory utilization of cooked potato starch and the use of maize starch was thoroughly studied by Naber and Touchburn (1969), Nitsan and Bartov (1972) and D'Mello and Whittemore (1975).

Halnan (1944) determined the digestibility of several processed types of potato in several ways, using hens (Table 6.28). His work shows that raw potato starch (indicated, however, by nitrogen-free extract) is indeed poorly digested by the fowl, but that the digestibility improves with cooking. In the case of potato slices, the degree of improvement in digestibility depends on the amount of heat applied. A similar positive effect of heat treatment is shown for the digestibility of nitrogen in a comparison of raw potato and cooked flake dried at low and high temperature. Crude protein and amino acid percentage from several processed samples in comparison with raw potatoes is shown in Table 6.29.

(d) Potato starch wastewater

The filterable solids in potato starch wastewater were explained under section 6.4.3 (c) (Potato starch). After concentrating by heating up to 71.1 °C and the consequently partially inactivation the trypsin inhibitor (37 000 down-6430 units per g). This waste product is called dehydrated soluble potato solids (DSPS).

This product still contains substantial levels of trypsin inhibitor activity as in the case of raw soybeans, which create difficulties when used as a feedstuff for broilers. Moreover it contains 35% sugars, it takes up moisture (hygroscopic) and it will cake solidly if exposed to humid air. When added to rations it will tend to form lumps in troughs. Calcium oxide can be added to eliminate the latter tendency (Gerry, 1977).

Strolle *et al.* (1975) analyzed DSPS and showed it to crude protein of 33-41%, total sugars 35%, reducing sugars 28-32%, sucrose 1-7%, organic acids 4%, minerals 20%. Gerry (1977) showed DSPS to have a lysine content of 1.51% and methionine 0.47%, these two essential limiting amino acids being abundant and comparable with meat and bone meal. If DSPS is used, care should be taken to adjust the inclusion of

Table 6.28. Composition and digestibility of several potato products determined with hens. From Halnan (1944)

	Raw	*Boiled*	*Cooked flake*	*Dried slices*		*Dried meal*
				low temp.	*high temp.*	
N in DM (%) [a]	1.15	1.13	1.33	1.15	1.17	1.10
GE in DM (MJ/kg) [b]	16.90	17.00	16.90	16.80	16.90	16.90
Digestibility coefficients						
Nitrogen	0.54	0.52	0.81	0.39	0.69	0.57
Organic matter	0.22	0.83	0.91	0.55	0.85	0.89
Nitrogen-free extract	0.18	0.90	0.95	0.59	0.90	0.95
ME value (MJ/kg DM)	2.90	13.60	15.00	8.60	13.70	14.50

[a] Nitrogen in Dry matter
[b] Gross energy in Dry mattter

Table 6.29. Amino acid percentage (DM basis) from several processed whole potato samples in comparison with raw potato. From Whittemore (1977)

%	*Raw*	*Cooked*	*Cooked flour*	*Cooked flake*
Crude protein in DM	9.50	9.80	10.30	9.50
Aspartic acid	2.20	1.78	1.92	1.97
Threonine	0.30	0.33	0.34	0.37
Serine	0.37	0.34	0.37	0.34
Glutamic acid	1.57	2.24	2.30	1.73
Glycine	0.27	0.29	0.31	0.28
Alanine	0.28	0.33	0.33	0.31
Valine	0.37	0.45	0.48	0.41
Cystine	0.10	0.10	0.09	0.09
Methionine	0.10	0.09	0.12	0.10
Isoleucine	0.28	0.33	0.34	0.30
Leucine	0.42	0.55	0.52	0.46
Tyrosine	0.30	0.35	0.38	0.32
Phenylalanine	0.33	0.36	0.41	0.34
Lysine	0.51	0.51	0.51	0.50
Histidine	0.24	0.17	0.18	0.21
Arginine	0.43	0.34	0.51	0.53

limestone and dicalcium phosphate by including calcium oxide (against hygroscopicity). Emphasis should also be put on the inactivation of the trypsin inhibitor by heat without affecting the protein and amino acids quality (Gerry, 1977).

(e) Potato shreds

Potato shreds is a typical German product. The processing is based on heating the potatoes with their own fluids, thus no water is added during treating. The process is based on washing the whole potato, cutting, first drying in a drum dryer (650C down to 120C) and second drying in a drum dryer (850C down to 60-70C) to a final moisture level of 9.5-12%. Finally the product will be cooled until 24C and shredded (Vogt and Stute, 1969[a]). Table 6.30 shows the chemical analysis and amino acid composition of potato shreds. Potato shreds are carbohydrate feedstuff which is not processed (cooked) with water. Therefore its starch is not completely gelatinized and it has low digestibility coefficients for nitrogen ranging from 0.36-0.40 (Vogt and Stute, 1969[a]), while Halnan (1944) reported higher digestibility coefficients for nitrogen with cooked potatoes being 0.81 and 0.69 (Table 6.28).

Table 6.30. Chemical analysis and amino acid composition in percentages of potato shreds (air dry basis). From Vogt and Stute (1969[a])

Composition %		*Amino acids %*	
Dry matter	91.60	Threonine	0.25
Crude protein	9.50	Serine	0.25
Crude fat	0.30	Glutamic acid	0.91
Crude fibre	2.50	Proline	0.26
N-free extract	74.40	Glycine	0.25
Combined Carbohydrates	76.90	Alanine	0.25
Ash	4.90	Valine	0.34
Starch	62.00	Cystine	0.15
Sugar	3.20	Methionine	0.09
Lignine	1.50	Isoleucine	0.26
Pentosans	4.50	Leucine	0.42
Crude-Cellulose	0.03	Tyrosine	0.28
Calcium	0.04	Phenylalanine	0.29
Phosphorus	0.32	Lysine	0.31
Amino acids %		Histidine	0.11
Aspartic acid	1.42	Arginine	0.27

(f) Solanine in potatoes

Solanine, known as a normal constituent in all solanaceous plants, is the term applied to the steroidal alkaloid fraction of potatoes soluble in acidified alcohol and insoluble in slightly aqueous solution the aglycone of which is solanidine. This aglycone fraction contains free solanidine as well as a mixture of glycosides of which α-chaconine is the most important one. In α-chaconine the sugar sequence is glucose-rhamnose-rhamnose. The sugar sequences of other solanidine glucosides also present in potatoes are as follows: ß-chaconine, glucose-rhamnose; ϒ-chaconine, glucose; α-solanine, galactose-glucose-rhamnose; ß-solanine, galactose-glucose; and ϒ-solanine, galactose (Talburt, Schwimmer and Burr, 1987).

Solanine, as the total alkaloid fraction, is present in the normal whole tuber to the extent of 0.01-0.10% of the dry weight. There is at least twice as much in the peel as in the flesh, and it is concentrated in the vicinity of the eyes. Exposure of freshly harvested potato tubers to ultraviolet light causes a several-fold increase in the solanine content. Sprouts contain much more solanine (4.3% solanine of the dry weight) than tubers. Potatoes containing more than 0.1% solanine are considered to be unfit for human consumption, owing to the solanine poisonous alkaloid (Burton, 1948; Kröner and Völksen, 1950; Talburt, Schwimmer and Burr 1987).

Vogt and Stute (1969[b]) studied the effect of solanine content in dried ground potatoes on broiler performance. They concluded that the dried ground potatoes contained 100-150 ppm solanine, a level which had no side effects on broiler performance. They reported also that inclusion of α-solanine in levels of 12.5-50 ppm in broiler diets had no negative effects on broiler performance as far as weight gain, feed conversion or water content of the faeces are concerned. α-Solanine supplemented caused a significant increase in final body weight at 45 days of age.

Maga (1980) reviewed the significance of glycoalkaloids in potatoes and potato products. This work explains the types and distribution of glycoalkaloids identified in potatoes, the factors affecting them, control measures to minimize their formation, method of analysis, and the health implication. Bushway and Pennampalam (1981) found that the glycoalkaloids are quite stable during the processing potatoes by boiling, frying or baking.

6.4.8 Feeding Potatos and Their Waste

Processed potatoes such as dehydrated potato flakes and potato starch if they are in surplus and potato waste meal and potato starch waste are very modern and relevant feedstuffs of a carbohydrate origin. They may replace corn or wheat in the poultry diets and save the cereals for human consumption.

(a) Potato waste meal

Hulan, Proudfoot and Zarkadas (1982[b]) investigated the use of potato waste meal in broiler diets to replace maize. Two experiments were carried out in which potato waste meal was included in diets with 5, 10, 15, 20 and 30% in practical diets of broilers fed up to 49 days old. Potato waste meal is limiting with respect to methionine and cystine, arginine and aromatic amino acids. Supplementation with methionine (0.5 g/kg) revealed test mixtures to have a synergestic effect compared with controls. They concluded also that potato waste meal can be considered a good substitute ingredient for 20%

of the ground maize in a practical diet for broiler chickens. The inclusion of up to 30% of potato waste did not show any negative performance as far as feed conversion is concerned (Table 6.31). Up to 30% inclusion did not increase wetness of the litter or hardness of the pellets.

(b) Potato flakes

Whittemore, Moffat and Mitchell-Manson (1974) and Whittemore, Moffat and Taylor (1975) studied the performance of broilers fed on diets containing cooked potato flakes as a replacer for maize meal. They formulated starter and finisher diets containing 10, 20 and 40% potato flakes and a control diet. All the diets were pelleted and fed ad libitum to broiler chicks (Ross I) from one-day old to 28 days and from 29-56 days. No significant differences were noticed between the performance of broilers given diets containing 0, 10 or 20% potato flakes. Broilers with 40% inclusion of potato flakes grew at a slower rate and ate significantly less feed in comparison with control and 10 and 20% potato flakes inclusion. However, feed conversion was not changed and was similar to that in the control and the other diets. Table 6.32 shows the performance of the broilers and the significance of the parameters used. Wetness of the litter and hardness of the pellets both increased as the proportion of the potato flakes in the diet increased. These are suggested as the main factors limiting the 40% inclusion in the diet. However, these factors could be overcome by attention to the litter and by feeding diets containing potato flakes in smaller particles.

D'Mello and Whittemore (1975) reported similar results to the above mentioned authors. They concluded from their trials with male broiler chicks (Marshall LB 9 strain) from 7 days old to 21 days, that growth and efficiency of feed conversion were lower ($P < 0.05$) in broilers fed the experimental diet with 40% potato flakes compared to those fed a diet with 40% maize or fed the control diet.

(c) Potato starch

Starch. Naber and Touchburn (1969) evaluated maize and potato starches for their effects on chick growth and energy utilization after treating them with water, heat and mechanically. Maize starch was selected as a representative of starches readily digestible in the raw state, whereas potato starch was used to represent starches poorly digestible in the raw state. Untreated potato starch was poorly utilized by the chick when growth, feed conversion and metabolizable energy value are considered as criteria of performance. Water treatment of potato starch at 58 °C without inducing a remarkable amount of gelatinization resulted in a significant improvement of chick performance and an increase of 15% in starch utilization. Complete gelatinization of potato starch at 68-72 °C greatly improved chick performance and starch utilization. However, water treatment of potato starch at 16 or 41 °C had no effect on its utilization by the chick. Mechanical fracture of the potato starch granules increased the starch utilization but chick growth was poor due to low feed consumption caused by hydration of the starch on the beaks and in the mouths of the chicks.

Table 6.31. Effects of inclusion of potato waste meal in broiler (50% males and 50% females) diets on body weight, feed consumption and conversion. From Hulan, Proudfoot and Zarkadas (1982[b])

	Treatment Live body wt (g)		*Feed Consumption (g)*		*Feed Conversion*	
	Starter	*Finisher*	*Starter*	*Finisher*	*Starter*	*Finisher*
First experiment [a]						
Control	902	1906	1416	3869	1.57	2.03
" + 5% PWM	899	1916	1402	3832	1.56	2.00
" + 10% "	905	1927	1403	3758	1.55	1.95
" + 15% "	917	1948	1421	3838	1.55	1.97
Second experiment [b]						
Control	924	2148	1469	4360	1.59	2.03
" + 10% PWM	917	2117	1403	4128	1.53	1.95
" + 20% "	936	2116	1441	4126	1.54	1.95
" + 30% "	911	2026	1412	3951	1.55	1.95

[a] First experiment: Starters were fed from one-day old to 28 days and finishers from 29 days to 46 days.
[b] Second experiment: Starters were fed from one-day old to 27 days and finishers from 28 days to 49 days.
PWM = Potato Waste Meal

Table 6.32. Effect of inclusion of potato flake in broiler diets on body weight, feed consumption and feed conversion. From Whittemore, Moffat and Mitchell-Manson (1974)

Treatment	*Body weight (g)*		*Feed Consumption (g/bird/day)*		*Feed Conversion g feed/g live-weight gain*	
	Starter[c]	*Finisher*[d]	*Starter*[c]	*Finisher*[d]	*Starter*[c]	*Finisher*[d]
Control	726[a]	1804[a]	45.9	100.4[a]	1.87	2.61
Cooked potato flake 10%	711[a]	1781[a]	40.4	105.2[a]	1.68	2.76
" " " 20%	698[a]	1764[a]	41.3	102.9[a]	1.73	2.70
" " " 30%	636[b]	1555[b]	38.3	79.0[b]	1.79	2.41
SE of treatment means	11.3	27.7	2.13	5.16	0.098	0.164
Significance	***	***	NS	**	NS	NS

[a,b] Means with different superscripts in a column differ significantly (P < 0.05)
[c] Starter: 1-28 days old
[d] Finisher: 29-56 days old
NS: Not significant
** : (P < 0.01)
*** : (P < 0.001)

On the other hand water treatment of maize starch failed to improve chick performance when compared to untreated maize starch unless water treatment induced gelatinization. Gelatinized maize starch stimulated growth and feed conversion without significantly increasing the metabolizable energy value of the diet: Tables 6.33 and 6.34 show the effect of water, temperature and mechanical treatments on corn and potato starches, and their results on metabolizable energy and meat type performance of the chicks.

Table 6.33. Effect of water and temperature treatment on the utilization of corn and potato starches by meat type chicks from 2 to 4 weeks of age (5 replicates of 10 chicks each/treatment). From Naber and Touchburn (1969)

Starch component in basal ration	*Water and temperature treatment*	*Body weight gain (g)*	*Feed/unit gain*	*ME-Value of diet*
Corn starch	None	247^b	2.45^c	13.18^b
Corn starch	16C	241^b	2.48^c	13.22^b
Corn starch	78C	292^c	2.24^d	13.43^b
Potato starch	None	176^c	5.07^a	6.65^a
Potato starch	16C	196^a	4.51^b	6.49^a
Potato starch	72C	301^c	2.21^d	13.26^b

a,b,c,d Values bearing different superscript letters in a column are significantly different from each other ($P < 0.05$) using Duncan's multiple range test.
ME = Metabolizable energy (MJ/kg).
The experimental processed starches were incorporated into the basal ration where they comprised 50% of the diet.

Table 6.34. Effect of ball-milling, water and heat treatment on chick performance at 2-4 weeks of age (5 replicates of 10 chicks each/treatment). From Naber and Touchburn (1969)

Treatment of potato starch component in basal ration	*Body weight gain (g)*	*Feed/unit gain*	*ME-Value of diet*
Not treated	176^a	4.76^a	6.99^a
Ball-milled	161^a	3.63^b	10.08^c
Water at:			
16C	175^a	4.45^a	7.03^a
41C	168^a	4.80^a	6.69^a
58C	209^b	3.89^b	8.12^b
68C	263^c	2.14^c	13.18^d

a,b,c,d Values bearing different superscript letters in a column are significantly different from each other ($P < 0.05$) using Duncan's multiple range test.
ME = Metabolizable energy (MJ/kg)
The experimentally processed starches were incorporated into the basal ration where they comprised 50% of the diet.

Nitsan and Bartov (1972) reported similar conclusions from their trials comparing the nutritional value of potato and corn starch for chicks. In their first trial when potato or corn starch supplied 50% of the dietary carbohydrates, no significant difference in growth rate was noticed. However, the potato starch group showed a higher feed consumption and a lower feed utilization. The digestibility of the corn diet was higher than that of the potato diet. In their second trial when potato or corn starch made up to 100% of the dietary carbohydrate, average body weight gains over 46 days were 352 g for chicks fed corn starch, and 304 g for those fed potato starch, this difference was significant. The amylase activity in the pancreas of the chicks fed potato starch was higher than in those fed corn starch. They concluded that in comparison with maize starch potato starch depressed the growth rate of chicks significantly when it was the main source of carbohydrate.

Hakansson and Lindgren (1974) studied the ability of laying hens to digest raw potato starch. They concluded that there are genetic differences in the hen's production of enzymes used in carbohydrate digestion or by individual differences in the manner of feed consumption. The time that the feed stays in the crop may be of importance for the gelatinization of the starch granules which, in turn, affect their digestibility. The use of potato starch in feeds for laying hens may help in finding the factors affecting the digestibility of carbohydrates by poultry and the importance of the different segments of the digestive tract.

Potato starch wastewater. Gerry (1977) studied the effect of Dehydrated Soluble Potato Solids (DSPS; a waste product from potato starch manufacture) on broiler performance. In his first experiment samples of DSPS containing trypsin inhibitor at high (58 800 units/g) and low trypsin inhibitor (4700 units/g) were fed in mixed diets at several levels to broilers till 54 days old. The 2 and 4% inclusion of DSPS, high and low had little or no effect on growth and feed efficiency of the broilers. The same trend was observed with levels of 2, 4 and 8% inclusion of high trypsin inhibitor against control. The inclusion of higher levels of DSPS (12 and 16%) depressed body weight and increased feed efficiency significantly. Mortality was low and not associated with the rations fed. The high sugar content of DSPS resulted in a very hygroscopic mixed feed mainly with the high inclusion. Moreover, the mixed feed tends to lump in storage. The 16% inclusion in the feed caused a brown sticky mass below the beak of 5 day old chicks. This brown mass was composed of chick down feathers and parts of feed. The droppings of the chicks were very loose and sticky, and the water troughs contained considerable feed, and the edges of the feed troughs were covered with a gummy substance. The feed with the 16% inclusion of DSPS was like glue and sticky when it was wet (Table 6.35).

The second experiment was aimed to inactivate the trypsin inhibitor by evaporating (spraydrying) the DSPS at 71C and lowering the hygroscopicity by adding calcium oxide to DSPS. The inclusion of the free DSPS with trypsin inhibitor at 3, 5 and 7% showed no significant difference against control as far as body weight and feed efficiency of chicks at 54 days of age. Owing to the inactivation of the trypsin inhibitor by heat it was suggested that the availability of amino acids in the protein were negatively affected and this was reflected on the slight decrease in body weight and increase in feed efficiency (Table 6.35). In the treatment with the addition of calcium oxide to DSPS (to obtain lower hygroscopic properties), a lower temperature was used in this process to prevent destruction of the amino acids. The results showed no significant difference between the mean weight of the birds at 3, 5 and 7% inclusion of DSPS and that of the controls. However, feed/gain was significantly lower in controls than in the 5% inclusion (Table 6.35). It was concluded that standard broiler diets containing 5% of DSPS treated with heat and calcium oxide produced growth and feed efficiency (significantly) equal to control diets.

Table 6.35. Effect of dehydrated soluble potato solids (DSPS) in several forms on the performance of combined 54-days-old male and female broiler chicks. From Gerry (1977)

Treatments	*Average body weight (g)*	*Feed/gain*
First experiment		
Control	$1770^{a,b}$	1.97^{a}
2% DSPS High T I [d]	$1744^{a,b}$	1.94^{a}
4% " " "	$1771^{a,b}$	1.97^{a}
8% " " "	$1757^{a,b}$	2.02^{a}
12% " " "	$1704^{b,c}$	2.11^{b}
16% " " "	1612^{c}	2.20^{c}
2% DSPS Low T I	$1755^{a,b}$	1.93^{a}
4% " " "	1817^{a}	1.95^{a}
Second experiment		
Control	1941^{a}	1.84^{a}
3% DSPS Free from T I [e]	$1883^{a,b}$	1.88^{a}
5% " " " "	1836^{b}	1.92^{a}
7% " " " "	1735^{b}	1.89^{a}
Control	1808^{a}	1.85^{a}
3% DSPS + Calcium oxide [f]	1772^{a}	1.90^{a}
5% DSPS + " "	1723^{a}	1.90^{a}
5% DSPS + " "	1775^{a}	1.97^{b}

[a,b,c] Means with each test column without a common superscript are statistically different (experiment 1 is separate from experiment 2).

[d] T I = Trypsin inhibitor; high = 58 800 units/g; low = 4 700 unit/g.

[e] DSPS evaporated at 71.1C to deactivate trypsin inhibitor.

[f] Calcium oxide added to DSPS to lower hygroscopic properties.

(d) Potato shreds

Vogt and Stute (1969[a]) studied the feeding value of dried potato shreds included in broiler diets and the performance. Their report showed that inclusion of 30% potato shreds in broiler diets, which were formulated to be similar in digestible protein and energy to the control diet, resulted in weight gains and feed efficiency equal to those of the controls. However, the inclusion of dried potato shreds in rations caused a higher water content of faeces produced by the broilers.

(e) Potato waste effluent

Studies on the effect of potato waste effluent as a feedstuff in the nutrition of poultry have yet to be completed. An alternative was reported by Meister and Thompson (1976) who evaluated the protein quality of precipitate from waste effluent of potato chip processing with field mice. They replaced several potato effluents by casein (control) in synthetic diets composed of mineral salts, vitamin mixture, corn oil, fibre, honey, sugar and starch. The preparation of the material used (tubers, heat precipitate and acid precipitate) took place as follows: Average-sized tubers of 3 varieties were selected, washed, autoclaved at 2.2 kP for 15 min., peeled, sliced, and freeze-dried. To simulate wastewater, tubers were washed and ground for 10 min. and blended. Mash was diluted with water to 5 times its volume, filtered through several layers of cheesecloth, and left for 30 min. to settle the starch. Half of each diluted extract and half of a water sample received from potato chip processing plant were acidified with HCl to pH 3.0. Samples were stirred for 15 min. and then subjected to sedimentation for 90 min. The supernatant was drained off. The other half of each sample was heated to 98C after adjustment to pH 4.5. This was followed by immidiate cooling in ice water and settling. The peel and the acid precipitates were heated in a water bath for 15 min. at 60C to gelatinize the starch. All samples were freeze-dried.

The results of Meister and Thompson (1976) showed that the protein quality of whole tubers, heat and acid precipitates of the three varieties, and a sample of abrasive peel was compared with that of casein. The analysis of variance of the three treatments of the three varieties indicated that treatment effects were less pronounced than variatal differences. Feed intakes, weight gains, and Protein Efficiency Indices (PEI) are summarized in Table 6.36. Feed consumption was the same for all diets while weight gain showed differences: casein control diet showed the highest weight gains and peel diets were the lowest. PEI values were highest for casein and varied between potato varieties. Protein recovered by either method of precipitation gave equal or higher PEI's than crude protein of the whole tubers. They concluded that a potato chip plant could reduce discharged waste and water use by relatively simple means, and obtain a high quality feed containing approximately 30% protein. This research was based on field mice as target animals and only plant protein from potato waste effluent was considered; starch, as a main source of potato carbohydrate was not put into consideration. This field is in great need of a scientific support using practical poultry nutrition diets as a basis of evaluation.

Table 6.36. Feed intake, weight gain and protein efficiency index (PEI) of varieties [1], potato processed products and a casein control of synthetic diets in meadow vole (mice) feeding. From Meister and Thompson (1976)

Inclusions to synthetic diet [2]	*Inclusion (%)*	*Protein % of inclusion*	*Feed intake (g/6 days)*	*Weight gain (g/6 days)*	*PEI*
Casein	8.0	87.0	30.0	6.25[a]	3.02[a]
First variety					
Tubers	64.8	10.8	35.8	4.75[c,d]	1.89[e]
Heat precipitate	25.7	27.2	35.5	5.31[b,c]	2.19[c]
Acid "	24.1	29.0	28.4	4.13[d,e]	2.07[c]
Second variety					
Tubers	58.3	12.0	35.3	5.98[a,b]	2.51[b]
Heat precipitate	20.0	35.0	40.3	6.26[a]	2.21[b,c]
Acid "	23.5	30.0	28.2	5.82[a,b]	2.96[a,b]
Third variety					
Tubers	46.7	15.1	32.2	3.93[e]	1.81[e]
Heat precipitate	21.6	32.4	33.8	4.93[c]	2.09[c]
Acid "	24.6	28.4	35.4	5.41[b,c]	2.18[c]
Potato peel	58.3	12.0	34.7	0.06[f]	0.02[d]

[a,b,c,d,e,f] Values with the same letter are not significantly different at the 5% level.

[1] Variety of potatoes, Burbank, Sebago, unknown and potato peel. Average values of 6 voles per diet.

[2] Synthetic diets composed of mineral salt, vitamin, mixture, cornoil, fibre, honey, sugar and starch.

6.5 GRAPE AND WINE RESIDUE

6.5.1 Introduction

Grapes belong to the grape family (*Vitaceae*), and the known European grape is *Vitis vinifera* L. which is spread in all Mediterranean Countries. Viticulture for wine production was known to the Assyrians and Egyptians by 3500 BC and to the Greeks by about 1400 BC. Grape culture had spread to the river Rhine by 200 AD. and had reached its widest extension in Europe by the 15th Century. The vine was cultivated and wines were made throughout Europe during the post-Roman and pre-modern period. For general information on wines, there are several articles and books published in this field (Jeffs, 1971; Schoonmaker, 1973; Amerine and Singleton, 1977; Lichine, 1979).

Grapes are round berries, smooth, blue-black, red or yellow-green, two-loculed, four-seeded or less. The seed is pear-shaped with two grooves, the endosperm is horny and the embryo is minute. Figure 6.15 and 6.16 show the microscopic structure of the grape seed which is composed of, firstly, spermoderm with five well marked layers: outer epiderm, parenchyma, stone cells, lattice cells and the inner epiderm layer.

Also, the prisperm and the endosperm layer are present. The bulk of the seed consists of thin-walled aleurone cells of the endosperm. The aleurone grains show well-developed crystalloids of calcium oxalate rosettes and fat. The microscopic structure gives an idea of the fat content of the seed which varies between 8.5-14.8%, which is formed in the seed (endosperm) (Winton and Winton, 1949).

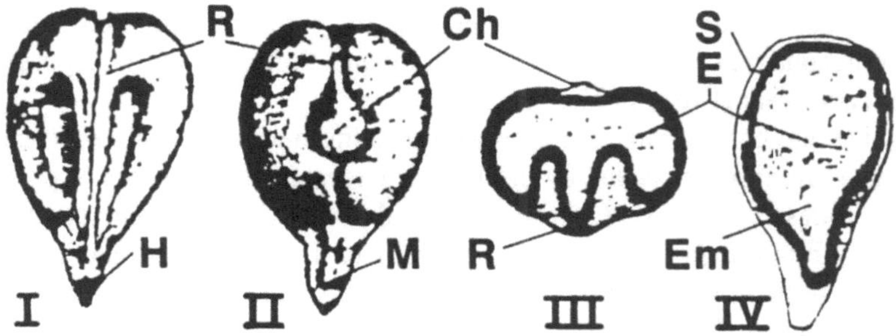

Figure 6.15. A grape seed: I, ventral side; II, dorsal side; III, cross section; I,V longitudinal section; *H,* hilum; *M*, micropyle; *Ch*, chalaza; *R*, raphe; *S*, spermoderm; *E*, endosperm; *Em*, embryo. X 15. Reproduced with permission from Winton and Winton (1949).

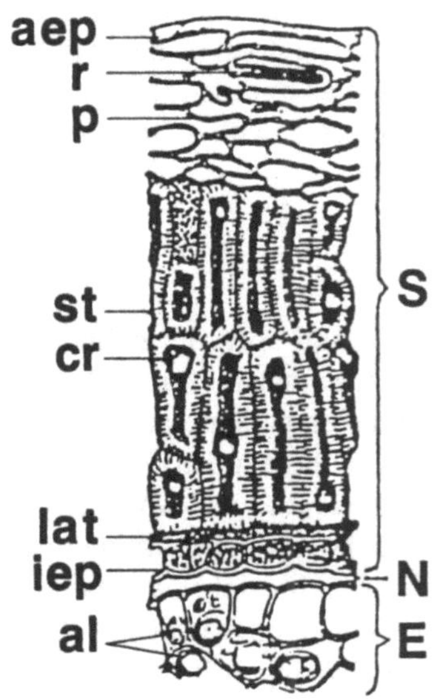

Figure 6.16. Grape seed in cross-section. *S*, spermoderm; *aep*, outer epiderm; *p*, parenchyma with raphides *(r)*; *st*, stone cell layer with crystals *(cr)*; *lat*, lattice cells; *iep*, inner epiderm; *N*, perisperm; *E*, endosperm with aleurone grains *(al)*. X 160. Reproduced with permission from Winton and Winton (1949).

Not every species of *Vitis* is suitable for wine production because of deficiencies in chemical composition or production, or possibly its undesirable flavour. Europe is the most important wine-producing area with more than 70% of the total acreage and wine production. The highest producers of wine in Europe are France and Italy followed by Spain, the Commonwealth of the Independent States (Russia), Germany, Portugal, Romania and former Yugoslavia. Moderate producers are Argentina and United States. In the third place as far as production is concerned in Europe are Hungary and Greece and outside Europe Chile, South Africa and Algeria can be mentioned. Other important wine producing countries are Australia, Bulgaria, Brazil, Uruguay, Cyprus, Israel and Turkey. Besides this many countries produce wine just for local consumption. The highest consumption/caput/year is found in France, Italy and Portugal, and is estimated to be 100.5 l per caput/year (FAO, 1991).

Grape juice and concentrates are used as refreshing beverages but the processing of wine with its several products and by-products is widespread. The types of products are white, red and desert wines, port, sherry, vermouth, brandy, liquors and champagne.

The winery by-products are numerous such as tartrates, seed oil and tannin, stock feedstuffs and fertilizers from grape pomace. Other possible by-products are charcoals, pigments and vinegar (Rice, 1976).

Grape or wine grape processing (total world wine production 29.2 million t) produces large quantities of waste products estimated to be 3.6 million t (FAO, 1991). These waste products are about 6-10% by weight of the total grape fruit but this varies somewhat from variety to variety but also with degree of maturity (Amerine *et al.*, 1980). The waste from grape processing is known as grape or winery (pomace), which consists of skin and seeds, with or without stalks. The dried skin and pulp, without seeds is called marc and after grinding, it may be used as a feedstuff for livestock. The grape seeds contain 22% edible oil which can be pressed with solvent to produce seed cake that can be used as a feedstuff (Göhl, 1975).

6.5.2 Processing

Wine production is a very wide and specialized field where several table wines are processed with a special techniques that vary according to the tradition of the country and the region. Climate as far as temperature and the type or species of *Vitis vinifera* are concerned, soil and consumers' taste are all factors playing an important role in the wine making. The type of end product as red, white, rosé, or desert wines, port, sherry, vermouth, brandy or champagne, however, determines the type of processing. The most important and relevant references in this field are Amerine and Joslyn, 1970; Cooke and Berg, 1973; Ough and Berg, 1974; Guymon and Crowell, 1977 and Amerine *et al.*, 1980. An example of the red wine processing will be discussed to show the various stages from harvesting to filling or bottling.

(a) Processing of red wine

Red grapes contain pigments which are localized in the red skin of the berries. In processing red wine most of the fermentation takes place on the skins particularly to extract the colour. On the other hand, with the processing of white wine the juice is fermented free of the skins to avoid the colour being extracted from the skins and to avoid tannin formation as much as possible.

For the processing of red wine special varieties are recommended to assure a short fermentation period, a high tannin level and a dark colour and early maturity of the wine variety which should be resistant to bunch rot and sunburn, and which ripens very early in the season, is of a great importance to the starting season. Figure 6.17 shows the schematic outline of the red wine processing.

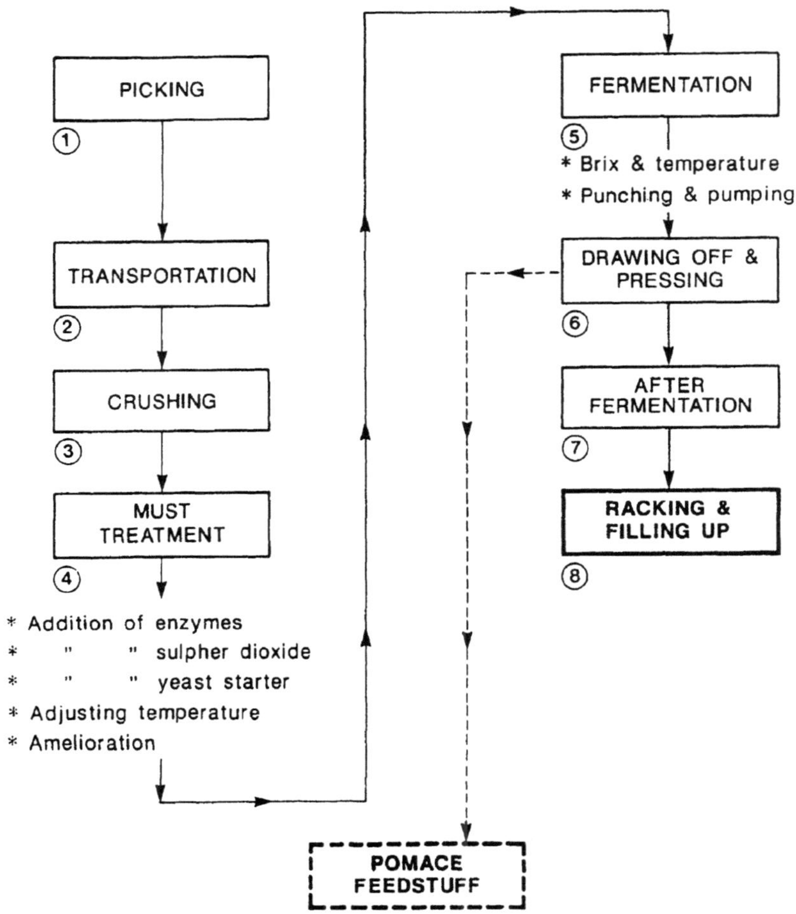

Figure 6.17. Schematic outline of red wine processing.

Before picking, the grapes should be tested for ripeness, sugar content, acidity, tartaric acid content and the Brix level (soluble solids measured as sugar). It is also desirable to follow the pH during ripening as musts (unripe wine) of high pH are unsuitable for making high quality wines.

Picking

Hand harvesting is usually carried out, often by contract, by crews of pickers who will pick at several vineyards in succession. Curved knives or short-bladed shears are used to cut the bunches from the vines. Care is taken to cut off rotten berries. Grape bunches are laid in buckets and, when they are full, transferred carefully into small gondolas (small trucks) and transported directly to the winery. Owing to the crushing effect and formation of bacterial growth of grapes during transportation with gondola trucks, mechanical harvesting is being introduced which will generally be used in the future.

Only sound grapes (not entaminated with fruit flies or mould) should be harvested. Most grapes do not ripe at the same time and therefore picking has to be planned for the second crop that ripens 2 or 3 weeks later. Proper ripening is of great importance as far as the acidity is concerned. The first crop to ripen contains less acidity than the later ripe. Therefore the second ripe crop will support much more acidity during the fermentation of must, depending on the temperature of the region.

Transportation

The location of wineries near the vineyards is of great importance for the quality of grapes. If the winery has a suitable location, transportation of the grapes will be quick and convenient. Transportation of grapes in large gondolas for long distances, causes the grapes will be damaged, crushed, fermented and bacterial growth will occur causing a lower quality. In the case of grapes being transported over large distances, it is advisable to process the grapes into wine in a winery near the vineyards and then transport the new wine. If this is not possible, the grapes can be to crushed at the vineyard into closed tanks and sulphur dioxide added prior to transporting.

Crushing

The crushing units are mostly located outside the winery for many reasons. To facilitate the delivery of the grapes, easiness of cleaning, avoidance of the spread of flies in the winery and ease of unloading from the gondolas. The roller crusher consists of two fluted, horizontal rolls of rubber coated-steel or stainless steel, operated by gears, which turn towards each other during operation. The berries are thoroughly crushed without breaking the seeds or grinding the stem. The crushed grapes and stems fall through the rollers into the stemmer, consisting of a stationary horizontal cylinder perforated with holes large enough to allow the crushed grapes to fall through and small enough to retain most of the stems. Revolving metal paddles hammer the grapes through the holes and the stems are carried out of the end of the cylinder. Stems and leaves should be removed as they contain undesirable compounds. Stems and leaves should be washed and ground for fermenting and distilling.

Beneath the crusher is a sump of metal or concrete into which the crushed grapes fall. From the sump crushed grapes are pumped through pipes of large diameter to the fermentation tanks.

Must Treatment

The variety of grape and the climate of a region determine the acidity of the must and the sugar content. If the acidity is too high and the sugar content too low, addition of both water and sugar may be essential for production of a palatable wine. These additions are a tradition per region, country and season.

Addition of enzymes. Addition of enzymes is of a great importance in the production of red wine. Musts of red grapes treated with pectic enzymes increased wine yield by 3.5% and colour by 12%. The aroma of wines was greatly enriched when the musts were treated with enzymes such as cellulase, protease and pectinase (Ough and Berg, 1974; Montedoro and Bertuccioli, 1976).

Addition of sulphur dioxide. Sulphur dioxide is added to sound grapes in a concentration of 75 mg/l and 150 mg/l for mouldy or soured grapes. The addition and mixing take place after grapes are crushed to protect them from wild yeasts and spoilage bacteria.

Addition of yeast starter. Within 2 h after the addition of sulphur dioxide which is a reasonable time to inhibit the wild yeast and spoilage bacteria, yeast is added. Pure yeast starter of a desirable strain is added in a concentration of 2% to the crushed grapes. Several hours after addition of the starter and after multiplication of the yeast, the crushed grapes are pumped over to mix the yeast with the entire contents of the vat. The use of fresh starter of pure yeast for inoculation is a special technique and tradition.

Amelioration. In very hot weather the grapes will ripen very quickly and they will become over-ripe. Addition of little water is permissible to adjust or lower the Brix. Over-ripe grapes which exceed 25 Brix and are not suitable for dry wine production. However, grape concentrate may be used for amelioration.

Temperature. For the optimal reaction of yeast, warming is necessary in cold areas. Limited quantities of crushed grapes are warmed till 60 °C and this is pumped and circulated to the vat until the mass will be heated to about 21 °C. There is evidence that this temperature will not only stimulate the yeast growing, but also that the colour and flavour extraction will be improved.

Fermentation

During the fermentation process of the must, some chemical reactions occur, sugars will be converted to alcohol and carbon dioxide will be produced, and owing to this reaction heat is formed. The carbon dioxide gas escapes from the vat. The sugar in the must that is converted to alcohol will be lower in density than water and accordingly the specific gravity (or the Brix) of the must decreases. The decrease in Brix is an approximate measure of the amount of sugar that has been fermented.

Brix and temperature recording. During wine fermentation it is of great importance to record Brix and temperature regularly and accurately. Temperature observations will indicate the increase of temperature which will inhibit the fermentation process and therefore cooling should take place. The Bix reading will show the progress of fermentation and indicating the approximate time of the end of the fermentation. At this stage drawing of the free-run of the pomace and processing it will start. Also punching and pumping over is carried out to assure a complete fermentation of the vat contents and to equalize the fermentation.

Punching and pumping over. The result of fermentation is the production of carbon dioxide gas which will bring to the surface skins, pulp and seeds to form a thick layer or cap. This cap will increase in thickness and become very dense. Fermentation is extremely rapid in the cap and a much higher temperature is formed here than the must. This will result in the destruction of the favourable yeast culture and increase the number of unfavourable thermophelic bacteria. Therefore, it is of great importance to mix the pomace of the cap thoroughly with the must. Mixing takes place by drawing the must from the bottom and pumping it back over the cap 1-10 times a day (Cook and Berg, 1969, 1973).

The progress of fermentation, the physical movements of the skin and the enzyme action lead to disintegration of the pulp. The production of the alcohol owing to fermentation leads to extraction of colour and tannin from the skins. Pectic enzymes hydrolize the pectic substances and destroy the slime of the crushed grapes. Pectines and other organic matters are precipitated by the alcohol produced. The finally fermented pomace is not slippery and can easily be pressed.

Drawing off

When the fermented must reaches the required amount of colour and tannin, pomace is drawn off. This step is mostly reached in a week or more according to the desired level of Brix (0-12), colour, tannin and alcohol 6-12%. The separation of pomace from wine may take place by drawing off or by pressing.

In drawing off, wine will mix with air, and the yeast will be invigorated. Accordingly fermentation will run smoothly to completion in the storage tank. In the drawing off system the free-run seeds and particles of pulp are strained out with stainless steel screens. The free-run wine is less astringent, smoother and contains less colour than press-wine. Grape seeds contribute a reasonable quality of tannin in the red wine estimated to be 0.2-0.4%. It was also noticed that from the tannins of red wine, 35-50% is originate from skins, 15-20% from seeds and 20-40% from stems (Singleton and Draper, 1964; Ribéreau-Gayon and Milhé, 1970).

In the case of pomace pressing, a grinding screw press or horizontal baskets, with only minimal grinding, are used. In some cases both systems are combined. In this stage of both systems, wine pomace is produced as a waste which will be discussed in detail under the processing of winery pomace.

The after fermentation

It is of great importance in this phase that the fermentation continues in the storage tank because the wine still contains unfermented sugar. The alcohol formed will show low readings of 0 Brix or even less. This stage after fermentation requires attention owing to the lower temperature which stops fermentation (sticking). Supplying oxygen may result in renewed unnecessary yeast growth. For these reasons wine should be sampled every 2 or 3 days to measure the Brix.

During this period wine tanks should be protected against oxygen and vinegar bacteria by means of a fermentation bung (a stopper for barrels). The fermentation bung builds up a slight pressure of carbon dioxide gas in the tank which prevents the aerobic growth of vinegar bacteria on yeast films. Through use of the barrel bung a sufficient atmosphere of carbon dioxide will be formed over the surface of the wine.

The cessation of bubble formation around the fermentation bung indicates that fermentation is nearly complete. The fermentation bung will be then be replaced by a plain, solid, cellar bung. However, some fermentation will still proceed and gas pressure will develop in the tank which has to be released frequently. Finally, the tanks will be completely filled.

Racking and filling up

Wine is mostly fermented and perfectly dry within six weeks after crushing. The tanks are kept completely full by regularly filling up with new wine and they are kept tightly bunged. Four to six weeks after drawing off from the fermentation, the wine is mostly well settled. Wine should be tasted at this time and accordingly appropriate blends can be made.

After racking, a sediment of yeast, pulp and tartrates (lees) is formed, which may be used in combination with sediments from other tanks, and allow to settle for recovery. These recovered lees will be used for the production of additional clear wine. By pressing the lees, a waste will be added to the wine pomace.

In some countries the second fermentation takes place known as the malo-lactic fermentation. This fermentation is produced from lactic acid bacteria and converts malic to lactic acid with a release of carbon dioxide. This fermentation reduces the titratable acidity and raises the pH. This fermentation is required in the case of acid wines and it may be prevented by early racking, cool storage and adding 10% sulphur dioxide (Ough and Amerine, 1960, 1961, 1962).

Finally, care of the wine as far as laboratory examination, aging, cellar operations, blending and bottling are concerned, requires very detailed specifications, found in the references in the beginning of this section.

(b) Processing of winery pomace

Besides the production of wine as a basic product from grapes there are many winery by-products such as tartrates or wine stone (argols) produced or recovered from pomace or from lees. The wine pomace is not only used to produce tartrates but also oil, tannin and pigments by extraction. Winery pomace is also known to be a very good fertilizer. The waste stems can also be distillated for brandy production. For example, 57 l of brandy can be obtained per t of stems.

The most important waste from wine is grape pomace to be used for livestock feed. After the fermentation of wine and its recovery, pomace may be discharged through the gate of the tank either into a conveyor or a pomace pump. Alternatively, pomace may be sluiced out with press wine (1 l wine / 9 kg pomace) into a 20 cm, double-action piston pump attached to the bottom tank valve. Pomace is usually sluiced out with water into the pomace conveyor. Continuous chain conveyors and screw-conveyors are commonly used.

Wine or grape pomace is mostly dried in rotary, direct-fired drum driers with dimensions of 1.7-2.2 m in diameter and about 17-20 m length. At the highest part of the drum, pomace is introduced through an entry part and gathered after processing from an exit lower part. Burners of gas or crude oil blow a flame into the upper end of the drier by means of a large fan. The blown hot air mixes with the pomace and carries gases and moisture vapour out of the drum drier. The wet pomace drops into the hot air combustion gases zone at the furnace end of the tunnel. The estimated temperature at this point is above 535 °C. Evaporation is very rapid, cooling the pomace and preventing scorching, although some skins dry instantly and then scorch.

The dried pomace (6% moisture) will pass through a hammer mill to be ground to a fine mash. Sometimes the mash is mixed with small amounts of molasses to improve its energy content and its palatability. Others improved the dried pomace meal by adding lime to bind the pectines and to raise the pH (Agostini, 1964; Amerine *et al.*, 1980).

Grape seed oil. From grape, seed oil may be recovered by drying the grape pomace and separating the seeds, and finally oil can be extracted. Pomace without seeds contains skin and pulp, and is called marc. The pomace is dried as discussed before in rotary, direct gas-fired drum driers, and seeds are separated by threshing and sieving. Oil may be extracted from the seeds by means of pressing the seeds, or by grinding the seeds and additional extraction by solvents (Kinsella, 1974).
Tannin may be extracted from the grape seeds to be added to wine making and for tannery in the leather industry.

Winery wastewater. Winery wastewater is produced from cooling fermentation must and from condensing the alcoholic distillate from the stems during brandy processing. Also the disposal of stillage (still slops) and condensation produced from the grape pomace, presents problems. Several methods are used for disposal such as emptying into streams (which causes pollution and may be harmful to fish) or irrigation to improve the soil structure. Other methods are the addition of commercial polyelectrolyte preparation followed by centrifugation resulting in a sludge from the effluent water. This by-product has no relevant nutritive value in animal nutrition, therefore it will not be discussed in details. For more detailed information about wine and grape wastewater, the reader is referred to the articles of Tofflemire (1972) and Ryder (1973).

Yields of grape residues

The residue produced after pressing the grape berries is estimated to be about 6-10% by weight of the total grape input. If the grapes are stripped from the stalks before processing, the residue will consist of about 40% seeds, 60% skin and pulp. Pomace produced with stalks will contain about 30% stalks, 30% seeds and 40% skin and pulp. On a basis of 63% moisture one t fresh pomace will yield about 350 kg dried pomace of 5% moisture content. One t of pomace on the basis of 50% moisture would yield about 470 kg dried pomace of 5% moisture content (Göhl, 1975; Amerine *et al.*, 1980).

6.5.3 Chemical Analysis and Nutritive Value

Grape and wine residues (pomace) may be classified as oil from grape seeds, seed oil cake (after pressing the oil), winery pomace with stems (stalks) or without, and grape marc (skin and pulp). Its chemical composition is shown in Table 6.37.

Table 6.37. Chemical composition (% in dry matter) of grape or winery seeds, marc, cake and pomace in several components. From (1) Winton and Winton (1949); (2) Amerine *et al.* (1980); (3) Göhl (1975) and (4) Bath (1981)

%	*1*			*2*				*3*			*4*	
	Seed	*Marc* *a*	*Cake*	*Dried pomace*				*b*	*Pomace* *c*	*d*	*e*	*Pomace*
Moisture	14.08	6.0	10.0	-	5.4	-	5.1	59.4	53.50	11.2	-	9.0
Crude protein	10.00	8.9	12.0	11.1	11.9	-	13.5	11.7	13.70	14.9	14.40	12.7
Crude fat	11.70	7.6	5.0	6.0	5.1	4.52	7.4	9.9	7.00	5.0	8.30	-
Crude fibre	34.90	19.8	36.0	40.0	36.1	34.00	26.9	25.5	23.60	35.8	22.30	30.0
N-free extract[f]	30.10	52.8	33.0	-	-	-	-	45.2	42.90	35.4	48.70	-
Reducing sugars	-	-	-	-	0.9	-	-	-	-	-	-	-
Starch	-	-	-	-	7.7	-	-	-	-	-	-	-
Ash	3.60	4.9	3.0	8.0	6.4	7.30	6.4	7.7	12.80	8.9	6.30	-
Calcium	0.54	-	-	-	-	-	-	-	0.82	-	0.79	-
Phosphorus	0.27	-	-	-	-	-	-	-	0.20	-	0.27	-
Potassium	0.50	-	-	-	-	-	-	-	-	-	-	-
Sodium	0.05	-	-	-	-	-	-	-	-	-	-	-
Chlorine	0.01	-	-	-	-	-	-	-	-	-	-	-
Iron	0.01	-	-	-	-	-	-	-	-	-	-	-

[a] Marc (grape skin and pulp)
[b] Stalks, skin and seed (Italy)
[c] Skin and seeds (Italy)
[d] Stalks and skin (Germany)
[e] Skin (Italy)
[f] Carbohydrate
ME (MJ/kg) = 7.3 for pomace (Farrell, Rose and Warren, 1983)

Oil from grape seeds

Grape seed oil is characterized by a very high content of linolic acid which is considered to be the most important fatty acid in human nutrition. In this respect grape seed oil is similar to safflower and somewhat superior to oils produced from corn, soybean and cottonseed. It can be safely used as an edible oil and its fatty acid composition makes it very desirable for its inclusion in human diets designed to decrease serum cholesterol (Kinsella, 1974; Amerine *et al.*, 1980).

The chemical analysis of grape seeds shows an average oil percentage of 11.7% (8.5-14.8%). The yield of oil that may be obtained by extraction ranges from 5-22% depending on the raw material and the extraction method. The glycerides of grape seed oil contain the following fatty acids: stearic acid 2.3%, palmetic acid 6.5%, oleic acid 32.4%, -linolic acid 37.5%, ß-linolic acid 8%, linolenic acid 0.2%, the rest be made up of hydroxy acids and unsaponifiable matter (Winton and Winton, 1949).

Seed oil cake

Some 7% of the oil can be removed from the seed which contains about 12% oil. After extracting the oil from the seed, the crude protein level in the seed cake will increase from 10%-12% owing to the loss of oil.

Seed oil cake, however, is very high in fibre content (36%). Its nutritive value is low due to its high tannic acid content, therefore, it is not used normally in poultry diets. Its use in animal nutrition is rather limited, for example, as a carrier for molasses in cattle feed (Winton and Winton, 1949).

Grape pomace with stems

This is the standard grape winery pomace and its chemical composition is as follows (Amerine *et al.*, 1980): crude fibre 27-40%; crude protein 11.1-13.5%; crude fat 4.5-7.4% and nitrogen free extract 35.4-48.7% (Göhl, 1975). Due to the high fibre content the use of grape pomace with stems in poultry rations is not recommended but is used for ruminants (Göhl, 1975).

Grape pomace without stems

With the new harvesting and winery techniques, other qualities of grape pomace can be produced which contain little or no stems. This stemless grape pomace has been successfully used with levels of 15-20% in feed lot rations. Its crude fibre content is lower than grape pomace with stems (23.6%) but it has a low level of potassium tartrate which can cause digestive failures (Table 6.37; Göhl, 1975; Bath, 1981; Boa, 1990).

Grape marc

Grape marc is the best by-product of the wine industry owing to its relatively low fibre content (19.8%) and its high level of nitrogen-free extract (52.8%). Its digestibility can be increased by soaking the marc in hot water (90 °C) for about 20 minutes to remove the tartrates. After this treatment the product has an acceptable nutritive value in horse diets at inclusion rates up to 10% (Göhl, 1975) or it is used and mixed during pelleting in rations of ruminants (Göhl, 1975; Boa, 1990).

All the by-products of wine processing such as winery pomace with or without stems, seed cake and marc are by-products which are produced on a large scale. If all research institutes paid attention to improve these winery by-products by means of special processing techniques, by various additions (addition of lime), or by mixing these by-products with other waste products like potato peeling or animal slaughtery waste, it will result in a product having a higher energy, protein and fat level and lower fibre content.

In general, the nutritive value of grape residue products depends on the type of by-product (grape pulp, marc, extracted grape seeds and presence or absence of stems) and it is dictated by its fibre content. Treatment with hot water affects the nutritive value as far as the tartrates are concerned. Although no toxins are reported to be present in grape residues the content of iron may be high and may affect phosphorus absorption.

6.5.4 Feeding Grape Pomace

There is little literature concerning the use of winery pomace in the nutrition of poultry. Farrell, Rose and Warren (1983) concluded that the inclusion of dried winery pomace up to 20% in well-balanced broiler diets showed an excellent and curvilinear growth ($P < 0.01$) indicating that this feedstuff may contain a growth promoting factor. The inclusion of winery pomace in levels of 10, 15, 20 and 25% in rat diets up to 14 days showed no significant differences in weight gain but some increase (not favourable) in feed conversion was noticed.

The aim of the study of Farell, Rose and Warren (1983) was to compare several waste by-products such as biscuit waste, citrus pulp, dried potato offal, potato, apple and winery pomace used separately and in several inclusion levels in one experiment on broilers and rats. This research needs further work to detect the exact effects of the winery pomace with a reasonable number of replicates and several inclusion levels of winery pomace as a sole waste product.

McKeen (1984) reported results of his trial in which grape pomace was fed to leghorn hens as an alternative to starvation to induce malt. Force malting is of a great economic value for a second-cycle egg production after a complete year's production (North, 1978; Scott, Nesheim and Young, 1982). He concluded that feeding grape pomace did not cause as much weight loss as 10 days with no feed, and further that grape pomace has a certain feed value and it can be used as an alternative to a 10-day starvation to force malt in leghorn hens.

This feedstuff needs further examination to evaluate its worth as a source of essential fatty acids and growth promoting factors and as a means to replace barley during the moulting period of layers after one year of production. In addition, its disadvantages of high fibre content and level of tartrates urge examination of processing methods to diminish these effects.

6.6 DATE RESIDUE

6.6.1 Introduction

Dates and palm belong to the fruits of the palm family (*Palmaceae*). Palm nut is known for its oil seed. Dates contain storage material in the seed as starchy endosperm. Date is known as *Phoenix dactylifera* L. Palm trees may become very large, a strong palm tree may reach 30 m in height. The leaves are very long and stiff, arching upwards with greyish green leaflets. The date fruits are 2-4 cm long, cylinder shaped, and grow profusely on long hanging strands. The stony long seeds (pits) are sometimes removed from the edible flesh before the fruit is dried.

The most essential requirements for the growth of date trees are the use of suitable varieties, careful pollination, hot dry climate and sufficient water at the roots. The proper conditions of intense heat above ground and moisture below are met in the Gulf regions (oasis in the desert) where a tree may yield as many as 20 clusters and as much as 100 kg of fruit.

Varieties of dates are usually grouped according to the nature of the fruit, being soft, dry or having a low sugar content.

Varieties giving soft fruits are rich in saccharine juice, and are mostly exported to Europe and America. The fruits of date varieties are hard, non-sticky, and have good keeping properties. These are used for local consumption in the Gulf region. Varieties producing low sugar fruits include a special type of fruit which has a low sugar content and is very juicy. Owing to the low sugar content the fruit is quickly spoiled during drying in the sun, therefore it should be quickly consumed in its fresh form.

All date varieties are diœcious, and they are commonly propagated by root suckers rather than seed. Pollination takes place by a male tree which can pollinate an orchard of hundred trees if nature assists in distributing this pollen.

Dates, free from the stem and pit, are commonly packed in the Gulf region in boxes or grass mates, and by the Americans in various modern packs. The colour of the dried date depends partly on the colour of the fresh fruit, which may be red, yellow or intermediate. However, the care taken during harvesting and drying plays an important role as far as the quality and the colour are concerned.

Dates stuffed with nuts or rolled in sugar are prepared on a commercial scale for human consumption. Besides that there is a very large industry for date processing to end products including date paste, juice concentrate, artificially dried date granules, date syrup, jelly, marmalade, soft drinks, compôte, pickled dates and alcohol. However, the by-products are of a great importance to the animal and poultry feed industry owing to the waste produced from this industry. From this waste the most known by-product is the seed or pit meal, discarded dates as dried, whole dates as well as the sugar-extracted fruit pulp. This is in addition to the other benefits the palm tree offers its cultivators: the tree trunk can be used to make, for example, walls, rafters, doors, shutters and stairs, in house construction.

The fronds (leaves) are used in the construction of complete houses, fences, roofs, crates, chicken cages, bedsteads, boats and as fuel, while the leaflets are woven to make mats and baskets. The fibres from the round part of the heart of the palm, are also used to make many products such as ropes, stuffing of pillows and mattresses. The fibre from the leaflets is used in furniture making. Beside those products palm trees have an important function in improving shade and providing protection from the wind for crops growing below such as citrus trees. (Dawson and Aten, 1962; FAO, 1965; Göhl, 1975; Rygg, 1975).

Figure 6.18 shows a longitudinal section of date with the skin, flesh, envelope of the seed, the cap and the seed (pit or stone). Figure 6.19 and 6.20 show that the date fruit is a one-seeded berry which is hard, elongated, with a groove on the ventral side and with a spot marking position of minute embryo on the dorsal side. The endosperm shows storage material in thickened walls with obvious oil in the cells. The oil percentage varies in the pits or stones from 8.1-11.2% (Winton and Winton, 1949; El-Shazly, Ibrahim and Karan, 1963; Rygg, 1975).

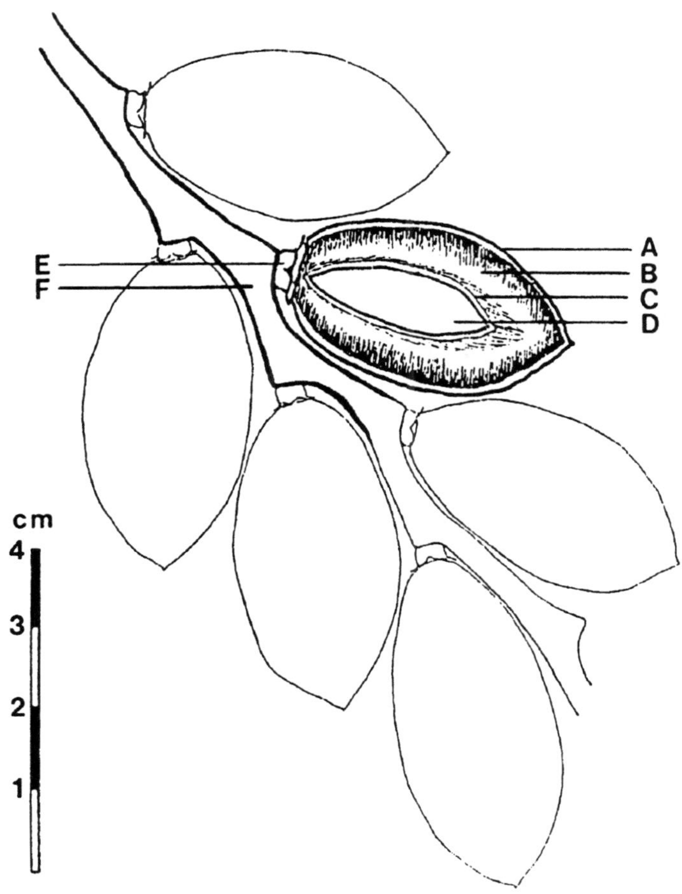

Figure 6.18. Longitudinal section of date, showing its parts: (A) epicarp, skin; (B) mesocarp, flesh; (C) endocarp, white, papery envelop of seed; (D) seed, pit or stone; (E) perianth, calyx, or cap; (F) stalk or spikelet. The heavily shaded outer part of the flesh represents that part which has been almost entirely converted to sugar. Reproduced with permission from Dowson and Aten (1962).

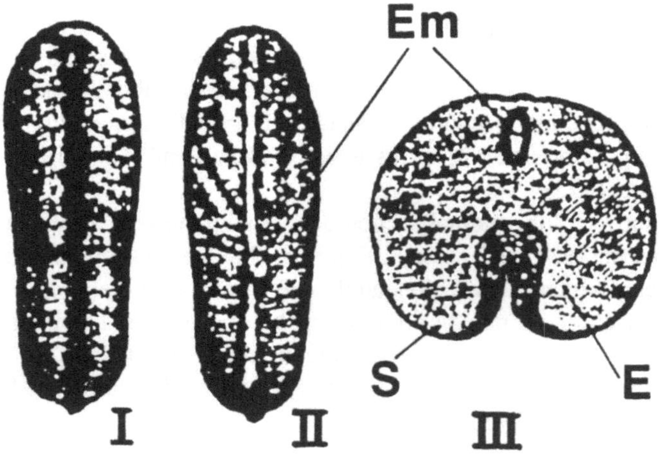

Figure 6.19. A date seed: I, ventral side; II, dorsal; III, cross-section. *S*, spermoderm; *E*, endosperm; *Em*, embryo in cavity. X 15. Reproduced with permission from Winton and Winton (1949).

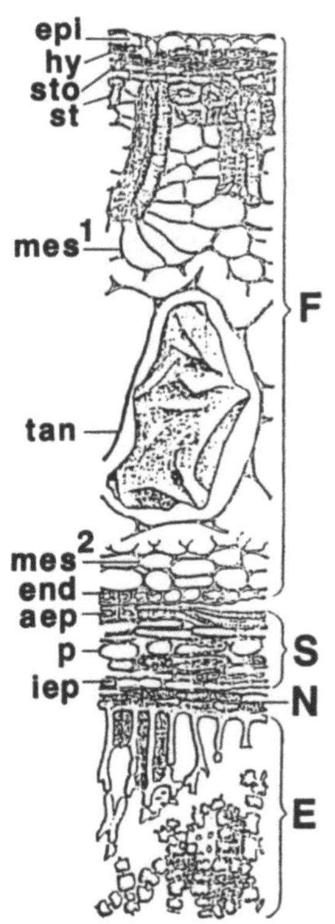

Figure 6.20. A date fruit and seed in cross-section. *F*, pericarp; *epi*, epicarp with stoma *(sto)*; *hy*, hypoderm; *st*, stone cells; *mes*^{1}, outer mesocarp; *tan*, tannin cell; *mes*^{2}, inner mesocap; *end*, endocarp; *S*, Spermoderm; *aep*, outer epiderm; *p*, paranchyma; *iep*, inner epiderm; *N*, perisperm; *E*, endosperm. X 160. Reproduced with permission from Winton and Winton (1949).

The main date-producing countries in the world are Egypt, Iran, Saudi Arabia, Iraq, Qatar and Algeria. Other main producers of dates are Tunisia, Oman and Marocco and, at lower rate, Spain, Israel and North America.

The main market for dates is in Egypt, Iran, Iraq, and other Asian countries, and some dates are shipped to Europe and the United States. The world production of dates amounts to probably over 3.4 million t / year (FAO, 1991). Assuming that 50% of this production will be processed, this amount will produce 0.17 million t of date pits which can be processed as a feedstuff instead of being discarded.

Ground date seeds can be used in rations of ruminants up to 20% of the total ration with good results. They have also been used for pig and poultry rations to replace carbohydrate in such feedstuffs as maize, wheat and barley with no side effects. Dried pulp as well as sugar-extracted pulp are reasonable carbohydrate feedstuffs for all farm animals (Göhl, 1975).

Several research workers have described the value of date waste as a feedstuff and detailed information has been reported (Tamimie, 1958, 1959; El-Shazly, Ibrahim and Karam, 1963; Jumah, Al Azzawi and Al Hashini, 1973; Al-Hiti and Rous, 1978, Kamel *et al.*, 1979, 1981; El Moghazy and El Boushy, 1982[b]; Sawaya, Khalil and Safi, 1984; Gualtieri and Rapaccini, 1990).

6.6.2 Processing

(a) Processing of fresh dates

Processing of dates and its generated waste has been discussed in detail by many research workers (Dawson and Aten, 1962; FAO, 1965; Rygg, 1975). The processing of dates differs from one country to another according to the weather conditions, the variety of date used (soft or hard), the curing required, dehydration, or glazing (glossy shape) and preservation. Those are all details related to the style in which the dates will be conserved and presented in their final packing form for export.

The harvesting method affects the quality of the end product of dates. If dates are harvested in a clean way, the labour and efforts of cleaning later will be saved. If dates are allowed to fall on the ground after shaking the tree, they will pick up soil and sand, and extra work will be needed to remove the dirt. However, the time and method of harvesting and the care given will be reflected on the type of processing needed. The poor conditions of harvesting in developing countries will considerably increase the processing costs and this will reflect on the quantity of waste produced (less effort is required to separate the seeds as a waste).

Hand harvesting. The dates which mostly hang very high on the tree, may be reached by climbing the tree barefoot with the aid of a rope, or by a light high ladder. These dates are harvested in a traditional manner by picking individual dates into a bucket. In addition methods are based on collecting bunches of the dates in containers or collecting dates fallen on the ground where they are mostly gathered in field containers or on palm leaflet mates. In most cases, field grading takes place by hand followed by pitting (removing the pit) and capping (removing the cap).

Field pitting is mostly done by women on the roofs of houses where the dates are drying in the sun. The methods of removal of the pits is also a tradition which differs from country to country. Generally they are removed by squeezing out the date between the finger and the thumb where the skin of the soft date is also slipped off. Crushing the date between two stones or moving the pit with a small knife is also practised. Finally pits are also removed by a blunt-ended needle set in a wooden handle, where the pit is pushed through the date endwise.

Capping (perianth removal or deperianthing, or calyx removal) are terms used for the removal of the cap. These caps vary in the manner in which they are joined to the date depending on the variety. This step improves the appearance and makes the date more attractive; caps have no nutritional value as a waste.

Mechanical harvesting. This new mechanical aid changed the handling procedures as far as the time of harvesting, maturity and damage to the dates are concerned. Only the dried kinds are suited to mechanical harvesting, as softer kinds are too tender and would be damaged. The mechanical harvesting is based on a container attached to a hydraulic, telescopic lift which is used to position a worker in a basket, near the bunches, where he can cut the mature bunches. They are equipped with a shaking device for removing the dates from the bunches and the dates drop into a bin beneath the device.

In the storing house dates go through a fumigation process (methyl bromide, carbon disulfide, hydrocyanic acid or etlylene oxide) which is used to destroy all insect life. Storing mostly takes place in a cool temperature (0-4 °C) to complete the destroy of insects and minimize bacterial growth (Rygg, 1975).

Cleaning takes place by means of spraying or brushes with some detergents or germicides, and then drying by blown air. After cleaning dates will be sorted to remove culls and are graded. Sometimes artificial maturation takes place using temperatures varying from 35-46 °C or chemical treatments such as the use of acetic acid, carbon dioxide, ethylene, hot lye or salt.

The dates are dehydrated and pitted. Hand pitting has been described above and mechanical operated pitters such as masher pitters or plunger-type pitters are semi-automatic and quick methods to separate the pits from the dates.

Finally, pitted dates are packed in large packages for wholesale or in small decorated plastic containers ready for shipment. Storage after packing is of great importance to avoid damage to the fruit, owing to deterioration of taste, or colour and to bacterial development. The required temperature for storage, therefore, is 1-2 °C for 5-6 months. Transportation also requires low temperatures.

(b) Processing of the date pits

Date pits are known to be very hard with a low moisture percentage 9.3% (Winton and Winton, 1949) which facilitates processing without any treatment. El-Shazly, Ibrahim and Karam (1963) reported that date stones are difficult to grind in an ordinary hammer mill. Firstly, they crushed the pits in a disc crusher followed by fine grinding in a grain-grinding stone mill. Their trials took place in the sixties where up-to-date developments of hammer mills were not completely established. Nowadays, however, heavy duty high rotation hammer mills are available for this purpose: the Retsch mill facilitates and accelerates the processing of grinding this waste.

Kamel *et al.* (1981) reported that whole Zahdi dates were dried in a forced draft oven at 65 °C for 24 h and then grounded in a hammer mill.

(c) Yields of date pits

The consumption of pitted dates is increasing because they offer greater consumer convenience and lower the transportation costs due to their lower weight compared with whole dates. In the United States it was reported by the California Date Growers Association that 80% or more dates were pitted. Date pits constitute about 10% of the weight of the whole date (Rygg, 1975).

6.6.3 Chemical Analysis and Nutritive Value of Date Pits and Waste

Kamel *et al.* (1981) and El Moghazy and El Boushy (1982[b]) classified date waste as pits and date waste without pit. Göhl (1975) came to the same classification with the addition of sugar-extracted fruit date pulp.

(a) Date pits (stones)

Date pits were found to have a low but acceptable protein content and a nitrogen-free extract. Their variability in chemical composition can be attributed to processing of the date as well as the variety. Chemical analysis and amino acid pattern of date pits in comparison with maize and barley is shown in Table 6.38. Date pits showed a protein percentage varying from 4.5-10.6%, crude fat from 6.2-10.4%, nitrogen-free extract from 47.7-62.5%. Those nutrients contents are comparable to maize and barley the only remarkable difference being the crude fibre content which varies in date pits from 16.2-30% while in the maize and barley the level is 2% and 7.5%, respectively. This high crude fibre percentage will limit the diet inclusion level of date pits to 10-15%. Macro elements as far as calcium and phosphorus are acceptable. The most limiting amino acids in a practical poultry ration (methionine and lysine) in date pits are similar to those of maize and barley (Table 6.38) (Göhl, 1975; Rygg, 1975; Kamel *et al.*, 1981; Scott, Nesheim and Young, 1982; Sawaya, Khalil and Safi, 1984; Gualtieri and Rapaccini, 1990).

Kamel *et al.* (1981) analysed the amounts of fatty acid in Zahdi pits to be capric 0.03%, caprillic 0.03%, lauric 1.27%, myristic 0.52%, palmetic 0.60%, stearic 0.02% and oleic and linoleic acid 2.85%. Linoleic acid and oleic acid contents constituted a substantial percentage of the lipid fraction in the pits and accounted for 52% of the total fat. These fatty acids could have a significant nutritional importance in the diet. Sawaya, Khalil and Safi (1984) analyzed the total minerals of date pits and showed a higher concentration (in ppm) of K 31.7, followed by Mg 6.3, Na 3.1, Fe 0.7, Zn 0.2, Cu 0.1 and Mn 0.3. The major fatty acids of date seeds were analyzed to be oleic 44.3%, lauric 17.4%, myristic 11.5%, palmitic 10.3% and linoleic 8.5%. They concluded that the average *in vitro* protein digestibility is 63.5% whereas that of casein is 90%.

(b) Whole date and date waste

Kamel *et al.* (1981) analyzed the Zahdi whole dates to have a protein percentage of 2.9% which is much lower than for date pits, maize and barley. This is due to the fruit flesh which is rich in carbohydrate N-free extract (76.2%) which is similar to that of barley (79.0%) but lower than that of the pits. The amino acid contents mainly methionine (0.06%) and lysine (0.10%) is lower than their contents in the pits, in maize and in barley. This is also due to the flesh of the fruit which originally had a lower protein percentage. Calcium and phosphorus level (0.76 and 0.52%, respectively) were shown to be much higher than those in maize and barley (Table 6.38).

As far as date waste without pits is concerned, the results of Al-Hiti and Rous (1978) showed that the protein content (8.1%) was higher than that of whole date and was nearly the same as that of maize (8.6%). Crude fat level was 1.8%, lysine was 0.32% and crude fibre was 9.1% and was shown to be higher than in barley and maize but much lower than in date pits. Calcium (0.31%) and phosphorus (0.62%) levels were far higher than that of maize and barley (Table 6.38).

Göhl (1975) reported the chemical composition of sugar-extracted fruit date pulp. This pulp contains a crude protein level of 5.5%, crude fibre 11.8%, fat 0.4%, nitrogen-free extract 79.6% and ash 2.7%. This product is rich in carbohydrates, and low in fibre and protein content. He reported that this feedstuff may be added to broiler diets in levels up to 20%.

6.6.4 Feeding Date Residues

(a) Feeding date pits

Tamimi (1959) carried out experiments and studied the use of date pits at levels of 2, 4, 8 and 16% in chick diets. He concluded that the inclusion of 4, 8 and 16% date pits in the diet had no depressing effect on body weight of the chicks till 3 weeks of age.

Jumah, Al Azzawi and Al Hashini (1973) reported that broilers fed diets with ground date pits at levels of 5, 10 and 15% in comparison with a control diet. They concluded that date pits may be used till 10% inclusion level. They noticed that body weight was significantly lower and feed consumption was significantly higher with date pits in the 15% inclusion level. Feed conversion was the highest with 15% inclusion of date pits. In the diet with 15% inclusion 6.1% crude fibre was reached. This level will increase the passage of ingesta in the intestines, resulting in a lower feed utilization and a lower body weight and an increase in feed consumption resulting in a high feed conversion (which is not economic). However, if the price of the ration with 15% inclusion of date pits is cheaper than the control with yellow maize, as in this case the inclusion level will be acceptable.

Kamel *et al.* (1979, 1981) carried out a study in which date pits were included at 5, 10 and 15%, with and without antibiotic zinc bacitracin, in broiler diets to replace wheat bran, corn and alfalfa meal. Their diets were formulated to be isonitrogenous and isocaloric. Their results indicated that the inclusion of date pits supported chick growth as efficiently as the control diets at all dietary levels tested. Inclusion of zinc bacitracin

Table 6.38. Chemical composition and amino acid analysis (air-dry product basis) of several date wastes in comparison with barley and yellow corn. From [(1)] Rugg (1975); [(2)] Göhl (1975); [(3)] Kamel *et al.* (1981); [(4)] Gualtieri and Rapaccini (1990); [(5)] Sawaya, Khalil and Safi (1984); [(6)] Al-Hiti and Rous (1978) and [(7)] Scott, Nesheim and Young (1982)

%	*Date pits, stones or seeds (%)*					*Whole date (%)*	*Date waste (%)*	*Maize (%)*	*Barley (%)*
	(1)	*(2)*	*(3)*[a]	*(4)*[b]	*(5)*[c]	*(3)*[a]	*(6)*[d]	*(3 and 7)*	*(3 and 7)*
Moisture	7.1	14.7	6.50	12.90	5.00	15.30	4.80	-	-
Crude protein	5.2	5.8	10.60	4.50	6.50	2.90	8.10	8.60	9.50
Crude fat	8.6	7.1	6.20	6.20	10.40	1.10	1.80	3.90	1.90
Crude fibre	16.2	30.0	26.50	19.20	22.00	2.40	9.10	2.00	7.50
N-Free extract	62.5	55.3	47.70	56.20	60.00	76.20	-	83.00	79.00
ME (MJ/kg)	-	-	-	-	-	11.80	-	14.10	9.70
Ash	1.1	1.8	2.50	0.96	1.10	2.10	3.50	1.50	2.50
Calcium	-	-	0.57	0.57	0.32	0.76	0.31	0.02	0.05
Phosphorus	-	-	0.31	0.31	0.05	0.52	0.62	0.30	0.32
Amino acids %									
Aspartic acid	-	-	0.67	-	0.71	0.24	0.77	0.57	0.61
Threonine	-	-	0.36	-	0.26	0.12	0.36	0.40	0.33
Serine	-	-	0.43	-	0.20	0.14	0.44	0.51	0.41
Glutamic acid	-	-	1.02	-	1.50	0.38	0.81	1.75	2.46
Proline	-	-	0.29	-	0.29	0.09	0.42	0.74	1.01
Glycine	-	-	0.32	-	0.36	0.12	0.46	0.40	0.30
Alanine	-	-	0.44	-	0.34	0.14	0.46	0.68	0.42
Valine	-	-	0.33	-	0.39	0.13	0.46	0.40	0.46
Methionine	-	-	0.17	-	0.15	0.06	-	0.18	0.14
Cystein	-	-	-	-	0.13	-	-	0.18	0.15
Isoleucine	-	-	0.31	-	0.21	0.10	0.36	0.40	0.40
Leucine	-	-	0.59	-	0.48	0.20	0.65	1.10	0.60
Tyrosine	-	-	0.15	-	0.14	0.05	0.17	0.41	0.19
Phenylalanine	-	-	0.36	-	0.27	0.11	0.39	0.50	0.48
Histidine	-	-	0.10	-	0.10	0.03	0.21	0.20	0.22
Lysine	-	-	0.31	-	0.33	0.10	0.32	0.20	0.30
Arginine	-	-	0.30	-	0.91	0.09	0.24	0.50	0.43
Tryptophan	-	-	-	-	0.05	-	-	0.10	0.13

[a] Zahdi dates; [b] Tunisian Deglit dates; [c] Mean of Ruzeiz and Sipri dates; [d] Date waste without pits

improved bird performance for both control and treatment diets. Examination of various organs from birds receiving date pits revealed no abnormalities (Table 6.39).

Table 6.39. Effect of diet date pit inclusion with or without zinc bacitracin on broiler performance up to 4 weeks of age. From Kamel *et al.* (1981)

	Body weight gain (g/bird)	*Feed consumption (g/bird)*	*Feed efficiency (feed/gain)*
Inclusion of zinc bacitracin [d]			
Control	578[a]	1013	1.75[a]
" + 5% Date pits	583[a]	998	1.71[a]
" + 10% "	573[a]	983	1.70[a]
" + 15% "	579[a]	1047	1.81[a]
No zinc bacitracin			
Control	535[b]	984	1.81[a]
" + 5% Date pits	547[b,c]	937	1.71[a]
" + 10% "	552[c]	965	1.75[a]
" + 15% "	540[b]	953	1.77[a]

[a,b,c] Means with each column followed by the same superscript are not significantly different at 5% probability level.

[d] Zinc bacitracin was used at a 50 ppm level.

Gualtieri and Rapaccini (1990) fed broilers up to 6 weeks of age with isocaloric and isonitrogenous diets. The diets were different in their cereal component (maize or low tannin sorghum) and in their inclusion level of ground date stones (0 or 10%). These results indicated that date stones are suitable as a feedstuff for broilers up to a maximum of 10% of the diet.

(b) Feeding whole dates and date waste

Kamel *et al.* (1981) studied the nutritional value of whole dates added at levels of 5, 10, 30 and 47.7% to replace ground yellow maize in broiler diets up to 3 weeks of age. The inclusion of 5, 10 and 30% of whole dates did not result in significant differences in responses obtained in body weight gains and feed utilization. However, when the inclusion level of the dates was increased to 47.7% thereby replacing all the maize in the diet, body weight gain was significantly depressed. This was accompanied by significantly poorer feed efficiency while feed consumption was not affected (Table 6.40).

Organ weights measured (liver, heart and spleen) varied among treatments owing to individual differences and the date as an energy source. When dates replaced maize totally (47.7% inclusion level) organ weights were not different from those of birds fed the control diet. In diets containing 30% whole date, these ground dates may effect the structure of the diet and cause granularity and stickiness owing to the high sugar content (65-75%; mostly glucose and fructose of dates). Al-Yousef and Vandepopuliere (1985) found similar results with inclusion levels of 8, 16 and 24% whole dates; no significant differences were observed between performance characteristics between birds fed experimental and control diets.

As far as the nutritive value of date waste without stones in broiler diets is concerned Al-Hiti and Rous (1978) have covered this item thoroughly. They reported that diets with date waste at inclusion levels of 5, 10 and 15% substituting shredded maize against a control diet showed promising results for weight gain of broilers at 7 weeks of age. Feed consumption of birds fed diets with the waste included in the three levels was significantly higher than that of birds fed the control diet. Feed conversion was not favourable for the birds fed the experimental diets. These unfavourable results for the feed conversion were acceptable owing to the low price of the date waste in comparison with the maize; the feed conversion figures were 2.22, 2.26, 2.26 and 2.07 for inclusions of 5, 10, 15 and 0% of date waste, respectively.

Table 6.40. Performance of broilers up to 3 weeks of age fed various levels of whole ground Zahdi dates and its effect on some organs. From Kamel *et al.* (1981)

Observations	*Inclusion % of whole dates*				
	0	*5*	*10*	*30*	*47.7*
Body weight gain (g)	507.00^{a}	525.00^{a}	535.00^{a}	501.00^{a}	454.00^{b}
Feed consumption (g)	795.00	814.00	801.00	783.00	795.00
Feed efficiency (feed/gain)	1.57^{a}	1.55^{a}	1.50^{a}	1.56^{a}	1.75^{b}
Organs (% of body weight)					
Liver	3.11^{a,b}	2.99^{a,b}	2.88^{a}	3.12^{a,b}	3.24^{b}
Heart	0.72^{b}	0.70^{a}	0.70^{a}	0.72^{b}	0.73^{b}
Spleen	0.12	0.12	0.13	0.14	0.11

^{a,b} Means within rows followed by the same superscripts are not statistically different ($P > 0.05$).

6.7 APPLE RESIDUE

6.7.1 Introduction

Apple is belonging to the fruits of the rose family (*Rosaceae*); other species belonging to this family yielding edible succulent fruits are: pomes (*Pomeae*), strawberries (*Potentileae*), bramble berries (*Rubeae*) and drupes (*Pruneae* and *Chrysobalaneae*). Apple is known as *Pyrus* (or *Pirus*) *Malus* L. = *Malus Communis* DC. = *M. Malus* Brit. Most cultivated varieties of apples are believed to have been derived from the wild hairy apple (var. *Pumila* Henry = var. *Mitis* Waller.) of western and central Europe. Only a few varieties owe their origin to the smooth wild apple (var. *sylvestris* L. = var. *austera* Waller) of southern and eastern Europe and of south-western Asia.

Apple trees grow in several parts of the world, particularly in areas where there is a temperate climate. Apple growing needs specific climatic conditions as far as temperature and water are concerned. A good apple orchard should yield about 10 t of apples per hectare. Apples are mostly grown in temperate climates as found in Europe. However, at the higher altitudes of some of the countries with a less temperate climate such as Mexico, apples are also grown. Apples may be classified into three major types in relation to the processed product:

- Dessert and culinary apples: This type is usually consumed directly either fresh or cooked.
- Cider apples: They can only be utilized by extracting the juice or by fermentation into alcoholic cider.
- Apple juice: This type is only used for producing fresh juice.

Apple products are numerous such as: dried, canned, frozen and sliced apples as well as sauce, juice, syrups, jelly, jam, butter, candy and sweets. Using fermentation, other products are manufactured such as cider, wine, brandy and vinegar. For general information on apples, the following articles and books are recommended: Hadorn and Högel (1945); Smock and Neubert (1950); Duman (1957); Moyer and Aitken (1980).

Apples vary in size, shape and colour, and in depressions at both the calyx (hairy on both sides) and stem end. The core of the fruit consists of the parchment-like endocarp about the five locules, each with two seeds or less. The endocarp often splits at the center structures, thus connecting the locules with each other and the central cavity. Figure 6.21 shows the microscopic structure of the apple seed which is composed of the spermoderm, perisperm, endosperm with aleurone cells, and embryo with its fleshy cotyledons and small radicle (Winton and Winton, 1949).

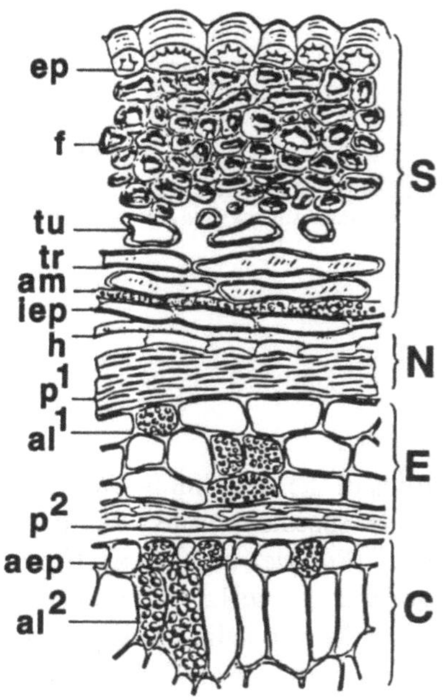

Figure 6.21. An apple seed in a cross-section. *S*, spermoderm; *ep*, outer epiderm; *f*, fibres; *tu*, tube cells; *tr*, cross cells; *am*, starch cells; *iep*, inner epiderm; *N*, perisperm; *h*, hyaline layer; P^1, obliterated cells; *E*, endosperm; al^1, aleurone cells; P^2, compressed parenchyma; *C*, cotyledon; *aep*, outer epiderm; mesophyl containing aleurone grains (al^2). X 160. Reproduced with permission from Winton and Winton (1949).

Apple is the most important and probably the most widely grown tree fruit in the world. While most apples are grown for dessert or cooking purposes throughout the temperate regions of the world, many in Germany, France, The Netherlands, and the United Kingdom are produced for cider (fermented apple juice) and are too sour for fresh eating. The total annual apple production is estimated to be 40.3 million t per year with a waste estimated to be 1.6 million t per year (FAO, 1991). The highest producers of the world are U.S.A., China and the Commonwealth of Independent States (USSR) followed by Iran, Japan, France, West Germany, Italy and Argentina. Medium producers are Korea, The Netherlands, Spain, Romania, Poland, Yugoslavia, Chile, Canada, South Africa and others.

Apple processing produces large quantities of waste products estimated to be 7.6% by dry weight of the total processed apples, but this figure varies somewhat according to the processing conditions and purpose of the final product (Smock and Neubert, 1950; Tressler and Joslyn, 1961).

During the processing of apples for juice or other by-products such as jelly, jam, butter and dried, canned, frozen, sliced, or candid apples, press cake composed of peel, core and seeds are produced known as apple pomace. Apple pomace is available in a fresh form during the normal processing season, beginning in September and continuing into the following year. Its value is mainly known as a source of carbohydrates mostly used as a stock feed and in the preparation of pectin. Apple pomace contains a high percentage of moisture (about 72%); it should therefore be used as soon as possible as silage or a fresh feedstuff for ruminants. Its use as a poultry feedstuff is limited owing to its high content of pectines (16%), fibre (18%), and low protein content (4.5%). Its advantage is its acceptable carbohydrate percentage 62% and metabolizable energy 8.20 (TME MJ/kg) (Sibbald, 1977).

A Few research workers evaluated the value of apple pomace as a feedstuff for poultry (Cicogna and Saibene, 1969; Sibbald, 1977; Farrell, Rose and Warren, 1983).

6.7.2 Processing

(a) Processing of fresh apples

Processing of apples to juice, syrup or other by-products such as dried, canned, frozen or sliced apples, and fermentation to cider, wine, brandy and vinegar has been discussed in detail by many research workers (Smock and Neubert, 1950; Tressler and Joslyn, 1961; Moyer and Aitken, 1980). Some processing methods have been selected to show how the waste is produced.

Juice processing

Before the apples are processed the quality control must be rigorous because any defective or damaged fruits have a large impact on the juice quality.

Culling and washing

Apples are carefully inspected to cull decayed and damaged fruits; small decayed spots are trimmed with care. After this step they are washed to remove dirt and to reduce the number of micro-organisms on the surface which may lead to fermentation after the juice is produced. Washing (mostly soaking) is followed by light spraying or by high-pressure water spraying. It is very important to spray out the residues of insecticides, fungicids or germicids attached to the fruit such as lead or arsenic.

Grinding

Grinding is carried out by means of roller, grater or hammer mills. The grinders are mostly made of stainless steel to avoid oxidation. All these treatments are based on the principle of breaking up the apple tissues and mechanically disrupting them into as many individual cells as possible.

Maceration

In this step the milled apples are kept for a period of 12 h or more before pressing to obtain a more pronounced fruit flavour. This method of maceration is only used for fermented ciders and not for juice processing.

Pressing

In this step several pressing systems are used such as the hand-operated barrel press, rack- and cloth press, hydraulic press and continuous press. They are all based on separating or dividing the pulp by passing it through screens where the juice will be filtered and the coarser material will be ejected as waste or apple pomace; this will be combined with the culled and trimmed parts.

After the juice is separated it will be deaerated and pasteurized for further processing as cider.

Apple processing for drying, canning or freezing

Most of the apples used for the process are delivered from fruit packing or cool cells. They are mostly washed to remove spray insecticidal or germicidal residues and they must be free from decay.

Washing

When apples are washed they pass a more careful hand selection procedure. They are washed again under high pressure to assure no trace of spray residues remains. Precautions are made to lower the spray residues, especially when the pomace will be used as animal feed or in the case pectin manufacture. In both cases, apples will be washed in an acid or diluted lye at an elevated temperature followed by a complete rinsing.

Peeling and coring

This process may be achieved by mechanical, or chemical and steam peeling.

The mechanical peeling is based on rotating the apple about its core on a spindl while a floating peeling knife follows the contour of the rotating fruit thus removing a narrow strip of peel. A circular coring knife may be used to remove the stem, calyx and seeds in the same operation. Grading apples by size is of great importance for the speed and efficiency of peeling.

Chemical and steam peeling are based on using hot lye solution. Lye peeling saves peeling loss, and reduces the immersion time and temperature required owing to the effect of lye as a wetting agent. After the lye treatment, apples are thoroughly washed with high-pressure water sprays to clean the lye residue and remove the loosened cooked skins and soft flesh. Apples peeled by the lye or steam process are cored in a separate operation to remove stem, calyx and seeds.

Trimming and slicing
In this step care is taken to remove by hand the remaining unpeeled parts such as skin, bruises and defects. Apples will then be sliced or cut. Apples cut into segments or slices are usually passed through radial knives. Rings may be cut by machines equipped with a series of knives. After slicing, the apples are checked again to remove residual core material and defects missed in the first inspection.

Preparation line
At this stage sliced or cut apples will be transported by means of a roller conveyor. They are submerged in a dilute salt solution for a short period to inhibit darkening. Finally the clean cut apples are transported to their final destination for additional processing, such as drying, canning or freezing.

(b) Processing of apple pomace waste

The waste remaining after apple processing for juice or other by-products is called apple pomace. This consists of the skins and fibrous parts of the stem, core, stalk and the seeds or pips. This apple pomace may be fed to ruminants fresh or as silage, and is sometimes dried. Another waste, the residue remaining after pectin extraction, is known as pectin pulp. Apple pomace can be utilized directly as animal feed after ensilation. Pomace can be chemically preserved by impregnating it with a solution of phosphoric acid (concentration 5-10%) at a level of 6% by weight.

Anaerobic fermentation of apple pomace is more economically than drying owing to high moisture content (79%; Morrison, 1961). The pomace may be ensiled alone or mixed with alfalfa, corn, silage, grass chippings or other material.

Drying
Apple pomace is mostly dried in continuous, rotary driers. Pomace is subjected to agitation and heat (steam) as it is moved progressively from the wet to the dry end of the large drum. Heat may be supplied by steam pipes within the drum or by hot gasses passed through the drying material. Care is taken that the pomace temperature does not exceed 71 °C during the drying operation to avoid decomposition of carbohydrate material or deterioration of the pectin (Smock and Neubert, 1950).

(c) Yields of apple pomace

About 225-315 kg of wet pomace is obtained from each ton of apples pressed in the manufacture of juice, juice concentrate, cider, wine, brandy, vinegar, *etc.* This will be

22.5-31.5% of wet pomace with a moisture percentage about 66.4-78.1% (Smock and Neubert, 1950). The figure of 7.6% waste based on dry matter will be a reasonable estimate for pomace as a waste from the apple industry.

6.7.3 Chemical Analysis and Nutritive Value

Apple pomace known as a wet product has no relevant value to poultry since wet feedstuffs cannot be mixed in rations. Dried apple pomace has a low protein content varying from 4.3-5.1%. Its crude fibre is high varying from 15.2-17.9% and it has also a high pectin level of 16.5%. Its nitrogen-free extract is also high (5.7-63.5%) and its metabolizable energy is acceptable (8.20 MJ/kg; Table 6.41) (Smock and Neubert, 1950; Morrison, 1961; Titus and Fritz, 1971; Sibbald, 1977; Huber, 1981).

Table 6.41. Chemical composition (%) of apple pomace. From (1) Smock and Neubert (1950); (2) Morrison (1961); (3) Huber (1981) and (4) Titus and Fritz (1971)

%	*Wet Pomace*		*Dried Pomace*				*Wet Pectin Apple Pulp*
	1	*2*	*1*	*2*	*3*	*4*	*2*
Moisture	72.3	78.9	11.8	10.4	11.00	10.70	83.3
Crude protein	1.4	1.3	5.1	4.3	4.90	4.50	1.5
Crude fat	1.1	1.3	4.2	4.6	-	4.80	0.9
Crude fibre	7.4	3.7	17.9	15.2	17.00	15.50	5.8
N-free extract	-	13.9	57.0	63.5	-	61.00	7.9
Pectin	2.0	-	16.5	-	-	-	-
Ash	1.4	0.9	5.6	2.0	-	3.50	0.6
Calcium	-	-	-	-	0.13	0.10	-
Phosphorus	0.2	-	-	-	0.12	0.10	-
Potassium	0.2	-	-	-	-	0.45	-

Metabolizable energy of dried apple pomace is 8.20 (MJ/kg) dry matter. (Sibbald, 1977)

This feedstuff needs a lot of care to improve its nutritive value by lowering its pectin content by soaking or boiling in water (Patil, Netke and Dabadghao, 1982). Mixing this waste up to 20% with other by-products such as rice polishing, tapioca and restaurant and potato wastes will enrich its nutritive value as far as metabolizable energy is concerned. Apple pomace is considered to be a carbohydrate feedstuff and, with this mixing possibilities its high crude fibre and pectine will be lowered in the final mixture.

6.7.4 Feeding Apple Pomace

There is little literature concerning the use of apple pomace as a feedstuff for poultry. Cicogna and Saibene (1969) reported the possibility of reducing the cost of concentrate mixtures for broiler poultry by replacing maize with a mixture of hard wheat, fine bran, carobs (fruit from the pea family) without seeds, dried apple pomace and fat. In an experiment with broilers they studied the partial or total substitution of maize (which was 61% in a control diet) compared with an experimental, low-cost diet containing wheat shorts, carob beans without seed, dried apple pomace, corn gluten meal, alfalfa meal and fat. Chicks fed the control diet showed the best results for weight gain, feed efficiency, dressing percentage, subcutaneous fat deposits, "free water" in the pectoral muscles, skin pigmentation and feeding costs. In this trial the authors made a mixture of the above-mentioned ingredients including apple pomace and substituted it for maize at 20, 40 and 100%, respectively (maize in the basal diet was 61%). The inclusion of this mixture, however, will never give a complete picture of the potential of apple pomace as a feedstuff. Further research is needed to test the real effects of inclusion of apple pomace as a sole energy source.

Farrel, Rose and Warren (1983) reported from their preliminary nutritional evaluation of some agricultural waste- and by-products for non-ruminants. From their trials with rats up to 14 days of age fed 20% apple pomace in a diet, there was no significant difference ($P > 0.05$) between the treatment and control groups as far as gain weight or feed conversion are concerned. A reasonable evaluation of apple pomace as a carbohydrate feedstuff in poultry diets is greatly needed.

Apple pomace may be fed fresh, as silage or dried or in a processed form as pectin pulp, for ruminants. There are initial benefits for pigs but it proved to be unsuitable for horses (Smock and Neubert, 1950).

6.8 MANGO SEED

6.8.1 Introduction

Mango belongs to the fruits of the Cashew family dicotyledon (*Anacardiaceae*). This family includes not only the mango but also pistachio and cashew nuts. Mango mesocarp is edible; the cotyledons from the other two are edible. Mango is known as *Magnifera indica* L. and was a native of India more than 4000 years ago, being among the first fruits to be cultivated. Mango includes several varieties which differ in form, size, colour and flavour. It is a fruit rich in sugar (sucrose; 8.8-12.6%) and citric acid (0.27-1.44%), and has a peculiar aromatic flavour due to the volatile (essential) oil. This oil is found in special oleores in cavities in conjunction with fibrous strands distributed through the flesh. This gives the fruit a distinctive character. When the fruit is ripe it is a very delicious dessert fruit and when green it can be used as a good salad fruit. It retains reasonably its flavour after processing for canning and freezing. The mango was introduced towards the end of the nineteenth century into Florida and California where it is now successfully grown for local consumption and export (Jagtiani, Chan and Saki, 1988).

Figure 6.22 shows a logitudinal section through the fruit with edible mesocarp (fruit flesh) and the endocarp (stone). Figure 6.23 shows a microscopical surface view of the seed with spermoderm, fleshy cotyledon with epiderm and narrow parquetted cells and mesophyl containing elongated starch grains; its nitrogen-free extract is copious (Winton and Winton, 1949; Patil, Netke and Dabadghao, 1982).

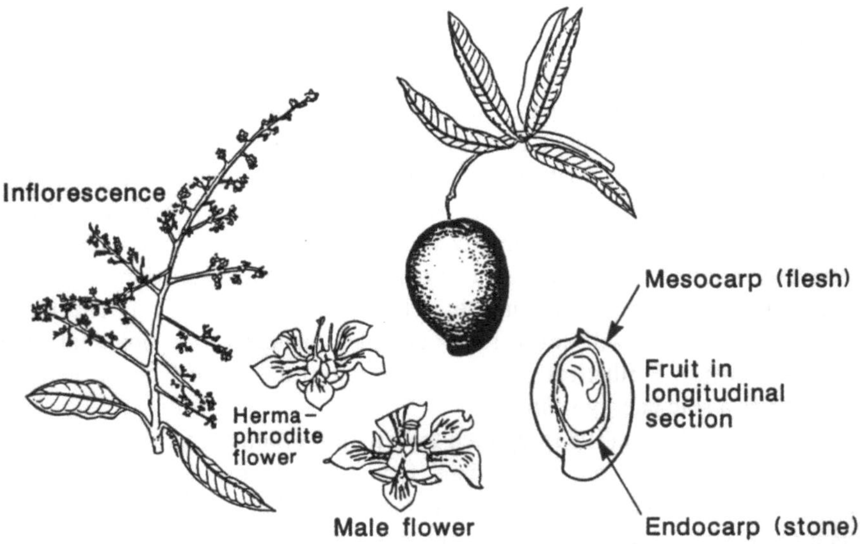

Figure 6.22. The mango fruit, its flowers, and a longitudinal section through the fruit showing the edible mesocarp (fruit flesh) and the endocarp (stone). Reproduced with permission from Jagtiani, Chan and Saki (1988).

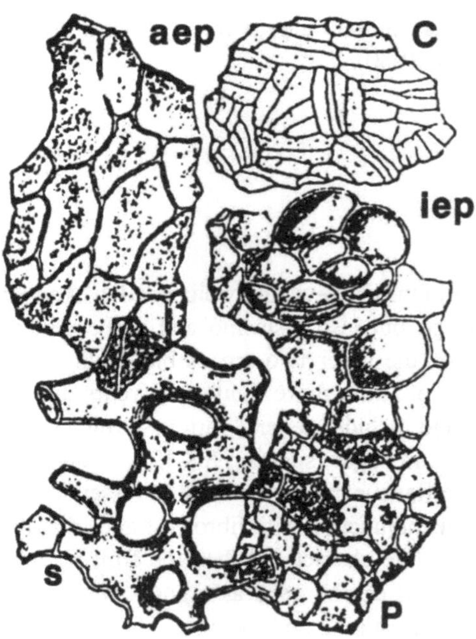

Figure 6.23. A microscopical surface view of mango seed. Spermoderm: *aep*, outer epiderm; *s*, spongy outer parenchyma; *p*, inner parenchyma; *iep*, inner epiderm; *C*, epiderm of cotyledon. X 160. Reproduced with permission from Winton and Winton (1949).

The highest production of mango fruit is in India which is estimated to have 62% of the total world production. Asia produces 79% of the total world production, while Africa produces 7% (FAO, 1991). The highly producing countries other than India are Mexico,

Brazil and Pakistan. The medium producers are Philippines, Indonesia, China, Haïti and Bangladesh. Low scale producers are Dominican Republic, Tanzania, Madagascar, Zaïre, Venezuela and Egypt. Other countries involved in a lower scale of production are: Peru, Sri Lanka, Sudan, Cuba and Saint Lucia (FAO, 1991).

Mango serves in human nutrition as a fresh product or processed as juice and nectar, slices in syrup, pickles, chutneys and slices in brine. Beside these products there is a wide variety of by-products such as dried mango slices, cubes and cheeks, mango squash, jam, toffee, powder and pickles

An estimation of yields showed that a tree of 6-year-old produces some 50-75 fruits and at the maturity (of 15 years old) it produces 1000 to 1500 fruits or some 45 t/ha. Mango trees can resist temperatures in excess of 48 °C, they survive in very dry areas and produce good crops if irrigated (Jagtiani, Chan and Saki, 1988).

Large quantities of mango seeds are available after processing in large factories and from small shops where fresh mango juice is sold during the fruiting season. The seeds are mostly wasted, but if care is taken, a seed kernel can be obtained after the removal of the hard seed coat and it can be processed to produce oil and residual kernel meal which is rich in carbohydrate (Salunkhe and Desai, 1984). Mango seed kernels have been used as a human food in India. The kernels, after soaking in water to lower the tannin level, are roasted or boiled and made into flour (FAO, 1988). In addition, treated mango seed kernel meal has been successfully used to replace maize up to 15-20% levels in poultry diets (Reddy, 1975).

The waste produced from the mango processing is estimated to be 40.6-44.9% of the whole fruit and this is a very significant amount of waste (Jagtiani, Chan and Saki, 1988). If it is assumed that 50% of the total world production is processed and 50% freshly consumed, with a world production of 15.7 million t/year (FAO, 1991) then the total waste is estimated as 3.4 million t/year. This amount is worth processing and using as an energy source in poultry diets.

Several research workers explored the nutritive value of mango waste as a feedstuff for poultry (Reddy, 1975; El Alaily, Anwar and El Banna, 1976; Patil, Netke and Dabadghao, 1982; Ravindran and Blair, 1991).

6.8.2 Processing

(a) Processing of fresh mangoes

Processing of mango and its generated waste has been discussed in detail by many research workers (Lynch and Mustard, 1955; Srivastava, 1967; Pantastico *et al.*, 1975; Lakshminarayana, 1980; Young and Sauls, 1980; Stafford, 1983; Jagtiani, Chan and Saki, 1988).

The processing methods applied to mango fruits differ according to the final product such as canned, sliced, dried, frozen, dehydrated mango and juice, nectars and purée. As an example to show the processing of fresh mangoes and the subsequent separation of waste, the process of canning mango slices will be explained. In general, several important products of commercial interest have certain common unit processes such as fruit selection, sorting, washing, peeling, blanching, slicing, dicing, acidification, pasteurization and packing.

Harvesting

Mango fruit is picked from the trees by means of a long-handled pole with a canvas bag attached to a ring that has a sharp-edged blade located on its inner side. Fruits are cut by the blade and fall into the bag. When the bag is full, with 2-4 mangoes the pole is lowered and the fruit is transferred to baskets, buckets or crates. Sometimes picking takes place from a platform which is attached to a tractor and can be raised or lowered by a hydraulic lift. Mangoes are mostly processed locally and if they are to be shipped, the ripening period during shipping has to be taken into consideration. This period is very important for the determination of the final quality of the fruit as far as taste, colour, firmness and sweetness are concerned. Storage temperatures are of great importance as far as the ripening period is concerned and vary accordingly; they are also related to the shipment period. Storage for the green stage, which is not completely ripened, is at 15-17 °C and for the ripe fruit is at 7 °C.

Fruit selection

Mangoes that are fully ripe and free from rot are usually selected for processing. Fully ripe fruits not only provide stronger flavoured and more highly coloured product, but also provide higher yields. Defected or rotten fruits should be discarded.

Washing

Mangoes should be thoroughly washed to remove dirt, adhering latex (honey) and other foreign matter. It is preferable to wash them in a soaker-washer fitted with brushes using a detergent, followed by rinsing in a rotary washer.

Peeling

Owing to low cost of labour in developing countries most of the peeling is carried out with a thin knife. However, blanching or lye peeling has been introduced in many regions. Water cooling followed by steam blanching for 2 min and then slitting the peel results in easy removal. Lye peeling efficiency depends on the maturity, ripening, variety and thickness of the peel. After the lye treatment, the peeling process is completed in a rotary-rod washer. The combination of water sprays and the tumbling action of the fruit in the washer removes more than 90% of the peel. Peel percentage as a (wet) waste produced is estimated to be 8-22%.

Slicing

Stainless steel knives are used to slice the peeled fruit. Two broad slices are taken from the flat sides of the mango, and two narrow slices from the remaining narrow sides. The flesh remaining on the seed is removed in a pulper for purée. In this step the relevant waste (seed) is produced which is estimated to be 7-23% depending mostly on the variety, maturity and ripeness.

After this step the slices will be put into cans, then syrup will be added (35-40 Brix syrup with 0.20-0.25% citric acid) and finally the open cans with their enclosures will be sterilized and closed mechanically, cooled, labelled and stored for export.

(b) Processing of mango waste

There are few reports detailing the processing of mango waste (seeds) as a feedstuff. None of these reports mentions the nutritive value or the processing of the peel from the mango fruit. El Alaily, Anwar and El Banna (1976) and Patil, Netke and Dabadghao (1982) described the processing of mango seed kernels on a laboratory scale. Unfortunately, no further large-scale processing was done. Mango seeds were collected immediately after juice processing and sealed in plastic bags. The seeds were air dried, dehulled and the kernels were ground to a powder known as Mango Seed Kernels (MSK).

Patil, Netke and Dabadghao (1982) reported that MSK was found to contain cyanogenetic glucosides and tannins. Therefore, the de-oiled MSK has to pass various processes to remove these antinutritional constituents. The processing treatments which were used for elimination and their effects are shown in Table 6.42 and may be summarized as follows:

Soaking in water. A slurry of MSK was made with four times its weight of water and allowed to stand for 24h at room temperature with frequent stirrings. After the suspended particles had settled, the supernatant liquid was decanted and the residue was sun dried. This method removed 61.4% of the tannin and 84.3% of HCN.

Boiling in water. A slurry prepared as above was boiled for 30 min, cooled at room temperature, and after decanting the supernatant, the residue was sun dried. This method removed 84.3% of the tannin and 69.8% of HCN.

Treatment with HCl. MSK was mixed with five volumes of 0.3*M* HCl, allowed to stand for 24h with occasional stirring, and, after decanting the supernatant, the residue was washed with two volumes of water. The process was repeated 5-6 times. The residue was then sun dried. This method removed 92.8% of tannin and 59.1% of HCN.

Treatment with NaOH. MSK was treated as above except that 2.5 *M* NaOH was used in place of HCl. This method removed 80.7% of tannin, but HCN was not determined.

Autoclaving. at 149 kPa (about 111 °C) for 5, 10 and 15 min: the removal of tannin and HCN was improved by longer processing times during autoclaving of MSK (Table 6.42).

Process KN. MSK was treated with HCl as described above. After removal of HCl, the residue was suspended in water (2 l/kg residue) and to the slurry 60 g $Ca(OH)_2$/kg residue was added to the slurry. The slurry was mixed thoroughly and kept overnight. The supernatant was then removed and the residue was washed five or six times with water. This process completely removed the tannins and removed 61.4% of the HCN. The resulting MSK product thus produced was termed processed mango seed kernel (PMSK).

Table 6.42. Effect of processing on the removal of tannins and cyanogenetic glucosides (HCN equivalent) from mango seed kernel. From Patil, Netke and Dahadghao (1982)

		Tannin		HCN	
Processing		Content (g/kg)	Removed (%)	Content (mg/kg)	Removed (%)
None		44.5	-	82.4	-
Soaking in water		17.2	61.4	12.9	84.3
Boiling in water		7.0	84.3	24.9	69.8
Treatment with HCl		3.2	92.8	33.7	59.1
Treatment with NaOH		8.6	80.7	-	-
Autoclaving (149 KPa) for	5 min.	32.0	28.1	31.2	62.1
	10 min.	18.0	59.6	21.2	74.2
	15 min.	7.0	84.3	20.0	75.7
Process KN [a]		0.0	100.0	31.8	61.4

[a] Process KN = Treatment of Mango Seed Kernel (MSK) with HCl followed by $Ca(OH)_2$

(c) *Yields of mango waste*

Most of the reports that were mentioned under processing of fresh mango reported that the composition of the mango varies considerably with variety and season. This is most evident from the wide ranges reported for the edible portion (55-75%), seed (7-23%) and peel (8 -22%). Another general estimation was that 55.1-59.4% is edible and 40.6-44.9% is waste, including peel and seed.

6.8.3 Chemical Analysis and Nutritive Value of Mango Kernel

Table 6.43 shows the chemical analysis of several samples of mango seed kernel (de-oiled) and processed mango seed kernel (PMSK) in comparison with maize (El Alaily, Anwar and El Banna, 1976; Patil, Netke and Dabadghao, 1982; Scott, Nesheim and Young, 1982; Ravidran and Blair, 1991). It may be concluded from this table that the crude protein content varies between 4.61-7.5% which is lower than that of maize (8.6%). Crude fat is between 1-1.5% in case of the de-oiled MSK while mango seed kernel with oil contains 12.5%. In general, the de-oiled MSK is lower in fat than is maize (3.9%). Crude fibre increases from 1.3-3% when MSK is de-oiled. Nitrogen-free extract varies between 77.5 and 84.0% which is similar to that of maize (79.6%). Metabolizable energy varies between 7.62 and 13.42 MJ/kg for MSK and PMSK, respectively while maize is 14.10 MJ/kg. Roy and Mitra (1970) concluded that mango seed kernels contain 75% carbohydrates and they are considered to be a rich energy source in animal diets. From the above-mentioned analysis it seems that MSK is a reasonable source of energy, its main disadvantage is the high HCN equivalent to cyanogenetic glucosides (82.4 mg/kg) and the high tannin content (4.45%). The PMSK meal showed very promissing results, as tannins were removed completely while HCN was lowered to about 32 mg/kg which is considered acceptable for a feedstuff.

Table 6.43. Chemical analysis of Mango Seed Kernel (MSK) and Processed Mango Seed Kernel (PMSK) in comparison with maize. From [1] El Alaily, Anwar and El Banna (1976); [2] Patil, Netke and Dabadghao (1982); [3] Ravidran and Blair (1991) and [4] Scott, Nesheim and Young (1982)

%	*(1)*	*(2)*		*(3)*	*(4)*
	MSK	*MSK*	*PMSK*	*MSK* [a]	*Maize*
Moisture	8.45	6.350	8.820	-	9.50[d]
Crude protein	6.74	5.490	4.610	7.50	8.60
Crude fat	12.53	1.150	1.000	1.50	3.90
Crude fibre	1.30	2.370	2.800	3.00	2.00
N-Free extract	77.46	80.600	78.600	84.00	79.63[d]
ME (MJ/kg)	7.62	10.910	13.420	10.04	14.10
Ash	1.97	4.040	4.100	4.00	2.00[d]
Calcium	-	-	-	0.20	0.02
Phosphorus	-	-	-	0.25	0.03
HCN [b]	-	0.008	0.003	-	-
Tannins	-	4.450	ND[c]	-	-
Glycosides	-	Positive	Traces	-	-

[a] MSK = Mango Seed Kernel de-oiled, soaked, dehydrated, ground.
[b] HCN= Equivalent to cyanogenetic glucosides 0.008% = 82.40 and 0.003% = 32.77 mg/kg.
[c] ND = Not detected
[d] values from author (1).

Several research workers reported that the treated (de-oiling, soaking or boiling) MSK has been successfully used to replace maize in levels up to 15-20% in poultry diets (Reddy, 1975; Patil, Netke and Dabadghao, 1982; Ravindran and Blair, 1991). If this waste is processed in a commercial way it will produce a very important feedstuff as a source of energy in poultry diets.

6.8.4 Feeding of Mango Waste or Seed Kernel

Welkins (1943) reported that mango seed kernels, processed as flour, give comparable feeding values to those of rice. Reddy (1975) reported that mango seed kernel flour has been successfully used to replace maize up to a 20% level in poultry diets. However El Alaily, Anwar and El Banna (1976) found that substituting mango seed kernels for maize at levels of 10 or 20% led to a reduction in growth rate, a poor feed consumption and a reduced efficiency of feed utilization in the period 0-4 weeks. The reason of the poor chick performance in their experiment was discovered by Patil, Netke and Dabadghao (1982). They concluded from trials in which de-oiled mango seed kernel meal was used, that the antinutritional factors were affecting chick performance causative factors, negatively. These factors include tannins, cyanogenetic glucosides and traces of alkaloids and gums. They concluded also that substitution of 14.1% of mango seed kernel for maize in a practical diet did not affect performance, whereas adverse effects occurred with 28.2, 42.3 and 56.3% as an inclusion level (Table 6.44). In addition, they concluded also that using the processed mango seed kernel, where various toxins were removed, (Table 6.42), improved chick performance but it was still not equivalent to that achieved by birds fed maize (control). Obviously, there was discrepancy with the processed mango kernel meal which may be due to the presence of other antinutritional factors that cannot be ruled or are difficult to rule out. This feedstuff needs further research to investigate more thoroughly the role of its antinutritional factors and the impact of residual levels after processing. It would also be worthwhile to study suitable methods for elimination these antinutritional factors. The use of the treated mango kernel meal should be limited and an inclusion level of not more than 14% in chick diets is advised.

6.9 GUAVA SEED

6.9.1 Introduction

Guava belongs to the fruits of the myrtle family (*Myrtaceæ*). This family includes two succulent fruits, *Psidium* (guavas) and *Eugenia* (rose apple, Jambolan, Surinam cherry, *etc.*).

Guava is known as *Psidium guajava* L. and several species and varieties are grown as a native of Brazil, Costa Rica, Mexico and Central America.

The guava fruits are collected from small tropical trees or shrubs which are yielding luscious, agreeable musty fruits which can be eaten or processed after removal of the seeds. Guava fruits are known as an excellent product known all over the world.

Guava grow well in most types of soil and in most climates in the tropics and subtropics. In some countries they have escaped from cultivation and grow wild and in some areas they are considered a weed. Guava can grow in areas of average maximum temperatures of 32 °C and average minimum temperatures of 3 °C. Guava fruits also produce shoots from the root if the shoots above-ground are killed by frost. These relevant characteristics make guava one of the easiest crops to grow commercially.

The chief structural character is that the fruit varies in colour, shape (peer like) and size (up to 10 cm long). Calyx lobes of this fruit are irregular and several (usually four) locules contain seeds. Seeds are numerous, hard triangular or kidney shaped (Figures 6.24 and 6.25). Mesocarp with oleresin cavities are found and grotesque stone cells are present, often in groups. The spermoderm largely consists of elongated stone cells. The endosperm has a single layer of aleurone cells while the embryo (largely radicle) contains aleurone grains of diameter up to 22 m. (Figure 6.25; Winton and Winton, 1949).

Table 6.44. Effects of inclusion several levels of Mango Seed Kernel (MSK) and Processed Mango Seed Kernel (PMSK) on the performance of chicks [e]. From Patil, Netke and Dabadghao (1982)

Diet	*Inclusion (%)*	*Crude protein (%)*	*ME (MJ/kg)*	*Weight gain (g/chick/day)*	*Feed intake (g/chick/day)*	*Gain/feed (ratio)*	*N retained (g) (per g N intake)*	*ME/GE (ratio)*
MSK not treated till 14 days of age								
Control	0.0	20.0	10.41	3.710^{a}	$11.600^{a,b}$	0.319^{a}	0.540^{a}	0.722^{a}
MSK	14.1	19.8	8.71	3.350^{a}	12.500^{a}	0.264^{b}	0.457^{a}	0.606^{b}
"	28.2	19.6	9.20	2.100^{b}	10.400^{b}	0.202^{c}	0.543^{a}	$0.654^{a,b}$
"	42.3	19.1	8.62	2.400^{b}	$10.800^{a,b}$	0.225^{c}	0.404^{a}	$0.615^{a,b}$
"	56.3	18.6	7.73	0.800^{c}	7.400^{c}	0.111^{d}	0.378^{a}	0.557^{b}
Pooled SE	-	-	-	0.130	0.460	0.010	0.047	0.030
PMSK treated till 12 days of age								
Control	0.0	20.8	11.30	6.300^{a}	14.000^{a}	0.450^{a}	0.622^{a}	0.769^{a}
PMSK	14.1	20.2	10.98	4.700^{b}	13.500^{a}	0.350^{b}	0.583^{a}	$0.752^{a,b}$
"	28.2	19.1	9.92	$4.500^{b,c}$	13.100^{a}	0.343^{b}	0.471^{b}	0.695^{b}
"	42.3	18.6	10.52	$3.800^{c,d}$	13.600^{a}	0.278^{c}	0.476^{b}	0.704^{b}
"	56.3	18.6	10.79	3.000^{d}	12.700^{a}	0.234^{d}	0.496^{b}	$0.716^{a,b}$
Pooled SE	-	-	-	0.225	0.517	0.006	0.017	0.106

[a,b,c,d] Values bearing different superscripts differ significantly from each other ($P < 0.05$).

[e] The experiment started with chicks of 7 days old selected for equal body weight.

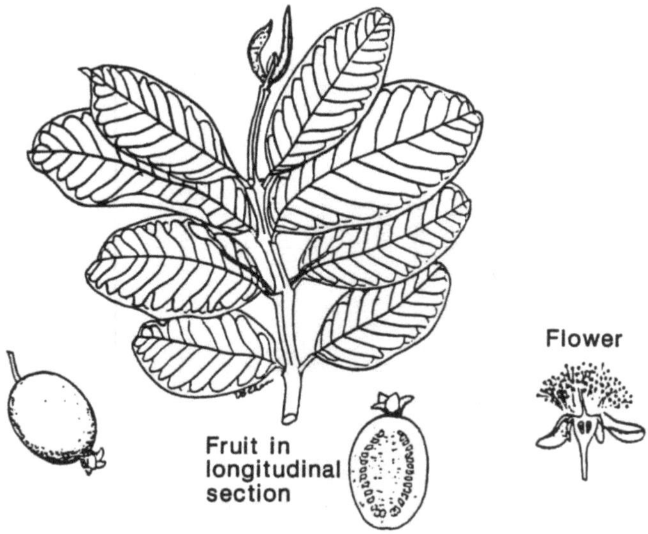

Figure 6.24. The guava fruit, its flower and a longitudinal section through the fruit showing the edible part and the pulp with the stones. Reproduced with permission from Jagtiani, Chan and Saki (1988).

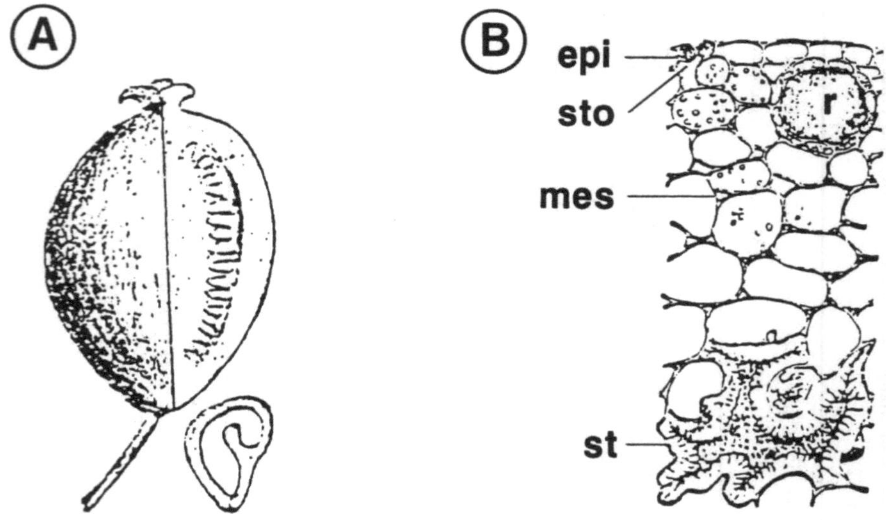

Figure 6.25. The guava fruit and its seed. (A) The central placenta and seed. The enlarged seed below shows horny spermoderm and curved embryo. (B) Outer pericarp in cross section. *epi*, epicarp with stoma *(sto)*; *r*, oleoresin cavity; *mes*, mesocarp with stone cell *(st)*. X 160. Reproduced with permission from Winton and Winton (1949).

The countries with high production of fresh or processed guava are India, which leads the world followed by Mexico and Pakistan. The medium-level producers are Colombia, Egypt, Brazil and South Africa. Other countries involved in lower-scale production are Venezuela, Dominican Republic, Puerto Rico, Jamaica, Kenya, Australia and Hawaï. The world production is estimated to exceed 0,5 million t per year. The total waste of seeds and stone cells is estimated to be 43 750 t per year on the basis of 50% guava fruits being processed (Jagtiani, Chan and Saki, 1988). This waste may create problems in the above mentioned countries characterized by a high production, such as India producing 165 000 t guava per year (Jagtiani, Chan and Saki, 1988).

Guava is used in the human nutrition not only as fresh fruit but also purée, canned or frozen fruit, juice (cloudy or clarified), syrup, concentrate, jelly, jam and guava spread, cheese, toffee and chutney. However, there are several additional canned types of guava products such as halves in syrup, pieces, powder, dried flakes and wines.

Unfortunately few research workers have evaluated the potential of guava waste by chemical analysis (Winton and Winton, 1949; Jagtiani, Chan and Saki, 1988). Its use as a feedstuff, however, has not been investigated.

6.9.2 Processing

(a) Processing of fresh guava

Processing of guava and its generated waste has been discussed in details by many research workers (Knight, 1980; Wilson, 1980; Chan, 1983; Jagtiani, 1985, Samson, 1986; Jagtiani, Chan and Saki, 1988).

Harvesting

Guava fruits are most commonly harvested by hand. Experimental orchards designed for harvest by shaking, have trees with one main trunk to facilitate attachment of the shaker.

Firm, yellow or half-yellow, mature fruit with no signs of insect attack or fungus disease should be harvested. It is important not to harvest over-ripe fruits, as they are easily damaged in transport and handling. Fruit-harvest intervals should not be more than 3-4 days; this prevents harvesting of overripe fruit. Fruits that have fallen and green fruits should be discarded. Guava fruits are usually harvested into field containers, then brought to trucks or trailers, and dumped into boxes for transport to the processing plants.

For long-distance shipments, soft fillings or cushioning material should also be used. Good ventilation is important to prevent build up of heat and humidity, which promotes microbial spoilage. If ripe fruit needs to be shipped, it is better to hold this in a cool or shady area and transport it during the night when the temperature is lower. The best practice would be to ship unripe fruit and ripen it under controlled conditions at the processing plant.

Fully ripe guava should be processed without delay, but if necessary, they can be held for about a week at 2-7 °C with only a small loss in vitamin C content. Storage at 0 °C for up to 2 weeks has also been practised commercially.

Processing

An example of fresh guava processing to be discussed, is the guava purée. Guava purée is also known as guava pulp, which is a liquid product prepared by pulping whole guavas. Purée is most commonly manufactured into nectars, various juice drink blends, syrups, ice cream toppings, jams and jellies. The fresh or previously stored fruit is transported by belts; it is then placed in a dump tank, which serves to soak and wash the fruit and also to separate out the over-ripe and immature fruit. This fruit tends to sink to the bottom of the tank. The fruit which floats is picked up on a moving conveyor belt, where it will be inspected and sorted for decay, insect damage and foreign materials such as leaves, dirt, and other trash. The washed fruit will pass through a chopper or slicer to be broken up and then fed to a pulper. The pulper removes the seeds and fibrous tissue (wastes) and forces the remainder of the product through a perforated stainless steel screen with holes of about 0.08-1.1 mm. The pulper is fed at a constant rate to insure efficient operation. The puréed material coming from the pulper passes through a finisher, which removes the stone cells from the fruit, and imparts the optimum consistency to the product. The finisher is equipped with a screen containing holes of approximately 0.05 cm. Finally the purée will be pasteurized, canned and cooled.

(b) Processing of guava waste and its yields

The guava waste is composed of some fibrous tissue such as stone cells and seeds. The further processing and use of these by-products is not very common.

Since the kidney-shaped seeds are 3-5 mm long and the stone cells are approximately 0.1 mm it will be difficult to use a hammermill since the matrix has large openings which would allow the waste to pass through the holes without grinding. It has been suggested that the waste is dried in the sun in small layers and then ground by a mill stone (El-Shazly, Ibrahim and Karam, 1963). They first crushed the dried products in a disc crusher and then powdered it using grain-grinding stone mills (see processing of date pits).

Yields. Yields are based on the variety, season and processing by the screener of the pulp and the finisher. The waste is estimated to be 12% of the seeds and 5.5% as stone cells (Jagtiani, Chan and Saki, 1988).

6.9.3 Chemical Analysis and Nutritive Value

There has been little research reported on this subject. It is clear that the ratio of guava seed to stone cells is about 2:1 and these two wastes will be produced as one combined waste product. Seed contains a rather high content of oil with high levels of essential fatty acids varying from 5-13.4% (Wilson, 1980). The fatty acid composition of the seed oil is as follows: myristic, 1.2%, palmitic, 8.9%, stearic, 4.8%, oleic, 53.9%, linoleic, 29.2% and linolenic 1.1% (Subrahmanyam and Achaya, 1957). The other nutrient contents are: crude protein 15.3%, crude fat, 14.3% and crude fibre 42.4% (Table 6.45). The most limiting factors in seeds are the high fibre content (42.4%) and the presence of tannin (1.4%). As far as stone cells are concerned the high lignin (37.1%) and cellulose content (53.9%) are further limiting factors for the inclusion of guava wastes in diets for monogastrics. The only suggested inclusion of these products is to mix them with other feedstuffs such as rice polishings or wheat middlings in low percentages up to 5%. They may be also mixed with kitchen waste or poultry manure up to 10% without no side effects as far as fibre percentage is concerned.

This waste should be nutritionally evaluated at several inclusion levels in broiler and layer diets.

Table 6.45. Chemical composition of guava seeds and stone cells. From (1) Winton and Winton (1949) and (2) Jagtiani, Chan and Saki (1988)

%	*(1)*	*(2)*
	Seeds	*Stone cells*
Moisture	10.3	-
Crude protein	15.3	1.5
Crude fat	14.3	0.9
Crude fibre	42.4	-
Ash	3.0	1.1
Glucose	0.1	-
Carbohydrate (soluble) -	5.5	
Starch	13.3	-
Lignin	-	37.1
Cellulose	-	53.9
Tannin	1.4	-

6.10 BREWERS' DRIED GRAINS

During the several processes involved in producing beer, distilled liquors and alcohol, various by-products are recovered after the fermentation of cereal grains. These by-products include spent grains, slops and brewers' yeast. All the by-products of the fermentation industries contain reasonable quantities of proteins, however, the carbohydrate content is lower than for the raw material owing to the losses caused by fermentation.

Brewing and the manufacture of distilled liquors, alcohol, vinegar and yeast are processes which convert starches to alcohol and carbon dioxide by the action of fermentation through yeast or other micro-organisms resulting in various by-products which have a reasonable nutritive value as a feedstuff for animals and poultry. The value of these by-products depends on the raw materials used and the channels they will pass through in the processing to the main final product and the processing of the by-product.

Through fermentation, much of the starch and the soluble nutrients from carbohydrate-containing raw materials such as cereal grains will be degraded leaving a spent-grain residue. The result of this fermentation will retain practically all the fibre, and much of the protein, fat, linolic acid, vitamins and mineral of the original grains. The spent grains retain their form, but the total solids and digestible constituents are greatly reduced. The great variation in these by-products can be seen from their chemical analysis and digestible protein levels shown in Table 6.46.

Table 6.46. Chemical composition (%) of brewers' grains and several by-products of the fermentation industries. From Ewing (1963)

Feedstuff	%							
	Moisture	*Protein*	*Fat*	*Crude fibre*	*Ash*	*Nitrogen-free extract*	*Digestible protein*	*Total digestible nutrients*
Brewers' dried grains -								
Below 23% protein	7.7	21.1	6.9	17.6	4.0	42.9	15.0	62.0
23.0-25.9% protein	7.4	24.6	6.4	16.2	4.2	41.2	18.0	-
Above 26% protein	7.2	27.5	6.7	15.3	3.9	39.4	21.0	66.0
Brewers' wet grains	75.6	5.6	2.0	4.3	1.0	11.5	4.3	16.7
Brewers' spent hops	6.2	23.0	3.6	24.5	5.3	37.4	6.7	29.0
Distillers' wet grains	77.4	4.5	1.6	2.8	0.6	13.1	3.2	19.0
Distillers' whole slop	93.8	1.9	0.6	0.5	0.3	2.9	1.3	5.3
Distillers' strained slop	95.9	1.4	0.7	0.2	0.3	1.5	1.0	3.7
Corn distillers' dried grains	7.1	28.2	9.0	12.2	2.6	40.9	22.6	80.0
Rye distillers' dried grains	6.8	17.0	6.0	15.6	2.4	52.2	11.0	57.0
Corn distillers' dried centrifuge sludge [a]	8.0	42.0	9.0	7.0	3.0	31.0	-	-
Corn distillers' dried solubles [a]	8.0	27.0	7.0	0.8	8.0	49.2	-	-
Corn distillers' semi-solid solubles [b]	60.0	12.0	3.0	0.4	3.5	21.1	-	-
Corn distillers' dried grains with solubles [a]	8.0	28.0	9.0	7.0	5.0	43.0	-	-
Corn distillers' oil-extracted dried grains with solubles [a]	8.0	31.0	1.0	8.0	5.0	47.0	-	-
Malt	7.7	12.4	2.1	6.0	2.9	68.9	-	-
Malt sprouts	7.6	27.2	1.6	13.1	5.9	44.6	19.7	74.4
Vinegar dried grains	6.8	19.5	7.0	17.3	2.9	46.5	12.5	62.3
Yeast dried grains	6.3	20.8	6.3	16.1	2.8	47.7	13.3	61.1
Yeast	4.3	50.0	0.5	0.5	10.0	34.7	40.0	75.0

[a] Values calculated to an 8% water content - actually samples ranged from 7 to 12% water.
[b] Semi-solid solubles calculated to a 60% water content.

The largest production of the above mentioned by-products (Table 6.46) is the brewers' dried grains in terms of quantity and as this process is used throughout the world, this is an important by-product.

6.10.1 Introduction

Brewing is the process of producing beer which is mostly based on barley and malt. Malt is barley germinated for a limited period of time and then dried. Barley belongs to the cereals of the grass family (*Gramineae*). The two-rowed barley is known as *Hordeum sativum* Jess. var. *distichon* (L.) Hackel = H. *distichon* L. The wild form of barley is known as

H. *sponteneum* Koch., found growing in various parts of Western Asia. This form appears to be the parent from which all the cultivated types of barley have been derived whether through evolution or cultivation. In central and northern Europe and also in certain regions of the United States, two-rowed barley is the type most commonly cultivated, being well adapted for malting.

Modern varieties of barley have relatively short stems (60-90 cm). Each flowering head has an axis with short internodes. From a nod three simple flowers arise, grouped on one side of the stem. At the next node, the grouping is on the opposite side. Looking vertically down the flower axis, six rows of flowers should be apparent. In some varieties all the flowers are fertile, in other varieties only the central flower of the three will form a fruit. Thus one can distinguish between 6-rowed and 2-rowed barley. In the 6-rowed barley, the fruits or grains that develop have less space for development than the 2-rowed. Therefore the central flower tends to produce a normal grain but the two lateral flowers (which are sterile in 2-rowed varieties) form rather twisted, thin grain (Figure 6.26).

More recently, 2-rowed barleys have become important because they have on average larger, more uniform grains and contain more starch than 6-rowed barleys. Their enzyme activity may be lower but nevertheless in modern varieties it is more than adequate for an equal mix with cereal starch. However, 6-rowed barley has been rejected because of the lack of uniformity of their grains, their high protein, high husk and low starch levels. The only advantage is its high enzyme activity and partly because the husks may improve the filtering quality of the mash bed during processing.

Figure 6.27 shows the barley grain in a longitudinal and transverse section. The bracts, called the lemma and palea are shown where the lemma is extended to form an awn. At their base is the former attachment of the flower to the mother plant and close to it is a region called the micropyle through which air and water can permeate to the embryo. The embryo is arranged mainly on the rounded or dorsal side of the grain. Its root sheath is close to the micropyle so that it can readily penetrate through this region when germination starts. In contrast the embryonic shoot points towards the distal end of the grain. Separating the embryo from its food store or endosperm is a shield shaped structure called the scutellum, regarded by some as the seed-leaf of this monocotyledonous plant. On the periphery of the endosperm is a layer called the aleurone. This region does not extend over the scutellum. The microscopic structure is gives an idea of the enormous starch accumulation in the endosperm which forms the basis of malting. (Figure 6.27) (Winton and Winton, 1950; Cook, 1962; Hough, 1985).

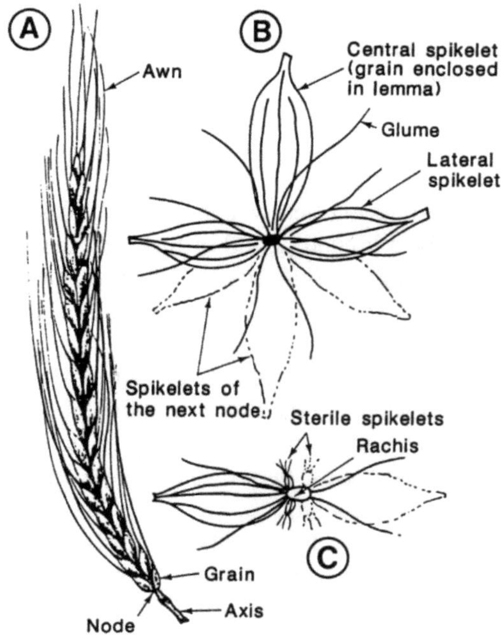

Figure 6.26. Detail of the flowering head (ear) of barley: (A) the ear of a two-rowed barley; (B) a six-rowed barley ear seen from above; and (C) a two-rowed barley ear seen from above. The dotted outline represents the florets attached at the next node or joint. Reproduced with permission from Hough (1985).

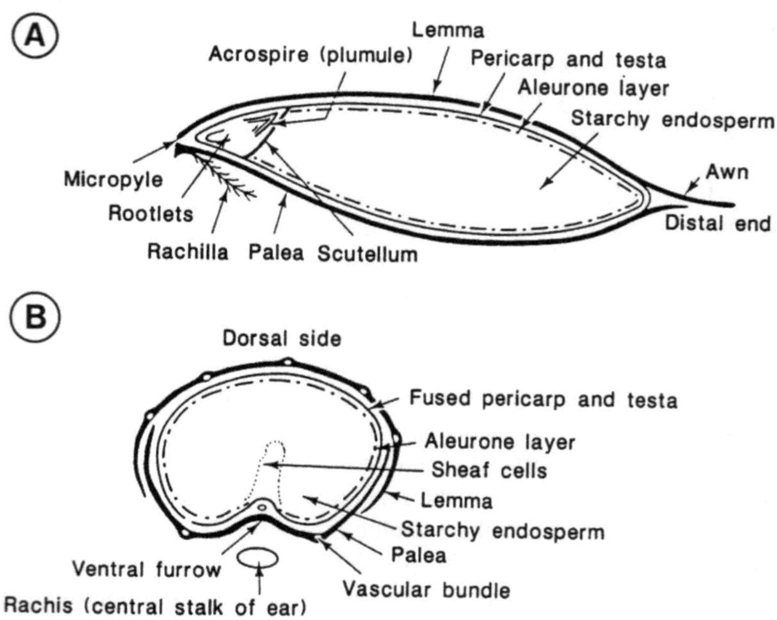

Figure 6.27. Detail of the barley grain: (A) Longitudinal (vertical) section; (B) Transverse (vertical) section. Reproduced with permission from Hough (1985).

The world production of barley is estimated to be 180.4 million t per year and the highest producers of the world are the Commonwealth of Independent States (formerly USSR) followed by the principal producers Canada, USA, France and UK (FAO, 1991).

Hough (1985) estimated for the years 1971, 1976, 1981 that the locally used barley in the UK, 14.6, 18.3 and 22.8%, respectively, was used for malting purposes. From these data it can be estimated that an average of 81.4% from the world barley production is used as animal feed. This means that this production will deliver 147 million t as feed and 33 million t as malting (FAO, 1991).

The world beer production for 1990 was estimated to be 1.17 milliard hectolitres (1 hl = 100 l). The highest consumption of 143 l per/caput/year was found for West Germany (CBK, 1991).

The production of waste from beer processing or the wet spent grains from malt is 60% on a wet basis which will be 15% on a dry basis (Hough, 1985). This equation will lead to a world production of wet spent grains estimated to be about 20 million t per year based on the total world production of barley (FAO, 1991).

Several research workers have reported the nutritive value of this waste as the brewers' dried grains or the brewers' spent (exhausted or spawned) grains as a food for human consumption and as a feedstuff for poultry (Eldred, Damron and Harms, 1975[a,b]; Damron, Eldred and Harms, 1976; Jensen, Chang and Maurice, 1976; Ward, 1977; Sullivan, Kuhl and Holder, 1978; Din, Sunde and Bird, 1979; Kissell and Prentice, 1979; White *et al.*, 1979; López and Carmona, 1981; López, Carmona and Pascal, 1981; Onwudike, 1986).

6.10.2 Processing

(a) *Malting and brewing*

Malting and brewing and its generated waste has been discussed in details by many research workers (Cook, 1962; Harris, 1962; Pollock, 1962; Schuster, 1962; Göhl, 1975; Broderick, 1977; Pollock, 1979; Hough *et al.*, 1982; Hough, 1985) and the process of malting, brewing and spent drying will be summarized.

The main process for beer or ale is the production of malt. Malt is a germinated barley which is soaked for 2 or 3 days in warm water and then dried. The process of malting or malt making is called malster. This process is involved in finding suitable stocks of barley, sorting them until required and steeping the grain in water to germinate. At the appropriate time, the germination is arrested by drying the grains in a stream of warm air (kiln-drying). The malted barley grains represent a package for the brewer that will keep stable for months, if not years. During the germination, the food store or endosperm of the grain is partly degraded by enzymes of several types such as:

- The amylolytic enzymes (ß-amylases and -amylases).
- Other carbohydrases (cystase, invertase, transfructosylases and maltase).
- Enzymes dealing with nitrogen (proteases, peptidases, nucleases and phosphatases).
- Lipolytic enzymes.

All those enzymes are attacking the cell walls, starch granules and the protein matrix. The package of enzymes provided in the malted grains is able to degrade the endosperm and the attendant enzymes are able to complete the degradation.

During arresting the germination malt by cool air, a pale colour malt will be produced which is very rich in enzymes. However when the arresting and drying temperature is high the malt will be darker in colour with a depleted enzyme content. Figure 6.28 shows a flow diagram of the malting process.

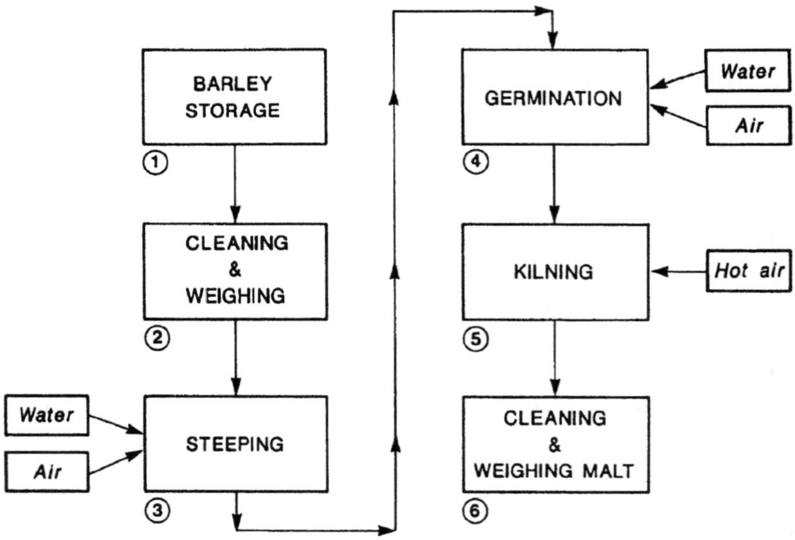

Figure 6.28. Flow diagram of the malting process. Reproduced with permission from Hough (1985) with modifications.

The growth of the sprouted grains is arrested by kiln-drying when the young shoot and rootlets are about 2 cm long. The dried shoots and rootlets are screened off and sold as a feedstuff under the name of "malt culms".

Brewing is a continuation of malting but on base of fermentation, the brewing process could be summarized as follows (Figure 6.29):

Grist. In this process of crushing, the malted barley is converted to a very coarse flour.

Warming. By adding warm water to the grist, a porridge-like mash is formed. Owing to the temperature, malt enzymes are encouraged to solubilize the degraded endosperm of the ground malt.

Wort. The aqueous extract known as wort is separated in suitable vessels. From spent (exhausted) solids in the wort more material is extracted by spraying hot water on to th mash.

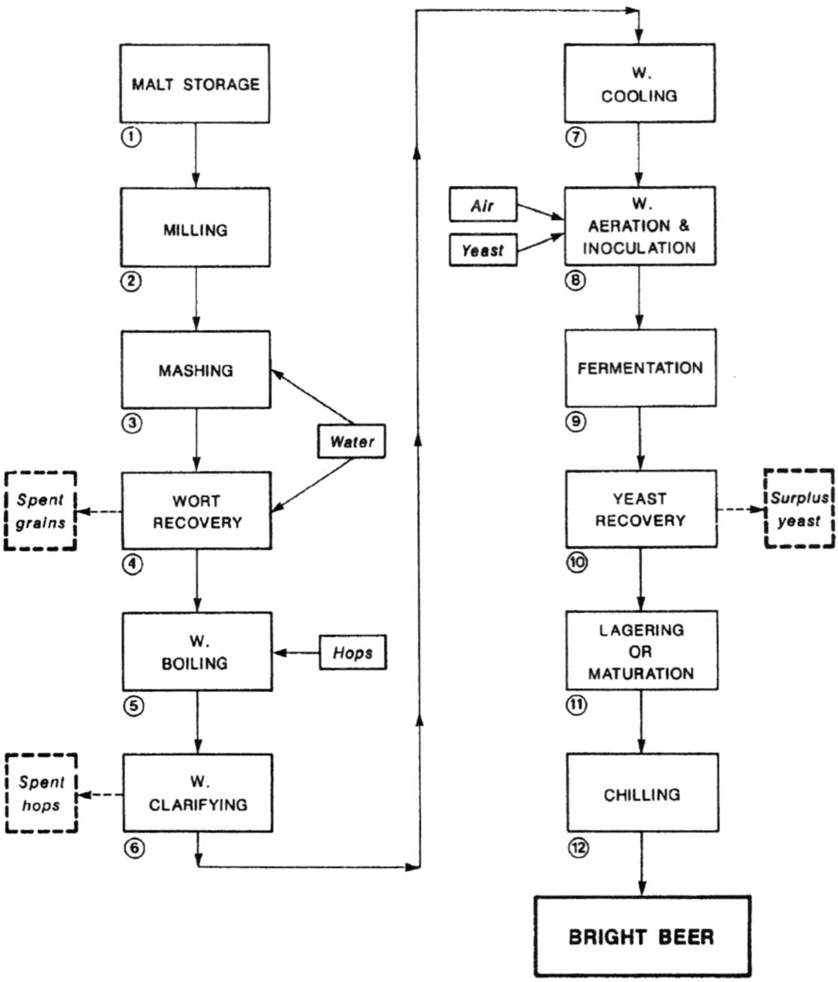

Figure 6.29. Flow diagram of the brewing process. Reproduced with permission from Hough (1985) with modifications.

Boiling. In this step hops are added to the wort and boiling takes place. This stops theaction of the enzyme. The wort is sterilized and some proteins coagulate. The addition of hops imparts distinctive flavours and aromas to the wort.

Addition of yeast. Clarifying, cooling and aerating the wort creates an ideal medium for yeast growth and fermentation.

Fermentation. By adding the yeast to the wort, fermentation takes place and much of the carbohydrate will be converted into alcohol and carbon dioxide. Other yeast metabolites contribute to flavour and aroma.

Maturing. In this step maturing and clarifying take place resulting in modification in flavour, aroma and keeping qualities of the beer.

Packing. After filtration, sterilization or pasteurization, packing is carried out.

Alternatively, for small packages like bottles or cans, the beer can be pasteurized within the package.

Several waste products generated from the malting and brewing process. From the malting, the malt culms or malt sprouts are produced after screening off the shoots and the rootlets. From the brewing, the spent grains of barley, spent hops and brewing yeast are generated. The most relevant by-products for poultry as far as nutritive value and useful quantity are concerned are the brewers' dried grains or spent brewers' grains.

(b) Processing of waste

Brewers' dried grains

The extracted malt, or brewers' or spent grains contain 75-80% moisture when filtered off. In large breweries the spent grains are dried to about 10% moisture. Wet spent grains spoil rather quickly and should be used fresh or stored out of contact with air. It can be stored up to two weeks quite successfully by heaping, treading and covering it with wet sacks. Wet grains stored in this way should not be produced in excessive amounts, as it is meant only for ensilage and fermentation purposes for ruminants.

For poultry it should be dried and this takes place by means of mechanical treatments as hydraulic pressing to lower the moisture content from 75-80% to 65% which will be cheaper due to less moisture to evaporate. The pressed brewers' spent grains are transferred to a rotary steam drier or a steam-tube drier with a temperature about 100-106 °C for 8 h. The end product will contain 10% moisture when it may be ground since it is bulky as a fine mash known as dried brewers' spent or grains (Göhl, 1981; Hough, 1985).

Brewers' spent grains are known to be a very relevant feedstuff for ruminants, calves and pigs, but they are not commonly included in commercial poultry feeds. However, several investigators showed that up to 20% dried brewers' grain can be included in poultry rations with excellent results concerning growth and feed conversion of broilers, and increased fertility and hatchability of the eggs. This point will be discussed in detail under feeding of spent grains.

Drying brewers' grains for human consumption

The first stage is the mechanical de-watering of the wet brewers' grains which saves energy and separates a proportion of the protein-rich fraction. As this stage is performed at the brewery, the fluids can be returned to the brewerhouse for further water purification.

Drying and pre-milling then takes place simultaneously in a drier where the brewers' grains are fed in a hot-air stream through a high speed rotor which continuously reduces the particle size of the material being dried. This method of drying has several advantages. The sudden evaporation of water in the hot-air stream effectively separates the husk and protein particles as the stream is formed. The rotor then reduces the protein particles to a very small size, while the more resistant husk remains in large elements, enabling easier subsequent separation.

The moisture content will be reduced to 5% without exceeding a temperature higher than 60 °C, which prevents any detrimental effect on the taste and colour of the brewers' grains.

The second stage of the process is a simple sieving operation to separate the fibre-rich material (about 70% of the total) from the protein-rich material (about 30% of the total).

The dietary-rich fraction will be milled to produce refined brewers' grains of suitable particle size for use by food processors (Anon, 1985).

Brewery liquid waste (effluent)

During the processing of beer as far as malting and brewing are concerned a lot of waste-water (effluent) is discharged. The components of this effluent are spent materials such as malt particles, hop fragments, excess yeast and lees (sediment of liquor at the base of a vessel) from beer tank. Weak worts or spoiled beer are also considered as effluents. Figure 6.30 shows the sources of effluent during beer processing.

Spent materials can be added to the spent grains after filtration and used as an animal feed, while weak worts and spoiled beer can be incorporated as a drinkwater for swine (swill).

The degree of polluted brewery wastewater or effluent is measured by means of the concentration of suspended solids (SS) and the concentration of the chemical oxygen demand (COD). This concentration of materials can be oxidized chemically by boiling with potassium dichromate and concentrated sulpheric acid. The reason for measuring COD is that when effluent enters a waterway, dissolved oxygen is consumed by aerobic micro-organisms as they metabolize the organic material. Thus, the more organic material that is present (in other words where there is a high COD), the more dissolved oxygen is used. High levels of organic material may completely deoxygenerate the water and would lead to the death of aerobic organisms. It is therefore necessary to restrict COD levels of effluents discharged into natural waterways, to say, 10-20 mg/l. Restriction of suspended solids is also required because they not only usually represent organic material but they may settle in natural waterways to give an anaerobic sludge (Metcalf and Eddy, 1979; Hough, 1985; Eilbeck and Mattock, 1987) (see chapter 3).

Typically, bulked brewery effluents have a suspended solid value of 240 mg/l and a COD value of about 1800 mg/l. Brewery effluents are usually in the pH range of 3.5-5.5. In large breweries the effluent is mostly treated by means of filtration which is roughly just bulking the outfalls and then followed by buffering. Finally, the effluent will be treated aerobically, or anaerobically.

Aerobic digestion

This process depends on a large population of micro-organisms that is able to take up both soluble and colloidal organic materials from the effluent and metabolize them to mainly carbon dioxide and water. The energy derived from the metabolism enables the microbial cells to grow (see Chapter 6, Citrus sludge and potato sludge). There are three basic types available for water purification under the aerobic digestion.

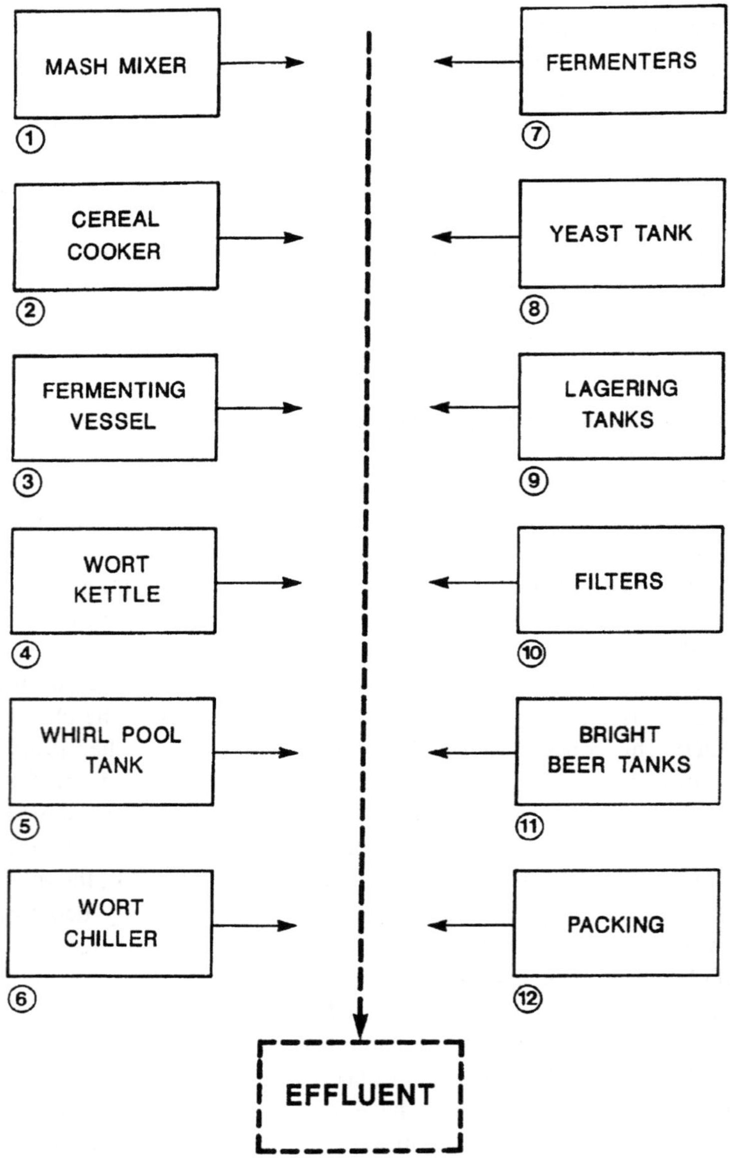

Figure 6.30. Flow diagram showing the production of effluent in a brewery. Reproduced with permission from Hough (1985) with modifications.

The percolating filter system. This comprises a 2 m bed of stones contained within a circular wall which is naturally ventilated. The effluent is sprayed from rotating distributor arms above the bed and trickles between the stones which are coated with a slimy population of micro-organisms. The percolating filter operates poorly in case of variable flow, composition and fluctuating pH of effluent, quite apart from the high level of suspended solids and COD.

Biotower. In this system a tower is very loosely filled with sheets of rigid plastic material on which the micro-organisms grow. The effluent trickles down the tower counter-current to rising air.

The activated sludge system. This system depends on a high concentration of micro-organisms that flocculate. These sludge organisms are kept in suspension by compressed air or mechanical agitation. In this process, the sludge organisms grow prolifically and the population has to be kept roughly constant by removing the surplus. The sludge is difficult to dewater and is not popular owing to its drawbacks of its bad odour and its high metal content. This activated sludge needs a lot of investigation as far as its nutritive value as a feedstuff is concerned in relation to the concentration of the micro-organisms and the metals.

Anaerobic system

This system of fermentation takes place in an enclosed container. The bacteria used for the digestion tend to be of two general types: one type producing acetic, propionic and other fatty acids from the metabolism of the effluent organic material; the other type yielding methan and carbon dioxide. This process yields gas which can be used in a gas engine, boiler or heat exchanger. The disadvantages of this process are the long time it takes to start up and its sensitivity to variations in loading and effluent composition. The effluent treatment is not very profitable and its expenses are much more than its output as a relevant sludge. Many breweries rely entirely on local authority sewage works for the treatment of their effluent or pooring the effluent into a lagoon system (Metcalf and Eddy, 1979; Eilbeck and Mattock, 1987; Hough, 1988).

Yields of brewery waste

It was mentioned before that spent grains are the most relevant waste by-product from breweries. Sometimes the spent grains are used as a filter for effluent slurries containing spent hops and sediments from tank bottoms. Thus, the nitrogenous content of the grains may be enhanced by a further discharge of unwanted materials and the effluent costs will be reduced. Spent hops which are very bitter and unpalatable are mostly added to the spent grains in the same proportion as they are produced; they will constitute about 5% of the total mixture. In this percentage, the palatability of the dried brewers' grain will not be negatively affected.

Each 100 units of weight of malt will produce 60 units of wet spent grains. After drying, the weight of spent dried grains will be 15 units. It has been calculated that 5 units of wet spent grains is equivalent to 1 unit of barley.

Spent grains are not only used as a feedstuff for animals and poultry (Göhl, 1981; Hough, 1985) but are also used for human consumption in bread, biscuits and breakfast cereals (Prentice *et al.*, 1978; Kissel and Prentice, 1979).

Sludge production is estimated to be 220-380 g dry matter per 100 liter beer produced (Breederveld, 1992).

6.10.3 Chemical Analysis and Nutritive Value

(a) Spent barley grains

Brewers' dried grains or spent grains are exhausted barley grains of which starch or carbohydrate are degraded owing to fermentation leaving all the fibre and much of the protein, fat and minerals of the original barley. They show a high protein percentage varying from 23.4-28% with a digestibility of 80%. The content of the essential most limiting amino acids lysine and methionine are 0.9 and 0.3-0.6%, respectively. The protein and methionine levels are higher than in barley and maize but lower than in maize gluten meal; lysine is higher than in barley and maize but nearly similar to that of maize gluten meal.

The metabolizable energy content of the spent grains varies from 7.3-10.8 MJ/kg which is lower than in maize and maize gluten meal, and in barley. Nitrogen-free extract varies from 41.3-46.7% which is lower than barley (68.1%), and maize (70.5%) and nearly the same as in maize gluten meal (39.4%). This is a clear picture of an exhausted grain which has lost a lot of starch during fermentation in brewing. As far as ash, calcium, phosphorus and sodium were shown to be higher in brewers' dried grain than in barley, maize and maize gluten meal. This is again a picture of a grain which has largely lost its starch and the other nutrient levels will be higher than in the original unfermented grain (barley). Crude fat varies between 3.7 and 8%, which is high compared to fat levels in barley, maize and maize gluten meal (1.9, 3.9 and 2.5% respectively). The fibre content varies between 12-18% which is much higher in comparison with barley (7.5%, maize 2%) and maize gluten meal (4.5%). The high fibre content limits the inclusion of brewers' dried grains to 20% in rations of starter broilers and growing chicks. Also its low energy content is a consideration when replacing the feedstuff with maize. The total digestible nutrient level (TDN: 63%) is similar to maize gluten meal (61%) much lower than maize (81%) and somewhat higher than barley (68%; Tables 6.47 and 6.48) (Crampton and Harris, 1969; Titus and Fritz, 1971; Sullivan, Kuhl and Holder, 1978; Göhl, 1981; Scott, Nesheim and Young, 1982; Hough, 1985; Onwudike, 1986).

(b) Sludge from brewery water purification

There is very little information available concerning the brewery sludge produced from wastewater. Breedeveld (1992) analysed the brewers' sludge and he found that it contains 29.4% crude protein; 2.5% calcium; 6.4% phosphorus and had a high content of ash (53.5%). This high ash content is due to its high silicon (Si) content of 56%. If the particles of silicon have a similar diameter to fine sea sand this will be a great advantage to the digestive system (mainly the gizzard) of poultry. It was reported by Onwudike (1986) that the use of fine sea sand at an inclusion of 4% improved feed conversion significantly. This point concerning the high silicon content of the sludge of brewers' wastewater needs further research. As far as the heavy metals are concerned the most striking element for dried brewers' sludge is iron, which was 5200 ppm in the sludge (Breedeveld, 1992), while the toxic levels are 4500 ppm according to

Table 6.47. Chemical analysis and amino acid composition (%) of brewers' dried grains in comparison with some feedstuffs. From (1) Crampton and Harris (1969); (2) Titus and Fritz (1971); (3) Sullivan, Kuhl and Holder (1978); (4) Göhl (1981); (5) Scott, Nesheim and Young (1982); (6) Hough (1985) and (7) Onwudike (1986)

	Brewers' dried grains							*Barley*	*Maize*	*Maize gluten M.*
%	*(1)*	*(2)*	*(3)*	*(4)*	*(5)*	*(6)*	*(7)*	*(5)*	*(5)*	*(5)*
Moisture	8.00	8.00	7.50	-	-	-	-	11.00[a]	8.30[a]	10.00[a]
Crude protein	25.90	25.90	28.00	27.80	27.00	23.4	23.7	9.50	8.60	42.00
ME (MJ/kg)[b]	10.50	7.30	-	10.80	10.10	-	-	9.70	14.10	13.80
Crude fat	6.20	6.20	6.00	8.00	7.00	7.7	3.7	1.90	3.90	2.50
Crude fibre	15.00	15.00	12.60	12.60	12.00	17.6	18.0	7.50	2.00	4.50
N-free extract	41.40	41.30	42.40	46.70	-	-	43.1	68.10	70.50[a]	39.40[a]
Ash	3.60	3.60	3.70	4.90	-	4.1	4.0	2.40[a]	1.70[a]	2.40[a]
Calcium	0.27	0.27	0.39	0.16	0.27	-	-	0.05	0.02	1.50
Phosphorus	0.50	0.50	0.71	0.65	0.50	-	-	0.32	0.30	0.40
Sodium	0.26	0.26	-	-	0.04	-	-	0.02	0.01	0.01
Potassium	0.08	0.08	-	-	0.13	-	-	0.50	0.38	0.03
Chloride	-	-	-	-	0.06	-	-	0.15	0.04	0.07
Amino acids %										
Arginine	1.30	1.30	-	0.12	1.30	-	-	0.40	0.50	1.40
Cystine	-	0.50	-	-	0.30	-	-	0.20	0.20	0.70
Glycine	-	1.00	-	-	1.00	-	-	0.30	0.40	1.50
Histidine	1.50	-	-	0.50	0.50	-	-	0.20	0.20	0.90
Isoleucine	1.50	-	-	1.50	1.50	-	-	0.40	0.40	2.10
Leucine	2.30	-	-	2.30	2.30	-	-	0.60	1.10	6.60
Lysine	0.90	0.90	-	0.90	0.90	-	-	0.30	0.20	0.80
Methionine	0.40	0.40	-	0.30	0.60	-	-	0.10	0.20	1.20
Phenylalanine	1.30	-	-	1.30	1.30	-	-	0.50	0.50	2.40
Threonine	0.90	-	-	0.90	0.90	-	-	-	0.40	1.30
Tryptophan	0.40	0.40	-	0.40	0.40	-	-	0.10	0.10	0.20
Tyrosine	1.20	-	-	1.30	1.20	-	-	-	0.40	1.00
Valine	1.60	-	-	1.50	1.60	-	-	0.50	0.40	2.20

[a] Reference (2)

[b] According to Sibbald (1977) ME for brewers' dried grains is 13.89 MJ/kg.

Table 6.48. Average chemical analysis and digestibility (%) of dried brewers' grain nutrients in comparison with barley, maize and maize gluten meal. From Titus and Fritz (1971)

%	*Spent dried grains*	*Barley*	*Maize*	*Maize gluten meal*
Crude protein	25.9	11.6	11.6	42.9
Digestibility	80.0	74.0	79.0	81.0
Crude fat	6.2	1.9	5.0	2.3
Digestibility	60.0	62.0	88.0	55.0
Crude fibre	15.0	5.0	2.9	4.0
Digestibility	12.0	9.0	6.0	11.0
N-Free extract	41.3	68.1	70.5	39.4
Digestibility	78.0	82.0	92.0	57.0
Total digestible nutrients	63.0	68.0	81.0	61.0

the NRC (1984). A high iron concentration may cause a colloidal suspension of insoluble iron phosphate which may absorb vitamins or trace inorganic elements, and this may cause growth depression in growing chicks and broilers (Scott, Nesheim and Young, 1982; El Boushy, Roodbeen and Hopman, 1984).

Also, the high concentration of aluminium (4300 ppm) exceeds the limits of toxicity reported by NRC (1984) which is 500-2200 ppm. Jones (1938) and Storer and Nelson (1968) reported that aluminium interferes with the phosphorus absorption by the formation of insoluble aluminium phosphate resulting in decreased serum phosphatase and bone mineral formation. Storer and Nelson (1968) reported that lower levels of aluminium (500-1000 ppm) as chloride or sulphate adversely affected growth, feed efficiency and bone mineralization. Water insoluble aluminium as the oxide and the phosphate had no adverse effects on performance.

As far as the heavy metals in brewers' sludge are concerned in conjunction with its use as a fertilizer, the permitted maximum levels were not reached (Table 6.49; Anon, 1991). It is safe to use brewers' sludge as a fertilizer and its heavy metals do not harm plants growing on fertilized soil with this sludge. If the sludge is included at low percentages in diets this will not create toxicity problems because an inclusion of 5% for example will never reach the toxic levels mentioned by NRC (1984). This relatively new feedstuff needs further research as far as the possibilities of lowering the level of the metals in the sludge and increasing the percentage of inclusion in diets of poultry are concerned. However, the analysis for the metabolizable energy and amino acid content is not known (Table 6.49).

Table 6.49. Chemical analysis of brewers' sludge produced from effluent in fresh and dry matter in comparison with toxic concentration in poultry diets and maximum permitted levels as a fertilizer. From [(1)] Breedeveld (1992); [(2)] NRC (1984) and [(3)] Anon (1991)

	(1)		*(2)*	*(3)*
	Fresh material (%)	*Dry matter (%)*	*Toxic levels* [a] *(ppm)*	*Maximum levels as fertilizer (ppm)*
Dry matter	19.900	100.00	-	-
Organic material	9.300	46.50	-	-
Crude protein (N x 6.25)	5.900	29.40	-	-
Ca	0.500	2.50	-	-
P	1.000	6.40	-	-
K	0.050	0.20	-	-
Mg	0.150	0.26	-	-
Si	11.260	56.00	-	-
Ash	10.600	53.50	-	-
COD [b]	13.700	69.00	-	-
Heavy metals (ppm)				
Cu	35.000	175.00	250-806	425.0
Cr	17.000	83.00	300	350.0
Zn				
Pb	14.000	68.00	320-1000	300.0
Cd	0.160	0.78	12-40	3.5
Ni	8.400	42.00	500	70.0
Hg (total)	0.040	0.20	5-400	3.5
As	0.700	3.50	100	25.0
Mn	12.000	61.00	6000-19600	-
Se	< 0.200	< 1.00	5-10	-
Fe	1035.000	5200.00	4500	-
F	0.760	3.80	400-1000	-
Al	856.000	4300.00	500-2200	-

[a] Toxic concentration may lead to growth depression or mortality for broilers or reduced egg production for layers.

[b] C O D = Chemical Oxygen Demand.

6.10.4 Feeding Brewers' Dried Grains (BDG)

(a) Broilers and chickens

The nutritional evaluation of Brewers' Dried Grains (BDG) in the diets of broilers and growing chickens has been reported by several research workers. Branckaert

(1967) evaluated BDG as a replacement for soybean meal and maize in broiler diets. The inclusion of 20% BDG slightly lowered body weight of broilers at 5 weeks of age while at 12 weeks of age, body weight was 11% higher than for control birds. Inclusion of 40% BDG decreased growth rate and increased feed intake significantly. However, Ademosun (1973) concluded that, on the basis of feed consumption, weight gain and feed to gain ratio, the inclusion of BDG in starter diets should not exceed 10%. For growing birds from 8-18 weeks of age, BDG should not exceed 30% of the diet. However, this level leads to increased feed consumption ($P < 0.01$) and feed to gain ratio ($P < 0.05$). He added also that the utilization of BDG for growing chicks is affected by its high crude fibre content.

López and Carmona (1981) evaluated the BDG in diets of broiler starter and finisher chickens with an inclusion of 0, 10, 20, 30 and 40%, diets which were formulated to be isonitrogenous and isocaloric (Table 6.50). There was a significant decrease in body weight gain of birds in the starter and finisher periods when the inclusion of BDG was 20% or more. However, final body weight was depressed significantly in the starter period only. The same trend was noticed with feed efficiency; feed intake was not affected and independent from the several inclusion levels of BDG. It was concluded that BDG should not be added in broiler diets in excess of 10% for starter and finisher (0-8 weeks) or 20% from 8-12 weeks if the broilers were kept to this age.

Table 6.50. Effect of inclusion levels of Brewers' Dried Grains (BDG) on the performance of growing chicks from 0 to 8 weeks. From López and Carmona (1981)

Inclusion	*Feed intake (g/day)*	*Body weight (g/day)*	*Final body weight (kg)*	*Feed efficiency*
Starter diet (0-4 weeks)				
Control	52.2	23.6^{a}	0.66^{a}	2.21^{a}
10% BDG	50.4	22.6^{a}	0.63^{a}	2.24^{a}
20% "	48.6	19.9^{b}	0.56^{b}	$2.43^{a,b}$
30% "	49.1	19.8^{b}	0.55^{b}	$2.48^{a,b}$
40% "	48.5	18.2^{c}	0.51^{c}	2.67^{b}
Significance	NS	**	**	**
Finisher diet (4-8 weeks)				
Control94.2	39.4^{a}	1.68	2.40^{a}	
10% BDG	92.5	$38.9^{a,b}$	1.67	2.38^{a}
20% "	92.0	$36.2^{b,c}$	1.59	2.57^{b}
30% "	93.0	36.4^{b}	1.59	2.56^{b}
40% "	93.5	34.6^{c}	1.55	2.70^{b}
Significance	NS	*	NS	*

^{a,b,c} Figures per diet in one column followed by the same letter vertically are not significantly different.

* $P < 0.05$

** $P < 0.01$

NS = not significant

Onwudike (1986) evaluated the effect of inclusion of BDG in starter and growers' pullet diets in levels of 0, 15, 20 and 30% and 0, 25, 30 and 40%, respectively, with the addition of 0 and 4% fine sea sand. It was concluded that the utilization of BDG by pullets depends on the age of the birds. Young starter pullets appear to derive the most benefit from the presence of sand in diets containing BDG. With starter birds, the use of sand allowed a greater quantity of BDG to be utilized than when no sand was fed. The rate of gain when 25% BDG was fed with the sand was similar to that rate of gain produced when 20% of BDG was fed without sand. This trend for growth was paralleled by feed conversion. The use of sand probably allowed a more efficient use of BDG through a better utilization of higher levels of crude fibre from BDG varying from 13-18% (Table 6.47). Several investigators came to similar conclusions in that sand and grit allow for a better digestion of coarse or fibrous feed in the domestic fowl (Otani, Yamatani and Sasaki, 1968; Norris, Norris and Steen, 1975; Hooge and Rowland, 1978; Oluyemi, Arafa and Harms, 1978). Onwudike (1986) added that the use of sand has no beneficial effect in the feeding of the growers owing to the percentage used and the size of the sand particles. It was concluded that growers may need more than 4% sand in the diet to obtain significant effects with high inclusions of BDG. It may be concluded that the presence of sand in the gizzard improved the grinding process of the BDG.

(b) Layers and egg quality

Several investigators have evaluated the Brewers' Dried Grains (BDG) in the diets of laying hens in relation to interior and exterior egg quality, liver fat and layer body weight.

Thornton and McPherron (1962) compared diets containing 20% BDG fed to pullets during the growing and laying phases to those fed a control ration without BDG. Pullets fed BDG grew faster, matured earlier and produced more eggs. It was noticed also that body weight gain and liver fat accumulation was significantly less in hens fed BDG. Further studies indicated that 40% BDG inclusion in a layer diet supported excellent laying performance, reduced body weight and liver fat (Kienholz, Thornton and Moreng, 1963) and improved fertility and hatchability (Kienholz and Jones, 1967). However, Kienholz, Darras and Caveny (1972) reported a marked decrease in egg weight associated with feeding 20-40% BDG. They concluded that the reason for egg size reduction may be due to the long period of storing the used BDG at a very high environmental temperature. The possible chemical damage of BDG upon these storage conditions, however, was not elucidated.

Eldred, Damron and Harms (1975[a]) studied the effect of inclusion of 10% BDG with or without brewers' yeast in layer diets. Egg production was significantly improved by adding 5% BDG and egg weight was improved with BDG in combination with yeast inclusion. Adding BDG to layer diets resulted in a significant improvement of egg Haugh unit values (Eldred, Damron and Harms, 1975[b]).

White *et al.* (1979) used BDG in levels of 0, 10, 25 and 40% with or without 1% lysine to replace maize and soybean meal in pullet layer diets. They reported that body weight gain was progressively less as the inclusion of BDG increased.

Eggs were somewhat lighter with 40% BDG. The addition of 1% lysine to 40% BDG gave better results than without for egg production percentage. Feed required to produce eggs increased with increasing the level of BDG. Haugh unit score was higher with 40% BDG, but this was probably related to the small size of clutches (sequence of a number of eggs).

Onwudike (1981) included BDG in diets for laying hens which had been laying for about 10 weeks and which were subsequently tested for 8 months. In the feeding trial, the inclusion level of BDG was 0, 10, 20, 30 and 40% in diets formulated to be isocaloric, isonitrogenous and equal in calcium and phosphorus. Average hen-day production values on the above-mentioned diets were 56.7%, 56.5%, 58.7%, 60.9% and 55.8%; average egg weight was 59.3 g, 59.7 g, 60.9 g, 62.4 g and 62.3 g and mean daily feed intake was 121.6 g, 130.8 g, 154.2 g, 166.4 g and 168.9 g, respectively.

López, Carmona and Pascual (1981) evaluated BDG in the diets of laying hens by replacing wheat, sorghum, barley and soybean meal by 0, 15, 30, 45, 60 and 90% (complete replacement) BDG. This experiment lasted for 75 days except for the inclusion of 90% BDG which was for two weeks. Diets with up to 45% BDG were isonitrogenous, while the higher levels of BDG had a higher level of protein. The productive characteristics

Table 6.51. Effect of Brewers' Dried Grains (BDG) on several layer production characteristics, external and internal egg quality. From López, Carmona and Pascual (1981)

Productive characteristics	*Control*	*Inclusion % of BDG*					*Significance*
		15	*30*	*45*	*60*	*90*[d]	
Feed intake (g/day)	122.20[a,b]	121.30[a,b]	123.20[a,b]	124.90[a]	115.50[b]	26.20	*
Feed intake (MJ ME/kg body weight)	0.66[a,b]	0.65[b]	0.68[a]	0.62[c]	0.59[d]	0.13	**
Egg production (%)	70.20[a,b]	71.40[a]	71.00[a,b]	70.70[a,b]	64.90[b]	30.70	*
Feed efficiency (kg/kg)	2.80	2.70	2.80	2.80	2.90	1.30	NS
Body weight gain (g/period)	16.20[a,b]	3.70[a,b]	30.90[a]	- 23.00[b]	- 75.00[c]	- 526.60	**
Defective eggs (%)	9.60	6.30	9.80	7.40	11.90	21.30	NS
Mean egg weight (g)	63.40	63.70	64.10	64.00	64.30	61.60	NS
External egg quality							
Shell weight (%)	8.90[a]	8.90[a]	8.60[a,b]	8.50[b]	7.90[c]	7.30	**
Shell density (mg/cm^2)	74.00[a,b]	75.00[a]	72.60[a,b]	71.60[b]	67.60[c]	62.00	**
Shell thickness (mm)	281.30[a,b]	286.90[a]	279.90[a,b]	269.60[b]	251.40[c]	227.40	**
Internal egg quality							
Yolk (%)	28.70	28.90	29.00	29.20	29.60	33.50	NS
Albumen (%)	62.40	62.20	62.50	62.30	62.50	59.20	NS
Haugh units	81.90[a]	86.30[b]	86.10[b]	85.30[b]	85.90[b]	77.40	*
Yolk index	48.40	48.60	48.70	48.80	48.80	45.10	NS

[a,b,c] Values followed by the same letter horizontally are not significantly different.

[d] Not statistically analyzed

* $P < 0.05$

** $P < 0.001$

NS = not significant

such as feed intake, egg production, feed efficiency, body weight gain, egg weight, and external and internal quality were studied (Table 6.51). The researchers concluded that diets containing 15 and 30% BDG appeared to be adequate for laying hens; 45% BDG was also acceptable, although the losses of body weight and egg shell quality suggested that this level should not be used without further research. Inclusion of 60% BDG caused a significant decrease in most of characteristics studied. The use of 90% BDG caused a reduction in feed intake, enormous body weight losses and, in the second week of the experiment, an inhibition of the lay.

Turkey diets

Few studies have been conducted to evaluate BDG in turkey diets. Sullivan, Kuhl and Holder (1978) evaluated BDG and Brewers' Dried Grains plus Yeast (BDGY) in turkey diets. Three trials were conducted to evaluate different inclusions of BDG and BDGY in corn-soy diets for starting and growing-finishing turkeys. The BDGY was composed of 95% BDG and 5% brewers' dried yeast, both of which were included in turkey diets at the expense of ground maize and soybean meal. All the diets were calculated to be isocaloric and isonitrogenous. The inclusion of 5% BDG was tolerated in turkey starter diets. Growing-finishing turkeys tolerated BDG and BDGY levels of 5-15% very well. However, levels higher than 15% depressed feed efficiency. Pelleting BDG diets with 5, 10 and 15% inclusion improved performance, however, no comparison was made between pelleted and unpelleted diets for each inclusion level. All BDG treatments without exception significantly ($P < 0.05$) increased body weight at 4, 8 and 12 weeks of age. In addition, all BDG treatments except 5% inclusion during the period, 0-20 weeks were associated with either statistically ($P < 0.05$) or numerically greater body weight at 16 and 20 weeks compared with the control. Turkey breeder hens receiving diets containing 10, 20 and 30% BDG or BDGY performed as well as or slightly better than hens fed the control diet. The metabolizable energy (ME) value of 10.5 MJ/kg used for BDG in the study of Sullivan, Kuhl and Holder (1978) was apparently an overestimate. Depressed feed efficiency of birds receiving the higher levels of BDG and BDGY support this conclusion. When levels of 20% or more of BDG replaced maize and soybean meal in diets for growing-finishing turkeys, lysine supplementation was obviously needed.

(d) Pheasant diets

Moreng, Pfaff and Kienholz (1989) evaluated BDG by replacing maize and soybean meal using 0, 15, 30, and 45% inclusion in diets of Chinese ringneck pheasant breeder hens. The birds were used over a 16 weeks breeding season, starting with one week prior to the onset of lay. Egg production and fertility from BDG fed hens were found to be lower than from hens fed a control diet. Hatchability of fertile eggs was higher ($P < 0.05$) in hens fed diets containing 30 and 45% BDG. Based upon the number of live chicks produced per hen, the economic advantage was highest in groups receiving 30% BDG in the diet. The inclusion of BDG in the diets of pheasant breeder hens obviously had a favourable influence on hatchability and total number of live chicks produced per hen.

(e) Use of brewers' dried grains in human nutrition

During fermentation in the brewering process, most of the grain starch is converted to alcohol, carbon dioxide and other fermented products; the remaining nutrients (protein, fat, fibre, minerals and vitamins) undergo an almost threefold concentration in the resulting products, spent grains or the brewers' grains (See Table 6.47).

In the developed market such as Europe and the USA, DBG are an inexpensive source of healthy, fashionable dietary fibre. The dried spent grains or dried brewers' grains can be blended with flour to replace 15-20% of the wheat flour in biscuits and 10-15% of the flour in bread. In developing countries this helps reduce expensive imports and will significantly increase the fibre, protein, fat, minerals and vitamin content of bakery products.

The combination of the large continuing supply, relative low cost and the desirable nutritional value makes BDG an attractive adjunct for the food processing industry. The research which has taken place in the USA, Canada, Germany, the UK and other countries proves that BDG is an excellent ingredient to be used in bakery products, snack foods and high fibre, protein-enriched biscuits. Some of this research is outlined below:

- In West Germany a patent has been granted covering the use of BDG in the preparation of bakery products. Also the production of refined BDG has been introduced to increase the fibre content of various snack foods (Anon, 1985).
- Prentice and D'Appolonia (1977) have investigated the use of BDG in high-fibre bread. Consumer pannels will accept favourable bread made with wheat flour of which 10% has been replaced by suitably ground BDG.
- Prentice *et al.* (1978) and Kissell and Prentice (1979) have investigated protein and fibre enrichment of biscuit flour with DBG and determined that 15% incorporation, corresponding to a 27% increase in nitrogen, a fourfold increase in crude fibre, was the upper limit for baking sugar biscuits.
- Tsen, Weber and Eyestone (1983) evaluated distillers' dried grain flour as an ingredient of bread and concluded that breads supplemented with 10% therefore was superior to whole wheat bread in loaf volume, crumb grain, and colour. Distillers' dried grains are nearly identical to BDG as far the end product as spent grains is concerned.

The benefit of BDG for the developing market

There are several studies on BDG in West Africa and other developing countries. Nigeria alone produces 11.5 million hectolitres of beer per year. Also, large quantities of wheat are imported by Nigerian flour mills. Replacing a portion of this relatively expensive imported wheat with BDG is of great importance for the following reasons:

- It helps reducing import requirements needing foreign currency.
- One ton of BDG costs only $ 87 while wheat flour costs $ 722 per t.
- The result of using BDG is of a high nutritive value; biscuits made with 20% BDG contained much more of the following: protein 55%, lysine 90% and fibre 220% and is of great demand for developing market economies.
- Organoleptic tests using critical consumer pannels showed detrimental alteration in taste when 10-15% of the wheat flour was replaced by BDG for use in bread

and biscuits (Prentice and D'Appolonica, 1977; Prentice *et al.*, 1978; Kissell and Prentice, 1979; Anon, 1985).

REFERENCES

Abou Akkada, A.R., Khalil, A., Kosba, M.A. and Khalifah, M.M. (1975). Effect of feeding residues of tomato canning on the performance of laying hens. *Alexandria J. of Agric. Res.*, **23**, (1), 9-14.

Adams, H.W., Hickey, F.D. and Willard, M.J., Jr. (1960). Lye-pressure steam peeling of potatoes. *Food Technol.*, **14**, 1-3.

Ademosun, A.A. (1973). Evaluation of brewers' dried grains in the diets of growing chicks. *Br. Poultry Sci.*, **14**, 463-8.

Agostini, A. (1964). Utilization of by-products of vines and wines. *Wijnboer*, **32** (395), 13-6.

Agricultural Research Service (1956). Chemistry and technology of citrus, citrus products, and by-products. *United States Department of Agriculture*. Washington D.C. *Agriculture Handbook*, No. **98**, p. 99.

Al-Hiti, M.K. and Rous, J. (1978). Date waste without stones in broiler diets. *Br. Poultry Sci.*, **19**, 17-9.

Al-Yousef, Y.M. and Vandepopuliere, J.M. (1985). Whole dates, date meat and date pits as ingredient in chicken broiler diet. *Poultry Sci.*, **64** (suppl.1)

Ambrose, T.W. and Reiser, C.O. (1954). Wastes from potato starch plants. *Ind. Eng. Chem.*, **46**, 1331.

Amerine, M.A., Berg, H.W., Kunkee, R.E., Ough, C.S., Singleton, V.L. and Webb, A.D. (1980). *The Technology of Wine Making*, 4th ed. AVI publishing Company, Inc. Westport, Connecticut. p. 794.

Amerine, M.A. and Joslyn, M.A. (1970). *Table Wines: The Technology of their Production*, 2nd ed. University of California Press, Berkeley and Los Angeles.

Amerine, M.A. and Singleton, V.L. (1977). *Wine. An Introduction for Americans*, 2nd ed. University of California Press, Berkeley and Los Angeles.

Angalet, S.A., Fry, J.L., Damron, B.L. and Harms, R.H. (1976). Evaluation of waste activated sludge (citrus) as a poultry feed ingredient. 2. Quality and flavour of broilers, egg yolk colour and egg flavour. *Poultry Sci.*, **55**, 1219-25.

Anon, (1985). Brewers' spent grains versatility at work. *Milling*, **168**, (6), 23, 24, 29.

Anon, (1991). Sludge indicator, slip wijzer. *N.V.A. Slibcommissie. In Dutch*, p. 48.

Anwar, A. El Alaily, H.A. and Diab, M.F. (1978). Nutritive value of tomato seed meal as a plant protein supplement for growing chicks. *Arch. Geflügelk.*, **42**, 56-8.

Bath, D.L. (1981). Feed By-Products and their Utilization by Ruminants. In: *Upgrading residues and by-products for animals*. (Ed. J.T. Huber) CRC Press, Inc. Boca Raton, Florida.

Ben-Gera, I. and Kramer, A. (1969). The utilization of food industries wastes. In: *Advances in Food Research*, vol. 17. (Ed. C.O. Chichester, E.M. Mark, and G.F. Stewart) Academic Press, New York. pp. 78-152.

Berry, R.E. and Veldhuis, M.K. (1977). Processing of oranges, grapefruit and tangerines. In: *Citrus Science and technology*. (Ed. S. Nagy, P.E. Shaw, and M.K. Veldhuis) The AVI

Publishing Company, Inc. Westport, Connecticut. pp. 177-367.

Boa, K. (1990). *Non Conventional Feedstuffs in the Nutrition of Farm Animals*. Elsevier, Amsterdam, Oxford-New York-Tokyo, pp. 168-204.

Bough, W.A. (1973). Composition of waste load of unit effluents from a commercial leafy greens canning operation. *J. Milk Food Technol.*, **36**, 547-53.

Branckaert, R. (1967). Utilisation des drΠches de brasserie dessechées dans l'alimentation du poulet de chair en régions tropicales. *Rev. Elev. Med. Vet. Pays Trop.*, **20**, 595-600.

Brautlecht, C.A. (1953). *Starch, its Sources, Production and Uses*. Reinhold Publishing Co., New York.

Breederveld, H. (1992). Analysis of the brewers' sludge from wastewater treatment. (*Personal Communication*).

Broderick, H.M. (1977). *The Practical Brewer*. Madison, Wisconsin, Master Brewers' Association of America.

Bryan, W.L., Anderson, B.J. and Norman, G.L. (1974). Mechanical grading of oranges based on dynamic behavior. *Proc. Fla. State Hortic. Soc.*, **87**, pp. 313-18.

Burdick, E.M. and Maurer, R.H. (1950). Removal of naringin from solutions containing same. *U.S. Patent* No. **2**, 510, 797.

Buriel, J.P., Criollo, M.L. and Rivera, O.M. (1976). Note on citrus meal in diets for grilling chickens. Agronomia tropical. *Escuela de Zootecnia*, Venezuela, **26** (3), 261-8.

Burton, W.G. (1948). *The Potato*. Chapman and Hall, London.

Bushway, R.J. and Pennampalam, R. (1981). - Chaconine and - solanine content of potato products and their stability during several modes of cooking. *J.Agric.Food Chem.*, **29**, 814-7.

Campbell, C.H. (1937). *Canning, Preserving and Pickling*. Vance Publishing Co. Chicago. p. 222.

CBK (1991). Facts and figures from the brewers' central office. The Netherlands, *Centraal Brouwerij Kantoor*, p. 5.

Chan, H.T., Jr. (1983). *Handbook of Tropical Foods*. Dekker, New York, p. 639.

Cicogna, M. and Saibene, G. (1969). Possibility of reducing the cost of concentrate mixtures for table poultry by replacing maize with a mixture of hard wheat fine bran, carobs (pea family) without seeds, dried apple pomace and fat. *Riv. Zootec.*, **42**, 363-74.

Coleman, R.L. and Shaw, P.E. (1977). Amino acid composition of dried citrus sludge and its potential as poultry feedstuff. *J. Agric. Food Chem.*, **25**, 971-3.

Cook, A.H. (1962). *Barley and Malt Biology, Biochemistry, Technology*. Academic press, New York, and London, p. 740.

Cook, G.M. and Berg, H.W. (1969). Varietal table wine processing practices in California. I. Varieties, grape and juice handling and fermentation. *American Journal of Enology and Viticulture*, **20**, 1-6.

Cook, G.M. and Berg, H.W. (1973). Table wine processing practices in the San Joaquin Valley. *American Journal of Enology and Viticulture*, **24**, 153-8.

Crampton, E.W. and Harris, L.E. (1969). *Applied Animal Nutrition*. The use of feedstuffs in the formulation of livestock rations. 2nd Edition, W.H. Freeman & Co, San Francisco, USA, p. 753.

Cruerss, W.V. (1958). *Commercial Fruit and Vegetable Products*, 4th Edition. McGraw-Hill Book Co., New York, p. 884.

Culp, R.L. and Culp, G.L. (1971). *Advanced Wastewater Treatment*. Van Nostrand Reinhold Co., New York.

Damron, B.L., Eldred, A.R. and Harms, R.H. (1976). An improvement in interior egg quality by the feeding of brewers' dried grains. *Poultry Sci.*, **55**, 1365-6.

Della Monica, E.S., Huhtanen, C.N. and Strolle, E.O. (1975). Stability of protein water concentrates from potato starch factory waste effluent. *J. Sci. Food Agric.*, **26**, 617-23.

Dickey, H.C., Brugman, H.H., Plummer, B.E. and Highlands, M.E. (1965). The use of by-products from potato starch and potato processing. *Proceedings of the international symposium on utilization and disposal of potato wastes*, New Brunswick, Canada. pp. 106-21.

Din, M.G., Sunde, M.L. and Bird, H.R. (1979). Effects of feeding plant by-product diets on growth and egg production. *Poultry Sci.*, **58**, 1274-83.

D'Mello, J.P.F. and Whittemore, C.T. (1975). Nutritive value of cooked potato flakes for the young chick. *J. Sci. Food. Agric.*, **26**, 261-5.

Douglass, I.B. (1960). By-products and waste in potato processing. *Proc. 15th Industrial Waste Conf., Purdue*

Univ. Engineering Ext. Series, No., **106**, pp. 99-106.

Douglass, I.B. (1966). The manufacture of potato starch. *Proc. Intern. Sym. Util. and Disposal of Potato Wastes*. New Brunswick (Canada) Res. and Productivity Council, 1965, pp. 122-34.

Dowson, V.H.W. and Aten, A. (1962). Date handling, processing and packing. Food and Agricultural Organization of the United Nations Rome. *FAO Agricultural Development Paper*, No. **72**, pp. 6-223.

Duman, K. (1957). *The Most Valuable Fruit Varieties*. Verlag G. Fromme, Wien. Echenfelder, W.W. Jr. and Mulligan, T.J. (1966). Basic concepts of the biological oxidation of organic wastes. *Proc. Intern. Symp. Util. and Disposal of Potato Wastes*, New Brunswick (Canada) Res. and Productivity Council, 1965, pp. 212-34.

Edwards, P.W., Eskew, R.K., Hoersch, A., Jr., Aceto, N.C. and Redfield, C.S. (1952). Recovery of tomato processing wastes. *Food Technol.*, **6**, 383-6.

Eilbeck, W.J. and Mattock, G. (1987). *Chemical Processes in Wastewater Treatment*. Ellis Horwood limited, Publishers Chichester. England, p. 331.

El Alaily, H.A., Anwar, A. and El Banna, I. (1976). Mango seed kernels as an energy source for chicks. *Br. Poultry Sci.*, **17**, 129-33.

El-Ashwah, E.T. (1963). Effect of detergents on the growth and thermal resistance of *Bacillus thermoacedurans*. *Ph.D. Thesis. Ohio State Univ.*, Columbus.

El Boushy, A.R., Roodbeen, A.E. and Hopman, L.C.C. (1984). A preliminary study of the suitability of dehydrated poultry slaughterhouse wastewater as a constituent of broiler feed. *Agricultural Wastes*, **10**, 313-8.

Eldred, A.R., Damron, B.L. and Harms, R.H. (1975[a]). Evaluation of dried brewers' grains and yeast in laying hen diets containing various sulphur amino acid levels. *Poultry Sci.*, **54**, 856-60.

Eldred, A.R., Damron, B.L. and Harms, R.H. (1975[b]). Improvement of interior egg quality from feeding of brewers' dried grains. *Poultry Sci.*, **54**, 1337.

Eldred, A.R., Damron, B.L., Harms, R.H. (1976). Evaluation of waste activated sludge (citrus) as a poultry feed ingredient. 1. Performance of chicks, broilers and laying hens. *Nutr. Rep. Int.*, **14**, 139-45.

El Moghazy, M.El.S.A. and El Boushy, A.R. (1982[a]). The effect of different levels of dried citrus pulp in isocaloric, isonitrogenous methionine and lysine supplemented rations. 1. On the performance of broilers. *Research Bulletin, Faculty of Agriculture. Ain Shams University Cairo*, Egypt, No. **2035**, p. 16.

El Moghazy, M.El.S.A. and El Boushy, A.R. (1982[b]). Some neglected poultry feedstuffs from vegetable and fruit wastes. *World Poultry Sci. J.*, **38**, 18-27.

El Shazly, K., Ibrahim, E.A. and Karam, H.A. (1963). Nutritional value of date seeds for sheep. *J. Anim. Nutr.*, **22**, 894-7.

Esselen, W.B., Jr. and Fellers, C.R. (1939). The nutritive value of dried tomato pomace. *Poultry Sci.*, **18**, 45-7.

Ewing, W.R. (1963). *Poultry Nutrition*. The Ray Ewing Company publisher Pasadonea, California, p. 475.

Fangauf, Von, R., Vogt, H. and Penner, W. (1961). Die Bedeutung der Kartoffelflocken als Geflügelfutter Kartoffelflocken im Mastfutter. *Arch. Geflügelk.*, **25**, 365-72.

F.A.O. (1965). Report of the second FAO Technical Conference on the improvement of date production and processing. *FAO meeting report* No. PL/1965/**16**, p. 23.

F.A.O. (1988). Traditional food plants. *FAO Food and Nutrition Paper*, **42**, 359-61.

F.A.O. (1991). *Production Yearbook*. Vol. **44**. Food and Agricultural Organization of the United Nations. Rome. Italy.

Farrell, D.J., Rose, C.J. and Warren, B.E. (1983). A preliminary evaluation of some agricultural waste- and by-products for non-ruminants. *Feed information and animal production proceedings of the second symposium of the international network of feed information centers*, Univ. New England, Armidal,

Australia, pp. 411-15.

Finks, A.J. and Johns, C.O. (1921). VIII. The nutritive value of the proteins of tomato seed press cake. *Am. J. Physiol.*, **56**, 404-7.

Flora, L.F. and Lane, R.P. (1978). Processing trials with Muscadine grapes. *Ga. Agric. Exp. Stn. Res. Rep.*, **289**.

Garcia, M.E. and Gonzalez, A. (1984). Preliminary study on the use of tomato and pepper seed meals and excreta meal as pigments for egg yolk. *Rev. Avicult.*, **28** (3), 155-63.

Gerry, R.W. (1977). Dehydrated soluble potato solids in broiler rations. *Poultry Sci.*, **56**, 1947-51.

Göhl, B. (1975). *Tropical Feeds*. Food and Agricultural Organization of the United Nations. Rome, Italy, p. 614.

Göhl, B. (1981). *Tropical Feeds*. Feed information summaries and nutritive value. Food and Agricultural Organization of the United Nations. Rome, Italy, p. 529.

Goose, P.G. and Binsted, R.C. (1973). *Tomato paste and other tomato products*. Food Trade Press Ltd, London, p. 266.

Gould, W.A. (1965). Effect of processing factors on the quality of fruits and vegetables. In: *Food Quality: Effects of Production Practices and Processing*. Adv. Sci., Washington, DC. pub. **77**, 57.

Gould, W.A. (1983). *Tomato Production Processing and Quality Evaluation*. AVI Publishing Company, Inc. Westport, Connecticut. pp. 112-32.

Gould, W.A., Geisman, J.R. and Sleesman, J.R. (1959). A study of some of the physical and chemical factors affecting the efficiency of washing tomatoes. *Ohio Agric. Exp. Stn. Res. Bull.*, **825**.

Graham, R.P., Huxsol, C.C., Hart, M.R. and Weaver, M.L. (1970). Process for peeling potatoes. *U.S. Pat.*, **3**, 547, 173.

Graham, R.P., Huxsoll, C.C., Hart, M.R., Weaver, M.L. and Morgan, A.I., Jr. (1969). "Dry" caustic peeling of potatoes. *Food Technol.*, **23**, 61.

Grames, L.M. and Kueneman, R.W. (1969). Primary treatment of potato processing wastes with by-product feed recovery. *J. Water Pollution Control Federation*, **41** (7), 1358-67.

Gualtieri, M. and Rapaccini, S. (1990). Date stones in broiler's feeding. *Tropicultura*, **8**, (4), 165-8.

Guymon, J.F. and Crowell, E.A. (1977). The nature and cause of cap-liquid temperature differences during wine fermentation. *American Journal of Enology and Viticulture*, **28**, 74-8.

Hackler, L.R., Newmann, A.L. and Johnson, B.C. (1957). Feed from sewage. III. Dried activated sewage sludge as a nitrogen source for sheep. *J. Anim.Sci.*, **16**, 125-9.

Hadorn, H. and Högl, O. (1945). Statistics for the stone fruit juices of 1944. *Mitt. Lebensmittelunters. u. Hyg.*, **36**, 216-31.

Hakansson, J. and Lindgren, E. (1974). The ability of laying hens to digest raw potato starch. *Swedish J. Agric. Res.*, **4**, 191-4.

Halnan, E.T. (1944). Digestibility trials with poultry. XI. The digestibility and metabolizable energy of raw and cooked potatoes, potato flake, dried potato shreds. *J. Agric. Sci. Camb.*, **34**, 139-54.

Harrington, W.O., Olson, R.L. and Nutting, M.D. (1960). Effects of glycerol monostearate on reconstituted potato granules. *Am. Potato J.*, **37**, 160-5.

Harrington, W.O., Olson, R.L., Weston, W.J. and Belote, M.L. (1959). Effects of process variables in the potato granule production. *Am. Potato J.*, **36**, 241-54.

Harris, G. (1962). The enzyme content and enzymic transformation of malt. In: *Barley and Malt, Biology, Biochemistry, Technology*, (Ed. A.H. Cook) Academic press. New York London. pp. 583-694.

Hendrickson, R. and Kesterson, J.W. (1951). Citrus by-products of Florida. *Fla. Agr. Expt. Sta. Bul.*, **487**, p. 56.

Hooge, D.M. and Rowland, L.O., Jr. (1978). Effect of dietary sand on feed conversion of broilers and laying hens. *Poultry Sci.*, **57**, 1145 (abstr.)

Hopper, T.H. (1958). Amino acid composition of foodstuffs. In: *Processed Plant Protein Foodstuffs*, (Ed. A.M. Altschul) Academic Press Inc., Publishers New York. Chapter 33, pp. 877-91.

Hough, J.S. (1985). *The Biotechnology of Malting and Brewing.* Cambridge University Press, Cambridge, London, New York, p. 168.

Hough, J.S., Briggs, D.E., Stevens, R. and Young, T.W. (1982). *Malting and Brewing Science* (2 Volumes). Chapman & Hall, London.

Howerton, W.W. and Treadway, R.H. (1948). Manufacture of white potato starch. *Ind. Eng. Chem.*, **40**, 1402-7.

Hubbell, C.R. (1981). Feedstuffs analysis table. *Feedstuffs*, **53**, (3), 4-45.

Huber, J.T. (1981). *Upgrading residues and by-products for animals.* CRC Press Inc., Boca Raton, Florida, p. 131.

Hulan, H.W., Proudfoot, F.G. and Zarkadas, C.G. (1982[a]). Potato waste meal. I. Compositional analyses. *Can. J. Anim. Sci.*, **62**, 1161-9.

Hulan, H.W., Proudfoot, F.G. and Zarkadas, C.G. (1982[b]). Potatao waste meal. II. The nutritive value and quality for broiler chicken. *Can. J. Anim. Sci.*, **62**, 1171-80.

Jagtiani, J. (1985). Report to the Kingdom of Tonga on the utilization and processing of Guava. *Special Project of the FAO/U.N.*

Jagtiani, J., Chan, H.T., Jr. and Saki, W.S. (1988). *Tropical Fruit Processing.* Academic press, Inc., Harcourt Brace Javanovich Publishers, New York, London, p. 184.

Jeffs, J. (1971). *The Wines of Europe.* Taplinger publishing Co., New York.

Jensen, L.S., Chang, C.H. and Maurice, D.V. (1976). Improvement in interior egg quality and reduction in liver fat in hens fed brewers' dried grains. *Poultry Sci.*, **55**, 1841-7.

Johns, C.O. and Gersdorff, C.E.F. (1922). The proteins of the tomato seed, *Solanium Esculentum. Journ. of Biol. Chem.*, **51**, 439-52.

Jones, C.R. (1940). The production of mechanically damaged starch in milling as a governing factor in the diastatic activity of flour. *Cereal Chem.*, **17**, 133-69.

Jones, J.H. (1938). The metabolism of Calcium and Phosphorus as influenced by the addition to the diet of salts of metals which form insoluble phosphates. *Am. J. Physiol.*, **124**, 230-7.

Jumah, H.F., Al Azzawi, I.I. and Al Hashini, S.A. (1973). Some nutritional aspects of feeding ground date pits for broilers. *Mesopotamia J. Agric.*, **8**, (2), 139-46.

Kalaisakis, D., Papadopoulos, G., Boufidis, B., Zacharioudakis, S and Gouraros, A. (1970). The use of tomato meal in broilers. *Poultry Sci. Rev.*, **1**, 1-9.

Kamel, B.S., Diab, M.F., Ilian, M.A. and Salman, A.J. (1979). Nutritional value of whole dates and date pits in the chick. *Poultry Sci.*, Vol. **58**, 1071-2 (abstract).

Kamel, B.S., Diab, M.F., Ilian, M.A. and Salman, A.J. (1981). Nutritional value of whole dates and date pits in broiler rations. *Poultry Sci.*, **60**, 1005-11.

Karunajeewa, H. (1978). Effect of rapeseed and dried citrus pulp meals on egg yolk colour and performance of crossbred hens. *J. Austr. Inst. of Agric. Sci.*, **44**, 208-9.

Kelley, E.G. (1958). Plant residues and pomaces. In: *Processed Plant Protein Foodstuffs*, (Ed. A.M. Altschul) Academic Press Inc., Publishers. New York, pp. 859-75.

Kienholz, E.W., Darras, C.A. and Caveny, D.D. (1972). Small egg size from brewers' dried grains. *Poultry Sci.*, **51**, 1825.

Kienholz, E.W. and Jones, M.L. (1967). The effect of brewers' dried grains upon reproductive performance of chicken and turkey hens. *Poultry Sci.*, **46**, 1280.

Kienholz, E.W., Thornton, P.A. and Moreng, R.E. (1963). The use of brewers' dried grains in some poultry rations. *Poultry Sci.*, **42** (2), 1280.

Kinsella, J.E. (1974). Grape seed oil: A rich source of linolic acid. *Food Technol.*, **28**, 58-60.

Kissell, L.T. and Prentice, N. (1979). Protein and fibre enrichment of cookie flour with brewers' spent grain. *Cereal Chem.*, **56**, (4), 261-6.

Klimenko, Yv.A. and Kaganskii, R.M. (1969). Utilization of waste-products from the manufacture of fruit preserves and from wine making plants in the USSR. *Maslo-Zhiro-Vaya Promystrlennost*, **35**, (7), 5-7.

Knight, R., Jr. (1980). Origin and world importance of tropical and subtropical fruit crops. In: *Tropical and Subtropical Fruits*, (Ed. S.J. Nagy, P.E. Shaw) AVI Publishing, Inc. Westport Connecticut, pp. 1-120.

Kröner, W. and Völksen, W. (1950). *The Potato*. Johann Ambrosius Barth, Leipzig.

Lakshminarayana, S. (1980). Mango. In: *Tropical and Subtropical Fruits*. (Ed. S. Nagy and P.E. Shaw) AVI Publ. Co., Westport Connecticut, pp. 184-257.

Landine, R.C. and Dean, J.R. (1973). Waste treatment and solids recovery system at a potato processing plant. *Industrial Wastes*, **19**, (1), 1-12.

Lichine, A. (1979). *Alexis Lichine's Guide to the Wines and Vineyards of France*. Knoph, New York.

López, J.D. and Carmona, J.F. (1981). Evaluation of brewers' dried grains in the diets of broiler chickens. *Animal Feed Science and Technology*, **6**, 179-88.

López, J.D., Carmona, J.F. and Pascual, J.L.M. (1981[a]). Evaluation of brewers' dried grains in the diets of laying hens. *Anim. Feed Sci. Technol.*, **6**, 169-78.

Lucas, L.L. (1967). Evaluation of lye peeling of tomatoes using high lye concentration and short time of exposure with wetting agents. Spec. Group Studies. *Rep. Dep. Hortic.*, Ohio State Univ., Columbus.

Lynch, S.J. and Mustard, M.J. (1955). *Mangoes in Florida*. University of Miami, Gainesville.

Maga, J.A. (1980). Potato glycoalkaloids. *Crit. Rev. Food Sci. Nutr.*, **12**, 371-405.

Maymone, B. and Carusi, A. (1945). Nutritive value of tomato seed cake for fattening pigs. *Inn. Ist. Sper. Zootecn. (Roma)*, **3**, 387-404.

McKeen, W.D. (1984). Feeding grape pomace to leghorn hens as an alternative to starvation to induce a malt. *Poultry Sci.*, **63** (suppl.1), 148-9.

Meister, E. and Thompson, N.R. (1976). Protein quality of precipitate from waste effluent of potato chip processing measured by biological methods. *J. Agric. Food Chem.*, **24** (5), 924-6.

Metcalf & Eddy, Inc. (1979). *Wastewater Engineering: Treatment Disposal Reuse*. McGraw-Hill Book Company. New York, p. 920.

Montedoro, G. and Bertuccidi, M. (1976). Essai de vinification en rouge avec l'emploi de différentes préparations enzymatiques. *Lebensm. Wiss. Technol.*, **9**, 225-31.

Moon, N.J. (1980). Maximizing efficiencies in the food system: A review of alternatives for waste abatement. *J. Food Protect.*, **42**, 231-8.

Moreng, R.E., Pfaff, W.K. and Kienholz, E.W. (1989). Feeding Chinese ringneck pheasant for efficient reproduction. In: *Recent Advances in Animal Nutrition in Australia*. (Ed. D.J. Farrell) pp. 305-10.

Morrison, F.B. (1961). *Feeds and Feeding*, abridged. The Morrison Publishing Company, Orangeville, Ontario, Canada. P.O. Box 130, p. 696.

Moyer, J.C. and Aitken, H.C. (1980). Apple juice. In: *Fruit and Vegetable Juice Processing Technology*, (Ed. P.E. Nelson and D.K. Tressler) AVI Publishing Company, Westport Connecticut, pp. 212-67.

Muller, G.J. (1941). Potato starch technology modernized. *Chem. Met. Eng.*, **48** (3), 78-81.

Naber, E.C. and Touchburn, S.P. (1969). Effect of hydration, gelatinization and hall milling of starch on growth and energy utilization by the chick. *Poultry Sci.*, **48**, 1583-9.

Nagy, S., Shaw, P.E. and Veldhuis, M.K. (1977). *Citrus Science and Technology*. Volume 2. Fruit production, processing practices, derived products and personnel management. The AVI Publishing Company, Inc., Westport Connecticut, p. 667.

Neal, W.M., Becker, R.B. and Arnold, P.T.D. (1935). *Florida Agr. Expt. Sta. Bull.*, **275**. In: *Plant Residues and Pomaces*, (Ed. E.G. Kelly) In: *Processed Plant Protein Foodstuffs*, (Ed. A.M. Altschul) Academic Press Inc., Publishers New York.

Nitsan, Z. and Bartov, I. (1972). Comparison between the nutritional value of potato and corn starch for chicks. *Poultry Sci.*, **51**, 836-40.

Norris, E., Norris, C. and Steen, J.B. (1975). Regulation and grinding ability of grit in the gizzard of Norwegian Willow Patramigan (*Lagopus lagopus*). *Poultry Sci.*, **54**, 1839-43.

North, M.O. (1978). Force molting. In: *Commercial Chicken Production Manual*. AVI Publishing Co. Inc, Westport Connecticut, pp. 692.

NRC (1984). *Nutrition Requirements of Poultry*. National Research Council. National Academy Press. Washington. D.C., pp. 33-4.

Olson, R.L. and Harrington, W.O. (1951). Dehydrated mashed potatoes - a review. *U.S. Dept. Agr., Agr. Res. Admin.*, AIC-**297**.

Oluyemi, J.A., Arafa, A.S. and Harms, R.H. (1978). Influence of sand and grit on the performance of turkey pullets fed on diets containing two concentrations of protein. *Br. Poultry Sci.*, **19**, 169-72.

Onwudike, O.C. (1981). The use of brewers' dried grains by laying hens. *Nutr. Rep. Int.*, **24**, (5), 1009-16.

Onwudike, D.C. (1986). The effects of dietary sand on the usage of diets containing brewers' dried grains by growing chicks. *Poultry Sci.*, **65**, 1129-36.

Otani, I., Yamatani, Y. and Sasaki, M. (1968). Fundamental studies on the digestion of the domestic fowl. 4. Effects of grit on the digestibility of feed. *J. Fac. Fish. Anim. Husb. Hiroshima Univ.* Japan, **7**, 281-9.

Ough, C.S. and Amerine, M.A. (1960). Experiments with controlled fermentation, IV. *American Journal of Enology and Viticulture*, **11**, 5-14.

Ough, C.S. and Amerine, M.A. (1961). Studies on controlled fermentation. V. Effects on colour, composition, and quality of red wines. *American Journal of Enology and Viticulture*, **12**, 9-19.

Ough, C.S. and Amerine, M.A. (1962). Studies with controlled fermentation. VII. Effect of anti-fermentation blending of red must and white juice on colour, tannins, and quality of Cabernet Sauvignon Wine. *American Journal of Enology and Viticulture*, **13**, 181-8.

Ough, C.S. and Berg, H.W. (1974). The effect of two commercial pectic enzymes on grape musts and wines. *American Journal of Enology and Viticulture*, **25**, 208-11.

Pader, M. (1962). Process for preparing potato flakes. *U.S. Pat.*, **3**, 067, 042. Dec.4.

Pader, M. (1964). Process for making dehydrated potatoes. *U.S. Pat.*, **3**, 163, 546. Dec.29.

Pailthorp, R.E. (1965). Potato waste treatment. *Oregon State Univ. Eng. Expt. Sta. Cir.* **29**, 101-9.

Pailthorp , R.E. and Filbert, J.W. (1966). Potato waste treatment in Idaho. Pilot unit study. *Proc. Intern. Symp. Util. and Disposal of Potato Wastes*, New Brunswick (Canada) Res. and Productivity Council, 1965, pp. 285-94.

Pailthorp, R.E., Filbert, J.W. and Richter, G.A. (1987). Treatment and disposal of potato wastes. In: *Potato Processing*, (Ed. W.F. Talburt and O. Smith) AVI Van Nostrand Reinhold Company, New York, pp. 747-88.

Pantastico, E.B., Mattoo, A.K., Murata, T. and Ogata, K. (1975). In: *Postharvest Physiology, Handling and Utilization of Tropical and Subtropical Fruits and Vegetables*. AVI Publ. Co., Westport Connecticut, pp. 339-62.

Patil, S.N., Netke, S.P. and Dabadghao, A.K. (1982). Processing and feeding value of mango seed kernel for starting chicks. *Br. Poultry Sci.*, **23**, 185-94.

Pendlington, S., Williams, T.P. and Lester, R. (1967). Dehydrated potato foodstuffs. *British Pat.*, **1**, 091, 159. Nov.15.

Petrenko, V.D. and Banina, N.N. (1984). Use of tomato wastes in diets for laying hens. *Nauchno- Usledovatel' Skogo Instituta Ptitsevodstava*, no. **16**, 13-6.

Pollock, J.R.A. (1962). The nature of malting process. In: *Barley and Malt Biology, Biochemistry, Technology*, (Ed. A.H. Cook) Academic Press, New York and London, pp. 303-98.

Pollock, J.R.A. (1979). *Brewing Science* (Volume 3). London, New York, Academic Press. Prentice, N. and D'Appolonia, B.L. (1977). High-fibre bread containing brewers' spent grain. *Cereal Chem.*, **54**, 1084.

Prentice, N., Kissell, L.T., Lindsay, R.C. and Yamazaki, W.T. (1978). High-fibre cookies containing brewers' spent grain. *Cereal Chem.*, **55** (5), 712-21.

Ratcliff, M.W. (1977). Citrus processing waste prevention handling and treatment. In: *Citrus Science and Technology*, (Ed. S. Nagy, P.E. Shaw, M.K. Veldhuis) The AVI Publishing Company, Inc., Westport Connecticut, pp. 546-57.

Ravindran, V. and Blair, R. (1991). Feed resources for poultry production in Asia and Pacific region. I. Energy sources. *World Poultry Sci. J.*, **47**, 213-31.

Rebeck, H.M. and Cook, R.W. (1977). Manufacture of citrus pulp and molasses. In: *Citrus Science and Technology*, (Ed. S. Nagy, P.E. Shaw and M.K. Veldhuis) The AVI Publishing Company, Inc. Westport Connecticut, pp. 368-81.

Reddy, C.V. (1975). Utilization of by-products in poultry feeding in India. *World Rev. of Anim. Prod.*, **11**, (2), 66-72.

Ribéreau-Gayon, P. and Milhé, J.C. (1970) Recherches technologiques sΩr les composés phénoliques des vins rouges. I. Influence des différentes parties de la grappe. *Connaiss. Vigne Vin*, **4**, 63-74.

Rice, A.C. (1976). Solid waste generation and by-product recovery potential from winery residues. *American Journal of Enology and Viticulture*, **27**, 21-6.

Rivoche, E.J. (1948). Improvements in and relating to the drying of vegetables. *Brit. Pat.*, **601**, 151.

Rivoche, E.J. (1950). Drying of starchy foodstuff. *U.S. Pat.*, **2**, 520, 891.

Roy, N. and Mitra, A.K. (1970). Amylose content in starches from some unfamiliar sources of food. *J. Food Sci. Technol.*, **7**, 164.

Rygg, G.L. (1975). Date developing, handling, and packing in the United States. Agriculture Research Service. United States Department of Agriculture. Washington D.C. *Agriculture Handbook*, No. **482**, p. 56.

Ryder, R.A. (1973). Winery wastewater treatment and reclamation. *Eng. Ext. Serv., Purdu Univ.*, **142**, 564-87.

Salunkhe, D.K. and Desai, B.B. (1984). *Postharvest Biotechnology of Fruits*. Volume I (pp. 168) and II (pp. 146). CRC Press, Boca Raton, Florida.

Samson, J.A. (1986). The minor tropical fruits. In: *Tropical Fruits*. Longman Scientific and Technical, pp. 270-5.

Sandstedt, R.M. and Gates, R.L. (1954). Raw starch digestion: a comparison of the raw starch digestion capabilities of the amylase systems from four alpha-amylase sources. *Food Res.*, **19**, 190-9.

Sawaya, W.N., Khalil, J.K. and Safi, W.J. (1984). Chemical composition and nutritional quality of date seeds. *J. Food Sci.*, **49**, 617-19.

Schaible, P.J. (1970). *Poultry: Feeds and Nutrition*. The AVI Publishing Company, Inc., Westport Connecticut, p. 636.

Schoonmaker, F. (1973). *Encyclopaedia of Wine*, 5th Edition. Hastings House, New York.

Schulte, W.A. (1965). Efficiency of chemical and physical tomato peeling systems and their effects on canned product quality. *Ph.D. Thesis. Ohio State Univ.*, Colombus.

Schultz, W.G., Graham, R.P., and Hart, M.R. (1976). Pulp recovery from tomato peel residue. In: Proceedings of the sixth national symposium of food processing wastes. Industrial environmental research laboratory, office of research and development, *U.S. Environmental Protection Agency*, Cincinnati, Ohio 45268. *Report No.* EPA-600/2-76-224, pp. 105-17.

Schultz, W.G., Graham, R.P., Rockwell, W.C., Bomben, J.L., Miers, J.C. and Wagner, J.R. (1971).Field protection of tomatoes: 1. Process and design. *J. Food Sci.*, **36**, 397-9. Schuster, K. (1962). Malting technology. In: *Barley and Malt, Biology, Biochemistry, Technology*. (Ed. A.H. Cook) Academic Press, New York and London, pp. 271-302.

Scott, M.L., Nesheim, M.C. and Young, R.Y. (1982). *Nutrition of the Chicken.* Scott, M.L. and Associates, Ithaca, New York, p. 562.

Shaw, R.L. and Shuey, W.C. (1972). Production of potato starch with low waste. *Am. Potato J.*, **49**, 12-22.

Shearon, W.H., Jr. and Burdick, E.M. (1951). Citrus fruit processing. *Ind. Eng. Chem.*, **40**, 370-8.

Sibbald, I.R. (1977). The true metabolizable energy value of some feedingstuffs. *Poultry Sci.*, **56**, 380-2.

Singleton, V.L. and Draper, D.E. (1964). The transfer of polyphenolic compounds from grape seeds into wines. *American Journal of Enology and Viticulture*, **15**, 34-40.

Slater, L.E. (1951). Quality peeling sets a new phase. *Food Eng.*, **23**, (12), 106-7.

Smith, O. (1987). Potato chips. In: *Potato Processing*. (Ed. W.F. Talburt and O. Smith). AVI - Van Nostrand Reinhold Company, New York, pp. 371-490. Smith, T.J. and Huxsoll, C.C. (1987). Peeling potatoes for processing. In: *Potato Processing*. (Ed. W.F. Talburt and O. Smith) AVI - Van Nostrand Reinhold Company, New York, pp. 333-69.

Smock, R.M. and Neubert, A.M. (1950). *Apples and Apple Products*. Interscience Publishers Inc., New York, London, p. 471.

Splittgerber, Von. H., and Gysae, M. (1962). Prüfung von Kartoffelpreßschrot als Komponente in Geflügelmast - Alleinfutter. *Arch. Geflügelk.*, **29**, 111-5.

Srivastava, H.C. (1967). In: *The Mango, A Handbook.* (Ed. C.G. Raghava, U. Narasinga Rao, P. Kachroo and S.N. Tata) Indian Council of Agricultural Research, New Delhi, India. pp. 99-149.

Stafford, A.L. (1983). In: *Handbook of Tropical Feeds*. (Ed. H.T. Chan Jr.), Dekker, New York. pp. 399-431.

Storer, N.L. and Nelson, T.S. (1968). The effect of various aluminium compounds on chick performance. *Poultry Sci.*, **47**, (1), 244-7.

Strolle, E.O., Cording, J., Jr. and Aceto, N.C. (1973). Recovering potato proteins coagulated by steam injection heating. *J. Agr. Food Chem.*, **21**, 974-7.

Strolle, E.O., Aceto, N.C., Gerry, R.W. and Cording, J., Jr. (1975). Utilization of soluble solids from potato starch factory waste effluents. *IFT, Chicago, Illinois*, June 8-12.

Subrahmanyam, V.V.R. and Achaya, K.T. (1957). Lesse-known Indian vegetable fats. 1. Oleic rich fats. *J. Sci. Food Agric.*, **8**, 657-62.

Sullivan, T.W., Kuhl, H.J., Jr. and Holder, D.P. (1978). Evaluation of brewer' dried grains and yeast in turkey diets. *Poultry Sci.*, **57**, 1329-36.

Talburt, W.F. (1987). History of potato processing. In: *Potato Processing*. (Ed. W.F. Talburt, and O. Smith) AVI - Van Nostrand Reinhold Company, New York, pp. 1-9.

Talburt, W.F., Boyle, F.P. and Hendel, C.E. (1987). Dehydrated mashed potatoes - Potato granules. In: *Potato Processing*. (Ed. W.F. Talburt and O. Smith) AVI - Van Nostrand Reinhold Company, New York, pp. 535-56.

Talburt, W.F. and Kueneman, R.W. (1987). Dehydrated diced potatoes. In: *Potato Processing*. (Ed. W.F. Talburt and O. Smith) AVI - Van Nostrand Reinhold Company, New York. pp. 613-46.

Talburt, W.F., Schwimmer, S. and Burr, H.K. (1987). Structure and chemical composition of the potato tuber. In: *Potato Processing*. (Ed. W.F. Talburt and O. Smith) AVI - Van Nostrand Reinhold Company,

New York, pp. 11-46.

Talburt, W.F., Weaver, M.L., Reeve, R.M. and Kueneman, R.W. (1987). Frozen French fries and other frozen products. In: *Potato Processing*. (Ed. W.F. Talburt and O. Smith) AVI - Van Nostrand Reinhold Company, New York, pp. 491-534.

Tamimie, H.S. (1958). The effect of feeding dates and date pits to chicks. *World Poultry Sci. J.*, **14**, 207-10.

Tamimie, H.S. (1959). Feeding graded levels of dates and date pits to chicks. *World Poultry Sci. J.*, **15**, 231-4.

Thornton, P.A. and McPherron, T.A. (1962). Controlling body weight and liver lipid accumulation in the chicken with dietary brewers' dried grains. *Fed. Proc.*, **21**, 397.

Titus, H.W. and Fritz, J.C. (1971). *The Scientific Feeding of Chickens*. The Interstate, Printers and Publishers, Inc. Danville, Illinois, p. 336.

Tofflemire, J.J. (1972). Survey of methods of treating wine and grape wastewater. *American Journal of Enology and Viticulture*, **23**, 165-72.

Tomczynski, R. (1978) Tomato seeds and skins for feeding of laying hens. *Zeszyty Naukowe Akademii Rolniczo - Technicznej W Olsztynic*, No. **189**, 153-64.

Treadway, R.H. (1987). Potato starch. In: *Potato Processing*. (Ed. W.F. Talburt and O. Smith) AVI - Van Nostrand Reinhold Company, New York, pp. 647-64.

Tressler, D.K. and Joslyn, M.A. (1961). *Fruit and Vegetable Juice Processing Technology*. AVI Publishing Company, Inc., Westport Connecticut, p. 1028.

Tsen, C.C., Weber, J.L. and Eyestone, W. (1983). Evaluation of distillers' dried grain flour as a bread ingredient. *Cereal Chem.*, **60**, (4), 295-97.

Veldhuis, M.K. and Rushing, N.B. (1954). Pasteurization and storage of sweetened and unsweetened lime juice. *Food Technol.*, **8**, 136-8.

Veldhuis, M.K., Berry, R.E., Wagner, C.J., Jr., Lund, E.D. and Bryan, W.L. (1972). Oil and water-soluble aromatic distilled from citrus fruit and processing waste. *J. Food Sci.*, **37**, 108-12.

Velloso, L. (1985). Use of citrus pulp in animal feeding. *Comunicacoes Cientificas da Faculdade de Medicina Veterinaria e Zootecnia da Universidade de Sao Paulo*, **9**, (2), 163-80.

Voetberg, J.W. (1986). Anaerobic wastewater purification in the potato processing industry. *Proceedings of the 19th annual conference of the research group of processors of potato food products.* Inst. for Storage and Processing of Agric. Produce. Wageningen. The Netherlands, pp. 81-91.

Vogt, Von. H. (1969). Kartoffelschrot im Legehennenfutter. *Arch. Geflügelk.*, **33**, 439-43.

Vogt, Von. H. and Stute, K. (1969a). Kartoffeltrockenschrot im Geflügelmastfutter. *Arch. Geflügelk.*, **33**, 323-31.

Vogt, Von. H. and Stute, K. (1969b). Kartoffeltrockenschrot im Geflügelmastfutter. *Arch. Geflügelk.*, **33**, 316-20.

Von Loesecke, H.W. (1952). Citrus fruits industry. *Industrial and Engineering Chemistry*, **44**, (3), 476-82.

Wagner, J.R. (1970). Field processing of tomatoes. *Proc. 1st National Symposium on Food Processing Wastes*. Portland. Ore. Federal Water Quality Admin., U.S. Dept. of Interior, Washington, D.C., pp. 350-4.

Ward, J.B. (1977). Brewers' dried grains in the diet of broiler breeder hens. *Poultry Sci.*, **56**, 1768.

Welkins, E.G. (1943). Mango kernels as food. *Indian Farm.*, **3**, 636-7.

White, J.W., Jr. (1973). Processing fruit and vegetable wastes. In: *Symposium Processing Agricultural and Municipal Wastes*. (Ed. G.E. Inglett) The AVI Publishing Company, Inc., Westport Connecticut, pp. 129-42.

White, W.B., Sunde, M.L., Bird, H.R., Burger, W.C. and Prentice, N. (1979). Brewers' dried grains in laying mashes. *Feedstuffs*, **51**, (8), 27-8.

Whittemore, C.T. (1977). The potato (*Solanium Tuberosum*) as a source of nutrients for pigs, calves and fowl - A review. *Anim. Feed Sci. Technol.*, **2**, 171-90.

Whittemore, C.T., Moffat, I.W. and Mitchell-Manson, J. (1974). Performance of broilers fed on diets containing cooked potato flake. *Br. Poultry Sci.*, **15**, 225-30.

Whittemore, C.T., Moffat, I.W. and Taylor, A.G. (1975). The effect of dietary cooked potato flake on performance of broilers and on litter quality. *Br. Poultry Sci.*, **16**, 115-20.

Willard, M.J. (1971). Systems used for waste disposal by potato processing in the United States. *Mitteilungen of the Inst. for Storage and Processing of Agric. Produce.* Wageningen, The Netherlands, **6** (12A), 1-9.

Willard, M.J., Cording, J. Jr., Eskew, R.K., Edwards, P.W. and Sullivan, J.F. (1956). Potato flakes: A new form of dehydrated mashed potatoes. Review of plant process. *Am. Potato J.*, **33**, 28-31.

Willard, M.J. and Hix, V.M. (1987). Potato flour. In: *Potato Processing*. (Ed. W.F. Talburt and O. Smith) AVI - Van Nostrand Reinhold Company, New York, pp. 665-82.

Willard, M.J., Hix, V.M. and Kluge, G. (1987). Dehydrated mashed potatoes - Potato flakes. In: *Potato processing*. (Ed. W.F. Talburt and O. Smith) AVI - Van Nostrand Reinhold Company, New York, pp. 557-612.

Willard, M.J., Pailthorp, R.E. and Smith, O. (1967). Waste disposal. In: *Potato Processing*. (Ed. W.F. Talburt and O. Smith) AVI Publishing Co., Westport Connecticut, pp. 551-79.

Wilson, C.W. III (1980). Guava. In: *Tropical and Subtropical Fruits*. (Ed. S. Nagy and P.E. Shaw) AVI Publishing Inc., Westport Connecticut, pp. 279-99.

Winton, A.L. and Winton, K.B. (1949). *The Structure and Composition of Food*. Volume II. Vegetables, legumes and fruits. John Wiley and Sons, Inc., New York. London: Chapman and Hall, Limited, p. 857.

Winton, A.L. and Winton, K.B. (1950). *The Structure and Composition of Food*. Vol. I. Cereals, starch, oil seeds, nuts, oils, forage plants. John Wiley and Sons, Inc., New York. London: Chapman and Hall, Limited, p. 710.

Woodman, H.E. (1945). (Ed. B. Göhl), *Bull. Minist. Agric. Fish Fd. London*, No. **124**. (Cited in Tropical Feeds. FAO of the United Nations Rome, 1975).

Woodroof, J.G. (1975). Plant sanitation and waste disposal. In: *Commercial Vegetable Processing*. (Ed. B.S. Luh and J.G. Woodroof) The AVI publishing Company, Inc., Westport Connecticut, pp. 603-38.

Woodroof, J.G. and Luh, B.S. (1986). *Commercial Fruit Processing*. AVI Publishing Company, Inc., Westport Connecticut, pp. 613-46.

Wright, R.C. and Whiteman, T.M. (1949). Tests show which potatoes have least peeling loss. *Food Ind.*, **21**, 69.

Yang, S.J. and Choung, C.C. (1985). Studies on the utilization of citrus by-products as livestock feeds. 4. Feeding value of dried citrus by-products fed to layers. *Korean J. Anim. Sci.*, **27**, (4), 239-45.

Yannakopoulos, A.L., Tserveni-Gousi, A.S. and Christaki, E.V. (1992). Effect of locally produced tomato meal on the performance and the egg quality of laying hens. *Anim. Feed Sci. Technol.*, **36**, 53-7.

Young, T.W. and Sauls, J.W. (1980). The mango industry in Florida. *Fla. Coop. Ext. Serv., Inst. Food Agric. Serv.*, University of Florida, Gainesville.

7. MUNICIPAL REFUSE

Contents

7.1 INTRODUCTION

Municipal refuse is a raw material, which varies in chemical composition, moisture and microbial contents. Other relevant points concerning the origin of municipal refuse are:

- Standards of living and lifestyle.
- Degree of industrialization.
- Size of the region, town, city or country.
- Season of the year and from one year to another.

Throughout the world, developed or developing, millions of tons of municipal wastes are produced; these are often collected and then disposed off. The potential value of this waste resource is largely ignored. It may be regarded as a valuable product for recycling, composting and also as an acceptable feedstuff for livestock and poultry.

In the developed market the separation of municipal refuse components was introduced more than 10 years ago. The separation of those components was stimulated and guided by the municipalities of many towns or cities. Often, this development is stimulated by an awareness of the need for recycling. The refuse is classified into the following major categories:

Household waste

In the ideal situation the consumer are encouraged to sort out their waste into eight categories:

Newspapers, magazines, journals and other written paper. This waste is collected and delivered for recycling and processing as carton, cardboard, packing paper *etc.*

Carton. Waste carton is gathered and delivered for recycling.

Glass bottles. Glass bottles and other glass receptacles, are selected for disposal according their colour, clear, brown or green. They are disposed off in separate containers per colour in convenient places near supermarkets and shopping centers. This glass can be recycled and used for bottles.

Metal. Cans and tins *etc.* are disposed off in special containers which are conveniently placed, where they may be disposed off for recycling as a metal for further processing purposes.

Plastics. Plastics such as bottles, packing material *etc.* are disposed off in special containers in convenient places where they may be collected for recycling and producing plastic bottles, containers, toys *etc.*

Fruit, vegetable and food waste and garden waste. This waste is gathered in a special plastic container with a cover which is easy to move with wheels. These containers are emptied periodically from individual houses by an automatic truck with hydraulic lift. For blocks of flats there are large containers for the whole building. This waste product is mostly used as a compost or a fertilizer for gardens and for enriching the soil structure.

Other household waste. This category is mostly in the form of paper products, wood, leather, rubber, cloth, carpets, simple plastics, dust and dirt *etc.* and it is gathered in a second plastic container, similar to the one used for fruit/garden waste but of another colour, which will be also emptied periodically. This waste will be incinerated (burned).

Chemical waste. This waste is also collected separately in special small containers and delivered periodically to a special vehicle or to a central place. The chemical waste is composed of old medicines, household chemicals as detergents, paints, cleaning agents, batteries *etc.* This waste mostly requires special knowledge and skill for its disposal. Hospital waste in particular belongs in this category as far as medicines, needles and chemicals *etc.* are concerned.

If this method of separation is used, very little waste needs to be incinerated and the final quantity of waste is minimized and mostly recycled to produce beneficial products such as paper, carton, glass, metal, plastic, and compost *etc.* Table 7.1 shows the composition of household refuse from several countries and over several years (Purdom *et al.*, 1966; Kaizer, Zeit and McCappery, 1968; Cardenas and Varro, 1973; Hays, 1973; Müller, 1976; Cornelissen, 1990), and shows the new development in separating the fruit, vegetable, food and garden waste which makes up about 50% of the household input (Cornelissen, 1990). Tables 7.1-7.3 show the percentages of the various ingredients in the household refuse, paper fractions and household garbage, respectively.

Table 7.1. Composition of household refuse from several countries with emphasis on garbage (food waste). From (1) Hays (1973); (2) Kaiser, Zeit and McCappery (1968); (3) Purdon *et al.* (1966); (4) Cardenas and Varro (1973); (5) Müller (1976) and (6) Cornelissen (1990)

Components of refuse	*Municipal household refuse in %*								
	(1)[a]	*(2)*[b]	*(3)*[c]	*(4)*[d]	*(5)*[e]	*(5)*[e]	*(5)*[e]	*(5)*[e]	*(6)*[f]
Garbage	25.9	12.0	5.0	13	64.0	60	56	45.0	50.5[g]
Cloth and synthetics	1.3	3.2	2.6	-	-	-	-	-	1.9
Paper products	45.5	43.5	54.4	70	18.0	22	17	28.0	26.5
Plastics	1.7	-	-	-	0.7	5	3	0.8	7.8
Leather and rubber	1.0	-	-	-	-	-	-	-	0.6
Rubber, plastic and leather	-	5.7	1.7	-	-	-	-	-	-
Yard waste	1.6	-	-	-	-	-	-	-	-
Wood	0.3	3.9	2.4	-	-	-	-	-	0.5
Glass	10.9	-	-	-	3.0	2	2	3.0	6.2
Glass and ceramics	9.5	9.1	10.0	-	-	-	-	-	-
Ceramics	-	-	-	-	-	-	-	-	1.5
Metal	10.8	8.9	8.4	10	7.0	5	3	8.0	3.1
Non metal	-	-	-	-	-	-	-	-	0.8
Brick, rock and dirt, *etc.*	1.0	-	-	-	-	-	-	-	-
Carpets	-	-	-	-	-	-	-	-	0.3
Grass, leaves and dirt	16.8	16.4	-	-	-	-	-	-	-
Soil	-	-	-	-	-	-	5	-	-
Unsorted	-	-	-	-	1.0	6	14	8.0	0.4
Moisture	-	-	-	-	45.0	38	43	35.0	-

[a] Johnson city, Tennessee USA.
[b] Ocean side, N.Y. USA.
[c] Philadelphia, USA.
[d] New York city, USA.
[e] Four municipalities in south east Asia.
[f] The Netherlands mean; of three provinces.
[g] Garbage is classified as a mixture of vegetables, fruit and garden wastes (50.5 is 45.3 garbage + 2.5 bread + 2.7 animal waste).

Table 7.2. Composition of paper fraction in garbage. From Müller (1975)

Paper type	*(%)*
Newspapers	18.7
Office paper (writing)	12.5
Printed paper	7.1
Corrugated paper	21.3
Packing paper	28.0
Other	12.4
Total	100.0

Table 7.3. Detailed composition of household waste in garbage. From Müller (1975)

Kind of Waste	*(%)*
Fruit waste	15
Bones	3
Egg shells	2
Fibrous shells	7
Food waste	10
Non-food waste (mostly leaves)	42
Unsortable	21
Total	100

A nutritionally relevant waste is the garbage which is also known as swill, kitchen refuse or food waste. Garbage is mostly gathered from hotels and restaurants, institutions (schools, homes of elderly people, hospitals *etc.*), and military and municipal establishments. This waste is relevant as a source of protein, energy and minerals which are the most important nutrients in poultry feed and also important with regard to the feed costs. The new developments in finding cheap resources of energy and protein to replace conventional feedstuffs is of great economic importance, not only for developing countries, but also for developed ones.

Beharrell (1940) estimated food waste to be 450 g per family per day on the basis of 4.5 persons per family. He calculated for the year 1940, that the waste of edible material in Great Britain would exceed 1.5 million t. He concluded from other sources that one t was produced per 4.000 persons per week, which gives 208 000 t per year. Assuming that 100 g food waste is produced per day, then the estimated world production using the population figure of 5,3 billion (FAO, 1990), will be 1932 million t per year.

As far as the food waste is concerned as a feedstuff, several investigators proved that this waste has a reasonable nutritive value for the nutrition of poultry, especially broilers and layers (Draper, 1945; Soliman *et al.*, 1978[a,b]; Hoshii and Yoshida, 1981; Lipstein 1984, 1985).

Immediately after the Second World War there was a tremendous shortage of grains and, later, in grain deficient areas, a lot of food waste was used as a livestock feed in its raw form. Cattle and swine, however, suffered from several diseases owing to this type of raw feeding. Accordingly, laws were passed which prohibited the use of untreated raw food waste. High temperature treatment was made compulsory to assure complete sterilization of the product (Baird and Young, 1973). Sterilization by heating and drying makes it possible to use this dry food waste as a feedstuff for poultry.

Municipal sewage sludge

Sewage is the total output of municipal drainage of domestic wastes and some agro-industrial wastes such as small breweries or slaughterhouses which are discharging their wastewater without treatment. The wastewater in the drainage is from a combination of lavatory, washing and bathing facilities. The agro-industrial effluent is rich in undigested nutrients with the addition of cellulose (paper) and other organic materials. This sewage then passes a water treatment system in several stages and results in a sewage sludge. The sludge is rich in total nitrogen mainly in organic form but is also very rich in (heavy) metals such as Zn, Fe, Cu, Cd and Cr. These (heavy) metals are present in high concentration which inhibit the microbial decomposition unless precipitatingted according to the population (5.3 billion; FAO, 1990). This will deliver 58 million t of dried municipal sewage sludge per year.

Sewage sludge, as an addition to the diet of ruminants and poultry after it has been proper dried and sterilized, has been studied by many research workers. Ammerman and Block (1964) evaluated the feeding value of sewage sludge, while Hackler, Newman and Johnson (1957) used dried activated sewage sludge as a nitrogen source for sheep. Chaney *et al.* (1978) studied the effect of sewage sludge on the cropland in relation to the heavy metals and Kienholz *et al.* (1979) fed dried sewage sludge to feedlot steers.

As far as feeding dried municipal sludge or sewage sludge to poultry is concerned, several investigators have included this material in experimental diets of layers and broilers. Wong and Leung (1979) used sewage sludge as a supplementary feed for chicks; Johnson and Damron (1980) used the same sludge for layers. Damron *et al.* (1982) used it in the diets of broilers and layers. Kienholz *et al.* (1981) used the anaerobically digested sewage sludge for broilers, while Ologhodo and Oyewole (1987) used dried activated sewage sludge in diets for broilers. Shoremi, Sridhar and Ekpenyoung (1990) tried other possibilities by using sewage effluents as drinking water for laying chickens.

7.2 PROCESSING OF MUNICIPAL REFUSE

The processing of municipal refuse is based on drying, sterilizing and grinding the final product. There are two types of municipal refuse, kitchen waste (garbage or food waste or swill) and sewage sludge.

7.2.1 Processing of Garbage or Swill

Raw garbage is not suitable for poultry because it contains numerous harmful bacteria owing to its storage and subsequent fermentation in relation to the high environmental temperatures; it cannot therefore be used immediately. Since poultry rations are always offered in a mash or pelleted form, so the mixing of a wet raw garbage is not possible.

Kornegay *et al.* (1965) and Baird and Young (1973) reported that garbage, collected from hotels and restaurants, and the military and municipal establishments, was first screened to exclude foreign bodies such as metal, glass, *etc.*, then cooked by bubbling steam into the garbage from perforated pipes in the truck bed. For another sample of hotel and restaurant garbage, an additional method of cooking in an open vat was used (it is required by law that garbage is cooked at 100 °C for at least 30 min.). The wet garbage was dried in a forced-air oven for 72 h at 65 °C and, finally, the dried product was ground in a hammermill and thoroughly mixed. Baird and Young (1973) reported the cooking of garbage in wagons equipped to pass steam through the garbage in order to bring the material to the boiling temperature for 30 min.

Baird and Young (1973) reported that the food waste of military origin consists of scrapings from plates with paper contamination at a minimum. The food waste used was one day old; it was emulsified to a slurry of about 80% moisture and 20% solids and then sterilized at 100 °C for 30 min. Following sterilization, feed additives were introduced and then the slurry was dehydrated to about 38% moisture, and passed through an extruder that formed 6 mm diameter pellets of about 28 mm in length. The pellets were then cooled and dried to about 12% moisture. These methods are based on adding water through the steam. This increases the moisture level and creates the necessity of drying (to about 10% moisture) with all its expenses.

Lipstein (1985) reported that samples of kitchen waste were dried at 80 °C and sterilized by gamma irradiation. This method is, however, not suitable for the complete drying and sterilization of large quantities of garbage.

The most relevant industrial method of drying kitchen waste is the use of a drum drier. The temperature in the drum is raised by an oil-fired burner up to 340-400 °C; with continuous agitation, this results in a sterilized dry product with a relative humidity of 8-14% (see Chapter 2, Figure 2.1). A disk-drier may also be used (see Chapter 4, Figure 4.15).

Several reports were published concerning the gathering and processing of swill in The Netherlands (Alten *et al.*, 1995; Kuiper, 1995 and Odekerken *et al.*, 1997). All the reports concluded that swill can be processed to reduce the moisture content (which varies between 60- 80%) to about 10% which is a normal moisture content in feedstuff.

However, after proper sterilization, swill is considered to be germ free, after processing at a temperature of 133 °C and a pressure of 300 kPa with a residence time of 20 min. according to the Dutch destruction requirements. The processing lay-out is demonstrated in Figure 7.1 according to Odekerken *et al.* (1997) as follows:

Processing of swill

Swill is delivered in a liquid trailer or in a container. Containers will be kept closed as much as possible to reduce an undesired smell. After quality control, the swill will pass several processes:

Pre-treatment. Paper, cardboard and/or strange materials is separated from the swill using a turning shaft with hitches. Periodically, these hitches will be cleaned. From the storage container, the swill is pumped to the dewatering unit. A magnet separates iron particles.

Pre-dewatering. A first dewatering step is carried out by a stepscreen, this uses a number of ladder-like lamella, which move independent from each other in a vertical way. By this movement, the solid particles are directed to the top while the fluids flow away through openings between the lamella. These fluids contain approximately 10% dry matter and are led to the fat separator and pre-sedimentation tank, respectively.

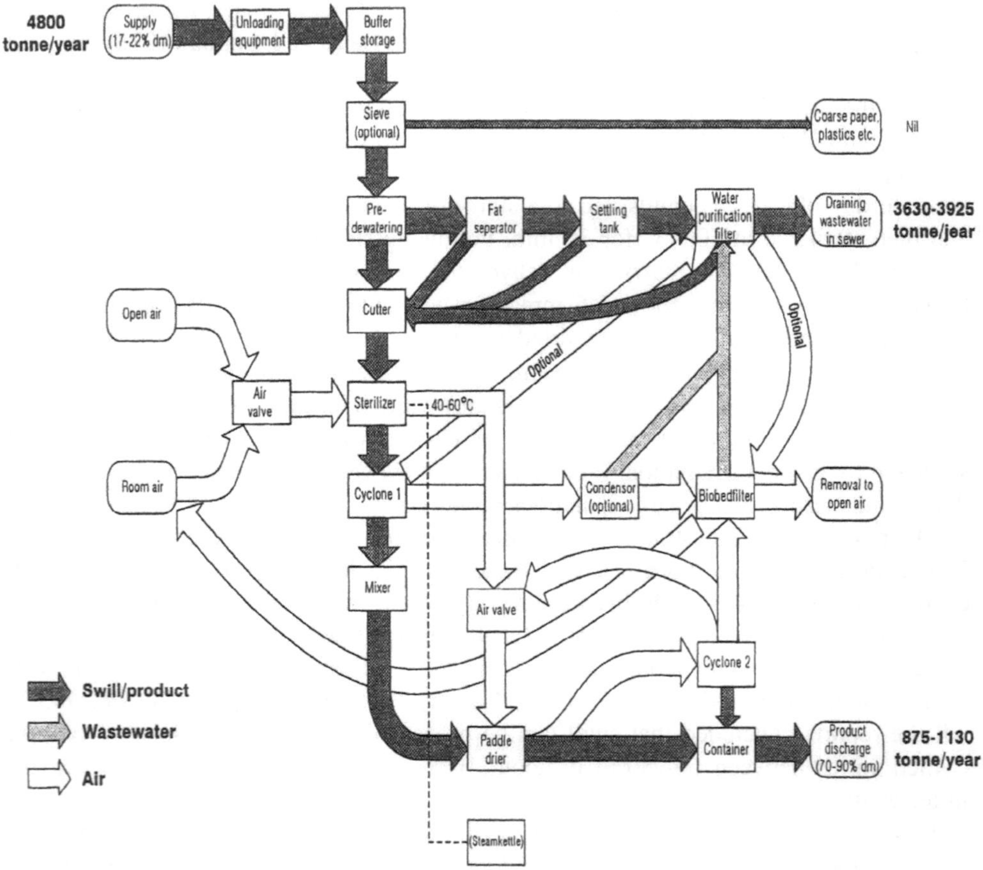

Figure 7.1. Flow diagram of the swill process. Reproduced with permission from Odekerken *et al.* (1997).

The separated fat as well as the sediment are returned to the process. Dewatering reduces the volume to about 40%, fat and sediment included. Dry matter content increases from 18-30%.

Sterilization. Sterilization process is a continuous process. The pre-dewatering mass is pumped through a double-walled tube, supplied with valves to maintain steam pressure in the system. Thermocouples control the pressure and the desired process conditions in the sterilization unit. In the barrel of the tube, steam is led to heat the swill mass. The sterilization tube ends in a cyclone, where surplus of steam is sucked off. This process takes place according to the Dutch destruction requirements, sterilization for 20 minutes, pressure of 300 kPa and a temperature of minimal 133 °C. Higher temperatures may create nutrients unavailability.

Thermal drying. After sterilization, the slurry will be transported from the cyclone to the mixer. The wet material from the sterilization unit and part of the dry end-product, will be mixed to a dry matter percentage of 45-50%. This mixing is done to avoid the critical drying phase, which is between 35 and 45% dry matter. In this critical phase (sticky phase) swill will stick and forms a vast and sticky "dough balls". This causes difficult evaporation of water, and increases energy costs. Mixture will pass into a barrel-heated paddle-drier. The material covers the warm barrel through the paddles, and is regularly scraped from the barrel. A low temperature air-stream in the driver dries the falling material; vapour drains away. The material is continuously subjected to particle size reduction to increase the drying surface. Storage and transportation of the end product takes place. Transportation by means of a conveyor belt or screw in an open drain. Finally it is delivered as swill meal or pellets. Transport can be carried out in bulk or in big-bags.

Wastewater treatment. Wastewater is processed by "pre-dewatering" and "thermal drying", will be transported to a biological purification installation. Prior to discharging to the sewer, the water is filtered; the sedimentation silt can be returned to the sterilization unit.

Air treatment. Air is extracted continuously from the storage room, water vapour and odours are present. This will be transported to a biofilter-bed, water is purified prior to drain off.

Quality assurance. Assurance of quality control of feed compounders is certified according to Good Management & Manufacturing Practices (GMP) or to ISO-9002. The Dutch Board of Feedstuffs is the authority to certify and controls the operation of the quality system regularly. The Dutch Board of Feedingstuffs has founded the commission "Directive quality requirements for moisty animal feeds". This commission has arranged concept-requirements concerning the processing of swill. These quality requirements comprise standards such as:

- veterinary aspects
- hygienic aspects
- undesired components
- contaminants by packing materials
- biological degradation products (e.g. protein quality)
- requirements for fat fraction: fat percentage and composition

The undesired components may form a bottleneck, concerning the non-detectable contaminants.

7.2.2 Processing of Municipal Sewage Sludge

At the present time, wastewater treatment combines methods (unit operations and processes) known as primary, secondary and tertiary (or advanced) treatments. In general, an almost infinite number of combinations of unit operations and processes are possible. In practice, therefore, sludge treatment processes can be divided into two general categories, depending on whether or not biological treatment is involved. We selected a conventional wastewater treatment as an illustration of the treatment of sewage waste water (Figure 7.2).

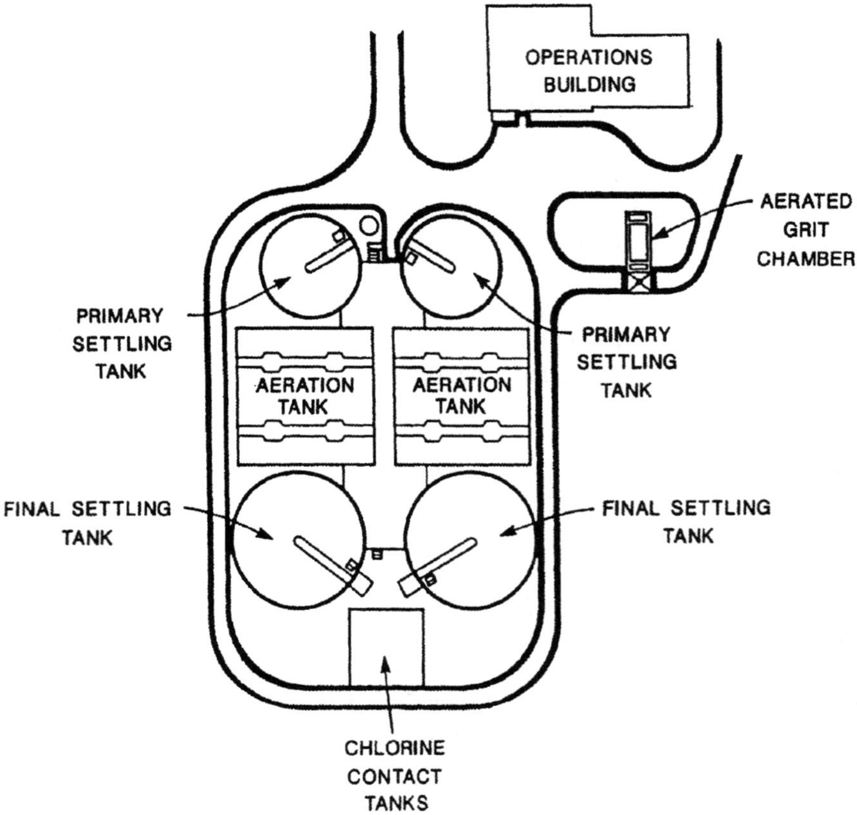

Figure 7.2. Simplified schematic diagram for the treatment of sewage into sludge. Reproduced with permission from Metcalf and Eddy (1979).

In primary treatments, physical operations such as screening and sedimentation are used to remove floating and settleable solids found in sewage. In secondary treatment, biological and chemical processes are used to remove most of the organic matter. This biological treatment process is an aerobic fermentation process in which the organic material, inherent in domestic sewage and the sewage from industries discharging organic wastes, is metabolized and assimilated by a variety of micro-organisms. The end product of this fermentation (activated sludge) is a gelatinous mass of these micro-organisms. Thus, biological processes are used to convert the finely divided and dissolved organic matter in wastewater into flocculant settleable solids. These solids consequently can be removed in sedimentation tanks. Biological processes therefore, are efficient in removing organic substances that are either soluble or in the colloidal size range.

Finally, in advanced treatments, additional chemical combinations are used to remove other constituents such as nitrogen and phosphorus, which are capable of stimulating the growth of aquatic plants and that are not removed by secondary treatment.

In Figures 7.3 and 7.4, a typical diagram and a layout of treatment plant are presented for the treatment of municipal sewage.

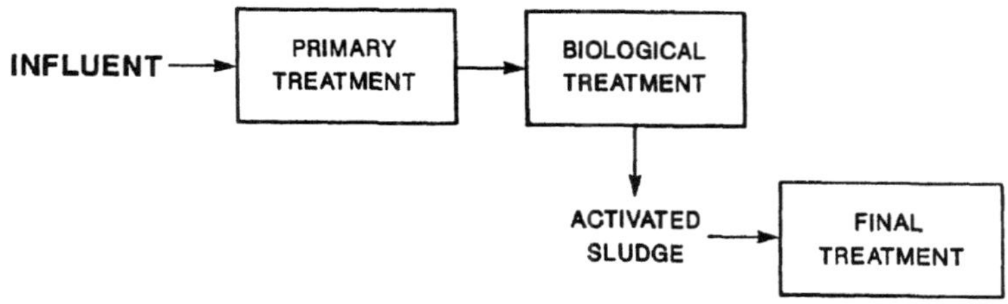

Figure 7.3. A typical flowsheet of municipal sewage treatment.

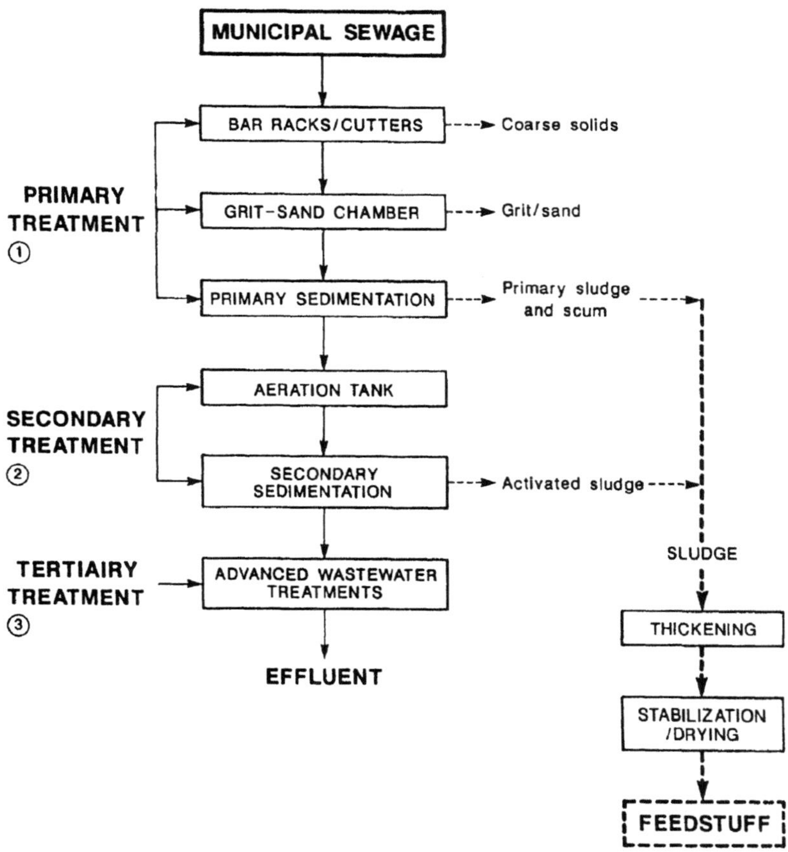

Figure 7.4. Layout of treatment plant for sewage processing. Reproduced with permission from Metcalf and Eddy (1979)

Primary treatment
In the primary treatment, racks and coarse screens remove coarse solids by interception and this is usually followed by a process of grinding of the remaining solids. Consequently, grit, sand and gravel are removed in so-called grit chambers. Settleable solids and floating materials are then removed by sedimentation, the principal operation used in the primary treatment of wastewater.

Secondary treatment
For biological waste treatment with the activated sludge process, organic waste is introduced into a reactor where a culture of a variety of micro-organisms, principally aerobic bacteria, is maintained in suspension. In the reactor, the micro-organism, culture carries out a conversion of colloidal and dissolved carbonaceous organic matter into various gases and into cell tissue. The aerobic environment in the reactor is achieved by the use of diffused or mechanical aeration, also used to maintain a completely mixed regime. Because bacterial cell tissue has a specific gravity slightly greater than that of water, the resulting tissue (biological solids) can be removed from the treated liquid by gravity settling (secondary sedimentation).

Advanced wastewater treatment (optional)
The conventional treatment operations mentioned before affect substances such as inorganic ions (Ca, K, S, N and P) or more complex organic compounds and inorganic compounds such as heavy metals (Fe, Cu, Cd, Zn, Mn, Al). It is anticipated that the treatment requirements in terms of the acceptable concentration of many of these substances in the effluent from wastewater plants will become more stringent. For this reason, advanced wastewater treatments including chemical precipitation, ion exchange, reverse osmosis (hyperfiltration) and electrodialysis are necessary which all will affect the final composition of the sludge. For more details the work of Fair and Geyer (1965) and of Metcalf and Eddy (1979) is recommended.

Treatment of sludge
The constituents removed in wastewater treatment plants are important by-products of the treatment process. They include screenings, grit, scum and (activated) sludge. The sludge is usually in the form of liquid or semi-solid liquid. which typically contains from 0.25-12% solids, depending on the operations and processes involved (Metcalf and Eddy, 1979). The operations involved with sludge processing include preliminary operations (grinding, degritting, blending and storage), thickening, stabilization and disinfection, conditioning/dewatering, drying (and eventually composting as well as thermal reduction for product recovery and volume reduction purposes), whereas sludge disposal methods include landfill, land application, and reclamation or reuse for example as a feedstuff in animal diets.

Thickening is a procedure used to increase the solids content of the sludge by removing a portion of the liquid fraction and which is generally accomplished by physical means such as gravity settling, flotation and centrifugation. Sludges are stabilized in order to reduce pathogens, to eliminate offensive odours, and to inhibit or reduce the potential for putrefraction. The technologies available for stabilization include chlorine oxidation, lime stabilization, heat treatments and anaerobic or aerobic digestion.

Sludge is conditioned to improve its dewatering characteristics. The two methods commonly used involve chemical and heat treatment conditioning, respectively. Chemical conditioning, with chemicals such as ferric chloride, lime, alum and organic polymers, results in a coagulation of the solids and in release of absorbed water. Heat treatment is both a stabilizing and a conditioning process that involves heating the sludge under high temperature, for a short time conditions. The treatment coagulates the solids, breaks down the cell structure, and reduces the water affinity of sludge solids. As a result, the sludge is sterilized and deodorized to a great extent and has a good potential for dewatering without the addition of chemicals.

The heat treatment is most applicable for biological sludges that may be difficult to stabilize or condition by other means. In addition to the reduction of pathogenic bacteria by stabilization, disinfection methods are in use such as sterilization, pH treatment, *etc.*

Dewatering is a physical/mechanical unit operation used to reduce the moisture content of sludge in order to reduce volume, and to allow proper handling and transportation. In addition, it is commonly required prior to its disposal on land or as a feedstuff. A number of processes for removal of the moisture are available including filter presses, horizontal belt filters and drying beds.

Heat drying, finally, is the unit operation that reduces water content by vaporization of water into the air. The most important processes used for drying sludge are flash driers, disc driers (for description see Chapter 4, Figures 4.15 and 4.16), drum driers (Chapter 2, Figure 2.1) and spray driers. The purpose of heat drying is to remove moisture from the wet sludge to less than 10 percent to permit grinding of the sludge, to reduce its weight and to prevent biological action. These are prerequisites for the processing of sludge into fertilizer or into a feedstuff. For example, Kienholz *et al.* (1981) reported air-dried anaerobically digested sewage sludge for incorporation into broiler diets. Sewage was passed through a leaf mulcher, screened for large particles, mixed with two parts of ground corn (wt/wt) and extruded at an operating temperature of 163 °C and with a processing time of about 10 s.

7.2.3 Yields of Municipal Refuse

(a) Garbage or swill

The amounts of swill produced from hotels and restaurants, institutions (hospitals, schools, homes of elderly people *etc.*), and military and municipal is not known. The amount will vary widely in developed markets from country to country according to the standard of living and lifestyle of the population. The degree of industrialization, and how are products introduced to the consumer (chicken and fish fillets, no bones; peeled fruit, no peel; milk products, packed or not), are factors affecting swill production. Seasons of the year (spring, summer, fall and winter) determine the percentage of swill as far as fruit and vegetable waste are concerned. In some countries fruits are found only at certain seasons while other countries import vegetables and fruits (bananas, oranges, grapes, *etc.*) all the year round.

In the developing market the composition of swill will vary according to the utilization of edible and non-edible food. In developed countries some non-edible food is rejected which in developing ones is regarded as being edible (rumens, lungs, brains, and spleens of some farm animals are used as food after cooking), mango pit kernels are used as flour for cooking, potatoes are consumed with peels, and tomatoes are processed with their seeds, *etc.*).

It was reported in The Netherlands that the total potential waste for composting purposes was estimated to be 14 342 000 t where 106 000 t were defined as swill or kitchen waste (Anon, 1992). It was also reported that 3 124 000 t waste was produced from hotels and restaurants, institutions and municipal establishments which was not recycled (Table 7.4). If more attention was paid to this waste, it could also be used as a potential feedstuff.

Table 7.4. Estimates of the total potential organic waste [a] produced by type and source per year. From Anon (1992)

Types and Sources	*Tonnes*
Hotel and restaurant	181 000
Institutional	2 619 000
Municipal	324 000
Military	Not estimated
Total	3 124 000

[a] The total potential waste is estimated to be 14 342 000 t of which 3 124 000 t is not recycled. Swill is mostly produced from both hotel and restaurant waste and is estimated to be 64 000 t with municipal waste estimated to be 42 000 t.

(b) Domestic sewage sludge

The yields of domestic sewage sludge depend on the components of wastewater flow and the proportion of solids in the sewage sludge, which is determined by the treatment process. The treatment process as discussed before results in products such as primary sludge, excess activated sludge, combined primary and activated sludge, digested primary sludge, and combined primary/excess activated sludge and digested sludge (Fair and Geyer, 1965).

As far as the components of wastewater flow with its solids are concerned they depend on the type of collection system used and may include the following:

Domestic sanitary wastewater. Wastewater discharged from residences and from commercial, institutional and similar facilities.

Industrial wastewater. Wastewater in which industrial wastes predominate.

Infiltration/Inflow. Extraneous water that enters the sewer system from the ground through various means, and storm water that is discharged from sources such as roof leaders, foundation drains and storm sewers.

Storm water. Water resulting from precipitation runoff.

In all cases, the percentage of wastewater components varies with local conditions, from country to country, standards and living habit, degree of industrialization and the time of the year.

Table 7.5 shows flows of wastewater in some residential and institutional sources. It is estimated roughly that the mean wastewater production per person per day is 75-200 l depending on the points mentioned before (Metcalf and Eddy, 1979).

Table 7.5. Daily wastewater flows from some residential and institutional sources. From Metcalf and Eddy (1979)

		Flow, l/day	
Source	*Unit*	*Range*	*Typical*
Apartment	Person	200-340	260
Hotel, resident	Resident	150-220	190
Average home	Person	190-350	280
Luxury home	Person	300-550	380
Hospital medical	Bed	500-950	650
Rest home	Resident	200-450	350
School, boarding	Student	200-400	280

As far as the municipal sludge yields are concerned, these depend on the processing system as far as the primary and secondary treatments are concerned, and also on the method of digesting the sludge and the method of fermentation used. In general, the quantities of sludge are based on the plain sedimentation (primary treatment), chemical coagulation (primary treatment) trickling filtration (secondary treatment), and activation (secondary treatment). For example, from the primary treatment, the daily production per caput of fresh sludge is 54 g, 39 g (or 72.2%) being volatile and 15 g (or 27.8%) fixed. The weight of dry solids originating from primary, secondary and advanced treatments is estimated to be about 30 kg per 1000 persons per day (Fair and Geyer, 1965). Table 7.6 shows the proportion of solids in municipal sewage sludge according to Metcalf and Eddy, 1979.

Table 7.6. Proportion of solids in municipal sewage sludge [a]. From Metcalf and Eddy (1979)

Type of Treatment	*Wet Sludge*		
	Dry solids (g/person/day)	*% solids*	*$m^3/10^3$ capita/day*
Primary treatment	72	5	1.44
Primary and trickling filter	108	4	2.70
Primary and activated sludge	114	3	3.80

[a] Based on 120 g of suspended solids per person per day in untreated wastewater.

7.3 CHEMICAL ANALYSIS AND NUTRITIVE VALUE

7.3.1 Garbage or Swill

Kornegay *et al.* (1965) studied the chemical composition of swill of several types and the sources of swill from hotels and restaurants, institutions, and military and municipal establishments. In addition, they analysed the swill per day and per season. Their results showed that there was no significant relation between days of the week (Monday till Friday) and the chemical composition of the various garbages. Unfortunately, Saturday and Sunday were not included in the product analysis and it is known that on these two days most people tend to eat more luxurious food and that there are plenty of left-overs. As far as the seasons are concerned, the chemical analysis showed that crude fibre and ash content were significantly higher during summer than during the other seasons ($P < 0.05$). This may be due to an increased consumption of fruit and vegetables which are known to contain relatively more minerals. Winter garbage had a higher gross energy content than summer garbage, which may be due to the high human requirements of energy in winter in the form of more carbohydrate (potatoes, pies, sweets *etc.*) and more fat (butter, margarine, oil etc). This is reflected in the food consumed and left-over as garbage (Table 7.7).

Kornegay *et al.* (1965) also analysed the chemical composition of garbage by type and source (hotel and restaurant, institutional, military and municipal). They concluded that all types of garbage were very low in dry matter content in comparison with normal food. Military garbage was significantly higher in dry matter, ether extract and gross energy than any of the other types. This may be due to the high energy food (carbohydrates and fats) offered to military personnel which is needed for their continuous physical training. Institutional garbage had a significantly ($P < 0.01$) higher nitrogen-free extract (NFE) content and lower ether extract and gross energy content than any of the other types. This may be due to the use of food with high NFE. With the exception of municipal garbage, the crude fibre content did not vary greatly among the different types of garbage (Table 7.8).

Table 7.7. Chemical composition of garbage by day of the week and season on a dry matter basis. From Kornegay *et al.* (1965)

Days of the week and seasons	*Samples* No.	*D.M.* (%)	*C.P.* (%)	*E.E.* (%)	*C.F.* (%)	*Ash* (%)	*NFE* (%)	*Gross energy* MJ/kg
Days of the Week								
Monday	29	20.2	15.6	22.9	2.7	5.5	53.5	21.55
Tuesday	31	19.6	15.9	23.2	3.5	5.6	51.8	22.00
Wednesday	30	18.7	14.9	24.4	4.2	6.0	50.5	22.22
Thursday	29	21.6	15.8	23.4	3.5	6.2	51.1	21.92
Friday	40	19.7	15.7	23.0	4.0	6.0	51.2	21.84
Seasons of the Year								
Spring	36	20.0	16.2[a]	23.3	4.0[b]	6.1	50.5	21.59
Spring and summer	25	18.8	14.3	21.9	3.8	6.1	53.7	21.67
Fall	32	20.3	14.5	23.5	3.2	5.4	53.3	22.26
Winter	36	20.4	16.2[a]	24.0	3.1	5.6	51.0	22.43[a]

[a,b] Significantly ($P < 0.05$) greater than the lowest mean.
DM = Dry matter; CP = Crude protein; EE = Ether extract; CF = Crude fibre; NFE = nitrogen-free extract.

Table 7.8. Chemical composition of garbages by type and source on a dry matter basis. From Kornegay *et al.* (1965)

Types and sources	*Samples* No.	*D.M.* (%)	*C.P.* (%)	*E.E.* (%)	*C.F.* (%)	*Ash* (%)	*NFE* (%)	*Gross energy* MJ/kg
Hotels and restaurants	30	16.0	15.3	24.9	3.3	5.7	50.7	22.30
Institutional	29	17.6	14.6	14.7	2.8	5.3	62.6	20.15
Military	28	25.7	16.0	32.0	2.8	5.6	43.8	23.70
Municipal	21	16.6	17.5	21.4	8.4	8.6	44.0	21.34
Mean	-	19.0	15.9	23.3	4.3	6.2	50.3	21.87

See ***Table 7.7*** for key to abbreviations

Soliman *et al.* (1978[a]) investigated the chemical composition of several types of food wastes from restaurants, mainly first, second, and third class restaurants besides restaurants of mixed classes. It was noticed that the first class restaurants had a higher crude protein (24.6%) in comparison with the second and the third (19.9 and 19.9%

respectively); other chemical items did not show a relevant difference. It seems that the first class menu contains more proteins such as fish, meat, chicken and eggs than the other classes which was reflected on the chemical composition of left-over food waste (Table 7.9).

Table 7.9. Chemical composition of restaurant food waste on an air dry matter basis. From Soliman *et al.* (1978) [a]

	Classes of restaurants (%)			*Mixed*
	First	*Second*	*Third*	*classes*
Moisture	7.2	6.8	7.1	7.5
Crude protein	24.6	19.9	19.9	21.6
Crude fat	17.1	18.1	15.3	16.6
Crude fibre	4.8	3.1	5.6	4.6
NFE	27.7	36.1	33.0	31.5
Ash	18.7	16.0	19.7	18.3
Ca	5.6	4.7	5.5	5.2
P	1.6	1.3	1.4	1.4

NFE = Nitrogen-free extract.

Table 7.10 shows the chemical composition of garbage as compared to several relevant feedstuffs. The crude protein of kitchen waste varies between 13.8-21% which is higher than in maize and rice polishings but lower than spent grains. Crude fat varies between 8-26.1% which is higher than maize and spent grains. Crude fibre varies between 5-8.1% which is higher than maize and rice polishing but lower than spent grains. Metabolizable energy content was 11.9 MJ/kg which is similar to rice polishing, lower than maize and higher than spent grains. Minerals were shown to be adequate as far as calcium, phosphorus, sodium, chloor are concerned, whereas trace elements were not present at a level to create problems in formulating diets. The essential limiting amino acids lysine and methionine both showed a high level (0.40-1.05% and 0.16-0.45% respectively) being higher than those in maize, spent grains and rice polishing (Table 7.11). From Tables 7.10 and 7.11 it is clear that the kitchen waste in column 3, gathered from self-service stores in which the waste consists of inedible parts of fresh fish and vegetables, is the richest in protein, fat, ash, phosphorus, lysine and methionine. This relatively higher nutritive value is due to the fish waste which was abundant in this garbage (Yoshida and Hoshii, 1979). The information in the tables shows that it is worthwhile encouraging not only the re-use of kitchen waste in the form of left-over food but also that of the waste from supermarkets which contains very valuable ingredients such as fish, meat, chicken, vegetable and fruit wastes.

Table 7.10. Chemical composition of kitchen waste (garbage) in comparison with some relevant feedstuffs. From [(1)] Baird and Young (1973); [(2)] Lipstein (1985); [(3)] Yoshida and Hoshii (1979); [(4)] Beharrell (1940); [(5)] Scott, Nesheim and Young (1982) and [(6)] Titus and Fritz (1971)

(%)	*(1)*[a]	*(2)*[b]	*(3)*[c]	*(4)*	*(5)*	*(5)*	*(5)*
	Kitchen waste				*Maize*	*Spent grains*[d]	*Rice polishing*
Moisture	5.90	-	12.70	7.3	12.00[e]	8.00[e]	11.00[e]
Crude protein	16.60	13.80	20.70	21.0	6.80	27.00	12.00
Crude fat	9.70	10.90	26.10	8.0	3.90	7.00	12.00
Crude fibre	8.10	5.00	5.00	7.2	2.00	12.00	3.00
ME (MJ/kg)	-	11.90	-	-	14.10	10.10	12.00
Gross energy(MJ/kg)	18.60	-	21.30	-	-	-	-
Ash	4.10	14.30	8.60	15.4	1.30[e]	3.60[e]	0.70[e]
NF Extract	55.30	-	27.10	41.2	71.80[e]	41.30[e]	78.60[e]
Ca	0.53	3.26	1.68	3.4	0.02	0.27	0.04
P	0.48	0.47	1.21	1.7	0.30	0.50	1.40
Na	0.25	0.86	0.35	-	0.01	0.04	0.07
K	1.99	-	-	-	0.38	0.13	1.10
Cl	-	1.34	-	-	0.04	0.06	0.07
Magnesium	0.17	-	-	-	-	-	-
Manganese (ppm)	51.00	-	-	-	5.00	38.00	13.00[e]
Iron (ppm)	805.00	-	-	-	35.00[e]	300.00[e]	8.00[e]
Copper (ppm)	17.00	-	-	-	3.00[e]	21.00[e]	2.00[e]

[a] Military food, waste consists of scraping from plates with a minimum paper contamination.
[b] Home kitchen, waste separated as edible and non-edible components.
[c] Self-sevice stores, waste consists of inedible parts of fresh fish and vegetable.
[d] Spent grains = Brewers' dried grains
[e] From reference (6)

Table 7.11. Amino acid content of kitchen waste garbage in comparison with some relevant feedstuffs. From [(1)] Baird and Young (1973); [(2)] Lipstein (1985); [(3)] Yoshida and Hoshii (1979) and [(4)] Scott, Nesheim and Young (1982)

(%)	*(1)*[a]	*(2)*[b]	*(3)*[c]	*(4)*	*(4)*	*(4)*
	Kitchen waste			*Maize*	*Spent grains*[d]	*Rice polishing*
Crude protein	16.80	13.80	20.70	8.70	26.0	7.50
Arginine	0.94	0.62	1.26	0.50	1.3	0.36
Cystine	0.20	0.18	0.24	0.18	0.3	0.09
Glycine	0.81	0.73	1.82	0.40	1.0	0.70
Histidine	0.35	0.40	0.50	0.20	0.5	0.18
Isoleucine	0.30	0.45	0.81	0.40	1.5	0.45
Leucine	0.36	0.73	1.32	1.10	2.3	0.71
Lysine	0.79	0.40	1.05	0.20	0.9	0.27
Methionine	0.16	0.22	0.45	0.18	0.6	0.17
Phenylalanine	0.64	0.48	0.75	0.50	1.3	0.53
Threonine	0.82	0.47	0.82	0.40	0.9	0.36
Tryptophan	-	-	0.25	0.10	0.4	0.09
Tyrosine	0.38	0.37	0.55	0.41	1.2	0.62
Valine	0.65	0.55	1.06	0.40	1.6	0.53

[a] Military food waste, consists of scraping from plates with a minimum paper contamination.
[b] Home kitchen waste separated as edible and non-edible components.
[c] Self-service stores' waste consists of inedible parts of fresh fish and vegetables.
[d] Spent grains = Brewers' dried grains

7.3.2 Sewage Sludge

The chemical composition of sewage sludge varies from country to country and from town to town in the same country. This variation is due to:

- The variation in systems used in the water purification of the sewage (eliminating the heavy metals or not). This is related to the degree of industrialization and development in the region.
- The processing of the sewage sludge (raw, aerobically, anaerobically, fermented or activated sludge). This affects the nutritive value of the sludge as far as its protein, energy and minerals are concerned.
- The Standards of living and lifestyles which affect the quality of the sewage sludge. The use of tissue paper increases the cellulose content, but, in many countries water is used instead of tissue paper. As far as the actual food is concerned, a high standard of living is related to food rich in protein (meat, fish, eggs, milk *etc.*) and lower standards of living to food less rich in protein (beans, rice, bread) (see chapter 1: energy and protein consumption in developed and developing countries).
- The intensity of use of wastewater treatment. If industries such as slaughterhouses, breweries, tanneries, wine processors *etc.* all use the public municipal sewage system, the sewage sludge will be very high in heavy metals and will be composed of low quality organic matter. This point will differ depending on to the level of industrialization in the region.
- The season of the year. In the winter, high energy food is consumed while in the summer, more fruit and vegetables are used. This is reflected in the quality of sewage sludge. This quality, of course, varies from one country to another depending on the climate of the region.

Several research workers have determined the chemical composition of the sewage sludge (Table 7.12). According to their results the following level of nutrients (range) can be reported for sewage sludge: crude protein 13-30.3%, crude fat 4.2-4.5%, crude fibre 6.9-19%, nitrogen free extract 15.2%, Ash 39.9-62.0%. The levels of macro elements are: Ca 0.48-2.6%, P 0.75-2.3%, Mg 0.07-0.9%, Na 0.05-0.5%, K 0.16-0.4%. Heavy metals in ppm are Fe 9250-36 600, Cu 228-2090, Zn 1500-2300, Cr 55-2500, Cd 8-200.

The large variation in the chemical composition, and particularly in the level of heavy metals due to the above-mentioned factors (Toth, 1973; Müller, 1975; Kienholz *et al.*, 1979; Damron *et al.*, 1982; Ologhodo and Oyewole, 1987).

Hackler, Newmann and Johnson (1957) analysed the amino acid contents of the activated sewage sludge and noticed that the levels of the most limiting essential amino acids lysine (1.3%) and methionine (0.5%) were acceptable (Table 7.13). These levels are comparable with those of hydrolized feather meal (lysine 1.5% and methionine 0.5%; Scott, Nesheim and Young, 1982).

In addition to the analysis of the heavy metals in the sewage sludge, the toxic levels for poultry (NRC, 1984) and the maximum levels permitted in fertilizer (Anon, 1991) are indicated in Table 7.12. There are three types of sewage sludge:

Table 7.12. Chemical composition of municipal sewage sludge in comparison with reported toxic concentration of heavy metals in poultry diets and maximum permitted levels in fertilizers. From (1) Toth (1973); (2) Müller (1975); (3) Kienholz *et al.* (1979); (4) Damron *et al.* (1982); (5) Ologhobo and Oyewole (1987); (6) NRC (1984) and (7) Anon (1991)

	(1)	*(1)*	*(2)[a]*	*(3)[b]*	*(4)[c]*	*(5)[d]*	*(6)*	*(7)*
(%)	***Municipal sewage sludge***						*Toxic levels in poultry diets*	*Maximum levels as fertilizer*
Dry matter	NR	NR	100.0	100	NR	94.0	-	-
Crude protein	20.8	13.00	28.0	14	30.30	27.7	-	-
Crude fat	-	-	4.5	-	-	4.2	-	-
Crude fibre	-	-	9.0	19[e]	-	6.9	-	-
Organic matter	-	-	68.0	-	-	-	-	-
NFE	-	-	-	-	-	15.2	-	-
Gross energy (MJ/kg)	-	-	-	-	-	18.2	-	-
Ash	-	-	9.0	62	-	39.9	-	-
Ca	0.48	1.02	2.6	-	1.95	-	-	-
P	1.40	0.75	1.9	-	2.30	-	-	-
Mg	0.15	0.07	0.9	-	-	-	-	-
Na	0.05	0.37	0.5	-	-	-	-	-
K	0.19	0.16	0.4	-	-	-	-	-
Cl	-	-	0.2	-	-	-	-	-
S	-	-	0.9	-	-	-	-	-
Heavy Metals (ppm)								
Mn	100	40	700	-	-	-	6000-19600	-
Fe	13500	9250	-	-	36600	-	4500	-
Cu	228	2090	-	710.0	1525	-	250-806	425.0
Zn	2000	2300	-	1500.0	2200	-	800-3000	1400.0
Ni	20	50	-	125.0	260	-	500	70.0
Cr	55	175	-	-	2500	-	300	350.0
Cd	8	14	-	21.0	200	-	12-40	3.5
Pb	108	150	-	780.0	670	-	320-1000	300.0
Hg	-	-	-	11.0	-	-	5-400	3.5
Mo	-	-	-	40.0	-	-	300-500	-
Se	-	-	-	5.4	-	-	5-10	-
F	-	-	-	200.0	-	-	400-1000	-
As	-	-	-	1.3	-	-	100	25.0

[a] Singapore (Asia); [b] Denver (USA); [c] Chicago (USA); [d] Activated sewage sludge; [e] as ADF
NR = not reported

- An activated sludge which contains no heavy metals, owing to the use of an expensivewastewater treatment system (advanced wastewater treatment). The absence of these metals, allows the sludge to be activated by micro-organisms growing on the organic matter (Hackler *et al.*, 1957; Ologhodo and Oyewole, 1987).
- A sludge with low levels of heavy metals which is produced using less expensive wastewater treatment (secondary treatments) (Singapore sludge, Müller, 1975; Denver sludge, Kienholz *et al.*, 1979).
- A sewage sludge very high in heavy metals, mainly Fe, Cu, Zn, Cd, which is produced when wastewater treatment is used only to separate course material and organic matter. This type of sludge will create problems as a feedstuff or as a fertilizer (Toth, 1973; Chicago sludge, Damron *et al.*, 1982).

Table 7.13. Protein and amino acid content of activated sewage sludge. From Hackler, Newman and Johnson (1957)

Amino acids	*(%)*
Crude protein	32.90
Arginine	1.10
Histidine	0.44
Isoleucine	1.40
Leucine	1.80
Lysine	1.30
Methionine	0.50
Cystine	0.20
Phenylalanine	1.60
Tyrosine	0.70
Threonine	1.50
Tryptophan	0.30
Valine	2.00
Glycine	1.60
Glutamic A	2.90

The third wastewater treatment will produce a sewage sludge with very high levels of heavy metals. The toxic levels represented by NRC (1984) for the heavy metals Fe, Cu, Zn, Cr, Cd and Pb are much lower than those in the sewage sludge reported by Toth (1973) and Damron *et al.* (1982).

Toxic levels of heavy metals affect the performance of broilers, growing chicks, layers and turkeys. Deobold and Elvehjem (1935) noticed bone abnormalities (rickets) in chicks with a level of 4500 ppm Fe. Kunishisa *et al.* (1966) found a reduced growth with 300 ppm of Cr. As far as Cu is concerned, levels from 250-800 ppm showed growth depression, mortality, exudative diathesis (severe oedema), muscular dystrophy (degeneration of muscle fibres) and gizzard erosion (Mehring *et al.*, 1960; Jensen, 1975; Poupoulis and Jensen, 1976; Robbins and Baker, 1980).

For Zn, toxic levels varying from 800-4000 ppm were reported and showed growth depression, exudative diathesis and muscular dystrophy in chicks (Roberson and Schaible, 1960; Johnson *et al.*, 1962; Vohra and Kratzer, 1968; Berg and Martinson, 1972; Jensen, 1975). Levels of Cd varying from 12-40 ppm were reported to be toxic and to result in reduced growth in chicken and decreased egg production for layers (Supplee, 1961; Hill *et al.*, 1963; Leach, Wang and Baker, 1979). For Pb levels of 1000 ppm were reported to cause reduced growth in chickens (Damron, Simpson and Harms, 1969).

7.4 MUNICIPAL REFUSE IN THE NUTRITION OF POULTRY

7.4.1 Garbage or Swill

(a) Growing chickens and broilers

Draper (1945) was one of the pioneers of the use of garbage meal in the nutrition of chicken. He concluded, at a time, when synthetic commercial amino acids such as methionine and lysine were not known, that soybean meal, fish meal and meat meal were essential ingredients for use together with garbage for optimal results. The inclusion of 30% processed garbage in diets with other essential ingredients gave satisfactory results for body weight, feed consumption and conversion in the presence of fish meal or soybean meal. The inclusion of up to approximately 20% dehydrated fat-extracted garbage in the chick ration gave satisfactory results when meat meal was used as a protein concentrate. It was reported that the processed garbage and edible waste from military camps was a worthwhile ingredient for chick rations.

Soliman *et al.* (1978[a]) studied the nutritive value of garbage classified into first, second, third and mixed classes of restaurant food waste in mixed feed fed to Alexandria chicks up to 12 weeks of age. The chemical composition of the experimental waste diets and the two control diets offered to the chicks and the performance of the chicks are shown in Table 7.14. The mixed class restaurant food waste in poultry feed resulted in a significant lower body weight at 4, 8 and 12 weeks of age in comparison with the standard diet. Feed consumption showed no significant differences and feed conversion thus was significantly higher than for the control birds. The comparison between the two controls against the average treatments showed an uneconomic feed conversion owing to the low body weight and similar feed consumption. The use of this waste is acceptable when representing 50% of the total diet due to the low price of the garbage.

Lipstein (1984) evaluated the nutritional value of treated kitchen waste in broiler diets. The inclusion of dried and sterilized (gamma irradiation) kitchen waste at levels of 5, 10, 15 and 20% compared with control diets, showed no significant differences in broiler performance when this was measured as weight gain, feed consumption and conversion up to 4 weeks of age. He also reported that abdominal fat and liver weights,

Table 7.14. Chemical analysis and average body weight, feed consumption and conversion of chicks fed standard or common diet in comparison with four diets with restaurant food (RF) waste added. From Soliman *et al.* (1978[a])

	Chemical analysis						*Body weight (g)*			*Feed consumption (g)*			*Feed conversion*		
	CP	*CF*	*C Fat*	*NFE*	*Ca*	*P*	*4W*	*8W*	*12W*	*4W*	*8W*	*12W*	*4W*	*8W*	*12W*
Standard diet	21.1	7.9	3.9	52.5	1.3	0.6	163.0[a]	602[a]	1076[a]	423[a]	1581[a]	3272[a]	2.61[d]	2.62[c]	3.04[d]
Common diet	21.2	7.5	4.0	48.7	2.0	1.0	152.0[a,b]	549[a,b]	961[b,c]	427[a]	1589[a]	3285[a]	2.82[c,d]	2.90[c]	3.42[c]
RF Waste 1st class[e] 50% inclusion	22.3	13.6	4.7	38.5	3.4	1.1	130.0[b,c]	451[a,b,c]	998[a,b]	434[a]	1620[a]	3404[a]	2.36[d]	3.00[c]	3.42[b,c]
RF Waste 2nd class[e] 50% inclusion	20.6	13.6	3.5	43.1	3.0	1.0	128.0[c]	478[c]	913[b,c]	443[a]	1589[a]	3308[a]	3.47[a,b]	3.32[b]	3.63[b]
RF Waste 3rd class[e] 50% inclusion	20.5	11.0	5.0	41.5	3.4	1.0	115.1[c]	480[c]	886[c]	438[a]	1702[a]	3544[a]	3.81[a]	3.62[a]	4.01[a]
RF Waste mixed[f] 50% inclusion	21.7	12.7	4.2	40.4	3.3	1.0	130.6[b,c]	493[b,c]	938[b,c]	434[a]	1692[a]	3423[a]	3.31[b,c]	3.43[a,b]	3.65[b]

[a,b,c,d] Average value having the same letter vertically are not significantly different.
[e] Restaurant food waste with an inclusion of 50% to the standard diet.
[f] A mixture of the 1, 2 and 3 class (mixed classes) restaurant food waste inclusion 50%.
CP = Crude protein; CF = Crude fibre; C fat = Crude fat; NFE = nitrogen-free extract

as percentages of body weight, did not show any significant difference between treatments. In a second experiment, kitchen waste was included at two levels (8 and 16%) and tested against control diets for starter and finisher chicks. Weight gain did not show great differences for the two levels of kitchen waste against the controls for both starter and finisher chicks, while feed consumption was higher and was reflected in feed conversion figures. In the finisher period, feed conversion was 2.65 for the control birds against 2.76 for both 8% and 16% kitchen waste, respectively. The results show that treated garbage is acceptable, since the differences in performance data were not significant, and food waste is cheap permitting economic results. It was concluded from these results that treated kitchen waste as a supplement for broiler diets is a suitable source of energy (11.9 MJ/kg) and protein (13.8%) with an in vivo protein absorbability of 77%.

Hoshii and Youshida (1981) concluded from their experiments by using garbage included in broiler diets, that the broilers' meat showed excellent palatability according to sensory pannel test on both flavour and taste. The percentage of the garbage included, however, was not reported in their summary.

(b) Layers

Soliman *et al.* (1978[b]) used restaurant food waste as a feedstuff for mature laying hens during a total experimental period of 7 weeks. The inclusion of the garbage in experimental diets was 10, 20, 30, 40 and 50% and was compared with a basal diet (Table 7.15). In this experiment, it was noticed that the feed consumption increased significantly, which could be attributed to high palatability or improved flavour, and low energy content of the diets with high levels (30, 40 or 50%) of garbage. Feed conversion showed no significant difference, however there was a numerical increase in the feed conversion with the increase of garbage level. Egg production showed no significant differences between the several treatments and the control. Egg weight, egg shape index, Haugh units, yolk colour, fertility and hatchability showed significant differences between birds fed the treatment and the control diets. Shell thickness and yolk index showed no significant differences. The significant differences detected are due to the variation in the quality of eggs produced by hens and are expected to be related to the pre-experimental genetic variation in these traits among birds allocated the different rations. However, these differences within the treatments are commercially acceptable characteristics. Results obtained for fertility and hatchability showed significant differences among treatments, which could be related to the differences in the physiological and genetical characteristics.

Yoshida and Hoshii (1979) studied the nutritive value of garbage produced from supermarkets (inedible parts of fresh fish and vegetables) with an inclusion of 20% in diets of laying hens. It was concluded that this inclusion had no significant effect on egg production of feed intake of the layers, but resulted in significant reduction of egg yolk colour. It was also reported from the panel test on the flavour of the eggs and hen meat that no differences could be detected.

Lipstein (1985) studied the nutritional value of dried, gamma irradiated kitchen waste in layer diets with inclusions of 7.5 and 15% of this waste, and tested this against control diets in linearly programmed rations for layers equal in crude protein (13.8%), metabolizable energy content (11.96 MJ/kg), methionine plus cystine (0.53%), lysine (0.60%), Ca (3.4%) and P (0.58%). In this experiment Leghorn x Rhode Island Red crossbred birds were used. The layers were selected and chosen according to a standard based on their equal individual daily egg record, egg weight and body weight and then allocated to each of the three treatments. The results of this experiment showed that there were no differences in production rate, egg weight, feed intake, feed conversion and body weight between the control hens and those receiving up to 15% of kitchen waste in the diet. It is concluded that the related kitchen waste is a suitable ingredient in layer diets.

Table 7.15. Average reproductive parameters of laying hens fed diets with several additions of restaurant food waste of various percentages after 7 weeks production and their egg quality, fertility and hatchability. From Soliman *et al.* (1978[b])

Parameters	*Control basal diet*	*Percentage inclusion of restaurant food waste to the basal diet*					*Sign-ificance*
		10%	*20%*	*30%*	*40%*	*50%*	
Feed consumption (kg/hen)	6.30	7.05	7.23	7.65	7.81	9.54	*
Feed conversion	2.67	2.78	2.95	3.23	3.02	3.49	NS
Egg production (egg/hen)	27.90	29.30	29.40	28.40	31.00	32.80	NS
Egg weight (g)	50.60	51.40	51.00	53.70	51.40	52.50	*
Egg shape index	75.00	75.00	76.00	74.00	76.00	75.00	*
Shell thickness (mm)	0.31	0.31	0.30	0.31	0.30	0.31	NS
Yolk index (ratio)	0.50	0.51	0.51	0.50	0.51	0.51	NS
Haugh units	93.10	90.00	91.30	82.80	88.10	94.80	*
Yolk colour	6.00	6.00	5.50	5.50	5.00	4.20	*
Fertility (%)	97.20	97.50	98.10	91.90	92.60	100.00	*
Hatchability (%)	59.70	80.10	80.50	66.00	72.70	82.30	*

* = significant differences at ($P < 0.05$)
NS = not significant

7.4.2 Sewage Sludge

(a) Growing chickens and broilers

Heavy metals

Damron *et al.* (1982) studied the effects of feeding dried municipal sludge or reagent grade minerals such as Cr, Cd, Cu and Fe (on the performance of broilers. Levels of 0, 3 or 6% Chicago dried municipal sludge were substituted into the basal diet while equivalent nutrient levels were maintained. Four additional treatments in experiment 1 and five in experiment 2 were tested. These treatments contained the levels of cadmium, chromium, copper and iron from reagent sources equivalent to the levels of these elements furnished in the diet when 6% sludge was added. In the second experiment, two different iron levels were used.

Table 7.16 shows the data for broiler performance from 0-3 weeks for chickens fed various levels of sludge or reagent grade minerals in relation to tissue (muscle, liver and kidney) mineral concentration. In experiment 1, the feeding of 2992 ppm of dietary iron or 150 ppm Chromium from reagent sources resulted in significant body weight depressions. Only the feed intake of the birds receiving the supplemental iron was significantly below that of the control group. In experiment 2, both iron levels (2993 and 2196 ppm) significantly depressed body weights and daily feed intake. The inclusion of 6% Chicago sludge had no effect upon bird performance in either experiments. The cadmium and iron treatments of both experiments resulted in elevated liver and kidney levels of these minerals in proportion to their levels of supplementation. There was also a trend of increased cadmium levels in the liver and kidney resulting from feeding increased sludge levels. However, the utilization rate of cadmium originating from sludge appeared to be only approximately 20%.

Anaerobically digested sewage sludge

Kienholz *et al.* (1981) studied the effect of feeding anaerobically digested Fort Collins sewage sludge on the performance of chicks up to 27 days of age. In this trial the inclusion of sludge was 20% and diets were balanced in protein and had the same vitamin and mineral supplement. The sludge diets contained an ME of approximately 10.46 MJ/kg compared to an ME of 12.55 MJ/kg in the control feed. Birds gained significantly less body weight on the sludge diet in comparison with the control. Also vitamin A content in chick liver was significantly reduced by feeding sludge. Extrusion of a corn-sludge mixture prior to its diet inclusion reduced growth, feed efficiency and lowered liver vitamin A levels. It was concluded that ingestion of sewage sludge impaired the utilization of vitamin A from the diet.

Dried activated sewage sludge

Ologhodo and Oyewole (1987) studied the effect of replacing groundnut meal by dried activated sewage sludge in diets for broilers up to 9 weeks of age. The performance of broilers was measured in terms of their growth rate and feed efficiency when 5 and 10% of groundnut meal were replaced by dried activated sewage sludge. All three diets contained 23% crude protein and 12.13% ME (MJ/kg). Inclusion of sewage sludge at a level of 10% significantly improved performance. However, inclusion of sludge at a level of 5% showed no significant improvement over the control group. Cost of feed per kg of diet and per kg body weight gain decreased with increasing levels of sludge. Nitrogen and lipid retention percentage and metabolizable energy content (corrected for nitrogen) were significantly ($P < 0.05$) higher at the 10% inclusion level of sludge. It was concluded from these results (Table 7.17) that replacing groundnut meal in broiler rations by 10% dried activated sewage sludge resulted in higher performance and economy of feed conversion compared to the 5% inclusion level and to the control diet.

Table 7.16. Performance of broilers chicks (from 0-3 weeks of age) fed various levels of sludge or reagent grade minerals in relation to wet-weight tissue mineral concentrations. From Damron *et al.* (1982)

Treatments	*BW* (g)	*FC* (g)	FCE (%)	*Muscle* [e] Fe	*Muscle* [e] Cd (%)	*Muscle* [e] Cu (%)	*Liver* [e] Fe (%)	*Liver* [e] Cd (%)	*Liver* [e] Cu (%)	*Kidney* [e] Fe (%)	*Kidney* [e] Cd (%)	*Kidney* [e] Cu (%)
First experiment												
Control	560[a]	809[a,b]	1.43[b]	0.73	ND	0.06	13.24	0.001	0.35	5.94	0.02	0.26
" + 3% sludge	541[a,b]	800[a,b]	1.48[b]	0.86	ND	0.06	19.00	0.003	0.34	8.48	0.06	0.24
" + 6% sludge	545[a,b]	802[a,b]	1.47[b]	0.77	ND	0.04	26.01	0.005	0.35	8.22	0.10	0.24
Cr 150 ppm	525[b]	762[b,c]	1.45[b]	0.79	ND	0.06	13.14	ND	0.31	5.25	ND	0.18
Cd 12 "	540[a,b]	823[a]	1.52[b]	0.60	ND	0.05	15.00	0.190	0.34	5.48	0.16	0·19
Cu 92 "	552[a,b]	815[a]	1.47[b]	0.58	ND	0.06	15.63	ND	0.36	6.36	ND	0.21
Fe 2992 "	450[c]	722[c]	1.63[a]	0.78	ND	0.04	63.54	ND	0.25	8.02	ND	0.21
Second experiment												
Control	576[a]	869[a]	1.50[d]	1.02	0.005	0.07	9.27	0.006	0.37	6.05	0.017	0.27
" + 3% sludge	563[a]	857[a,b]	1.53[b,c,d]	0.89	0.004	0.06	9.17	0.046	0.39	5.95	0.078	0.25
" + 6% sludge	553[a]	853[a,b]	1.54[b,c,d]	0.90	0.004	0.06	11.28	0.047	0.37	6.99	0.130	0.32
Cr 150 ppm	559[a]	869[a]	1.56[a,b,c,d]	0.98	0.004	0.05	5.95	0.004	0.33	6.02	0.008	0.24
Cd 12 "	528[a]	880[a]	1.68[a]	0.84	0.005	0.06	7.55	0.297	0.38	5.90	0.458	0.25
Cu 92 "	559[a]	853[a,b]	1.52[c,d]	0.87	0.003	0.06	8.01	0.003	0.39	5.90	0.006	0.26
Fe 2992 "	439[b]	737[c]	1.66[a,b]	1.02	0.002	0.06	64.05	0.002	0.36	7.84	0.005	0.25
Fe 2196 "	470[b]	775[b,c]	1.65[a,b,c]	0.84	0.002	0.06	39.25	0.002	0.34	6.27	0.004	0.22

[a,b,c,d] Means within columns without common letters are significantly different ($P < 0.05$).

[e] Chromium below *detection limits.*

BW = Body weight; FC = Feed conversion; FCE = Feed conversion efficiency.

However, the superior nitrogen retention and efficiency of energy utilization may be due to the favourable amino acid balance of the sludge. These findings are in agreement with the work of Firth and Johnson (1955) who found that 10% inclusion of activated sewage sludge in diets of chicks gave a substantial increase in growth and efficiency of chick performance.

Sova *et al.* (1980) studied the effect of cadmium in dried activated sludges used in poultry diets. It was concluded that the inclusion of 2% sterilized activated dried sludge, containing 35 ppm cadmium in the dry matter to replace 1 or 2% meat and bone meal resulted in 0.7 ppm Cd in the diet; no traces were found in liver, kidney or muscle of chicken after 7 weeks of age.

Table 7.17. Chemical composition (%) of sewage sludge, performance and nutrient utilization of broilers (0-9 weeks old) receiving dried activated sewage sludge in comparison with a control diet. From Ologhobo and Oyewole (1987)

Parameters	*Control*	*% Inclusion of sewage sludge*	
		5	*10*
Dry matter	88.100	89.000	93.500
Crude protein	23.100	23.100	23.100
Ether extract	4.400	4.000	6.100
Crude fibre	15.200	10.700	13.100
Gross energy (MJ/kg)	19.900	18.500	19.200
Feed intake (g/bird)	5.364[a,b]	5.433[a,b]	6.489[b]
Body weight gain (g/bird)	1.400[a,b]	1.425[a,b]	1.469[b]
Feed efficiency	2.610	2.620	2.260
Mortality (%)	0.220	0.560	0.330
Cost of production [c]			
Feed cost/kg diet	100.000	94.000	89.000
Cost of feed/kg liveweight gain	100.000	95.000	88.000
Nutrient utilization (%)			
Dry matter digestibility	74.160[a]	75.800[a]	78.220[b]
Nitrogen retention	80.250[a]	82.790[b]	84.480[b]
Lipid retention	90.620[a,b]	90.880[a,b]	93.660[b]
ME (MJ/kg)	14.480[a,b]	15.650[b]	15.980[b]
ME_n (MJ/kg)	13.470[a]	14.600[b]	15.020[b]

[a,b] Means within a row without a common letter are significantly different (P < 0.05).
[c] Control is 100%
ME_n = Metabolizable energy corrected for nitrogen.

Petkov *et al.* (1981) used dried activated sewage sludge as partial replacement of meat and bone meal in broiler diets. Inclusion of 1.6% for starter and 3.2% for finisher diets had no adversely effect on health, weight gain or efficiency of feed conversion for broilers.

Cibulka, Sova and Muizikar (1983) investigated the lead and cadmium in the tissues of broilers fed a diet with added dried activated sewage sludge. They used two sources of sludge (dried activated) that contained 30.5 and 18.4 ppm Pb and 2.9 and 2.1 ppm Cd, respectively. There was a relationship found between Pb contents in the sludge and in the tissues; however, the health standard limit for Pb (1.0 ppm) was not exceeded, while Cd contents exceeded the permitted limit (0.02 ppm).

Koucky and Adamec (1989) used the activated sewage sludge at an inclusion level of 3, 6 and 9% to replace soybean meal and compared this with a basal diet. Live weight of broilers at 8 weeks old was 1908 g for control birds. With increasing amounts of sludge, live weight was 1963, 1979 and 1875 g, respectively. Feed to gain ratio was 2.24 for the control group and 2.12, 2.05 and 2.17 for the three levels of sludge. Mortality was 7.2% for control and 0.5, 5 and 7.5% for the different inclusion levels, respectively. Carcass quality and chemical composition of the meat and liver were similar in all groups. The lead and cadmium content in tissue from birds fed the sludge diets did not exceed those of the control birds.

Sewage as a fertilizer in soil-plant-poultry systems

Several investigators have studied the application of sewage municipal sludge to fields on the content of heavy metals in crops such as maize, sorghum, wheat and soya, and on the performance and tissue of birds fed on these crops. Damron *et al.* (1980) used maize fertilized with municipal sludge to replace half of all the maize in the diets of broiler chicks from 1 day-old till 3 weeks of age and in diets of laying hens for 3 or 4 months of production. Replacement had no significant effect on growth, laying, hatching or body mineral composition. In one of the two trials with hens, the feed intake and the final body weight were significantly increased. In one trial with broilers the efficiency of feed conversion was significantly decreased. Damron *et al.* (1981) used maize fertilized with municipal sludge in the diet of chicks and hens. From their duplicate experiments, each of which lasted 21 days, they used 1 day-old broiler chickens to study the effect of replacing 50% or all of the normal dietary maize with maize grown on soil fertilized with municipal sewage sludge. Neither level of sludge-fertilized maize had any adverse effect upon final body weights or daily feed intake. The feed conversion values of experiment 1 were not significantly influenced by treatment, but efficiency of feed conversion was, significantly, for all sludge-fertilized maize of experiment 2. As far as the experiments with laying hens are concerned, the partial or total replacement of sludge-fertilized maize by maize produced with commercial fertilizer had no significant effects on any of the production values in experiment 1. In the second experiment, the 100% of the sludge-fertilized maize was associated with a significant increase in daily feed intake and final body weight. Hatchability and taste panel results for eggs indicated no significant relation with dietary treatment. Mineral values of liver, kidney and muscle tissue from hens and broilers were not affected by dietary treatment.

Hinesly *et al.* (1985) studied the transfer of sludge-born cadmium through plants to chicks. The maize hybrids and soybean cultivars, used were selected to accumulate high levels of cadmium, and were grown on strip-mine soil amended with sewage sludge. The maize grain and soybean meal were used to formulate starter, developer and layer diets for White Leghorn chickens. The diets contained 0.09 ppm 0.05 (low Cd), 0.57 ppm 0.11 (medium Cd), or 0.97 ppm 0.14 (high Cd) of biologically incorporated Cd. The highest dietary Cd level did not alter the concentration of Cd in brain, breast muscle, leg muscle or eggs. There was no indication that the highest dietary level of Cd affected feed consumption, body weight gains, the rate of mortality, egg production, egg quality, or the absorption of essential inorganic nutrients. The bodies of spent hens retained about 1-3% of the highest level of Cd ingested as a constituent of feed, of which about 60% was found in kidneys.

(b) Layers

Several investigators have studied the effect of the nutritional value of sewage sludge in the diets of layers on their performance. Johnson and Damron (1980) studied the performance of White Leghorn hens fed various levels of municipal sludge or selected minerals. The hens were fed diets containing 0, 3.5 and 7% municipal sludge in two experiments. Additional treatments consisted of selected levels of minerals from reagent sources comparable to those provided by the 7% sludge treatment (14 ppm Cd, 107 ppm Cu, 175 ppm Cr and 2562 ppm Fe). The results of these experiments showed that sludge diets have no adverse effects on egg production, feed/dozen eggs, daily feed, egg weights, specific gravity, Haugh units or body weights. A significant increase in Haugh unit values occurred in eggs from birds fed a diet with 7% sludge, 14 ppm Cd, 175 ppm Cr and 2562 ppm Fe. Supplemental Fe also significantly decreased daily feed intake. Hens fed 7% sludge had a significantly decreased hatchability of fertile eggs. Mineral composition of egg components had no influence by dietary treatment. Cadmium and iron levels of liver and kidney were elevated by sludge feeding.

Damron *et al.* (1982) have drawn more conclusions to the work of Johnson and Damron (1980) by repeating the same trial on a large scale. In their experiments Damron *et al.* (1982) used two levels of Chicago sludge and various levels of supplemental reagent grade minerals for feeding caged White Leghorn hens in two experiments for 12 weeks of production (first experiment) and 16 weeks (second experiment) (Table 7.18). Levels of 3.5 and 7% sludge were substituted into a basal diet. In addition, the amounts of cadmium (14 ppm), chromium (175 ppm), copper (107 ppm), or iron (2562 ppm) in the form of ferrous sulphate which were equivalent as an additional treatment in the second experiment equal to those found in the 7% sludge diet were fed as additional treatments using reagent sources.

In the first experiment, hen production criteria, egg weight and specific gravity were not significantly influenced by the dietary treatments, while, daily feed intake was significantly reduced by the iron level fed. Mineral analysis of the eggs from the hens in experiment 1 did not indicate any effect from any of the dietary additions. In contrast, the tissue mineral concentration found in experiment 1 indicated that increased mineral stores would result from Chicago sludge and reagent mineral feeding.

Table 7.18. Performance data of White Leghorn hens fed various levels of sewage sludge or reagent grade minerals in relation to wet-weight tissue mineral concentration. From Damron *et al.* (1982)

	Egg production (%) [d]	*Feed dozen (kg)*	*Daily feed (g)*	*Egg weight (g)*	*SG 1.0*	*HU*	*Final body weight (kg)*	*Muscle (%)*				*Liver (%)*				*Kidney (%)*			
								Cd	*Cu*	*Cr*	Fe	Cd	Cu	Cr	Fe	Cd	Cu	Cr	Fe
First experiment (84 days)																			
Control	76.0	1.56	99.2	62.1	$839^{b,c}$	$73.4^{a,b}$	1.63	ND	0.058	ND	0.90	0.020	0.24	ND	8.50	0.15	0.26	ND	6·70
" + 3.5%sludge	75.8	1.56	98.5	61.2	$846^{a,b,c}$	74.3^{b}	1.63	0.010	0.047	ND	0.76	0.520	0.30	ND	13.33	2.30	0.30	ND	13.90
" + 7% sludge	72.6	1.62	97.4	60.7	861^{a}	74.6^{b}	1.65	0.015	0.046	ND	0.61	0.950	0.31	ND	26.20	3.74	0.32	ND	18.55
Cd 14 ppm	73.6	1.56	95.7	61.5	828^{c}	74.1^{b}	1.53	0.020	0.040	ND	0.64	1.220	0.31	ND	9.55	8.06	0.37	ND	7.25
Cu 107 "	73.9	1.60	98.5	60.3	$856^{a,b}$	71.9^{a}	1.62	0.005	0.045	ND	0.60	0.075	0.29	ND	9.58	0.29	0.28	ND	8.68
Cr 175 "	73.0	1.60	100.4	61.5	$844^{a,b,c}$	$73.6^{a,b}$	1.71	ND	*0.038*	*ND*	0.78	0.020	0.27	0.04	7.48	0.13	0.27	0.04	6.88
Second experiment (112 days)																			
Control	$74.6^{a,b}$	$1.59^{a,b,c}$	95.7^{a}	63.0	770	76.2^{c}	$1.63^{a,b}$	-	-	-	-	-	-	-	-	-	-	-	-
" + 3.5% sludge	75.8^{a}	1.55^{c}	96.0^{a}	62.1	780	$78.8^{a,b,c}$	1.71^{a}	-	-	-	-	-	-	-	-	-	-	-	-
" + 7% sludge	77.3^{a}	1.52^{c}	96.2^{a}	61.2	760	81.3^{a}	1.68^{a}	-	-	-	-	-	-	-	-	-	-	-	-
Cd 14 ppm	77.3^{a}	1.52^{c}	96.6^{a}	62.0	770	$80.0^{a,b}$	1.58^{b}	-	-	-	-	-	-	-	-	-	-	-	-
Cu 107 "	76.0^{a}	$1.57^{b,c}$	97.2^{a}	62.7	760	$78.0^{b,c}$	1.72^{a}	-	-	-	-	-	-	-	-	-	-	-	-
Cr 175 "	73.4^{a}	$1.64^{a,b}$	96.3^{a}	63.1	770	$79.6^{a,b}$	1.68^{a}	-	-	-	-	-	-	-	-	-	-	-	-
Fe 2562 "	66.8^{b}	1.68^{a}	89.6^{b}	62.4	770	$79.9^{a,b}$	1.54^{b}	-	-	-	-	-	-	-	-	-	-	-	-

[a,b,c] Means without a row without common letters are significantly different ($P < 0.05$).

[d] % Hen-day egg production

SG= Specific gravity; it reads 1.0839; *ND = Below detection limits; HU = Haugh units.*

Shoremi, Sridhar and Ekpenyoung (1990) investigated the effect of sewage and sewage effluent as drinking water sources for point-of-lay chickens of 20 weeks of age. During this experiment four different water sources were used:

- Tap Water as a Control: The tap water was chlorinated at the source.
- Alum Effluent: The alum effluent was obtained by treating 40l of raw sewage with commercial grade aluminium sulphate at 400 mg/l. The clear water obtained after settling the coagulated effluent for 15 min. gave the alum effluent. The effluent was disinfected by using calcium hypochlorite to a residual chlorine level of 3 mg/l, to ensure that the pathogenic micro-organisms were destroyed.
- Secondary Effluent: Secondary effluent was obtained from sewage treated by trickling filters and activated sludge process. The effluent was chlorinated at the chlorine contact basin.
- Raw Sewage: Raw sewage was collected after screening and disinfected as the alum effluent. Layers in the point-of-lay or sexually mature were receiving a standard diet containing 16% crude protein and 12.03% ME (MJ/kg) during a 16 week treatment.

The conclusion of this trial was that the drinking water sources had no significant effect on water consumption, feed intake, egg production, final body weight, efficiency of feed utilization, or egg shell thickness but affected egg Haugh units significantly ($P < 0.05$). The mean egg production of birds receiving the raw sewage and effluent was higher than that for birds receiving tap water at the start of lay; birds receiving the raw sewage sustained this trend until 50% production was attained. Over the period of study, all the indices increased significantly. Table 7.19 shows the effect of various drinking water sources on the performance of layers from 20 weeks of age up to 16 weeks of production and egg quality indices.

Table 7.19. Effect of various drinking water sources on the performance of layers (Golden Hubbard) from 20 weeks old till 16 weeks of production and egg quality indices. From Shoremi, Sridhar and Ekpenyoung (1990)

Parameters	*Drinking water source*			
	Tap water (control)	*Alum effluent*	*Secondary effluent*	*Raw sewage*
Water consumption (l) [a]	1.200	1.130	1.230	1.200
Feed intake (kg) [a]	0.600	0.580	0.570	0.590
Feed utilization efficiency	0.700	0.730	0.660	0.630
Body weight (kg)	1.570	1.590	1.530	1.520
Egg production (%)	22.970	26.940	28.980	31.080
Egg weight (g)	47.030	48.600	49.160	48.360
Haugh units	47.010	74.290	70.500	70.300
Egg shell thickness (mm)	0.293	0.291	0.288	0.295
Mortality (%)	3.360	5.000	1.670	3.360

[a] Per week

In general, it may be concluded that the nutritive value of the municipal sewage sludge and its use as a feedstuff for broilers and layers largely depends on its processing and the condition of the sewage in relation to its content of (heavy) metals.

7.4.3 Upgrading Sewage Sludge

Several research workers have studied the nutritive quality of sewage sludge in producing a relevant feedstuff. Jiazhi, Fengjum and Hua (1988) cultivated earthworms on municipal sewage sludge, the dried worms and the rest of the sludge were used in diets of feeding laying hens with good results. Ocio *et al.* (1980) used housefly larva (*Musca domestica*) to grow and upgrade municipal sludge. The meal has a gross protein value of 89.5% which is greater than for good quality meat meal and less than that of fish meal. Gruhn (1990) estimated the value of sewage sludge (biosludge) treated with yeast which grow and upgrade the sludge. In this experiment, the spray-dried yeast grown on wastewater, biosludge or on pig slurry, was incorporated at a 0, 9, 18 or 27% level in isoenergetic and isonitrogenous diets at the expense of carcass meal for colostomized laying hens receiving 100 g feed per day. Biosludge yeast contained more protein and less ash and amino acids than that of slurry yeast. Apparent digestibility of protein from biosludge and slurry yeast, and from carcass meal was 67, 57 and 78%. Digestibility values of crude fat were 36, 33 and 97%, for lysine 75, 58 and 83% for methionine 66, 54 and 83%, and for cystine 93, 65 and 85%.

Calvert (1976) used several systems for indirect recycling by using animal and municipal sewage sludge as a substrate for protein production. Algae, yeast, bacteria, fungi, moulds, house-fly larvae, earthworms *etc.*, were grown on the sludge to upgrade its nutritive value. This point was discussed in detail in Chapter 2 in the sections describing dried poultry waste and its upgrading.

Finally, Muller (1976) discussed several methods of treating municipal sludge as ensilage, and other processes mostly used for ruminants and not suitable for monogastrics.

REFERENCES

Alten, E., Gevers, L. and Vormans, R. (1995). *Swill, een lekker kostje?* Agriculture High School, P.O. box 90108, 5200 MA 's-Hertogenbosch, The Netherlands, p. 71.

Ammerman, C.B. and Block, S.S. (1964). Feeding value of rations containing sewage sludge and sawdust. *J. Agr. Food Chem.*, **12**, 539.

Anon (1991). *Sludge indicator. Slip wijzer.* N.V.A. "Slibcommissie", p. 48 (in Dutch).

Anon (1992). *Potentieel composteerbaar bedrijfsafval in Nederland. Vrom* 92217-b.6-92, p. 47 (in Dutch).

Baird, D.M. and Young, C.T. (1973). Food waste makes high energy swine diet. *Feedstuffs*, **45**, (11), 20-3.

Beharrell, J. (1940). Kitchen waste for feeding farm stock. *Nature*, No. 3689, **146**, 47-8.

Berg, L.R. and Martinson, R.D. (1972). Effect of diet composition on the toxicity of Zinc for the chick. *Poultry Sci.*, **51**, 1690-5.

Calvert, C.C. (1976). Systems for the indirect recycling by using animal and municipal wastes as a substrate for protein production. In: New feed resources. Proceedings of a technical consultation held in Rome. *FAO, Animal Production and Health Paper*, No. **4**, pp. 245-64.

Cardenas, R.R. and Varro, S. (1973). Disposal of urban solid wastes by composting. In: *Symposium: Processing Agricultural and Municipal Wastes.* (Ed. G.E. Inglett). The AVI Publishing Company, Inc. Westport, Connecticut, pp. 183-204.

Chaney, R.L., Hundemann, P.T., Palmer, W.T., Small, R.J., White, M.C. and Decker, A.M. (1978).Plant accumulation of heavy metals and phytotoxicity resulting from utilization of sewage sludge and sludge composts on cropland. *Proc. 1977 Nat. Conf. Composting of Municipal Residues and Sludges.* Information Transfer Inc. Rockvill, MD, pp. 86-97.

Cibulka, J., Sova, Z. and Muizikar, V. (1983). Lead and Cadmium in the tissues of broilers fed a diet with added dried activated sewage sludge. *Environ. Technol. Lett.*, **4** (3), 123-8.

Cornelissen, A.A.J. (1990). *Fysisch onderzoek naar de samenstelling van het Nederlands huishoudelijk afval. Resultaten 1989.* (Data of sorted components of the domestic refuse in The Netherlands, data on the year 1989). Report No. 736201002. *National Institute of Public Health and Environmental Production*, De Bilt, The Netherlands, p. 55.

Damron, B.L., Hall, M.F., Janky, D.M. and Lutrick, M.C. (1980). Feeding corn fertilized with municipal sludge to broiler chicks and laying hens. *Poultry Sci.*, **59**, (7), 1561 (Abstract).

Damron, B.L., Hall, M.F., Janky, D.M., Wilson, H.R., Osuna, O., Suber, R.L. and Lutrick, M.C. (1981). Corn fertilized with municipal sludge in the diet of chicks and hens. *Poultry Sci.*, **60**, (7), 1491-6.

Damron, B.L., Simpson, C.F. and Harms, R.H. (1969). The effects of feeding various levels of Lead on the performance of broilers. *Poultry Sci.*, **48**, 1307-1509.

Damron, B.L., Wilson, H.R., Hall, F.M., Johnson, W.L., Osuna, O., Suber, R.L. and Edds, G.T. (1982). Effect of feeding dried municipal sludge to broiler-type chicks and laying hens. *Poultry Sci.*, **61**, (6), 1078-81.

Deobold, H.J. and Elvehjem, C.A. (1935). The effect of feeding high amounts of soluble iron and aluminium salts. *Am. J. Physiol.*, **111**, 118-23.

Draper, C.I. (1945). Processed garbage meal in the chick ration. *Poultry Sci.*, **24**, 442-45.

Fair, G.M. and Geyer, J.C. (1965). *Elements of Water Supply and Wastewater Disposal.* John Wiley & Sons, New York, p. 615.

FAO (1990). *Production Handbook*, vol. 42. Food and Agricultural Organization of the United Nations, Rome, Italy.

Firth, J.A. and Johnson, B.C. (1955). Sewage sludge as a feed ingredient for swine and poultry. *J. Agric. Food Chem.*, **3**, (9), 795-6.

Gruhn, K. (1990). Contents and digestibility of crude nutrients and amino acids in yeasts grown on biosludge and liquid slurry, compared with carcass meal. *Tierernährung und Fütterung*, **16**, 241-7.

Hackler, L.R., Newmann, A.L. and Johnson, B.C. (1957). Feed from sewage. III. Dried activated sewage sludge as a nitrogen source for sheep. *J. Anim. Sci.*, **16**, 125-9.

Hays, J.T. (1973). Composting of municipal refuse. In: *Symposium: Processing Agricultural and Municipal Wastes.* (Ed. G.E. Inglett). The AVI Publishing Company, Inc. Westport, Connecticut, pp. 205-15.

Hill, C.H., Matrone, G., Payne, W.L. and Barber, C.W. (1963). In vivo interactions of Cadmium with Copper, Zinc and iron. *J. Nutr.*, **80**, 227-35.

Hinesly, T.C., Hansen, L.G., Bray, D.J. and Redborg, K.E. (1985). Transfer of sludge-borne Cadmium through plants to chickens. *J. Agric. Food Chem.*, **33**, (2), 173-80.

Hoshii, H. and Yoshida, M. (1981). Variation of chemical composition and nutritive value of dried samples of garbage. *Jpn. Poultry Sci.*, **18**, (3), 145-50.

Jensen, L.S. (1975). Precipitation of a selenium deficiency by high dietary levels of Copper and Zinc. *Proc. Soc. Exp. Biol. Med.*, **149**, 113-16.

Jiazhi, J., Fengjum, S. and Hua, L. (1988). Effects of earthworms cultivated in municipal sewage sludge on feeding laying hens. *Environmental Sci. (China)*, **9**, (4), 10-13.

Johnson, W.L. and Damron, B.L. (1980). Performance of White Leghorn hens fed various levels of municipal sludge or selected minerals. *Poultry Sci.*, **59**, (7), 1565. (Abstract).

Johnson, D., Jr., Mehring, A.L., Jr., Savins, F.X. and Titus, H.W. (1962). The tolerance of growing chickens for dietary Zinc. *Poultry Sci.*, **41**, 311-17.

Kaiser, E.R., Zeit, C. and McCappery, J. (1968). Municipal incinerator refuse and residue. *Proc. National Incinerator Conf.*, Amer. Soc. Mech. Engrs., New York.

Kienholz, E.W., Ward, G.M., Johnson, D.E. Baxter, J., Braude, G. and Stern, G. (1979). Metropolian Denver sewage sludge fed to feedlot steers. *J. Anim. Sci.*, **48**, (4), 735-41.

Kienholz, E.W., Young, N.S.K., Smith, J.L. and Whiteman, C.A. (1981). Reduced vitamin A levels of chicks fed anaerobically digested sewage sludge. *Poultry Sci.*, **60**, (4), 884-6.

Kornegay, E.T., Vander Noot, G.W., Barth, K.M., MacGrath, W.S., Welch, J.G. and Purkhiser, E.D. (1965). Nutritive value of garbage as a feed for swine. I. Chemical composition, digestibility and nitrogen utilization of various types of garbage. *J. Anim. Sci.*, **24**, (2), 319-24.

Koucky, M. and Adamec, T. (1989). Use of activated sewage sludge in feeding broiler chickens. *Zivocisna Vyroba*, **34**, (11), 1027-36.

Kuiper, T.C. (1995). Market inventarisation of swill. *Report No. PK 9523/D01*, Groeneweg 39, 2718 AA Zoetermeer, The Netherlands, p. 22.

Kunishisa, Y., Yaname, T., Tanake, T., Fukuda, I. and Nishikava, T. (1966). The effect of dietary Chromium on the performance of chicks. *Jpn. Poultry Sci.*, **3**, 10-14.

Leach, R.M., Jr., Wang, K.W.L. and Baker, D.E. (1979). Cadmium and the food chain: The effect of dietary Cadmium on tissue composition in chicks and laying hens. *J. Nutr.*, **109**, 437-43.

Lipstein, B. (1984). Evaluation of the nutritional value of treated kitchen waste in broiler diets. *Proc. 17th World's Poultry Science Congr.* (Helsinki). pp. 372-4.

Lipstein, B. (1985). The nutritional value of treated kitchen waste in layer diets. *Nutr. Rep. Int.*, **32** (3), 693-8.

Mehring, A.L., Jr., Brumbaugh, J.H., Sutherland, A.J. and Titus, H.W. (1960). The tolerance of growing chickens for dietary Copper. *Poultry Sci.*, **39**, 713-19.

Metcalf & Eddy, Inc. (1979). *Wastewater Engineering: Treatment, Disposal, Reuse.* (Revisor: Tchobanoglous, G.). McGraw-Hill Book Company, New York, p. 920.

Müller, Z.O. (1975). Pre-investment study "Conversion of Jakarta city garbage into compost or organo-chemical fertilizer". *Asia Research Pte. Ltd., Singapore.* Olstoba, 1975.

Müller, Z.O. (1976). Economic aspects of recycled wastes. In: *Proceedings of a Technical Consultation on New Feed Resources.* 22-24 November, 1976. F.A.O. Rome, Italy, pp. 265-94.

NRC (1984). Nutrition requirements of poultry. *National Research Council.* National Academy Press. Washington, D.C. pp. 33-4.

Ocio, E., Vinaras, R., Rey, J.M. and Richelet, R. (1980). Biological value of house-fly larva (*Musca domestica*) estimated in chickens by the crude protein method. *Avances en Alimentacion Y Mejora Animal*, **21**, (5), 211-15.

Odekerken, G.C.M.M., Poelen, J.G.M. and Reeuwijk, A.F.A. (1997). Processing of swill. *Netherlands Patent No. 1004243*, Office of Industrial Property, Postbox 5820, 2280 HV Rijswijk, The Netherlands.

Ologhodo, A.D. and Oyewole, S.O. (1987). Replacement of groundnut meal by Dried Poultry Droppings (DPD) and Dried Activated Sewage Sludge (DASS) in diets for broilers. *Biol. Wastes*, **21**, (4), 275-81.

Petkov, S., Kacerovsky, O., Sova, Z., Pardus, J., Kalous, J., Cibulka, J., Jedlicka, Z. and Rajnochova, J.

(1981). Use of dried activated sewage sludge as partial replacement of meat and bone meal in feeds for broilers. *Sbornik Vysoke Skely Zemedelske V Praze, Fakulta Agronomica*, **B 33**, 197-210.

Poupoulis, C. and Jensen, L.S. (1976) Effect of high dietary Copper on gizzard integrity of the chick. *Poultry Sci.*, **55**, 113-21.

Purdom, P.W., Shoenberger, R., Michaels, A. and Bergstrom, A. (1966). Incinerator residue - a study of its characteristics. *Ann. Meeting, American Public Works Assoc.*, New York.

Robbins, K.R. and Baker, D.H. (1980). Effect of sulphur amino acid and source on performance of chicks fed high levels of Copper. *Poultry Sci.*, **59**, 1246-53.

Roberson, R.H. and Schaible, P.J. (1960). The tolerance of growing chicks for high levels of different forms of Zinc. *Poultry Sci.*, **39**, 893-96.

Scott, M.L., Nesheim, M.C. and Young, R.J. (1982). *Nutrition of the Chicken*. M.L. Scott and Associates, Ithaca, New York, p. 562.

Shoremi, O.I.A., Sridhar, M.K.C. and Ekpenyoung, T.E. (1990). Effect of sewage and sewage effluents as drinking water sources for point-of-lay chickens. *Biol. Wastes*, **32**, (1), 1-7.

Soliman, A.A., Hamdy, S., Khaleel, A.A., Abaza, M.A., Akkada, A.R. and El-Shazly, K. (1978[a]). The use of restaurant food waste in poultry nutrition. I. Effect on growing chicks. *Alex. J. Agric. Res.*, **26**, (3), 489-99.

Soliman, A.A., Khaleel, A.R., Hamdy, S., Abaza, M.A., El-Shazly, K. and Abou Akkada, A.R. (1978[b]). The use of restaurant food waste in poultry nutrition. II. Effect on laying hens. *Alex. J. Agric. Res.*, **26**, (3), 501-14.

Sova, Z., Petkov, S., Kristoufkova, V., Cibulka, J., Pardus, I., Nemic, Z., Fucikova, A. and Zidek, V. (1980). Cadmium in dried activated sludges used in diets for poultry. *Biologizace a Chemizace Zivocisne Vyroby, Veterinaria*, **16**, (2), 151-6.

Supplee, W.C. (1961). Production of zinc deficiency in turkey poults by dietary Cadmium. *Poultry Sci.*, **40**, 827-8.

Titus, H.W. and Fritz, J.C. (1971). *The Scientific Feeding of Chickens*. Vth Edition, De Interstate Danville, Illinois, USA, p. 336.

Toth, S.J. (1973). Composting agricultural and industrial organic wastes. In: *Symposium: Processing Agricultural and Municipal Wastes*. The AVI Publishing Company, Inc., Westport, Connecticut, pp. 172-82.

Vohra, P. and Kratzer, F.H. (1968). Zinc, Copper and manganese toxicities in turkey poults and their alleviation by EDTA. *Poultry Sci.*, **47**, 699-704.

Wong, W.H. and Leung, K.L. (1979). Sewage sludge and seaweed (*Ulva sp.*) as supplementary feed for chicks. *Environmental Pollution*, UK, **20**, (2), 93-101.

Yoshida, M. and Hoshii, H. (1979). Nutritive value of garbage of supermarkets for poultry feed. *Jpn. Poultry Sci.*, **16**, (6), 350-5.

8. PALATABILITY AND FEED INTAKE REGULATIONS

Contents

8.1 INTRODUCTION

As a result of the increase in the total world population, more grains and animal products have to be produced for food. The competition between the humans and the birds for consuming grains, as wheat and maize, has triggered much research to find suitable alternative feedstuffs for poultry (milling by-products, animal by-products and biological waste).

To obtain the optimum nutritive value from grains and protein sources, it is important to have a correct feed formulation. The increased application of linear programming techniques in feed formulation has enhanced the need for more detailed information about nutrient levels in feedstuffs. Evaluation of palatability is included in feed formulation only by setting limits on the maximum inclusion of raw materials on the basis of their assumed palatability. It would be more advantageous to set the limits as realistic figures (percentages) which can be directly attributed to palatability of feedstuffs. Unfamiliar combinations of ingredients such as waste may have a definite effect on ration palatability, which is reflected in the broilers' and layers' performance.

Since gustatory impulses (the sense of taste) arouse sensations of taste and receives much reinforcement from the other senses (for example chemical receptors), they form a part of the perceptive processes of a bird, thereby determining the palatability of any feedstuff.

It is a fact that broilers and layers have an acute sense of taste (gustation). They have the ability to differentiate between sweet, salt, sour and bitter tasting chemicals. Basically, their selection or rejection of feed is based on shape, colour, texture, viscosity, osmotic pressure, nutritive value and toxicity (El Boushy *et al.*, 1989[a,b]). It has been shown that chickens possess 218-499 taste buds spread over the whole oral cavity and the tongue (Saito, 1966). This number in comparison with pigeons, which have 50-75 (Bath, 1906) seems to be a relevant number for taste buds for chickens.

Most studies on the sense of taste in chickens have used flavoured aqueous solutions (Jacobs and Scott, 1957; Kare, Black and Allison, 1957; Kare and Medway, 1959; Kare and Pick, 1960; McNaughton, Deaton and Reece, 1978). It is generally accepted that the concentration of a flavour in solution that is needed to elicit a response, is lower than the concentration in dry feed (Kare, Black and Allison, 1957). Balog and Millar (1989), working with high sweetening agents, concluded that birds can detect differences in flavour and consumed flavoured feed in a specific order of preference: aspartame, saccharin, citric acid, salt and quinine. From this, it was concluded that it is economically feasible for a producer to increase feed intake of birds in order to improve performance. This will shorten the time needed for the birds to be ready for market and accordingly reduce the number of days for which the producer must pay for the birds' upkeep.

8.2 SENSORY INVOLVEMENT IN CONTROLLING FEED INTAKE

Olfactory and gustatory impulses arouse sensations of smell and taste. Several factors are involved with the response of birds to their environment. Chickens possess about 218-499 taste buds (Saito, 1966) which are found at the base of the tongue, the floor of the pharynx and the oral cavity. They are innervated by the fibres from the glossopharyngeal nerve. The sense of taste in the fowl may be analysed in terms of the classical categories such as sweet, salty, sour and bitter. The fowl is also very sensitive to the temperature of water, and will tend to reject water which is above the ambient temperature but will readily drink ice-cold water (Kare and Mason, 1986).

8.2.1 Feed Recognition

In selecting their feed, birds appear to rely to a large extent on vision. The refusal or acceptance of feed on its first introduction is determined by colour and general appearance as far as attractiveness and structure are concerned.

Visual properties, colour and surface texture of the feed have been reported to take precedence over all other qualities in the birds' selection of feed, and examples of this statement are numerous. As far as colour is concerned, Gentle (1971[b]) reported that chickens with lingual nerve section showed a marked reduction of feed intake by changing only the colour of the feed. Van Prooije (1978) noticed that the chicken selects for colour. It was concluded from a test with ornamental corn, that chickens prefer yellow-white seed, followed by yellow, orange and finally orange-red. The red, red-blue and blue seeds were only eaten in exceptional circumstances (severe hungriness). Hess and Gagel (1956) concluded that newly-hatched chicks preferred to eat orange and blue feed. Capretta (1969) reported a preference for green over red feed in domestic-fowl chicks and Cooper (1971) found the same results with one-day-old turkeys while Kennedy (1980) found a significant preference for red and natural coloured diets over black and green with adult chickens (Table 8.1). Kennedy (1980) concluded from his trials with hatching chicks that all chicks showed a preference for diets of the colour that was fed after hatching. This may explain the conflicting reports in feed colour preference.

Chickens have a strong bias to use colour in avoiding substances which produce illness after ingestion (Martin, Bellingham and Storlien, 1977; Gillette, Martin and Bellingham, 1980).

Table 8.1. Selection of coloured diets by adult chickens in paired choice feeding trials. From Kennedy (1980)

Diet Colour	*Feed intake/bird/day (g)*	*Number of occasions (days) preferred*
Light Red	92.4	62[a]
Black	83.8	44
Green	86.4	45
Control (Natural)	92.4	65[a]

[a] Red and natural-coloured diets were significantly preferred to black and green ones.

The development of feed recognition by young, newly-hatched chicks has been extensively studied by Hogan (1971, 1973[a,b], 1977) and by Hogan-Warburg and Hogan (1981). Young chicks given a mixture of feed and sand learn to ingest primarily feed but will still ingest some sand. The increase in feed ingestion is probably the result of an association between the visual-tactile-gustatory stimuli from the feed and the positive long-term effects of the feed ingestion. The fowl often initially rejects the unfamiliar feed by recognition. The chicken will select the largest seed when offered one large and one small corn-seed (Frantz, 1957; Schreck *et al.*, 1963; Dawkins, 1968; van Prooije, 1978).

Portella, Caston and Leeson (1988) studied the apparent feed particle size preference of broilers. They concluded that feed particles were selected by broilers according to size. They added that the disappearance from the trough of particles (soybean meal - maize based diet) larger than 1.18 mm occurred at all ages, however, at 8 and 16 days old, the disappearance rate of particles between 1.18 and 2.36 mm in size was the greatest. As chickens aged, the disappearance rate was greatest for particles larger than 2.36 mm. However, Hogan (1977) studied the development of feed recognition in young chicks in relation to associative and non-associative effects of experience. There was immediate discrimination between feed and sand, based on taste indicators.

Kare and Medway (1959) stated that birds' vision possibly exceeds that of humans and that birds can differentiate between the refractive index of solutions. A further prove of visual recognition is that when a cockerel is offered real and imitation peas, it selects the real ones without touching the others (van Prooije, 1978).

Gentle (1985) provided a complete review about the sensory involvement in the control of feed intake in poultry. It was reported as far as feed recognition is concerned, that the domestic hen is a nidifugous bird; because the chicks are not fed directly by the parents there is an elaborate system of innate behavioural patterns which protect the birds from ingesting noxious diets. The innate reflexes are subsequently modified by new experiences allowing the birds to exploit a variety of valuable feed sources. It was also reported that the visual factors can be separated into those which relate directly to the feed as colour, shape and structure, and those which are not related to feed such as directing attention to feed sources by the mother hen (Turner, 1965; Hogan, 1973[a]; Savory *et al.*, 1978). Young chicks eat more feed in the presence of an active companion than they do in isolation and this has been called social facilitation (Tolman, 1964, 1965; Tolman and Wilson, 1965). Social facilitation is responsible for the increase in growth rate of group-reared chicks (Schreck *et al.*, 1963) but Savory (1975) proved that these differences in growth rate were due to feed conversion efficiency. Social facilitation plays an important role in the initiation of pecking (Strobel and Macdonald, 1974). These social factors in chicks have the effect of synchronizing the feeding of the whole groups and similar behaviour is noticed in adult birds (Hughes, 1971).

8.2.2 The Sense of Smell (Olfaction)

Work over the last two decades has indicated that the idea that birds possess a sense of smell is a controversial one. The only clear example is that irritants such as ammonia and acids stimulate the free nerve endings of numerous surfaces, including those of the nasal chambers. Air breathing is usually thought to have a telereceptor association, enabling the reception of airborne chemical stimuli in extreme dilution over relatively large distances. The nasal cavity of birds is innervated by the trigeminal as well as the olfactory nerve. There is no direct evidence for chickens using this olfaction in feed intake, while some recent work suggests that they may regulate their behaviour in response to olfactory factors (Jones and Gentle, 1985).

Smell takes place in the olfactory organ, which consists of external nares (nostrils), nasal chambers (conchae), internal nares (choane), olfactory nerves, the peripheral terminals (taste buds which lie in the olfactory epithelium) and the olfactory bulbs in the brain (Bang, 1971). Chemicals used for olfactory investigation in laboratories were not biologically relevant nor presented at levels of sensitivity to chicks (Wurdinger, 1979).

Poultry lack the behaviour of sniffing, indicating that birds need moving air to effect contact between odour stimuli and receptors. An example of the most developed olfactory system is the pigeon which uses olfactory cues for navigation over long distances (e.g. 500 km) (Kare and Mason, 1986).

8.2.3 The Sense of Taste (Gustation)

The function of taste is to encourage the ingestion of nutrients, to select among feeds that which is palatable and to avoid those that are unpalatable or toxic. The sensation of taste is stimulated by saliva and drinking. The bird's sense of taste differentiates the taste qualities of sweet, salt, sour and bitter. Kare, Black and Allison (1957) demonstrated that the fowl has an acute sense of taste. The ability to taste, however, is not uniformly present in all chickens. Some of them are "taste blind". This observation explains differences between replicates in experimental trials (Kare and Pick, 1960; Ficken and Kare, 1961; Gentle, 1972). Williamson (1964) found significant sex differences indicating a genetic difference, in the ability of chicks to taste ferric chloride, which can be a base for the selection of special lines.

It is suggested that taste plays a major role in the initial selection of feed and possibly in the motivation to eat (Gentle, 1971[b]). Therefore many flavours have been studied to improve feed consumption, weight gain and feed conversion (Berkhoudt, 1985). The preference test is thereby the most common laboratory method used to measure sensitivity of birds to taste stimuli (Kare and Mason, 1986). However, special care must be taken in cafetaria-type tests in which more than three taste stimuli are presented: these seem to overwhelm the discriminatory ability of birds (Kare and Mason, 1986). In addition, the flavours used are much more effective in chicken when administered through water than through feed. Chickens have a much dryer oral cavity than mammals after intake of dry feed, so the taste buds are less stimulated (Weischer, 1953; Kare and Pick, 1960). Weischer (1953) also reported that for chickens, contrary to parrots and starlings, all taste receptions are accompanied by a negative protective reaction (the more negative the reaction, the more toxic the substance).

(a) Reaction to sweetness

As discussed earlier, that chickens avoid certain feedstuffs on the basis of taste, lack of visual appeal or because of adverse effects on metabolism or "sense of well being". Sweeteners may help to prevent "starve outs" in baby chicks, or to keep chickens "on feed" during periods of disease or other stress (Thaxton and Parkhurst, 1976; McNaughton, Deaton and Reece, 1978).

Contrary to the widely held supposition that sugars have a universal appeal, most avian species do not avidly select sugar solutions, when fed on a balanced diet (Jukes, 1938; Kare and Medway, 1959; Kare and Pick, 1960; Kare and Rogers, 1976). However Jacobs and Scott (1957) and Gentle (1972) reported sucrose to be the only sugar for which the chickens showed a significant preference at concentrations of 5 and 12%, respectively. Gentle (1971[b]) also reported that chickens had a preference for a diet with 20% sucrose. Gentle (1972) also reported a rejection of 30% sucrose, fructose or glucose solutions, and that glucose was rejected at concentrations up to 5%. The rejection of xylose was noticed by Kare and Medway (1959). Several investigators reported the reaction of chickens to several concentrations of sweeteners (Tables 8.2 and 8.3). Kare and Rogers (1976) and Weischer (1953) postulated that nectar or fruit-eating bird species are more likely to respond positively to sugars than are insectivorous birds which respond negatively or not at all.

Table 8.2. Reaction of chickens to different types of sweeteners. (Medium is water in each case). From [(1)] Kare and Medway (1959) and [(2)] Gentle (1972)

Kind of Sweetener (1)	*Concentration (%)*	*Reaction*
Dextrose	20	Indifferent
Sucrose	< 20	Indifferent
Sucrose	> 20	Moderate rejection
Xylose	> 2.5	Rejection
Lactose	5	Indifferent
Galactose	5	Indifferent
Mannose	5	Slight rejection
Raffinose	5	Indifferent
Fructose	5	Indifferent
Arabinose	5	Slight rejection
Kind of Sweetener (2)	*Concentration (%)*	*Reaction*
Sucrose	5	Preference
Sucrose	> 30	Rejection
Sucrose	10-20	Indifferent
Fructose	1, 5, 30	Rejection
Fructose	10-20	Indifferent
Glucose	>5	Rejection
Glucose	<2.5	Indifferent

Table 8.3. Reaction of chickens to different types of sweeteners.

Sweeteners	*Concentration (%)*	*Reaction*	*Medium*	*References*
Sucrose	> 20	Preference	Feed	Gentle (1971[b])
Sucrose	12	Preference	Water	Jacobs and Scott (1957)
Saccharine [a]	0.028	Moderate rejection	Water	Kare, Black and Allison (1957)
Saccharine [a]	0.09	Moderate rejection	Water	Deyoe *et al.* (1962)
Molasses	0.09	Moderate rejection	Water	Deyoe *et al.* (1962
Sugar	< 20	Indifferent	Feed	Jukes (1938)

[a] High intensity sweetener.

Table 8.4. Reaction to different kinds of salts by chickens.

Kind of salt	*Concentration*	*Reaction*	*Medium*	*References*
NaCl	>2 %	Rejection	Water	Gentle (1971[b])
NaCl	5 %	Rejection	Feed	Gentle (1971[b])
NaCl	0.9 %	Indifferent	Water	Kare and Rogers (1976)
NaCl	1.5 %	Indifferent	"	Engelmann (1934)
KCl	<0.1 N	Indifferent	"	Gentle (1972)
NaCl	<0.25 N	Indifferent	"	Gentle (1972)
$CaCl_2$	<0.05 N	Indifferent	"	Gentle (1972)

Cameron (1947) gave a warning concerning the use of glucose solutions that were not freshly prepared in tests. A freshly prepared glucose solution, containing -glucose is much sweeter than a solution which is 16 hours old.

A number of factors other than taste may be involved, individually or collectively, in the response of a bird to a sugar solution, e.g. osmotic pressure, viscosity, melting point, nutritive value, toxicity and visual characteristics (Kare and Mason, 1986); these factors will be discussed in detail later.

(b) Reaction to salt

Electrolytes are set the conditions in which living systems can exist. The electrolytes include the "cations" sodium, potassium, magnesium, calcium and the "anions" chlorides, bicarbonates and sulphates. These electrolytes are commonly used in rations to achieve optimum performance from the birds.

Birds are able to distinguish between different salts in different concentrations when given a free choice (Williamson, 1964). Rensch and Neunzig (1925) investigated sodium chloride thresholds (i.e. the lowest concentrations at which solutions are rejected) for 60 species. They found that the threshold levels were very low in parrots (0.35%) and pigeons (0.5-0.9%) and extremely high in partridge (20%) and siskin species (37%). Engelmann (1934) and Pick and Kare (1962) found a threshold level for fowl of 1.5% sodium chloride, and an even lower threshold level for baby chicks. Mariotti and Fiore (1980), using operant-conditioning techniques (key-pecking and other tasks), reported a discrimination threshold level for pigeon of 0.2% sodium chloride, a level four times lower than the level reported by Engelmann (1950) using the usual choice of technique.

Many birds are indifferent up to the concentration at which their kidneys can no longer tolerate the salt solution (Kare and Beily, 1948). A hypertonic solution is strongly rejected and many chicks will die of thirst rather than consume a toxic 2% salt solution (Kare and Mason, 1986). However, these authors added that birds kept on a salt-free diet, will eagerly pick pure salt, when it is made available to them. The different threshold levels for the various salt concentrations are given in Tables 8.4 and 8.5.

Table 8.5. Preference for sodium and chloride metallic solutions at various concentrations over distilled water for chicks [a]. From Kare and Mason (1986)

	Concentration (g/100 ml)				
Solution	*0.1*	*0.2*	*0.4*	*0.8*	*1.0*
Na acetate	55[a]	52	56	52	51
Na sulfate	54	52	52	53	50
Na phosphate [b]	52	53	52	52	54
Na succinate	49	52	54	50	56
Na citric	54	52	54	47	35
Na phosphate [c]	51	49	47	44	14
Na tungstate	50	46	48	-	-
Na bicarbonate	52	43	38	20	14
Na benzoate	49	41	23	15	10
Na bisulphate	38	23	35	17	23
Na pyrophosphate	46	37	20	3	4
Na perborate	42	29	10	9	4
Na carbonate	42	30	10	4	2
Na phosphate [d]	46	20	4	1	2
Na cholate	4	20	3	-	3
Sodium Cl	50[a]	50	55	50	45
Magnesium Cl	49	51	51	53	45
Choline Cl	51	48	49	50	51
Manganese Cl	49	51	46	16	-
Strontium Cl	50	38	44	18	9
Ammonium Cl	49	46	35	12	6
Barium Cl	36	48	41	-	15
Calcium Cl	43	45	27	15	5
Zinc Cl	33	24	10	2	2
Cobalt Cl	26	12	6	5	6
Tin Cl	30	7	1	1	2
Copper Cl	6	11	3	8	4
Iron Cl	2	4	2	3	4

[a] Preference = (salt solution consumed x 100)/total fluid intake.
[b] Monobasic.
[c] Diabasic.
[d] Tribasic.

In the formulation of broiler and layer rations, special precautions are needed for the addition of sodium chloride. To avoid wet droppings, sodium chloride was lowered from 0.5% till 0.25-0.30%. The effect of salt levels in pheasant diets was investigated by Scott, Nesheim and Young (1982; Table 8.6). It was concluded that increased salt levels in the diet from 0.25-4% increased the moisture content of faeces from 74-89%, respectively; levels higher than 4% caused high mortality.

Table 8.6. Effects of high salt intake in pheasants. From Scott, Nesheim and Young (1982)

Salt levels in diet (%)	*Average weight 4 weeks (g)*	*Mortality (%)*	*Moisture content of faeces (%)*
0.25	230	4	74
1.0	232	5	78
2.0	223	3	86
3.0	219	6	88
4.0	218	5	89
5.0	197	7	-
7.5	165	23	-

(c) Reaction to sourness (acid)

Acidity and alkalinity in drinking water have a wide range of tolerance by birds, with a threshold level for acid being below 1% (0.17 M) (Rensch and Neunzig, 1925; Engelmann, 1934; Kare and Mason, 1986). Fuerst and Kare (1962) reported that chicks will tolerate strong mineral acid solutions up to pH 2. Organic acids are less acceptable and the tolerance for the hydrogen ion is not equivalent to that for the hydroxyl ion (Kare and Mason, 1986) (Tables 8.7 and 8.8).

Table 8.7. Percentage intake in chickens of acids and bases at different pH levels[a]. From Kare and Mason (1986)

pH	1.0					2.0				3.0			4.0
Acids													
HCl		4		19		50				59			
H_2SO_4			15	35	54						56		
HNO_3		8				62				52			
Acetic									16				53
Lactic							15	61					

pH	10.0		11.0		12.0			13.0	
Bases									
NaOH		45		47			33		2
KOH				48		36			3

[a] Tabled values are the mean of replicate lots. The percentage intake = (volume of tested fluid/total fluid intake) x 100 (18 daily values were averaged). The position of the numbers is an indication of the pH of the test solution. For example, at pH 1.5 the average daily consumption of HCl was 19% of the total fluid intake. Distilled water was the alternative in every instance.

Table 8.8. Reaction of chickens to different kinds of acids.

Kind of acid	*Concentration*	*Reaction*	*Medium*	*References*
Butyric acid	a	Indifferent	Feed	Fisher and Scott (1962)
Citric acid	>2 %	Rejection	Water	Gentle (1971[b])
Citric acid	2 %	Preferation	Feed	" "
Acetic acid	>0.3 N	Rejection	Water	" "
Acetic acid + glucose	0.3 N + 10%	Moderate rejection	"	" "
Acetic acid	0.2 N	Slight rejection	"	" "
Hydrochloric acid	<0.1 N	Indifferent	"	" "
Acetic acid	<0.1 N	Indifferent	"	" "

[a] = Not reported.

Gentle (1971[b]) reported that chicks showed no differences in feed intake with an addition of 6% citric acid. Gentle (1972) added that the addition of glucose to an aversive sour solution, however, reduced the aversion.

Engelmann (1934) found that the pigeon is more sensitive than the duck or the domestic fowl to hydrochloric acid concentration is concerned (threshold level of 0.01 M in pigeon versus 0.02 M and 0.025 M in duck and fowl, respectively). Engelmann (1950) added for the same acid that the sensitivity in baby chicks was more pronounced than in adults (0.009 M and 0.025 M, respectively).

(d) *Reaction to bitterness*

There are offensive taste compounds which are, in low concentrations, bitter to man but quite acceptable to chickens, e.g. sucrose octacetate (Kare and Mason, 1986). Some bitter components are rejected by birds but accepted by man e.g. dimethylanthranilate. Other bitter components are offensive to both man and birds, like quininehydrochloride or -sulphate (Kare and Pick, 1960; Yang and Kare, 1968; Rogers, 1974).

Engelmann (1934) showed that the pigeon, which is most sensitive to salt and sourness, seemed to be indifferent for quinine hydrochloride and sulphate, standard bitter stimuli for man and rats. In contrast, the duck and the chicken showed a marked rejection at low concentrations. However, Duncan (1960) reported a marked rejection for the pigeon at 0.1% quinine. Gentle (1976) found that quinine-HCl acceptability changed significantly on water deprivation.

(e) Reaction to other flavours

Sizemore and Lillie (1956); Romoser, Bossard and Combs (1958), Fisher and Scott (1962) and Williams and Kienholz (1974) found that none of the flavours they included in feed was responsible for a marked improvement in weight gain or feed utilization of chicks over control birds (Tables 8.9, 8.10 and 8.11). The flavours which were reported most accepted by chicks were butter, and flavour made up of esters and organic acids: EBPA (Ethyl Butyrate Phenylacetic Acid) (Deyoe et al., 1962; Table 8.12). In this test, EBPA was the only flavour which was consumed by chicks in larger quantities than by chicks on the control diet. Deyoe et al. (1962) indicated also that most flavours (in solution) in combination with butter were more acceptable to the bird and had an effect that might be compared with the neutralisation of the aversion to acid by glucose.

Table 8.9. Influence of various flavours added to feed on chick body weight and feed utilization at 28 days of age. From Fisher and Scott (1962)

Flavour	*Body weight (g)*	*Feed utilization (g gain/g feed)*
None	358	0.520
None	345	0.501
Coumarin	347	0.507
Vanillin	339	0.519
Butyric acid	326	0.516
Geraniol	317	0.479
Anise imitation	342	0.514
Anise compound	369	0.520
Spicing compound	322	0.508
Apple	352	0.518
Caramel	338	0.518
Oil cassia	380	0.534
Black walnut	339	0.526
Maple	332	0.503
Butter arome	349	0.533
Olive arome	326	0.504
Coconut	351	0.511
Honey imitation	320	0.514

Table 8.10. Weight gain of New Hampshire x Columbian chicks fed with flavoured feeds. From Romoser, Bossard and Combs (1958)

Pen Flavour choice (fed in two separate hoppers)		Weight gain (g) Experiment 1 (0-3 wks.)	Experiment 2 (1-3 wks.)	Average weight gain (g)
1	None	203	280	242
2	Strawberry	210	275	242
3	Anise	194	277	236
4	Grape	203	278	241
5	None and strawberry	194	287	241
6	None and anise	213	278	246
7	None and grape	207	279	243
8	Strawberry and anise	227	286	257
9	Strawberry and grape	222	270	246
10	Anise and grape	227	291	259

Table 8.11. Effect of addition of spices to diets [g] on broiler chick body weight gain and feed efficiency. From Williams and Kienholz (1974)

Spice added to diet	*Level added to diet (%)*	*Average 27 day weight gain (g)*	*g Gain/g feed (x 100)*
None (control)	0.0	614[a]	63
Black pepper	1.5	317[c]	49
Black pepper	3.0	242[d]	46
Black pepper	6.0	163[e]	42
Black pepper	12.0	53[f]	31
Black Pepper [h]	12.0	55	26
Chili powder	1.5	584[a,b]	58
Chili powder	3.0	608[a]	57
Chili powder	6.0	599[a,b]	61
Chili powder	12.0	515[b]	52
Curry powder	1.5	588[a,b]	58
Curry powder	3.0	585[a,b]	57
Curry powder	6.0	512[b]	54
Curry powder	12.0	460[b]	50

[a,b,c,d,e,f] Data without common letters are significantly different ($P < 0.01$).

[g] The basal diet contained (% basis) the following: 31 ground corn, 22.6 ground milo, 30 soybean meal (44% protein), 3 fish meal, 5 meat and bone meal, 2.5 dehydrated alfalfa, 0.7 ground limestone, 0.5 dicalcium phosphate, 0.4 NaCl, 4 fat, 0.25 vitamin mix, 0.05 trace mineral mix, 0.05 methionine. The vitamin mix added to diets the N.R.C. level of each vitamin. The trace mineral mix supplied the following to the diet (as ppm): 50 Mn, 50 Fe, 50 Zn, 5 Cu, 1.5 I and 0.5 Co.

[h] That group received control diet until 9 days of age, then 12% black pepper to 27 days of age.

Table 8.12. Effect of flavour on water consumption, growth and feed conversion at four weeks of age. From Deyoe *et al.* (1962)

Experiment no.	*Flavour*	*Flavoured water consumed (%)*	*Plain water consumed (%)*	*Body weight (g)*	*Feed/gain (g/g)*
1	Butter	44.9	55.1	290.4	-
	Licorice	39.3	60.7	286.2	-
	Molasses	38.9	61.1	330.1	-
	Anise	39.5	60.5	275.7	-
	Anethole	36.4	63.6	291.0	-
	Rum	40.7	59.3	280.4	-
	Cinnamon	36.7	63.3	287.9	-
	Strawberry	41.1	58.9	292.9	-
	Milk	35.8	64.2	280.7	-
	Grass	31.9	68.1	286.1	-
	Saccharin	34.3	65.7	275.9	-
	EBPA [a]	53.9	46.1	286.2	-
2	Garlic	38.7	61.3	316.5	2.07
	EBPA	51.2	48.8	319.3	2.21
	Barley	49.5	50.5	301.5	2.01
	EBPA (double strength)	51.4	48.6	297.6	2.04
	Chocolate	45.3	54.7	296.4	2.38
	Vanilla	47.8	52.2	293.1	2.10
	Coffee	48.2	51.8	293.1	2.05
	Quince	48.9	51.1	287.6	2.23
	Onion	45.4	54.6	295.4	2.33
	Camphor	46.8	53.2	296.5	2.48
	Eugenol	22.7	77.3	304.8	2.10
	Coconut	38.2	61.8	323.3	2.01
3	Molasses butter	50.2	49.8	307.5	2.15
	EBPA	56.9	43.1	320.3	2.30
	Barley butter	49.0	51.0	319.7	2.10
	Anise butter	42.7	57.3	324.4	2.22
	Vanilla butter	46.7	53.3	306.0	2.23
	Orange butter	50.7	49.3	329.7	1.90
	Garlic butter	49.1	50.9	335.0	2.06
	Licorice butter	48.3	51.7	308.9	2.13
	Camphor butter	47.7	52.3	309.7	2.25
	Quince butter	51.1	48.9	305.2	2.32
	Chocolate butter	51.1	48.9	315.5	2.07
	Onion butter	52.2	47.8	297.0	2.20
	Coconut butter	51.0	49.0	322.1	2.19
4	Coconut butter	51.5	48.6	274.5	2.21
	EBPA (double strength)	54.0	46.0	279.3	2.20
	Indol	39.7	60.3	262.8	2.14
	Seatol	50.6	49.4	267.3	2.18
	Nerolin	27.0	73.0	249.0	2.24
	Coumarin	39.0	61.0	289.8	2.24
	Grape	35.9	64.1	271.0	2.38
	Safrol	50.3	49.7	278.0	2.22
	Lilac	49.0	51.0	305.0	1.76
	Walnut	43.1	56.9	269.8	2.19
	Fecal	41.0	59.0	245.5	2.21
	Orange	40.4	59.4	273.0	2.27

[a] The qualitative formula for flavour EBPA contains Ethyl Butyrate and other esters, Phenylacetic Acid and other organic acids, Juniper Oil and other essential oils, Diacetyl and Maltol.

8.2.4 Other Chemical Senses

Factors other than vision, olfaction and gustation may be involved, individually or collectively, in the response of the bird to feed intake. These factors include temperature, viscosity, osmotic pressure of water, salivary production, nutritive value of feed or toxicity of feed (components). These factors may affect the sense of taste to water and feed consumption accordingly.

(a) Thermo receptors

Warm receptors. Warm receptors responding to an increase in temperature, have been found in the pigeon (Necker, 1972, 1973; Necker and Reiner, 1980) but not in the chicken (Kitchell, Strom and Zotterman, 1959). Nevertheless, Kare and Mason (1986) reported that the domestic fowl possesses an acute sensitivity to the temperature of water. It rejects water with a temperature higher than the ambient temperature and suffer acute thirst, rather than drink water at 10 °C above its body temperature. The work of Henning (1921), Goudriaan (1930) and Hahn, Kuckulies and Taeger (1938) was in accordance with these findings. Gentle (1985) reported the rejection of water above body temperature was a result of nociceptors (thermal and mechanical stimulation) and not only of warm receptors.

Cold receptors. Cold receptors have also been identified in the beaks of poultry. They respond to cooling of the surface of their beak through the oral epithelium (Kitchell, Strom and Zotterman, 1959; Gregory, 1973; Necker, 1973; Leitner and Roumy, 1974). Finally, Kare and Mason (1986) reported that the chickens accepted water, that was nearly frozen.

(b) Viscosity and osmotic pressure

The fowl might react to a sugar solution on the basis of physical quality such as osmotic pressure or viscosity and be indifferent to the quality of sweeteners (Kare and Medway, 1959). These researchers added that the rejection of the very viscose sugar solution (25%) in water was remarkable, while Gentle (1979) reported no significant rejection of a solution of carboxy methyl cellulose (CMC) below a concentration of 2.5%, (168.2 viscosity).

Dissection provides an anatomical support for the suggestion that the bird has been endowed with innervation which could represent an acute oral sense of touch and pressure (Kare and Medway, 1959). In addition, Shurlock and Forbes (1981[a]) suggested that in relation to feed intake, there is a major osmotic control in the duodenum which may affect a secondary control system in the upper gastro-intestinal tract.

Post-ingestional factors (gastrointestinal activity)

Apart from the beak, there are other receptors such as the assumed xylose perception organs in the crop (Kare and Medway, 1959), the glucose-dependent mechanism for the control of feed intake in the hepatic area of the chicken (Shurlock and Forbes, 1981[b]) and the post-ingestional salt receptors (Gentle, 1971[b]). These post-ingestional factors are probably more important than taste sensation (Gentle, 1975[a]).

Birds eat to satisfy energy requirements and volume receptors, or to attain a state of fullness or satiety. The amount of feed consumed depends on many factors, including the size and age of birds, activity, stage in the reproductive cycle, appearance and taste of feed and availability of water and environmental temperature. Some investigators believe that energy requirements and, ultimately, adipose tissue stores are the main regulators of appetite or feed intake (Mu, Hamilton and Brobeck, 1968).

Hodgkiss (1981) has demonstrated the presence of two types of distension-sensitive receptors in the crop of the chicken, slowly-adapting receptors and rapidly-adapting receptors. The slowly adapting receptors are encountered about four times more frequently than the rapidly adapting ones, and are capable of signalling distension of the crop for prolonged periods of time, although small local changes in tension may modulate the firing rate. The most relevant evidence for gastrointestinal receptors were reported by Duke, Kuhlman and Fedde (1977) who reported mechanoreceptors present in the gizzard. The presence of thermoreceptors, chemoreceptors or osmoreceptors in the crop and duodenum were reported by various authors (Gentle and Richardson, 1972; Richardson and Gentle, 1972; Shurlock and Forbes, 1981[a]). Shurlock and Forbes (1981[a]) suggested that, in relation to feed intake, there is a major osmotic control in the duodenum which may affect a secondary control system in the upper gastrointestinal tract. Shurlock and Forbes (1981[b]) demonstrated a glucose-dependent mechanism for the control of feed intake which exists in the hepatic region of the chicken.

There is general agreement that the satiety and appetite centres are located in the ventromedial and lateral nucleus of the hypothalamus (Lepkovsky and Yasuda, 1966; Smith, 1969). Polin and Wolford (1973) concluded that the feed intake is controlled by volume receptors that are influenced by rate of filling, capacity and discharge of feed.

Steffens (1978) presented evidence concerning the role of both the ventromedial (VMH) and lateral hypothalamus (LH) in the regulation of feed intake. He added that these two areas of the brain receive information regarding the nutritional state of the individual. The information going to the VMH and LH is centrally and peripherally mediated. During a meal, a positive feedback mechanism is at work for the continuation of a meal through neural and humoral messages to the VMH and LH. Termination of a meal is probably brought about by an increase of blood insulin and glucose levels which acts as a negative feedback system. The duration of a meal is dependent on a balance between positive and negative feedback of the hypothalamic level.

(d) Chemoreceptors

Nutritional value

Nutrient requirements are presented in two ways:

- In amounts required per kg of diet.
- In amounts required per 4.18 MJ/kg metabolizable energy.

Scott, Nesheim and Young (1982) early expected the second method to be universally adopted in the near future since the chick consumes feed largely to satisfy an inner need for energy. Table 8.13 shows that broiler chickens offered diets with a different energy content have a similar metabolizable energy intake. In other words, the chicken eats until its need for energy is satisfied (Scott, Nesheim and Young, 1982). In addition, Kare and Ficken (1963) showed that when caloric intake is restricted, a chick selects a sucrose solution to which it is normally indifferent and increases fluid intake to overcome the energy deficiency.

Table 8.13. Effect of metabolizable energy content (ME) of diet [a] and feed consumption by broilers from 0 to 8 weeks of age. From Scott, Nesheim and Young (1982)

ME (MJ/kg)	*Feed consumption (kg)*					
	0-2 weeks		*2-6 weeks*		*6-8 weeks*	
	Males	*Females*	*Males*	*Females*	*Males*	*Females*
11.71	0.31	0.29	2.75	2.08	2.23	2.00
12.13	0.30	0.28	2.65	2.01	2.15	1.93
12.55	0.29	0.27	2.56	1.94	2.09	1.87
12.97	0.28	0.26	2.48	1.88	2.02	1.81
13.39	0.27	0.25	2.40	1.82	1.96	1.76
13.81	0.26	0.24	2.33	1.77	1.90	1.70

[a] Assuming a moderate environment and a diet adequate in all nutrients.

Not only energy is involved in feed intake, as stated by Kare and Pick (1960) and Fisher and Scott (1962) but also an adequate diet (without flavour) appeals more than an inadequate diet with flavour. Kare and Rogers (1976) showed that the domestic fowl on a diet very low in sodium or calcium will exhibit a specific appetite and will select, when given a choice, the diet or solution that corrects its deficiency.

As reported before, the chicken consumes feed to satisfy its need for energy. Therefore it is important to adapt the other nutrients to the energy content. Scott, Nesheim and Young (1982) showed this adjustment for protein (Table 8.14). In this table, the protein requirements and expected efficiencies of feed utilization for broilers receiving graded levels of energy and correspondingly adequate levels of protein are shown.

Kitchell, Strom and Zotterman (1959) reported that taste information was relayed to the brain exclusively along branches of the glossopharyngeal nerve. Recent work of Gentle (1983) proved that the chorda tympani branch of the facial nerve has an important role in taste preception. The chorda tympani relays taste information from those taste buds in the anterior mundibular area. Gentle (1985) added that some recordings were noticed from the geniculate ganglion showing that many of the taste buds in the palate also send information to the brain along other branches of the facial nerve.

Table 8.14. Protein requirements of broiler chickens in relation to the metabolizable energy content (ME) of the diet. From Scott, Nesheim and Young (1982)

ME (MJ/kg)	*Protein requirement (%)*	*Efficiency of feed utilization (kg feed/kg weight)*
	In pre-starting diet (0-2 weeks)	
11.71	23.2	1.20
12.13	24.0	1.16
12.55	24.8	1.12
12.97	25.7	1.08
13.39	26.5	1.05
	In starting-growing diets (2-6 weeks)	
11.71	19.5	2.00
12.13	20.0	1.93
12.55	20.6	1.87
12.97	21.3	1.80
13.39	22.0	1.75
13.81	22.7	1.70
	In finishing diets (6 weeks to market)	
11.71	18.1	2.27[a]
12.13	18.7	2.19[a]
12.55	19.3	2.13[a]
12.97	20.0	2.05[a]
13.39	20.5	1.99[a]
13.81	21.2	1.93[a]

[a] These are cumulative feed conversions at 8 weeks of age, assuming that the chicks received approximately corresponding energy levels in the starting diets.

Chickens have an acute sense of taste (Gentle, 1975[a]) and changes in the taste preferences readily occur following experimental manipulation. When fed a diet adequate in energy, the chickens did not exhibit any marked preference for a sucrose solution (100 g/l) but, when fed on a diet low in energy, they showed a marked preference for sucrose (Kare and Maller, 1967).

Taste sensitivity
Hughes and Wood-Gush (1971) reported that birds, offered Ca deficient diets, selected more calcium carbonate to compensate for the deficiency of Ca in this diet. They discussed that this behaviour was due to gustatory involvement. Birds rapidly form conditioned aversion to weakly-flavoured solutions (Lett, 1980; Gilette, Thomas and Bellingham, 1983) but not to strongly-flavoured feed (Gilette, Martin and Bellingham 1980; Gilette, Thomas and Bellingham, 1983).

Gentle (1976) concluded that after short periods of water deprivation, chickens will accept solutions which were unacceptable previously. He added that this increase in acceptability may be due to changes in taste sensitivity. Another example was reported by Gentle (1974), that changes in thirst (hydration) will affect the birds' response to water. After deprivation, water is positively rewarding while after too much drinking the water in the mouth will be stressful due to the loading of the crop with water.

Oral behaviour in response to gustatory stimulation
Mandibulatory movements of the beak coupled with movements of the tongue and swallowing are all oral behaviour related to gustatory stimulation (Gentle and Harkin, 1979). Gentle and Dewar (1981) noticed that the oral behaviour may change according to the vitamin A level in the diet. Gentle, Dewar and Wright (1981) noticed a change in oral behaviour in terms of the number of beak and tongue movements. They reported that all the chicks fed a zinc-deficient diet showed a significant increase in beak and tongue movements compared with control birds.

(e) Toxicity

The biological benefit of avoiding toxic feed is self-evident. The chicken shows a characteristic response to aversive oral stimulation typified by persistent tongue and beak movements, head shaking and beak wiping behaviour (Gentle, 1979). As reported under feed recognition, the chicken is able to avoid certain toxic substances which, after ingestion, would have produced illness (Martin, Bellingham and Storlien, 1977; Gillette, Thomas and Bellingham, 1980). Also El Boushy and Kennedy (1987[a]) reported that, if a feedstuff or diet has once caused digestive disturbances or discomfort, it usually will not be eaten with relish a second time. In many cases the toxic matter has a threshold value. Below this threshold value no harmful effects are noticed. Scott, Nesheim and Young (1982) reported an intolerance above 20% of Philippines coconut oil meal in the ration because of a toxic compound, contrary to Mexican and Jamaican coconut oil meal (toxic free) which was successfully included in rations up to 40%. In addition to smell and taste, Kare and Mason (1986) discussed the common chemical sense, with its major component, the trigeminal system, which is usually reserved for non-specific, often irritating stimuli.

Cheeke *et al.* (1983) studied the feed preference responses for poultry fed alfalfa meal (lucern meal) containing saponin and quinine sulphate. They concluded that at levels of 10% or more lucern meal, all bird species preferred the lucern-free diet. With the inclusion of quinine sulphate at 0.01, 0.05 and 0.1% of diet, discrimination was noted, indicating that poultry can detect substances in the diet perceived as bitter and toxic. Rotter *et al.* (1989) made a comparison of the effects of toxin-free and toxin-containing mould-contaminated barley on chick performance. Their results indicated that mould-contaminated barley can seriously affect chicken performance even when no toxins are detectable and that antinutritive agents in mould-contaminated barley decrease the utilization of all nutrients in the diet rather than only those of the affected barley. Factors affecting diet palatability may also be involved (Rotter *et al.*, 1989).

(f) Salivary production

The salivary production is of a great importance because saliva dissolves, amongst other things, the flavours of the feed so that the taste buds can be stimulated (van Prooije, 1978). Other important factors for salivary production are:

- Washing the taste buds so they are ready for a new sensation.
- Serving as a transporting medium.
- Defending against dehydration of the beak epithelium (Berkhoudt, 1977).

For the chicken Belman and Kare (1961) observed a salivary production greater than that of man relative to body weight, but less relative to the amount of feed consumed.

8.2.5 Receptors Leading to the Sense of Taste

It was mentioned previously that the ability to differentiate between several taste stimuli is present in chicken. Therefore, it should be possible to associate the sense of taste with definite sensory end-organs or taste buds.

(a) Number and distribution of taste buds

Gustation usually requires close contact of a high concentration of the chemical stimuli with the taste receptors. These receptors are normally found in the taste buds of the oral cavity.

Contrary to those of mammals, the taste buds of birds are not situated in papillae but in the epithelium, mostly in the neighbourhood of the salivary glands (Ganchrow and Ganchrow, 1985; Berkhoudt, 1985). The taste buds are divided over three regions of the oral cavity, the tongue, palate and lower beak.

Tongue (excluding the region behind the lingual spines).
For a long time, the region behind the lingual spines (papillae), including larynx (Figure 8.1), was thought to be the only site of the taste buds. Lindenmaier and Kare (1959), Gentle (1971[a]) and Kare and Rogers (1976) reported that the taste buds counted from serial sections, varied from 8-24 in both chicks and adults. They were all located on the dorsal surface of the tongue (just behind the papillae, including the larynx) and were associated with the salivary ducts.

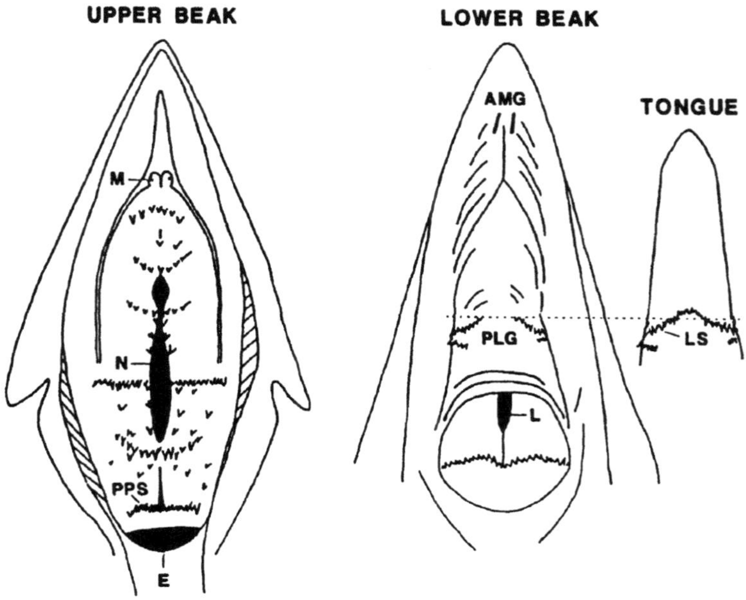

Figure 8.1. Schematic representation of three regions of the oral cavity in the one-day-old chick. Anterior is at the top. Upper beak: *M*, monostomatic opening of maxillary glands; *N*, nasopalatal opening; *PPS*, posterior palatal spines; *E*, oesophageal opening. Lower beak: *AMG*, region of anterior mandibular glands; *PLG*, region of posterior lingual glands; *L*, laryngeal opening. Tongue: *LS*, Lingual spines. Reproduced with permission from Ganchrow and Ganchrow (1985).

Other investigators have found that taste buds in other regions than the tongue, the palate and the lower beak (Bath, 1906; Saito, 1966). The real number of taste buds reported in the tongue, not including the region behind the lingual spines, varies from 0-12 according to Saito (1966) and Kurosawa *et al.* (1983). No taste buds were found near the highly cornified edges or sides of the tongue (Kare and Rogers, 1976). The numbers of taste buds found for several animal species are given in Table 8.15.

Palate

Of the taste buds located in the palate, there are far more in the anterior part of the oral cavity than in the posterior part (Saito, 1966).

The total number of taste buds in the anterior region varies between 87.4 and 219 according to van Prooije (1978) and Ganchrow and Ganchrow (1985), respectively. Table 8.16 shows the numbers detailed by each author. Saito (1966) also reported a difference between the left- and right part of the palate: the left part contained 87.2 and the right one 109 taste buds.

Table 8.15. Number of taste buds reported for different species by several authors.

Species	*Number of buds*	*References*
Chicken	30-70	- Bath (1906)
	24	- Lindenmaier and Kare (1959)
	218-499[a]	- Saito (1966)
	68	- Gentle (1975[b])
	124	- Prooije, van (1978)
	213	- Kurosawa *et al.* (1983)
	316	- Ganchrow and Ganchrow (1985)
Duck	200	- Bath (1906)
Mallard	375	- Berkhoudt (1977)
Pigeon	50-75	- Bath (1906)
	37	- Moore and Elliott (1946)
	24	- Gentle (1975[a])
	1285	- Kan, van (1979)
Parrot	300-400	- Bath (1906)
Human	9000	- Cole (1941)
Rat	1265	- Miller and Spangler (1982)
Barbary dove	54	- Gentle (1975[a])
Bath	800	- Moncrieff (1951)
Blue tit	24	- Gentle (1975[a])
Bull finch	46	- Duncan (1960)
Catfish	100 000	- Hyman (1942)
Hamster	723	- Miller and Smith (1984)
Japanese quail	62	- Warner, McFarland and Wilson (1967)
Kitten	473	- Elliott (1937)
Pig and goat	15 000	- Moncrieff (1951)
Rabbit	17 000	- Moncrieff (1951)
Snake	0	- Payne (1945)
Starling	200	- Bath (1906)

[a] Average 360.4.

Table 8.16. Distribution of taste buds over tongue, palate and lower beak.

Number in:			
Tongue	*Palate*	*Lower beak* [a]	*References*
-	-	5-12/24 [b]	- Lindenmaier and Kare (1959)
12	196.2	152.2	- Saito (1966)
-	-	8-15 + 56	- Gentle (1971[a], 1975[b])
4,6	87,4	32,2	- Prooije, van (1978)
0	112	101	- Kurosawa *et al.* (1983)
5-9	219	90	- Ganchrow and Ganchrow (1985)

[a] Including the area behind the papillae.

[b] 5-12 in one-day-old chicks; 24 in 3-month-old cockerels.

Lower beak
Taste buds in the lower beak (including the region behind the lingual spines) were reported in right and left parts by Ganchrow and Ganchrow (1985). They added an apparent asymmetry of bud counts between left (mean = 51 8, = 3) and right (mean = 40 11, = 3) lower beak. The total number of taste buds reported in the lower beak varies between 152.2 and 32.2 according to Saito (1966) and van Prooije (1978), respectively.

Berkhoudt (1977) and Ganchrow and Ganchrow (1985) defined the taste buds distribution in the chicken as follows: "Taste buds are situated in the areas of the beak, where there is prolonged contact with feed and thus will enable better gustatory determination".

Not all the salivary ducts possess taste buds. Kurosawa *et al.* (1983) found a large number of ducts without any adjacent taste buds on the larynx and pharynx, and others with 1-3 taste buds adjacent to them.

Lindenmaier and Kare (1959), Duncan (1960) and Saito (1966) reported an increased number of taste buds, especially in the lingual spines, with increasing age of the bird. The fact that other investigators were unable to find these doubled numbers in adults may be explained by differences in counting techniques. In serial sections (Saito, 1966) it is possible to count a taste bud more than once while with rapid surface staining technique (Berkhoudt, 1977), one might overlook taste buds with their opening in a salivary gland. The first method leads to an overestimation, the second to an underestimation (Berkhoudt, 1985).

(b) The morphology of taste buds

Kind of cells forming the taste buds
There is a debate as to how many different cell types, particularly of chicken taste buds, can be distinguished as singly or less frequently, in twos, threes or fours, as reported by several investigators (Bath, 1906; Lindenmaier and Kare, 1959; Saito, 1966; Gentle, 1971[b]; Dmitrieva, 1981; Kurosawa *et al.*, 1983 and Berkhoudt, 1985). Kurosawa *et al.* (1983) distinguished three different cell types (based on cytology): light cells (gustatory cells). dark cells (supporting cells). and flattened cells (peripheral cells). Other investigators additionally defined sensory, supporting, basal and follicular (perigemmal) cells (Bath, 1906, Dmitrieva, 1981; Berkhoudt, 1985) (Figures 8.2, 8.3 and 8.4).

The structure of the taste bud
The true taste buds are closely parallel in both structure and neural dependency to taste buds in other vertebrates (Gentle, 1971[a]). Their shapes lie between those of fish and mammals (Kare and Rogers, 1976). Bath (1906) discovered thee different types of taste buds in birds but only type I was found in chicken. This bud, type I, is ovoid and has a central core of sensory and supporting (sustentacular) cells surrounded by follicular (sheath) cells (Figure 8.5).

Saito (1966) and Gentle (1971[b]) reported only one kind of taste bud, including taste canals. Kurosawa *et al.* (1983) distinguished two types of taste buds (based on morphology), type I (pear shaped) and type II (fork shaped) (Figure 8.6).

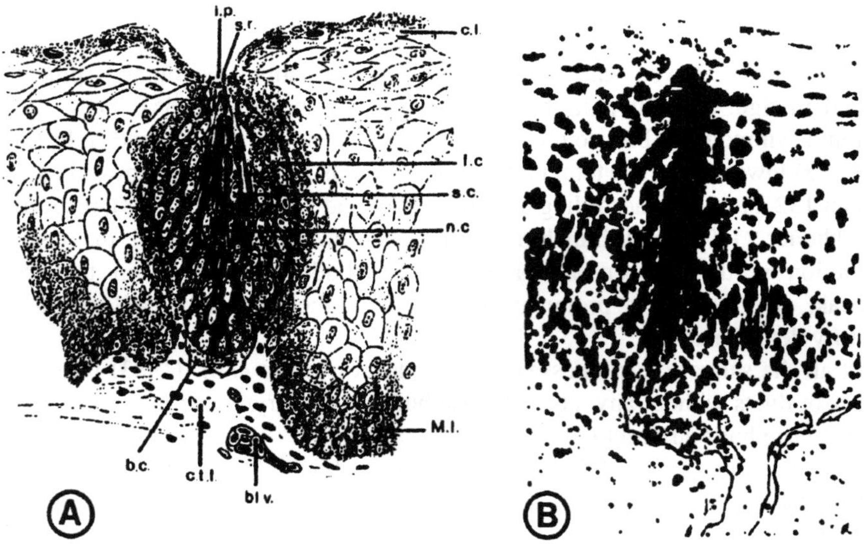

Figure 8.2. (A) Photograph of an original drawing by Bath (1906) of a taste bud in the pigeon *(Columba livia). b.c.,* Basal cell; *bl.v.,* blood vessel; *c.l.,* cornified layer; *c.t.f.,* connective tissue fibre; *f.c.,* follicular cells; *n.c.,* nerve cell; *i.p.,* inner pore; *M.l.,* Malpighian layer; *s.c.,* supporting cell; *s.r.,* sensory rods. x 420. (B) Photomicrograph of a taste bud in the mandibular mucosa of the pigeon *(Columba livia).* x 400. Reproduced with permission from Berkhoudt (1985).

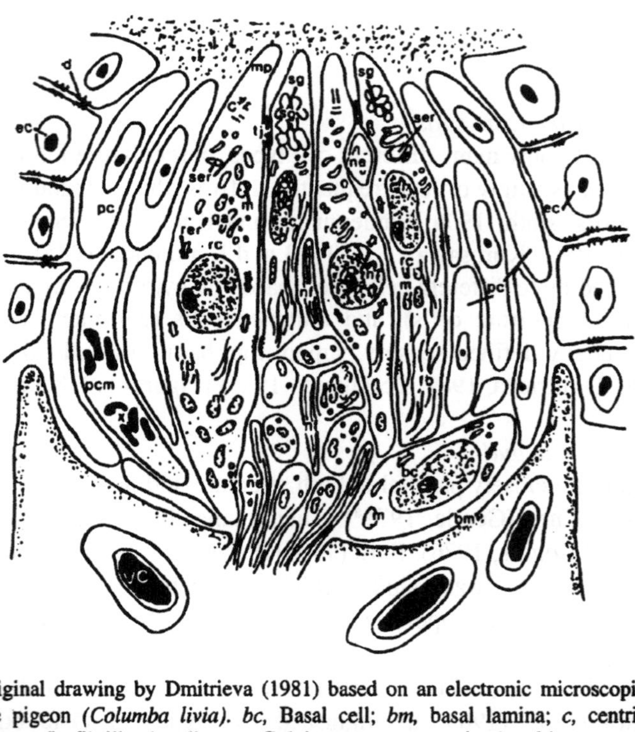

Figure 8.3. Original drawing by Dmitrieva (1981) based on an electronic microscopic study of lingual taste buds in the pigeon *(Columba livia). bc,* Basal cell; *bm,* basal lamina; *c,* centriole; *ec,* epithelial cells; *d,* desmosome; *fb,* fibrillar bundle; *ga,* Golgi apparatus; *m,* mitochondria; *mp,* microvillar process; *n,* nucleus; *ne,* nerve ending; *nf,* nerve fibres; *pc,* perigemmal cells; *pcm,* perigemmal cells in mitosis; *rc,* receptor cells; *rer,* rough endoplasmic reticulum; *sc,* supporting cell; *ser,* smooth endoplasmic reticulum; *sg,* secretory granules; *sy,* synaptic contact; *tj,* tight junction; *vc,* blood capillary; *x,* chromosome. Reproduced with permission from Berkhoudt (1985).

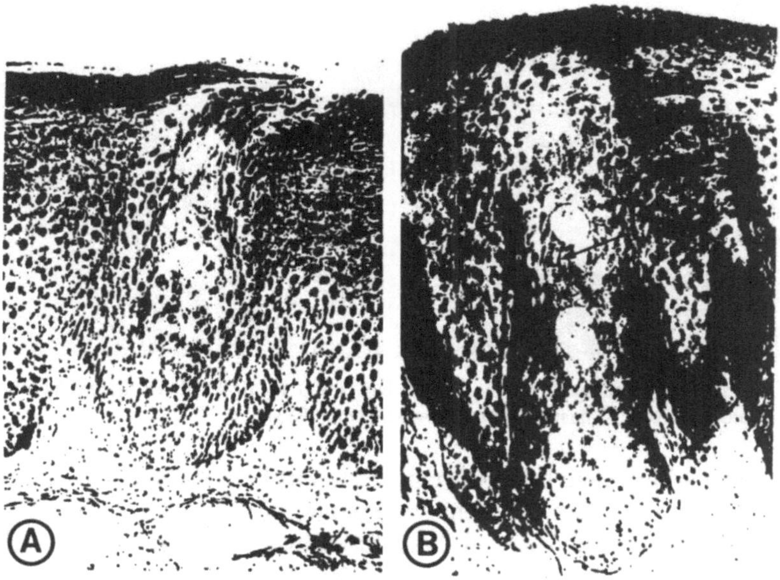

Figure 8.4. Photomicrographs of taste buds (A,B) from the maxillary region in the chicken. In (B) note lumen (empty spaces) inside the buds, and sensory cells (arrow). x 240. Reproduced with permission from Berkhoudt (1985).

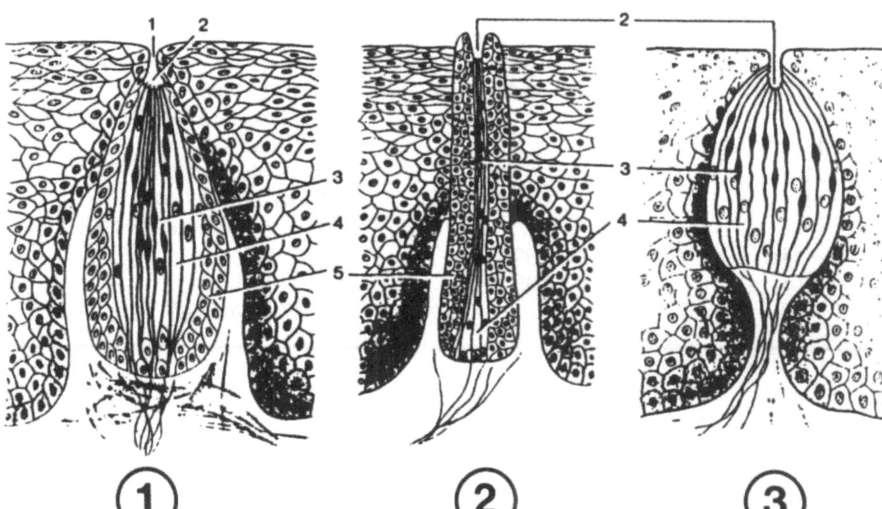

Figure 8.5. Schematic representation of the three different types of taste buds. Type 1 occurred in most songbirds (oscines) and also the chicken and pigeon, type 2 in ducks and flamingos, and type 3 in parrots. 1, Outer pore; 2, inner pore; 3, sensory cell; 4, supporting cell; 5, follicular cell. Reproduced with permission from Bath (1906).

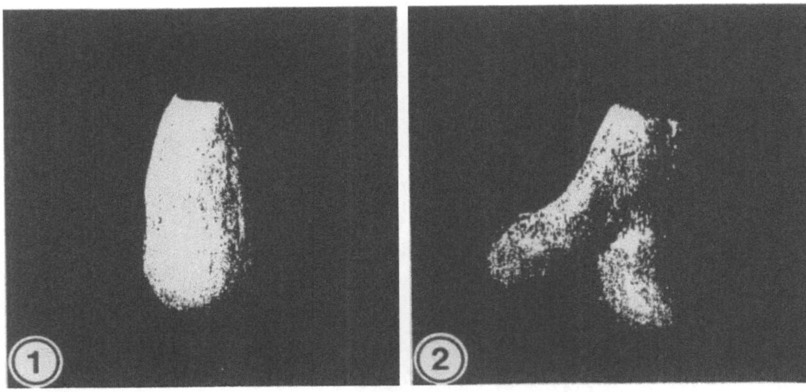

Figure 8.6. (1) External morphology of a type I taste bud. x 800. (2) External morphology of a type II taste bud. x 800. Reproduced with permission from Kurosawa *et al.* (1983).

The taste bud's dimensions have been reported to vary between 30 x 70 m (Lindenmaier and Kare, 1959) and 35-150 x 75-300 m (van Prooije, 1978) and are given in Table 8.17. An explanation for the differences in these dimensions, could be the use of shrinking fixation compounds by the investigators, who found low values (van Prooije, 1978). Also, the age and breeds of the chickens may play an important role in these measurements. Another explanation could be that the investigators who found large dimensions, measured a cluster of taste buds, instead of one (Berkhoudt, 1985).

Table 8.17. Dimensions of taste buds reported.

Dimensions		
Width (m)	*Length (m)*	*References*
40-53	73-99	- Bath (1906)
30	70	- Lindenmaier and Kare (1959)
40	110	- Saito (1966)
35	90	- Gentle (1971[b])
35-150	75-300	- van Prooije (1978)
		- Kurosawa *et al.* (1983)
55.2 (20-87.5)	140.8 (49-266)	Type I
36.9 (15-87.5)	159 (119-294)	Type II
40-49/60-69	-	- Ganchrow and Ganchrow (1985)

(c) Nerve fibre connection with taste buds

It was originally thought that taste information was relayed to the brain exclusively along branches of the glossopharyngeal nerve (Kitchell, Strom and Zotterman, 1959) but recent work has shown that the chorda tympani branche of the facial nerve has an important role in taste perception (Gentle, 1983). The chorda tympany relays taste information from the taste buds in the anterior mandibular region. Some recent recordings taken from the geniculate ganglion show that many of the taste buds in the palate also send information to the brain along other branches of the facial nerve (Gentle, 1985). At the same time Ganchrow and Ganchrow (1985) reported the very poorly developed afferent (or sensory) root of the facial nerve, one of the reasons of the poor sense of taste in the chicken in comparison to the mammal.

The nervous fibres which are distributed in the taste buds make a bundle in the membrana propria of the mucous membrane. These nervous fibres stretch and form the so-called taste bud surrounding fibres (basal nerve plexus) (Saito, 1966; Gentle, 1971[b]; Kurosawa *et al.*, 1983) (Figure 8.7). The synaptic regions of the taste bud cells show a high level of acetylcholinesterase activity and were found to degenerate according to denervation (Gentle, 1971[b]).

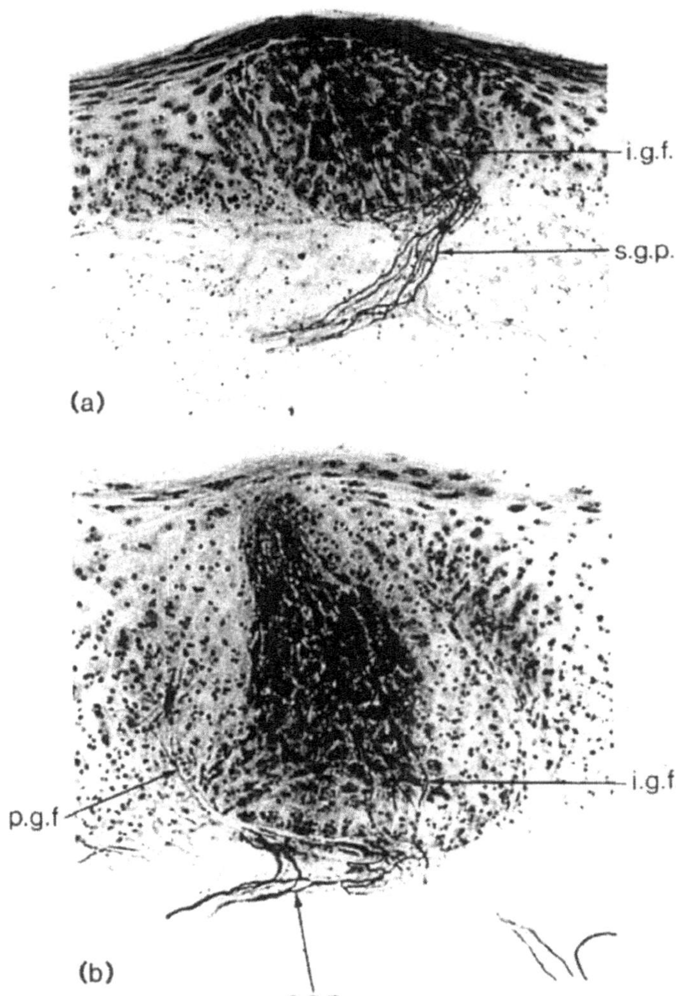

Figure 8.7. Photomicrographs showing innervation of taste buds located in the oral mucosa of the lower jaw of the Carrion Crow *(Corvus corone)*. Note the subgemmal plexus *(s.g.p.)* under the taste buds, and intragemmal fibres *(i.g.f.)* entering the bud between the sensory cells. *p.g.f.*, Perigemmal fibres. (a) x 225. (b) x 360. Reproduced with permission from Berkhoudt (1985).

(d) *The sensitivity of the taste bud*

The general acuity of taste discrimination diminishes in the order pigeon, duck, chicken (Berkhoudt, 1985). This is striking because the pigeon has only 128.5 taste buds (van Kan, 1979), in contrast to the duck and the chicken which have 375 (Berkhoudt, 1977) and 360.4 (Saito, 1966), respectively. According to above-mentioned studies, taste buds are obviously not the only sensory organ used for gustation. The receptor cells in the taste bud could differ in their reception pattern from one to another, depending on the distribution of the other receptor cells over the taste bud (van Prooije, 1978). Gentle (1985) divided the mechanoreceptors in the taste bud into two physiologically different types: one adapting rapidly and the other adapting slowly to mechanical stimuli.

(e) *Taste buds in relation to other processes*

Lindenmaier and Kare (1959) discussed the enzyme inhibition concept in which a flavoured substance inactivates one or more enzymes, causing an imbalance, which gives a characteristic sensation. Through the taste bud, the feed is selected and gives an impulse to the brain causing muscle contractions, secretion of gustatory fluids in the crop and stomach and salivary production in the oral cavity (van Prooije, 1978).

8.3 PALATABILITY AND FEED ACCEPTABILITY

The voluntary intake of feed is an extremely important factor, which often determines the quantity of nutrients that the birds obtain from their diets when fed *ad libitum*. Feedstuffs may have essentially the same nutritive value but differ in palatability. The recent development in using by-products and waste products may affect palatability of the ration. This should be kept in mind because a reduction in feed consumption will lead to a lower growth rate of broilers or to a decreased egg production of layers. In order to formulate diets which cover the needs of the birds as far as their feed intake capacity is concerned, it is of great importance to know the main factors affecting palatability.

8.3.1 Factors Affecting Acceptability

Memory. A feedstuff or a formula feed that has once caused digestive disturbances or discomfort usually will not be eaten with relish a second time. It will also be difficult to get adult fowls to eat yellow corn if they have been fed exclusively on wheat and oats (Card, 1956; Titus, 1961; El Boushy and Kennedy, 1987[a]).

Physical conditions. Feed consumption is affected by the physical conditions of the feed. Very fine mixtures are not eaten with such relish as are feeds which are more coarsely ground, pelleted or crumbled. Feedstuffs which are sticky are mostly swallowed with difficulty and not selected. The fibre content of a starting ration may easily be so high as to prevent chicks from eating as much as they would voluntarily consume, thereby reducing both weight gain and the efficiency of feed conversion (Card, 1956).

Savory (1974) studied the growth and behaviour of chicks fed pellets or mash. He concluded that all chicks on pellets were heavier after 40 days of age but had eaten no more than those on mash. All chicks on mash spent more time eating than those on pellets but the times spent on drinking and resting were similar. He concluded also that pelleted feed was converted more efficiently than mash mainly because the digestibility of feed was improved by pelleting, and the chicks spent less time on feed intake and so expended less energy.

Hijikuro and Takemasa (1981) studied the palatability and utilization of some whole grains for finishing broilers. They used whole or ground barley, wheat, milo or rice included at 63% as the only cereal in diets given to female broiler chickens from 6 until 8 weeks of age. Feed intake was not affected by the grain source, but chickens consumed much more whole grain than ground grain. The percentage of whole grain in the left-over diets was much less than in the original diets, indicating that finishing broiler chickens prefer whole grains.

Attractiveness. The colour and the sparkling quality of the feedstuffs are of some importance. Grits which shine and sparkle are usually chosen in preference to those which are dull. Green is a favourite colour, as determined by the attractiveness of feed components (green alfalfa or lucern). Chickens are able to distinguish between fresh and stale feed. They neglect the stale and select the fresh (Card, 1956).

Choice and selection. Kare and Medway (1959) observed that sucrose and xylose, which have toxic properties, were rejected by fowl. Kare and Scott (1962) concluded that buckwheat, barley or rye in combination with soybean meal, linseed cake meal or beans, reduced the acceptability as compared to that of a standard chick starter ration containing corn and soybean meal. They reported that palatability as judged by selection with two choices could not be correlated with nutritional adequacy of the diet.

Williams and Kienholz (1974) reported that the addition of spices such as curry, chili and black pepper powders at 1.5, 6 or 15% to the diet were all detrimental to the growth of the broiler-type chicks, when added to a corn-milo-soybean meal type diet (Table 8.11).

Balog and Millar (1989) studied the influence of the sense of taste on broiler chick feed consumption by inclusion of several flavours in diets for birds up to 21 days of age. Birds detected the differences in flavour and consumed the feed in a specific order of preference: aspartame, saccharin, citric acid, salt and quinine. This observation of preference and its effects is important for the use of a dietary combination that would decrease feed intake and slow weight gain while maintaining good feed efficiency. This will be a welcome alternative to the current practice of feed restriction.

8.3.2 Practical Factors Affecting Palatability

Additives. The addition of inorganic elements like calcium, phosphorus, manganese, iron, zinc *etc.* and high doses of some drugs may cause some "off flavours" in feed and they are not easily detected by the odour itself, since a feed can smell good and still taste bad. The addition of medicaments with their bitter taste, minerals, vitamins and antibiotics *etc.* is known to impair the natural good palatability of freshly prepared formula feeds.

Feedstuffs. The use of some feedstuffs may impair the palatability of formula feeds. These are feedstuffs such as old grain, old by-products from the milling industry (rice polishings), stale meat and blood meals, marine fats, rancid fats and oils. The use of dry poultry waste, all biodegradation products from the recycling of manure, algae meal, yeasts, tannery, sludge *etc.* are all examples of unpalatable feedstuffs which may reduce the feed consumption.

Nature. Wheat, hemp and sunflower seeds, polished rice, cooked potatoes, potato flakes and fresh fish are reported to be very palatable feedstuffs (Fangauf, Mackrott and Vogt, 1960; Feltwell and Fox, 1978).

Unpalatable feedstuffs as oats, rye, rough rice, buckwheat and barley, are only eaten when the birds are hungry (Fangauf, Mackrott and Vogt, 1960; Feltwell and Fox, 1978). Diets, containing large amounts of buckwheat, barley or rye, were found to be moderately rejected (Kare and Scott, 1962). The refusal of feedstuffs such as barley, rough rice and oats may be due to their hulls and awns. To improve their consumption they must be finely ground to break these fibrous hulls (Feltwell and Fox, 1978). Treatment of barley or pearled barley with water increased the acceptability of diets containing these grains (Ewing, 1963).

Fangauf, Mackrott and Vogt (1960) classified the grade of preference of several grains consumed by chicken in the order wheat, corn, barley, rye, and oats (the preference was graded from a complete acceptance to rejection). The best known unpalatable feedstuffs is linseed meal which is possibly toxic. It should not be used in poultry rations at levels above 2.5% (Fangauf, Mackrott and Vogt, 1960; Ewing, 1963). Some varieties of sorghum (low tannin varieties excluded) are also unpalatable to poultry owing to their bitter and stringent taste because of high condensed tannin content (Feltwell and Fox, 1978). One of the most unpalatable feedstuffs to poultry is blood meal, owing to its structure, colour and composition (Feltwell and Fox, 1978). Some by-products of raw potato starch also had a relatively low palatability owing to their small particle size (Narasaki, Ataku and Fujimoto, 1980).

Unidentified Growth Factors. Several research workers have come to the conclusions that some feedstuffs as manhaden fish meal, whey and its fermentation products, distillers' dried grains with solubles, fish liver and glandular meal contain an unidentified growth factor. This factor improves the palatability of these feedstuffs which, when fed to broilers, results in an improvement in growth (Combs and Alenier, 1979; Alenier and Combs, 1981; Cantor and Johnson, 1983[a,b]).

Staleness. Freshly prepared feeds should be fed within a short period, otherwise the danger of becoming stale will be greater. Staleness is mostly due to rancidity or oxidized fat in the diets, which will be mostly rejected by chickens (Schaible, 1970).
Growth of moulds in mixed feeds owing to long storage causes stale effects. Four types of fungal infestations have been shown to affect chickens and other animals.

They may be classified (according to Scott, Nesheim and Young, 1982) as:
- Feed ingredients in the field before harvest.
- Feed ingredients during storage after harvest.
- Mixed feeds in bulkbins and in feeding equipment.
- The gastrointestinal or respiratory tract of chickens.

Rosenberg and Tanaka (1956) concluded that chickens prefer fresh feed to old, stale or mouldy rations. They reported that chicks fed stale feed required 416 g more feed per day per kg gain than chicks on fresh feed. Bradley (1980) added that a properly formulated flavour in poultry diets or in drinking water will help in maintaining uniform feed intake and aid water intake regulation, and may improve feed conversion under certain circumstances.

Moisture content and texture. Moisture content plays an important role in feed consumption. Birds prefer to consume mash which contains the same moisture content as natural grains. It was also reported that water-treated barley was much more palatable than the dry barley (Ewing, 1963). It was proved that dusty mash is rejected while the wet mash is accepted. Finely ground wheat bran, corn or rice are readily accepted, while finely ground wheat and middlings are rejected (Ewing, 1963). Pelleting mash diets leads to a better feed consumption owing to treating the mash with steam and lowering the dust level of the feedstuffs (El Boushy and Kennedy, 1987[b]).

Spoilage factors. Feed mixtures are susceptible to quick spoilage especially in hot, humid climates and during shipping. The spoilage is manifested by the appearance of staleness, moulds and putrid odours, which indicate a breakdown in fat, carbohydrates and proteins. These changes are due to the chemical and biological reactions of oxidation induced by unfavourable enzymes, oxygen and microbial activities in the feed. The problems with these factors could be solved by improving shipping times and conditions of storage. Further solutions include the use of spraying or coating techniques for feed pellets, which mask the unpleasant taste of additives and medicines by flavouring agents and the use of antioxidants to protect the fats and oils from rancidity and oxidation (El Boushy and Kennedy, 1987[a]).

8.3.3 Regulation of Need Intake

As mentioned before, the other receptors, except those in the beak, are the post-ingestinal factors which are involved with the energy requirements, the volume receptors, which monitor the state of fullness. According to the present feed formulation techniques (linear programming), every ration should contain the following (Scott, Nesheim and Young, 1982):

Carbohydrates. Cereals as corn, oats, rice, sorghum, wheat and root crops (such as tapioca and potatoes).

Mill by-products. Wheat bran, rice bran and polishings, hominy feed and corn gluten feed and meal.

Protein of vegetable origin. Cotton seed meal, groundnut meal, treated rapeseed meal, soybean meal and sunflowerseed meal.

Protein of animal origin. Blood meal, meat meal, poultry by-product meal and feather meal. Also in this category are proteins of fish origin such as herring, menhaden and cod meal.

Fats and oils. Under this category, hydrolyzed animal fat and lard are known as animal fats. Vegetable oils are cottonseed oil, soy oil and coconut oil.

Minerals. The best known minerals are: calcium, phosphorus and NaCl. Both calcium and phosphorus are involved with building up the skeleton of growing chicks while calcium is also utilized for the forming of the eggshell. Examples of phosphorus are dicalciumphosphate and steamed bone meal; examples of calcium are limestone ($CaCO_3$) and oystershell.

Trace elements. The most essential trace elements are potassium, manganese, magnesium, iron, copper, zinc, selenium and iodine.

Vitamins. The most important vitamins are the oil-soluble ones such as vitamins A, D_3, E and K_3 and the water-soluble ones such as thiamine, riboflavin, pantothenic acid, niacin, pyridoxine, biotin, folic acid, choline and B_{12}.

Non-feed constituents. Antibiotics, xanthophylls, antioxidants, coccidiostats, pellet binders, flavouring agents, enzymes and synthetic amino acids.

Using new methods of linear programming, the lowest cost of rations can be established. However, the palatability of the final diet cannot be defined according to raw materials or nutrient restrictions using objective parameters. When waste- and by-products are used, special additives have to be included, to regulate and to improve feed intake. To summarize the effects of adding flavouring and sweetening agents on unpalatable feed, Figure 8.8 shows their benefits for the birds and their technological advantages. As mentioned before, the reaction to sweeteners has been reported by various authors.

Taste buds on the tongue and in the oral cavity of the fowl have been described by Gentle (1971[a]) and Berkhoudt (1985). Kare and Medway (1959) suggested two explanations for the reaction of the fowl to sweeteners.

- Dissection provides anatomical support for the suggestion that the bird has been endowed with enervation which could represent an acute oral sense of touch and pressure.
- The bird's vision exceeds our own and it can perceive differences in the refractive index of solutions.

They reported also that there is no clear-cut evidence for explaining the rejection of or indifference to certain sugars. Instead, there is a suggestion that the preference is based on the absolute specificity for the sugar involved. Kare and Pick (1960) added that feed and fluid intake can be regulated with the inclusion of certain flavours.

BENEFITS FLAVOR TREATMENT IN POULTRY AND TURKEY FEEDS

BIRD BENEFITS	TECHNOLOGICAL ADVANTAGES
A. Helps Improve Rudimentary Taste Perception	E. Neutralization of Unpalatable Ingredients
B. Aids Rudimentary Salivary Secretion	F. Permits Increased Use of By-Products
C. Helps Regulate Water Intake	G. Equalizes Feed Taste Variability
D. Helps Overcome Stress	H. Permits Use of Medicated Water

Better Feed Intake
Better Conversion (Conditional)
Less Mortality (Conditional)

Figure 8.8. Benefits to the birds and technological advantages of the inclusion of flavour in feed. Reproduced with permission from Bradley (1980).

(a) *The effects of sweetening agents and carbohydrates on feed consumption*

Kare and Medway (1959) noticed that chicks can discriminate between different carbohydrates. While it either fails to perceive or is indifferent to dextrose and sucrose, it rejects xylose. The results of six sugars at a 5% level showed that lactose, galactose, raffinose and fructose provided results of indifference or acceptance. There was no apparent discrimination between the sugar solutions and water. On the other hand, there was a moderate degree of rejection of the arabinose and the mannose solutions (Figure 8.9).

Feed consumption is regulated by physical factors (volume), physiological factors (hormones and nerve receptors) and by taste (appearance and palatability).

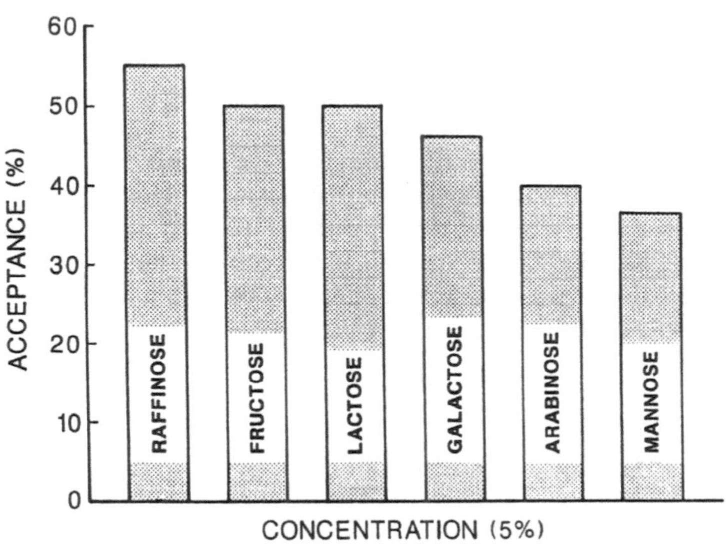

Figure 8.9. Acceptance of various sugars of a concentration of 5%. Reproduced with permission from Kare and Medway (1959).

Thaxon and Parkhurst (1976) reported a larger body weight and lower feed conversion ratio when newly hatched chicks were given a 10% sucrose solution instead of water before they were placed on feed. McNaughton, Deaton and Reece (1978) concluded that mortality of broilers was reduced when 8% glucose solution was supplemented in the first 15 hours of life.

El Boushy (1990) evaluated the effect of feeding palatable and unpalatable rations to broilers from 0-6 weeks of age. Both the palatable and unpalatable rations were formulated to be high or low in energy and protein. All diets were equally balanced as far as amino acids, trace elements, minerals and vitamins are concerned. Each of the four diets was tested: a control diet (no sweetener) and three diets which included a high intensity sweetener at levels of 150, 300 and 450 mg/kg, respectively. Table 8.18 shows the ingredients of the four rations and their chemical analysis. Results of the growth experiment (Table 8.19) illustrate that birds fed unpalatable diets showed a highly significant ($P < 0.01$) lower feed consumption compared to birds fed the palatable ration from 0-6 weeks of age. High energy/protein diets showed lower feed consumption ($P < 0.01$) than the low energy protein diets. The effect of the high intensity sweetener on feed consumption from 0-6 weeks was significant ($P < 0.05$) when combined levels of the three inclusions and control diets were compared.

From the same experiment, some birds were sacrificed to investigate the morphological and histological characteristics of their tongues as far as the papilla and the taste buds are concerned. Figure 8.10 shows a general view of the tongues of birds in the experimental groups: palatable control high protein, high energy (H/H), unpalatable control (H/H), palatable (H/H) + sweetener (450 g/kg) and unpalatable (H/H) + sweetener (450 g/kg).

The papillae from the birds fed with the unpalatable (H/H) diet seems to be more developed. This may be due to the weak pellet structure (due to composition) of this diet, which made the "catch and throw mechanism" more difficult. The relation between texture of feed and papilla development was investigated, but, no promising results were noticed (Berkhoudt, 1988). It is therefore difficult to draw a clear conclusion about the cause of the morphological difference in the tongues of birds receiving the palatable (H/H) and unpalatable (H/H) diets.

Figure 8.11 shows a histological section from the tongue of a bird fed with the unpalatable (H/H) diet. It was noticed during the preliminary microscopical comparison that the tongues of birds receiving the unpalatable (H/H) diet seemed to contain more developed taste buds in comparison with those receiving the palatable (H/H) diet. There were no differences in taste bud development between birds fed the control diet and diets with 450 mg/kg inclusion of sweeteners.

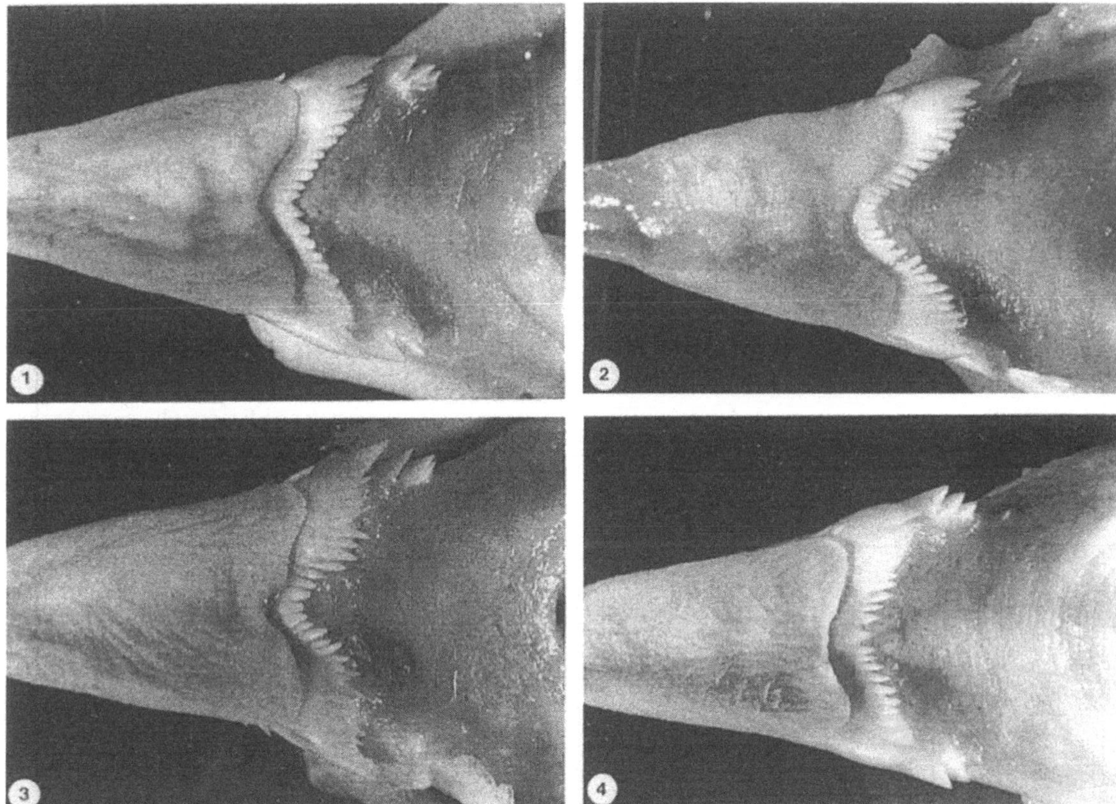

Figure 8.10. General view of four tongues from birds fed high energy/high protein diets. x 4. (1) Palatable diet (control); (2) Palatable diet + 450 mg/kg high intensity sweetener; (3) Unpalatable diet (control); (4) Unpalatable diet + 450 mg/kg high intensity sweetener. Reproduced with permission from El Boushy (1990).

Table 8.18. Composition of the experimental broiler standard rations [a] and their chemical analysis. From El Boushy (1990)

	Palatable		*Unpalatable*	
%	*HP/HE*	*LP/LE*	*HP/HE*	*LP/LE*
Crude protein	21.30	19.70	21.40	19.60
ME (MJ/kg) [b]	8.10	7.34	8.10	7.34
Ca	1.09	1.05	1.02	0.96
P available	0.45	0.45	0.45	0.45
NaCl	0.39	0.41	0.55	0.51
K	0.90	0.90	0.58	0.65
Lysine	1.15	1.12	1.16	1.11
Methionine + Cystine	0.96	0.80	0.91	0.89
Ingredients (g/kg)				
Maize yellow	612.2	546.7	469.7	353.3
Cassava [c]	-	-	140.0	187.6
Wheat middlings	-	150.0	-	150.0
Soybean meal (49%) [d]	247.5	211.9	59.7	40.0
Rapeseed meal [c,d]	-	-	114.8	100.0
Soybean oil	41.8	21.8	-	-
Animal fat [c]	-	-	65.0	41.5
Concentrate	98.6	69.6	150.8	127.6
Meat meal	45.7	30.5	-	-
Fish meal	30.7	10.0	-	-
Blood meal [c]	-	-	65.0	50.0
Feather meal [c]	-	-	55.0	45.0
Ca phosphate	7.1	9.9	14.1	13.6
Limestone	11.4	11.9	10.5	10.8
NaCl	1.0	1.2	1.5	1.4
Premix	1.3	1.3	1.3	1.3
Dl-Methionin	1.2	1.8	1.9	2.2
L-Lysine	-	2.3	0.8	2.7
Choline chloride	0.5	0.5	0.5	0.5
Ethoxyquin	0.2	0.2	0.2	0.2

[a] H = high; L = low; P = protein; E = energy.

[b] Calculated.

[c] Unpalatable feedstuffs.

[d] Solvent extracted.

Table 8.19. Average body weight gain (BW), feed consumption (FC), and feed conversion efficiency (FCE) of broilers at 6 weeks of age fed a basal diet[a] (either palatable or unpalatable) and the effect of high intensity sweetening agents[b]. From El Boushy (1990)

	Palatable						*Unpalatable*					
	HP/HE			LP/LE			HP/HE			LP/LE		
	BW (g)	FC (g)	FCE	BW (g)	FC (g)	FCE	BW (g)	FC (g)	FCE	BW (g)	FC (g)	FCE
1 Control	1902	3455	1.82	1900	3790	2.00	1353	2598	1.92	1558	3159	2.03
2 " + 150 ppm	1889	3472	1.84	1908	3816	2.00	1517	2849	1.88	1542	3234	2.10
3 " + 300 ppm	1908	3433	1.80	1953	3873	1.98	1481	2874	1.94	1461	3142	2.15
4 " + 450 ppm	1918	3488	1.82	1912	3801	1.99	1544	2825	1.85	1524	3113	2.04
5 Mean of 2, 3 and 4	1905	3464	1.82	1924	3830	1.99	1514	2858	1.89	1509	3163	2.10
Difference between 5 and 1 (in %)[c]	+ 0.2	+ 0.3	0.00	+ 1.3	+ 1.1	+ 0.5	+ 11.9	+ 10.0	+ 1.6	- 3.2	+ 0.1	- 3.3
				BW	FC	FCE						
Palatability effects	: Palatable/unpalatable			**	**	**						
Protein energy "	: High/low			NS	**	**						
Sweeteners "	: Levels			NS	*	NS						
Combined sweeteners	: 2, 3 and 4 against 1			*	*	NS						

a Abbreviations: see footnote to ***Table 8.18***.

b High intensity sweetening agent Thaumatin (Intake plus ®: International Additives Ltd., UK.).

c " + " means: improvement.

* : P<0.05

** : P<0.01

NS : Not significant

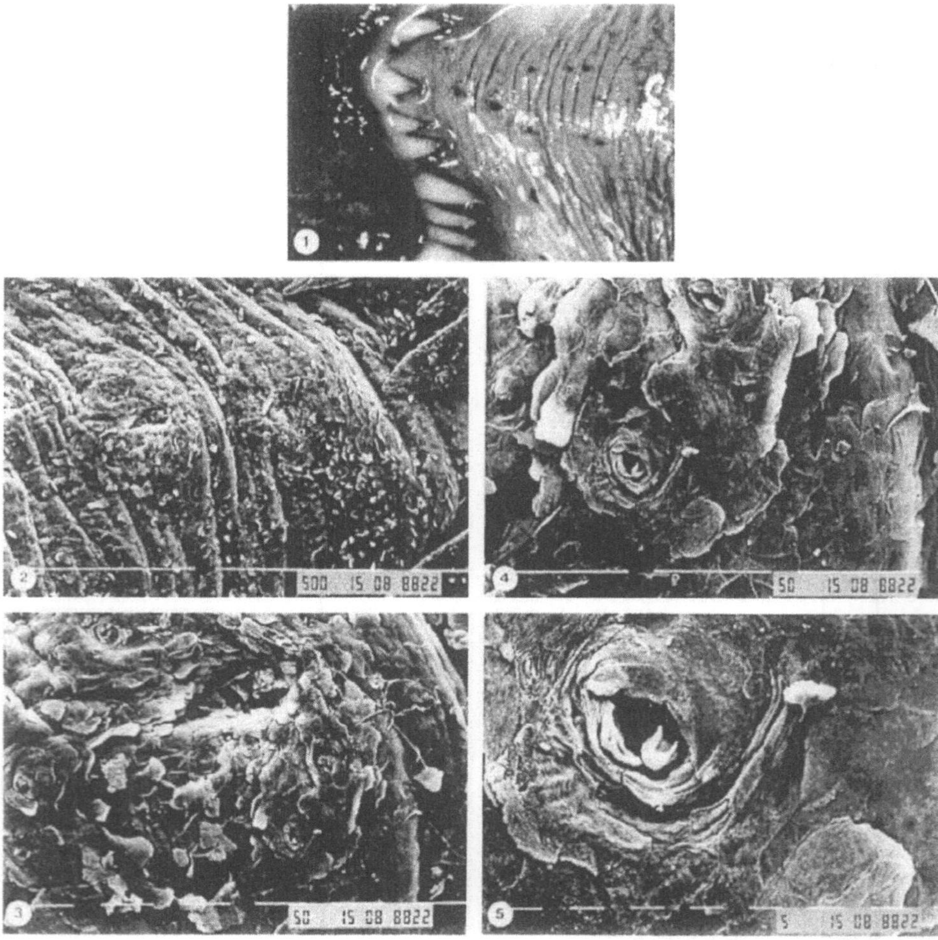

Figure 8.11. Tongue from a bird fed an unpalatable diet (high energy/high protein). Note salivary glans (a), taste bud pores (b) and papillae (c). (1) x 7; (2) x 60; (3) x 200; (4) x 400; (5) x 1200. Reproduced with permission from El Boushy (1990).

(b) Non-nutritive sweeteners

There are some sweet flavourings which are very concentrated. One gram of the high intensity sweetener is equivalent to 2-3 kg of sucrose (Figure 8.12, Tables 8.20 and 8.21). There are also other sweet flavours which are less concentrated than sucrose. If the sweetness intensity of a mixture is greater than the sum of the sweetness from the individual components then the sweeteners are acting synergistically. The formula for such a mixture, or how two or more sweeteners are combined, is mostly based on research. The breeder's benefits through the high production of birds and the price of the sweeteners added in the feed, are of a great importance for the feed industry (Kennedy, 1985).

Saccharin, cyclamate, aspartame, Acesulfam-K ®, stevioside, and thaumatin (Talin ®) protein sweetener are known as high intensity sweeteners. The characteristics of these sweeteners are:

- A sweetness quality and profile identical to those of sucrose.
- Sensory and chemical stability under the relevant food-processing and storage conditions.
- Compatability with other food ingredients and stability towards other constituents naturally present in or intentionally added to foods.
- Complete safety, shown as freedom from toxic, allergenic and other undesirable physiological properties.
- Complete freedom from metabolism in the body.
- High specific sweeteners intensity (Lindley, 1983).

The non-nutritive sweeteners or the high intensity sweeteners have been described in detail (Nicol, 1979; Higginbotham, Lindley and Stephens, 1981; Lindley, 1983; Grenby, Parker and Lindley, 1983; Kennedy, 1988; Higginbotham, 1990).

Saccharin

It is known that saccharin has a distinct aftertaste, described as bitter or metallic, which is an intrinsic characteristic of the saccharin molecule and not a property of impurities or decomposition products. The ability to taste the bitterness of the saccharin correlates with a genetically determined ability to perceive bitterness in compounds such as phenylthiocarbamide and -propylthiouracil. Therefore some consumers are particularly sensitive to the bitter/metalic aftertaste, of the saccharin while others hardly perceive it at all. To reduce the bitter metallic aftertaste, some formulations were introduced combining saccharin with lactose and with cream of tartar, or with maltodextrin and calcium chloride, or with aspartame. The results obtained did not eliminate the saccharin aftertaste completely. Its sweetness is 300 times that of sucrose.

Table 8.20. Relative sweetness of various non-nutritive sweeteners. From Lindley (1983)

Sweetener	*Sweetness intensity* [a]
Saccharin	300 x
Cyclamate	30-50 x
Aspartame	180 x
Acesulfame-K ®	150 x
Stevioside	300 x
Talin ®	2500 x

[a] Sweetness intensity relative to sucrose (weight basis). Quoted intensities should be used only as a guide since precise values depend on application.

Table 8.21. Relative sweetness of various organic chemicals. From Beck (1974)

Chemical	*Sweetness* [a] *(sucrose = 1)*
Sucrose	1
Lactose	0.4
Maltose	0.5
Galactose	0.6
D-glucose	0.7
D-fructose	1.1
Invert sugar	0.7-0.9
D-xylose	0.7
Sorbitol	0.5
Mannitol	0.7
Dulcitol	0.4
Glycerol	0.8
Glycine	0.7
Sodium 3-methylcyclopentyl sulfamate	15
P-anisylurea	18
Sodium cyclohexylsulfamate (cyclamate)	30-80
Chloroform	40
Glycyrrhizin	50
Aspartyl-phenylalanine methyl ester	100-200
5-Nitro-2-methoxyaniline1	67
5-Methylsaccharin	200
P-ethoxyphenylurea (dulcin)	70-350
6-Chlorosaccharin	100-350
N-hexylchloromalonamide	300
Sodium saccharin	200-700
Stevioside	300
2-Amino-4-nitrotoluene	300
Naringin dihydrochalcone	300
P-nitrosuccinanilide	350
1-Bromo-5-nitroaniline	700
5-Nitro-2-ethoxyaniline	950
Perillaldehyde anti-aldoxime	2000
Neohesperidine dihydrochalcone	2000
5-Nitro-2-propoxyaniline (P-4000)	4000

[a] Many factors affect sweetness, and different methods have been used to determine sweetness ratios. The sweetness of sucrose, the usual standard, will change with age due to inversion. Sweet taste depends upon concentration of the sweetener, temperature, pH, type of medium used, and sensitivity of the taster. The usual test methods are dilution to threshold sweetness in water and duplication of the sweetness of a 5 or 10% sucrose solution, although other techniques have also been employed. Where different sweetness values have been reported, the most commonly accepted ones have been cited in this table.

Physical properties. Several trials were carried out to examine the effects of subjecting saccharin to extremes of temperature and pH. Decomposition of saccharin took place at low pH and at high temperature up to 150 C. Results showed that saccharin is stable under a normal range of conditions employed in food processing and storage.

Cyclamate
Cyclamate is an acceptable sweetener but its sweetness is characterized as being a "sweet-chemical". It is used in combinations with saccharin in a weight-ratio 10:1. In this ratio the sweetness quality of the mixture is optimized and the cyclamate effectively eliminates the bitter aftertaste of saccharin. This mixture is not longer permitted as a food additive in a lot of countries. Its sweetness is 30 times that of sucrose.

Physical properties. Cyclamate is considered to be a chemically stable sweetener, although it is less stable than saccharin. A solution of cyclamate buffered to pH 2 and held at a temperature of 125 °C is stable for a very long time.

Aspartame
Aspartame is very similar to sucrose in flavour quality and profile. It is a sweetener with no bitter aftertaste. Its apparent sweetness is 200 times that of sucrose.

Physical properties. The great disadvantage of aspartame is its relative instability in aqueous systems. Under high temperature storage conditions and low PH, aspartame undergoes internal condensation producing diketopiperazine with the elimination of methanol, or it may be hydrolized to an unesterified dipeptide. Its stability in beverages and dessert mixes is excellent, however, aspartame is relatively unstable in aqueous environments, particularly at pH 2-3.

Acesulfame-K ®
Acesulfame-K ® is not similar to sucrose owing to its bitter metallic aftertaste. It is, however, superior to saccharin owing to the fact that this bitter metallic aftertaste is less prominent suggesting that it may be widely used in food systems. It is more expensive than saccharin which restricts its market. Its apparent sweetness is 130 times than that of sucrose.

Physical properties. Acesulfam-K ® is a stable sweetener, particularly over the range of temperatures and pH values that foods are subjected to. Hydrolysis takes place only under extreme conditions producing acetone, carbon dioxide and ammonium sulphate ions. A temperature of 120 °C and pH of 4 do not promote decomposition.

Stevioside
Stevioside is a diterpene glycoside extracted from the leaves of *Stevia rebaudiana*, the sweet herb of Paraguay. It is used frequently in Japan as a sweetening agent. It also has a bitter aftertaste and a clear tardy sweetness. However, new varieties have shown a cleaner sweetness quality. Its apparent sweetness is 300 times that of sucrose.

Physical properties. Stevioside is stable when added into all types of food systems. It is liable into enzymic hydrolysis, yielding its aglycone, steviol, which may play a role as an anti-androgenic substance.

Thaumatin (Talin ®) protein sweetener
Talin ® protein sweetener is the proprietary name given to the purified sweet extract obtained from the fruit of the West African plant *Thaumatococcus daniellii.* The sweetness profile is characterised by a delay in perception with a lingering sweet liquorice aftertaste which limits its usefulness as the sole sweetener in a feed or food system. It is currently permitted for use as a food ingredient in Japan. It is used widely there, both in low caloric foods and drinks and also as a flavour modifier at sub-sweetness threshold levels of addition.

The biochemical structure and physical properties are as follows:

Physical properties. Talin ® consists of closely related proteins known as the thaumatins. The amino acid content and the sequence of the major constituent proteins thaumatin I and thaumatin II are shown in Table 8.22. Both these proteins have 207 amino acid units linked in a single chain, cross-linked with eight disulphide bridges. It was suggested that its digestibility was easy confirmed by human digestibility and mammalian enzyme studies. Its apparent sweetness is 2500 times that of sucrose. Talin ® is freely soluble in cold water to give solutions of greater than 60%, whereas normal levels of use are 0.5-30 ppm by weight. In alcohols, 10% solutions can be obtained in 60% aqueous alcohol and prehydration in a little water will allow solubility in up to 90% glycerol, propylene glycol and the sugar alcohols. Gels are formed at high concentrations of Talin ® in alcohols. Talin ® is not soluble in the common organic solvents.

Talin ® protein sweetener, is unexpectedly stable to extremes of pH and temperature. Stability is enhanced at lower pH values, allowing heat treatment at 100 °C for several hours without loss of sweetness at pH values less than 5.5.

(c) Sweetness synergy

When two or more sweeteners are combined, the effective sweetness of the mixture is often different to that when they are tasted separately. If the sweetness intensity of the mixture is greater than the sum of the sweetness of the individual components, then the sweeteners are acting synergistically. Thaumatin sweetener has been shown to act synergistically with saccharin, acesulpham-K and stevioside, but not with cyclamate or aspartame. Consequently, there is a distinct improvement in the sensory properties of mixtures containing thaumatin sweetener and the sweetener that has an intrinsic, bitter component. Maximum synergy is observed when there is approximately equal sweetness contribution from each sweetener in the mixture. The masking of the bitterness of sweeteners, such as saccharin, is particularly relevant in this case.

Thaumatin/saccharin combinations, for example, can be further improved by the addition of carbohydrate taste modifiers to minimize the lingering thaumatin sweet aftertaste.

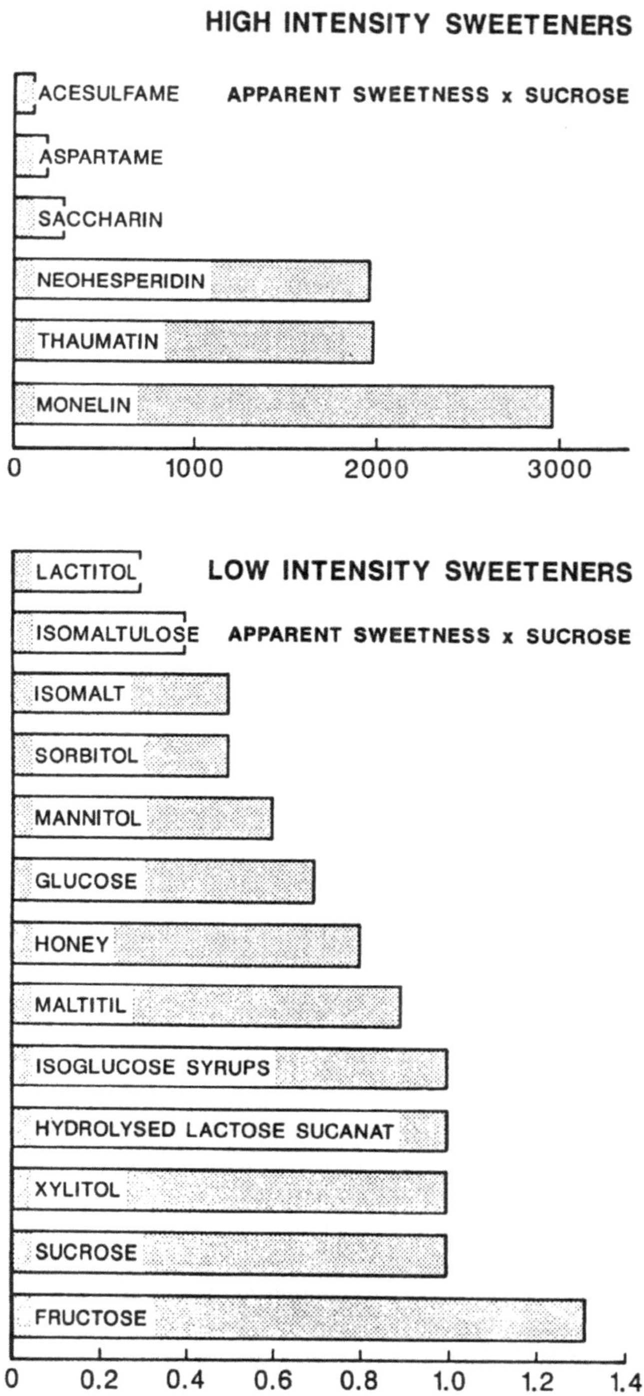

Figure 8.12. The difference between low and high intensity sweeteners in terms of apparent sweetness x sucrose. Reproduced with permission from El Boushy and Kennedy (1987[b]).

There is also a synergistic effect when high intensity sweeteners are mixed, and then blended with low intensity sweeteners, the main ones being sucrose and sugar alcohols, such as sorbitol. The most successful formula is to produce a high intensity sweetener, with a sweetness profile equal to that of sucrose and with a low price which permits its use in the feed industry.

Table 8.22. Amino acid composition of Thaumatin I (T_I) and Thaumatin II (T_{II}). From Kennedy (1988)

Amino acid	*No. of Amino acid residues*	
	T_I	T_{II}
Glycine	24	24
Threonine	20	20
Alanine	16	16
Half-cystine	16	16
Serine	14	13
Aspartic acid	12	13
Proline	12	12
Arginine	12	13
Phenylalanine	11	11
Lysine	11	11
Asparagine	10	8
Valine	10	10
Leucine	9	9
Iso-leucine	8	8
Tyrosine	8	8
Glutamic acid	6	6
Glutamine	4	5
Tryptophan	3	3
Methionine	1	1
Total	207	207

(d) Use of high intensity sweeteners in animal and poultry nutrition

Thaumatin has been widely used as an additive to rations of farm animals, poultry and pets. It is incorporated into the majority of canned catfeeds, a large number of piglet starter diets and into cattle diets. The inclusion of thaumatin in the several diets makes the feed more attractive and improves the weight gain owing to a better feed consumption (Higginbotham, 1990). As far as poultry is concerned it was proved that thaumatin improved broilers' (0-6 weeks) body weight gain, and lowered the feed conversion numerically and improved feed consumption significantly when added to unpalatable diets (El Boushy, 1990).

REFERENCES

Alenier, J.C. and Combs, Jr, G.F. (1981). Effects on feed palatability of ingredients believed to contain unidentified growth factors for poultry. *Poultry Sci.*, **60**, 215-24.

Balog, J.M. and Millar, R.I. (1989). Influence of the sense of taste on broiler chick feed consumption. *Poultry Sci.*, **68**, 1519-26.

Bang, B.G. (1971). Functional anatomy of the olfactory system in 23 orders of birds. *Acta Anat.*, **58**, (Suppl.), 1.

Bath, W. (1906). Die Geschmacksorgane der Vögel und Krokodille. *Arch. für Biontologie*, **1**, 5-47.

Beck, K.M. (1974). Synthetic sweeteners: past, present, and future. In: *Symposium: Sweeteners*. (Ed. G.E. Inglett), The AVI Publishing Company, Inc., pp. 240.

Belman, A.L. and Kare, M.R. (1961). Character of salivary flow in the chicken. *Poultry Sci.*, **40**, 1377.

Berkhoudt, H. (1977). Taste buds in the bill of the Mallard (*Anas platyrhynchos* L.). Their morphology, distribution and functional significance. *Neth. J. of Zool.*, **27**, (3), 310-31.

Berkhoudt, H. (1985). Special sense organs: Structure and function of avian taste receptors. In: *Form and Function in Birds*, vol. 3. (Ed. A.S. King and J. McLelland) Academic

Press Inc., London Ltd, pp. 463-96. Berkhoudt, H. (1988). *Personal communication*. Zool. Lab., Leiden, The Netherlands.

Bradley, B.L. (1980). Animal flavor types and their specific uses in compound feeds by species and age. In: *Palatability and flavor use in animal feed.* (Ed. H. Beckel), Verlag Paul Parey, Hamburg und Berlin, pp. 110-22.

Cameron, A.T. (1947). The taste sense and the relative sweetness of sugars and other sweet substances. *Scientific Report, New York, Sugar Research Foundation*, Ser. **9**.

Cantor, A.H. and Johnson, T.H. (1983[a]). Influence of dietary sources of unidentified growth factors upon feed preference of broiler chicks. *Nutr. Rep. Int.*, **28**, 1119-27.

Cantor, A.H. and Johnson, T.H. (1983[b]). Effects of unidentified growth factor sources on feed preference of chicks. *Poultry Sci.*, **62**, 1281-6.

Capretta, P.J. (1969). The establishment of food preferences in chicks, (*Gallus gallus*). *Animal Behaviour*, **17**, 229-31.

Card, L.E. (1956). *Poultry Production*. 8th edn. Lea and Febiger, Philadelphia. p. 409. Cheeke, P.R., Powley, J.S., Nakaue, H.S. and Arscott, G.H. (1983). Feed preference responses of several avian species fed alfalfa meal, high - and low - saponin alfalfa, and quinine sulfate. *Can. J. Anim. Sci.*, **63**, 707-10.

Cole, E.C. (1941). *Comparative Histology*. Blakiston, Philadelphia.

Combs, G.F., Jr., and Alenier, J.C. (1979). Palatability of poultry feeds. *Proceedings of Cornell Nutrition Conference for Feed Manufacturers*, Cornell University, New York, pp. 42-8.

Cooper, J.B. (1971). Colored feed for turkey poults. *Poultry Sci.*, **50**, 1892-3.

Dawkins, R. (1968). The ontogeny of a pecking preference in domestic chicks. *Z. Tierpsychol.*, **25**, 170-86.

Deyoe, C.W., Davies, R.E., Krishnan, R., Khaund, R.K. and Couch, J.R. (1962). Studies on the taste preference of the chick. *Poultry Sci.*, **41**, 781-4.

Dmitrieva, N.A. (1981). Fine structural peculiarities of the taste buds of the pigeon *Columba livia*. (In Russian) *Dokl. Akad. Nauk. SSSR*, **23**, 874-8.

Duke, G.E., Kuhlman, W.D. and Fedde, M.R. (1977). Evidence of mechanoreceptors in the muscular stomach of the chicken. *Poultry Sci.*, **56**, 297-9.

Duncan, C.J. (1960). The sense of taste in birds. *Ann. Appl. Biol.*, **48**, 409-14.

El Boushy, A.R. (1990). Influence of intake plus as a sweetening agent on broiler chick feed consumption. *Seminar, International Additives Ltd, U.K.* Held in Baden, Switzerland, 5 April. pp. 1-16.

El Boushy, A.R. and Kennedy, D.A. (1987[a]). Palatability, acceptability of feed influenced by senses. *Feedstuffs*, **59**, (25), 25, 27.

El Boushy, A.R. and Kennedy, D.A. (1987[b]). Non-nutritive additives improve palatability, consumption. *Feedstuffs*, **59**, (39), 16, 23, 24.

El Boushy, A.R., Van der Poel, A.F.B., Verhaart, J.C.J. and Kennedy, D.A. (1989[a]). Sensory involvement controls feed intake in poultry. *Feedstuffs*, **61**, (25), 16, 18, 19, 40, 41.

El Boushy, A.R., Van der Poel, A.F.B., Verhaart, J.C.J. and Kennedy, D.A. (1989[b]). Palatability, acceptability of feed by poultry explored. Feedstuffs, **61**, (30), 14-16.

Elliott, R. (1937). Total distribution of taste buds on the tongue of the kitten at birth. *J. Comp. Neurol.*, **66**, 361.

Engelmann, C. (1934). Versuche über den Geschmackssinn von Taube, Ente und Huhn. *Z. Vergl. Physiol.*, **20**, 626-45.

Engelmann, C. (1950). †ber den Geschmackssinn des Huhnes IX. *Z. Tierpsychol.*, **7**, 84-111.

Ewing, W.R. (1963). *Poultry Nutrition.* The Ray Ewing Company, publisher. Division of Hoffmann-La Roche, Inc. 2690, E. Foothill Blvd. Pasadena, California. p. 475.

Fangauf, R., Mackrott, H. and Vogt, H. (1960). Geflügelfütterung. Eugen Ulmer Stuttgart.

Feltwell, R. and Fox, S. (1978). *Practical Poultry Feeding.* Faber and Faber, London and Boston.

Ficken, M.S. and Kare, M.R. (1961). Individual variation in the ability to taste. *Poultry Sci.*, **40**, 1402.

Fisher, H. and Scott, H.M. (1962). Flavors in poultry rations. *Poultry Sci.*, **41**, 1978-9.

Frantz, R.L. (1957). Form preferences in newly hatched chicks. *J. Comp. Phys. Psychol.*, **50**, 422-30.

Fuerst, F.F. and Kare, M.R. (1962). The influence of pH on fluid tolerance and preferences. *Poultry Sci.*, **41**, 71-7.

Ganchrow, D. and Ganchrow, J.R. (1985). Number and distribution of taste buds in the oral cavity of hatchling chicks. *Physiol. Behav.*, **34**, 889-99.

Gentle, M.J. (1971[a]). The lingual taste buds of (*Gallus domesticus* L.). *Br. Poultry Sci.*, **12**, 245-8.

Gentle, M.J. (1971[b]). Taste and its importance to the domestic chicken. *Br. Poultry Sci.*, **12**, 77-86.

Gentle, M.J. (1972). Taste preference in the chicken (*Gallus Domesticus* L.). *Br. Poultry Sci.*, **13**, 141-55.

Gentle, M.J. (1974). Changes in habitation of the EEG to water deprivation and crop loading in (*Gallus domesticus*). *Physiol. of Behav.*, **13**, 15-19.

Gentle, M.J. (1975[a]). Gustatory behaviour of the chicken and other birds. In: *Neural and Endocrine Aspects of Behaviour in Birds.* (Ed. P. Wright, P.G. Caryl and D.M. Vowles), Amsterdam, Elsevier, pp. 305-18.

Gentle, M.J. (1975[b]). *Personal communication*, cited in Berkhoudt (1985).

Gentle, M.J. (1976). Quinine hydrochloride acceptability after water deprivation in (*Gallus domesticus*). *Chem. Sens. Flav.*, **2**, 121-8.

Gentle, M.J. (1979). Single unit responses from the solitary complex following oral stimulation in the chicken. *J. Comp. Physiol.*, **130**, 259-64.

Gentle, M.J. (1983). The chorda tympani nerve and taste in the chicken. *Experientia*, **39**, 1002-3.

Gentle, M.J. (1985). Sensory involvement in the control of food intake in poultry. *Proc. Nutr. Soc.*, **44**, 313-21.

Gentle, M.J. and Dewar, W.A. (1981). The effect of vitamin A deficiency on oral gustatory behaviour in chicks. *Br. Poultry Sci.*, **22**, 275-9.

Gentle, M.J., Dewar, W.A. and Wright, P.A.L. (1981). The effects of zinc deficiency on oral behaviour and taste bud morphology in chicks. *Br. Poultry Sci.*, **22**, 265-73.

Gentle, M.J. and Harkin, C. (1979). The effect of sweet stimuli on oral behaviour in the chicken. *Chem. Sens. and Flavour*, **4**, 183-90.

Gentle, M.J. and Richardson, A.J. (1972). Changes in the electroencephalogram of the chicken produced by stimulation of the crop. *Br. Poultry Sci.*, **13**, 163-70.

Gillette, K., Martin, G.M. and Bellingham, W.P. (1980). Differential use of food and water cues in domestic chicks (*Gallus gallus*). *J. Comp. Physol.*, **130**, 259-64.

Gillette, K., Thomas, D.K. and Bellingham, W.P. (1983). A parametric study on flavoured food avoidance in chicks. *Chem. Sens.*, **8**, 41-57.

Gomez, G., Tellez, G. and Caicedo, J. (1987). Effects of the addition of vegetable oil or animal tallow to broiler diets containing cassava root meal. *Poultry Sci.*, **66**, 725-31.

Goudriaan, J.C. (1930). Über den Einfluß der Temperatur auf die Geschmacksempfindung. *Arch. Néerl. Physiol.*, **15**, 252.

Gregory, J.E. (1973). An electrophysiological investigation of the receptor apparatus of the duck's bill. *J. of Physiology*, **229**, 151-64.

Grenby, T.H., Parker, K.J. and Lindley, M.G. (1983). *Development in sweeteners - 2.* Applied Science Publishers, London and New York, p. 254.

Guttridge, D.G.A. and Morgan, J.T. (1959). Methionine supplementation of broiler diets at varying protein levels. *World Poultry Sci. J.*, **15**, 381-6.

Hahn, H., Kuckulies, G. and Taeger, H. (1938). Eine systematische Untersuchung der Geschmacksschwellen, 1. *Mitt. Z. Sinnesphysiol.*, **67**, 259.

Henning, H. (1921). Physiologie und Psychologie des Geschmacks (*Asher u. Spiros*). *Erg. Physiol.*, **19**.

Herstad, O. (1976). Effect of dietary protein and energy utilization by chicks. *Arch. Geflügelk.*, **2**, 47-51.

Hess, E.H. and Gagel, W. (1956). Natural preferences of chicks and ducklings for objects of different colours. *Psychological Reports*, **2**, 477-83.

Higginbotham, J.D. (1990). Development of Talin ® for use in animal feeds. *Seminar, International Additives Ltd*, Baden, Switzerland, 5 April, 1990. pp. 1-6.

Higginbotham, J.D., Lindley, M.G. and Stephens, J.P. (1981). *The Quality of Foods and Beverages.* (Ed. G. Charalambous and G. Inglett), Academic press, New York, pp. 91-111.

Hijikuro, S. and Takemasa, M. (1981). Studies on the palatability and utilization of some whole grains for finishing broilers. *Japanese Poultry Sci.*, **18**, 301-6.

Hodgkiss, J.P. (1981). Distension-sensitive receptors in the crop of the domestic fowl (*Gallus Domesticus*). *Comp. Biochem. and Physiol.*, **70**$^{\mathbf{A}}$, 73-8.

Hogan, J.A. (1971). The development of hunger system in young chicks. *Behaviour*, **39**, 128-201. Hogan, J.A. (1973^{a}). Development of food recognition in young chicks: I. Maturation and nutrition. *J. Comp. Phys. Psychol.*, **83**, 355-66. Hogan, J.A. (1973^{b}). Development of food recognition in young chicks: II. Learned associations over long delays. *J. Comp. Phys. Psychol.*, **83**, 367-73.

Hogan, J.A. (1977). Development of food recognition in young chicks. 4. Associative and non-associative effects of experience. *J. Comp. Phys. Psychol.*, **91**, (4), 839-50.

Hogan-Warburg, A.J. and Hogan, J.A. (1981). Feeding strategies in the development of food recognition in young chicks. *Anim. Beh.*, **29**, 143-54.

Hughes, B.O. (1971). Allelomimetic feeding in the domestic fowl. *Br. Poultry Sci.*, **12**, 359-66.

Hughes, B.O. and Wood-gush, D.G.M. (1971). A specific appetite for Calcium in domestic chickens. *Anim. Beh.*, **19**, 490-9.

Hyman, L.H. (1942). *Comparative Vertebrate Anatomy*. Chicago, University of Chicago press.

Jacobs, H.L. and Scott, M.L. (1957). Factors mediating food and liquid intake in chickens. I. Studies on the preference for sucrose and saccharine solutions. *Poultry Sci.*, **36**, 8-15.

Jones, R.B. and Gentle, M.J. (1985). Olfaction and behavioral modification in domestic chicks (*Gallus domesticus*). *Physiol. Behav.*, **34**, 917-24.

Jukes, C.L. (1938). Selection of diet in chicks as influenced by vitamins and other factors. *J. Comp. Psychol.*, **26**, 135-56.

Kan, S. van (1979). Touch and taste in the pigeon (*Columba livia domestica*). *Internal report, Zool. Lab., Leiden*, The Netherlands.

Kare, M.R. and Beily, J. (1948). The toxicity of sodium chloride and its relation to water intake in baby chicks. *Poultry Sci.*, **27**, 751-8.

Kare, M.R., Black, R. and Allison, E.B. (1957). The sense of taste in the fowl. *Poultry Sci.*, **36**, 129-38.

Kare, M.R. and Ficken, M.S. (1963). Comparative studies on the sense of taste. In: *Olfaction and Taste*, Vol. I. (Ed. Y. Zotterman), Pergamon Press, New York.

Kare, M.R. and Maller, O. (1967). Taste of food intake in domesticated and jungle fowl. *J. Nutr.*, **92**, 191-6.

Kare, M.R. and Mason, J.R. (1986). The Chemical Senses in Birds. In: *Avian Physiology*, 4th edition. (Ed. P.D. Sturkie), pp. 59-73.

Kare, M.R. and Medway, W. (1959). Discrimination between carbohydrates by the fowl. *Poultry Sci.*, **38**, 1119-27.

Kare, M.R. and Pick, H.L. (1960). The influence of the sense of taste on feed and fluid consumption. *Poultry Sci.*, **39**, 697-706.

Kare, M.R. and Rogers, J.R., Jr., (1976). Sense of Organs. In: *Avian Physiology*. (Ed. P.D. Sturkie), Springer Verlag, New York, pp. 29-52.

Kare, M.R. and Scott, M.L. (1962). Nutritional value and feed acceptability. *Poultry Sci.*, **41**, 276-8.

Kennedy, D.A. (1985). A new generation of sweeteners for animal feeds. *Internal report, International Additives Limited*, Merseyside, L44 4AH U.K. p. 9.

Kennedy, D.A. (1988). *Personal communication from director of International Additives Limited*, Merseyside L44 4AH U.K.

Kennedy, J.M. (1980). The development of dietary preferences in pigs and poultry. In: *Palatability and Flavour Use in Animal Feeds*. (Ed. H. Bickel), Verlag Paul Parey, Hamburg und Berlin, pp. 141-7.

Kitchell, R.L., Strom, L. and Zotterman, Y. (1959). Electrophysiological studies of thermal and taste reception in chickens and pigeons. *Acta Physiologica Scandinavica*, **46**, 133-51.

Kurosawa, T., Niimura, S., Kusuhara, S. and Ishida, K. (1983). Morphological studies of tastebuds in chickens. *Jap. J. Zootech. Sci.*, **54**, 502-10.

Leitner, L.M. and Roumy, M. (1974). Short communications and technical notes: Thermosensitive units in the tongue and in the skin of the duck's bill. *Pflugers Archiv für die gesamte Physiologie des Menschen und der Tiere*, **346**, 151-5.

Lepkovsky, S. and Yasuda, M. (1966). Hypothalamic lesions, growth and body composition of male chickens. *Poultry Sci.*, **45**, 582-8.

Lett, B.T. (1980). Taste potentiates color-sickness associations in pigeons and quail. *Animal Learning and Behaviour*, **8**, 193-8.

Lindenmaier, P. and Kare, M.R. (1959). The taste end-organs of the chicken. *Poultry Sci.*, **38**, 545-50.

Lindley, M.G. (1983). Non-nutritive sweeteners in food systems. In: *Developments in sweeteners-2*. (Ed. T.H. Grenby, K.J. Parker and M.G. Lindley), Applied Science Publishers, London and N.Y, pp. 225-46.

Mariotti, G. and Fiore, L. (1980). Operant conditioning studies of taste discrimination in the pigeon (Clumba livia). *Physiol. Behav.*, **24**, 163-8.

Martin, G.M., Bellingham, W.P. and Storlien, L.H. (1977). Effects of varied color experience on chickens' formation of color and texture aversions. *Physiol. Behav.*, **18**, 415-20.

McNaughton, J.L., Deaton, J.W. and Reece, F.N. (1978). Effect of sucrose in the initial drinking water of broiler chicks on mortality and growth. *Poultry Sci.*, **57**, 985-8.

Miller, I.J. and Smith, D.V. (1984). Quantitative taste bud distribution in the hamster. *Physiol. Behav.*, **32**, 275-85.

Miller, I.J. and Spangler, K.M. (1982). Tastebud distribution and innervation on the palate of the rat. *Chem. Sens.*, **7**, 99-108.

Moncrieff, R.W. (1951). *The Chemical Senses.* London, Hill. p. 172.

Moore, R.A. and Elliott, R. (1946). Numerical and regional distribution of tastebuds on the tongue of the bird. *J. Comp. Neurol.*, **84**, 119-31.

Mu, J.Y., Hamilton, C.L. and Brobeck, J.R. (1968). Variability of body fat in hyperphagic rates. *Yale J. Biol. Med.*, **41**, 133-42.

Narasaki, N., Ataku, K. and Fujimoto, M. (1980). Feeding value of dried waste produced as a by-product of potato starch industry for poultry. 2. Nutritive value of potato pro-feed in cocks. *J. of the College of Dairying*, **8**, 363-70. (Japan).

National Research Council, (1977). *Nutrient requirements of poultry*, 7th ed. Washington DC,

National Research Council, National Academy of Sciences.

Necker, R. (1972). Response of trigeminal ganglion neurons to thermal stimulation of the beak in pigeons. *J. Comp. Physiol.*, **78**, 307-14.

Necker, R. (1973). Temperature sensitivity of thermoreceptors and mechanoreceptors on the beak of pigeons. *J. Comp. Physiol.*, **87**, 379-91.

Necker, R. and Reiner, B. (1980). Temperature-sensitive mechanoreceptors, thermoreceptors and heat nociceptors in the feathered skin of pigeons. *J. Comp. Physiol.*, **135**, 201-8.

Nicol, W.M. (1979). Sucrose and food technology. In: *Sugar: Science and Technology*. (Ed. G.G. Birch and K.J. Parker), Applied Science Publishers, Ltd, London, pp. 211-30.

Payne, A. (1945). The sense of smell in snakes. *J. Bomb. Nat. Hist. Soc.*, **45**, 507.

Pick, H. and Kare, M.R. (1962). The effect of artificial cues on the measurement of taste preference in the chicken. *J. Comp. Physiol. Psych.*, **55**, 342-5.

Polin, D. and Wolford, J.H. (1973). Factors influencing food intake and caloric balance in chickens. *Fed. Proc.*, **32**, 1720.

Portella, F.J., Caston, L.J. and Leeson, S. (1988). Apparant feed particle size preference by broilers. *Can. J. Anim. Sci.*, **68**, (3), 923-30.

Prooije, A. van (1978). The distribution, morphology and functional significance of taste buds in the chicken (*Gallus domesticus* L.) (in Dutch). *Internal Report, Zool. Lab., Leiden*, The Netherlands.

Rensch, B. and Neunzig, R. (1925). Experimentelle Untersuchungen über den Geschmackssinn der Vögel II. *J. Ornithol.*, **73**, 633-46.

Richardson, A.J. and Gentle, M.J. (1972). Changes in the electroencephalogram of the chicken produced by stimulation of the duodenum. *Br. Poultry Sci.*, **13**, 171-3.

Rogers, J.G. (1974). Responses of caged red winged blackbirds to two types of repellents. *J. Wildl. Management*, **38**, 418-23.

Romoser, G.L., Bossard, E.H. and Combs, G.F. (1958). Studies on the use of certain flavours in the diet of chick. *Poultry Sci.*, **37**, 631-3.

Rosenberg, M.M. and Tanaka, T. (1956). Effect of length of storage of mixed feed on growth rate of chicks. *Feedstuffs*, **28**, (31), 29, 30, 32.

Rotter, R.G., Frohlich, A.A., Marquardt, R.R. and Abramson, D. (1989). Comparison of the effects of toxin-free and toxin-containing mold-contaminated barley on chick performance. *Can. J. Anim. Sci.*, **69**, 247-59.

Saito, I.C. (1966). Comparative anatomical studies of the oral organs of the poultry. V. Structure and distribution of the fowl. (In Japanese). *Bulletin of the Agriculture Univ. of Miyazaki Daigaku Nogakubu, Kenkyu Hokoku*, **13**, 95-102.

Savory, C.J. (1974). Growth and behaviour of chicks fed on pellets or mash. *Br. Poultry Sci.*, **15**, 281-6.

Savory, C.J. (1975). Effects of group size on the feeding behaviour and growth of chicks. *Br. Poultry Sci.*, **16**, 343-50.

Savory, C.J., Wood-Gush, D.G.M. and Duncan, I.J.H. (1978). Feeding behaviour in a population of domestic fowls in the wild. *Applied Animal Ethology*, **4**, 13-27.

Schaible, P.J. (1970). *Poultry: Feeds and Nutrition.* The AVI publishing company, Inc. Westport, Connecticut. p. 636.

Schreck, P.K., Sterritt, G.M., Smith, M.P. and Stilson D.W. (1963). Environmental factors in the development of eating in chicks. *Anim. Beh.*, **11**, 306-9.

Scott, M.L., Nesheim, M.C. and Young, R.J. (1982). *Nutrition of the Chicken.* (Ed. M.L. Scott and associates), Ithaca, New York.

Shurlock, T.G.H. and Forbes, J.M. (1981[a]). Factors affecting food intake in the domestic chicken: The effect of infusions of nutritive and non-nutritive substances into the crop and duodenum. *Br. Poultry Sci.*, **22**, 323-31.

Shurlock, T.G.H. and Forbes, J.M. (1981[b]). Evidence for hepatic glucostatic regulation of food intake in the domestic chicken and its interaction with gastro-intestinal control.

Br. Poultry Sci., **22**, 333-46.

Sizemore, J.R. and Lillie, R.J. (1956). Lack of effect of a synthetic poultry feed flavor on chick growth and feed efficiency. *Poultry Sci.*, **35**, 360-1.

Smith, C.J.V. (1969). Alterations in the food intake of chickens as a result of hypothalamic lesions. *Poultry Sci.*, **48**, 475-7.

Steffens, A.B. (1978). Models for the regulation of food intake. *Proc. of the Zodiac Sym. on Adaptation*, Wageningen, The Netherlands, pp. 105-9.

Strobel, M.G. and MacDonald, G.E. (1974). Induction of eating in newly hatched chicks. *J. Comp. Phys. Psychol.*, **86**, 493-502.

Thaxton, J.P. and Parkhurst, C.R. (1976). Growth, efficiency and liveability of newly hatched broilers as influenced by hydration and intake of sucrose. *Poultry Sci.*, **55**, 2275-9.

Titus, H.W. (1961). *The Scientific Feeding of Chickens.* Fourth edition. The Interstate Printers and Publishers, Inc. Danville, Illinois. p. 297.

Tolman, C.W. (1964). Social facilitation on feeding behaviour in the domestic chick. *Anim. Beh.*, **12**, 245-51.

Tolman, C.W. (1965). Emotional behaviour and social facilitation of feeding in domestic chicks. *Anim. Beh.*, **13**, 493-6.

Tolman, C.W. and Wilson, G.F. (1965). Social feeding in domestic chicks. *Anim. Beh.*, **13**, 134-42. Turner, E.R.A. (1965). Social feeding in birds. *Behaviour*, **24**, 1-46.

Warner, R.L., McFarland, L.Z. and Wilson, W.O. (1967). Microanatomy of the upper digestive tract of the

Japanese quail. *Am. J. Vet. Res.*, **28**, 1537-48.

Wurdinger, I. (1979). Olfaction and feeding behavior in juvenile geese (*Anser a. anser* and *Anser domesticus*). *Z. Tierpsychol.*, **49**, 132.

Weischer, B. (1953). Untersuchungen über das Verhalten von Eidechsen und Vögeln gegenüber "süssen" Stoffen. *Z. für vergl. Physiologie*, Bd. **35**, S. 267-99.

Williams, N. and Kienholz, E.W. (1974). The effect of chili, curry and black pepper powders in diets for broiler chicks. *Poultry Sci.*, **53**, 2233-4.

Williamson, J.H. (1964). Genetic Differences in the ability of chicks to taste Ferric Chloride. *Poultry Sci.*, **43**, 1066-1068.

Yang, R.S.H. and Kare, M.R. (1968). Taste response of a bird to constituents of arthropod defence secretions. *Ann. Entomol. Soc. Am.*, **61**, 781-2.

Index

MIX
Papier aus verantwortungsvollen Quellen
Paper from responsible sources
FSC® C105338

FSC
www.fsc.org

If you have any concerns about our products,
you can contact us on
ProductSafety@springernature.com

In case Publisher is established outside the EU,
the EU authorized representative is:
Springer Nature Customer Service Center GmbH
Europaplatz 3, 69115 Heidelberg, Germany

Printed by Libri Plureos GmbH
in Hamburg, Germany